# Wild Mammals
of New England

# Wild Mammals of New England

## ALFRED J. GODIN

WITH DRAWINGS BY THE AUTHOR
AND A FOREWORD BY DANA P. SNYDER

The Johns Hopkins University Press
Baltimore and London

The Johns Hopkins University Press, Baltimore, Maryland 21218
The Johns Hopkins Press Ltd., London

Library of Congress Catalog Card Number 77-4785     ISBN 0-8018-1964-4
Library of Congress Cataloging in Publication data will be found on the last
printed page of this book.

For my parents and sisters

# Contents

# Foreword

*Wild Mammals of New England* is the first extensive account of mammals found in the New England states and their offshore waters. Although short summaries and check lists have been published, other comprehensive works either deal with a more restricted area or, as Hamilton's *The Mammals of Eastern United States*, cover a much broader section of the country. Alfred Godin has accomplished the painstaking task of searching the scientific literature, examining thousands of specimens in museum collections, and presenting his findings in a very readable style. He has included both marine and land mammals and has given special attention to the preparation of range maps showing the known distribution, within New England, of each species. His artistic and accurate pencil drawings, portraying each of the mammals, are a unique contribution and they add greatly to the esthetic quality of the work.

This book will be welcomed by the interested layman as an aid to identifying specimens and as an authoritative source of information on the habits, life history, and appearance of the wild mammals of New England. Mr. Godin's inclusion of introduced species that have become established will help avoid the confusion that can occur if one is unaware of these forms.

The compilation of published ecological and distributional data along with citations from the literature will be a handy reference for professional wildlife biologists, especially as their responsibilities are now coming to include small non-game mammals as well as the larger game species. For the mammalogist, the book provides a guide to what is known—and what is unknown—about the distribution, food habits, reproduction, habitat selection, and other ecological and behavioral characteristics of New England mammals. This information, in conjunction with the tabulation of museum specimens, should stimulate further research on the mammalian fauna of the region.

This work will prove valuable to those engaged in environmental impact studies in New England—studies in which time is frequently not available to search for the data that have now been assembled in one volume. As pressures on the natural environment continue to grow, this timely publication can serve to acquaint the public with our heritage of wild mammals and can awaken and stimulate an interest in furthering the protection and conservation of this heritage. Only by such increased public awareness can we hope to avoid adding to the list of the some half dozen native species that, as the author notes in his discussion of extirpated mammals, are now missing from the New England fauna.

Dana P. Snyder
Department of Zoology
University of Massachusetts
Amherst, Massachusetts

# Preface

MANY PEOPLE are interested in mammals and their conservation, and much has been written on the subject. But to date, no comprehensive work on the classification, distribution, ecology, and behavior of the wild mammals of New England, based upon specimens and supplemented by field data, has been published. A few brief accounts have appeared, including some excellent state publications and checklists. However, there is a dearth of information concerning New England wild mammals, and a need exists for a book that provides technical information on their lives.

The aims of this book are to identify the species of wild mammals occurring in New England and to provide up-to-date factual information and references on their classification, distribution, ecology, and behavior. It is intended that the book might serve both as a text supplement and reference for courses in the study of mammals and as a reference for the layman, naturalist, and fieldworker. The book is also intended as a contribution to the knowledge of New England mammals and as a stimulant to further studies. Much information is available on some species and very little on others.

The book is basically a compendium of information found in scientific literature that seldom is accessible to any but biologists. In preparing the book I have examined more than a thousand scientific publications and some papers not yet published and have drawn much from them. I do not vouch for the correctness of all the statements made by the various authors whose papers I cite or quote. I have also consulted with many biologists and have examined over 20,500 specimens of wild mammals of New England. The information presented here is as up to date as possible; data collection was terminated on 31 December 1975. But the book is not definitive because new research findings constantly supplement current information. Of course, as the book was prepared some errors have crept in. I will appreciate having them called to my attention.

The book is composed of three sections. Chapter 1 is introductory and deals briefly with the general characteristics of mammals, adaptation, and classification, including a key to the orders and keys to the species of adult wild mammals of New England. Chapter 2 treats the physiographic features, original forest, current forest zones, and climate of New England. Chapters 3–12, the main text, cover the general characteristics of each order, family, species, and subspecies, found in New England, giving their original and current classifications, description, distribution, ecology, behavior, methods of age and sex determination, specimens examined, and references. The book includes species that are extinct, extirpated, or in danger of extirpation, and those that have been introduced and have become established in New England. The glossary lists many of the technical terms used in the book. I have prepared all the illustrations and maps.

I have not illustrated the skull of each species because of space limitations. Other books have been devoted to characteristics of skulls. I have therefore constructed artificial dichotomous keys to identify the current wild mammals of New England on the bases of whole adult specimens, omitting keys for the extinct and extirpated species.

The body measurements and weights stated for the mammals described are based on random samples of specimens I have examined and on data recorded in the literature. The measurements and weights of adults of both sexes are given in both the metric and the English systems and may be used for classification and for comparison between individual animals and species. The measurements given usually include total length, tail length, and hind-foot length. The general term "measurement" refers to the total length of the animal from the tip of the nose to the tip of the fleshy part of the tail beyond the last vertebra, measured with the animal on its back. The tail length is measured from the base of the tail to the tip of the fleshy part of the tail beyond the last vertebra when the tail is bent at a right angle to the body. The hind-foot length is measured from the heel to the end of the claw on the longest toe.

The information in the distribution maps is based on the specimens I have examined and on authenticated records from various sources. To put a separate symbol for every specimen examined would result in broadly overlapping dots and would in some instances obliterate county and state borders. Therefore the shaded areas on the maps represent the distribution of the species. The total number of specimens examined from each locality, county, and state is listed in

Specimens Examined; I have arranged alphabetically the counties within each state and the localities within each county. When only the county is recorded on the specimen label, I have listed the specimen under the known county as "no location." I have used the following symbols to indicate the museum collections of specimens I have examined.

| | |
|---|---|
| AMNH | American Museum of Natural History, New York, New York |
| DC | Dartmouth College Museum, Hanover, New Hampshire |
| HMCZ | Harvard Museum of Comparative Zoology, Cambridge, Massachusetts |
| UCT | University of Connecticut, Storrs |
| UMA | University of Massachusetts, Amherst |
| UME | University of Maine, Orono and Farmington |
| UNH | University of New Hampshire, Durham |
| URI | University of Rhode Island, Kingston |
| USNM | U.S. National Museum of Natural History, Washington, D.C. |
| UVT | University of Vermont, Burlington |

Since this book deals with nine orders and one hundred species of mammals there are likely to be some minor errors in scientific names. I have attempted to avoid this by following the classification of terrestrial mammals by Hall and Kelson (1959) and the classification of aquatic mammals by Scheffer and Rice (1963) except where otherwise noted.

## Acknowledgments

I COULD NOT have completed this book without the help of many people, agencies, and institutions. I enjoyed working with them. I acknowledge specifically the many contributions of the biologists whose findings serve as the basis of this book. The following biologists have read parts of the manuscript or have rendered valuable assistance: Albert A. Barden, Jr., Joseph A. Chapman, Malcolm W. Coulter, Wendell E. Dodge, Harold B. Hitchcock, Harry E. Hodgdon, Joseph S. Larson, Richard H. Manville, Robert D. McDowell, James G. Meade, Edward D. Mitchell, Mark Mowatt, John L. Paradiso, David T. Richardson, David E. Sergeant, William G. Sheldon, Hellenette Silver, Dana P. Snyder, Ralph M. Wetzel, and Howard A. Winn.

The following curators have let me use their facilities and specimen collections, and to each I am very grateful: Sydney Anderson and Karl F. Koopman, American Museum of Natural History, New York; Robert G. Chaffee, Dartmouth College Museum, Hanover, New Hampshire; Barbara Lawrence, Harvard Museum of Comparative Zoology, Cambridge, Massachusetts; Robert E. Dubois and Ralph M. Wetzel, University of Connecticut, Storrs; Albert A. Barden, Jr., Malcolm W. Coulter, and Robert L. Martin, University of Maine, Orono and Farmington; Wendell E. Dodge and Dana P. Snyder, University of Massachusetts, Amherst; Edward N. Francq, University of New Hampshire, Durham;

Robert K. Chipman, University of Rhode Island, Kingston; Robert W. Fuller and Charles A. Woods, University of Vermont, Burlington; and Don W. Wilson, U.S. National Museum of Natural History, Washington, D.C.

Grateful acknowledgment is made to the staff members of the American Museum of Natural History and the National Museum of Natural History for allowing me to use their libraries, and to the state fish and game departments and other state agencies of New England for their generous cooperation and use of their photographs.

Special thanks are due to the following people who have helped me in one way or another: Rene M. Bollengier, Jr., Colton H. Bridges, James E. Cardoza, James Chadwick, James E. Forbes, John S. Gottschalk, Francis J. Gramlich, Richard E. Griffith, William C. Hickling, Edward R. Ladd, James J. McDonough, John W. Peterson, Alberta V. Rowley, James D. Stewart, Alice M. Swayne, and Salvatore A. Testaverde.

My sincere thanks are extended to my former professor, Dr. William G. Sheldon, Professor Emeritus of Wildlife Management, University of Massachusetts at Amherst, for his generous guidance and assistance during the preparation of the book.

I thank my uncle, Mr. Gerard A. St. Pierre, who pointed me toward the profession of fish and wildlife resources conservation.

# Wild Mammals
# of New England

# Introduction

**Characters of Mammals.** Mammals are the only class of vertebrates in which females possess mammary glands, producing milk with which they nourish their young, and the word "mammal" is thus derived from the Latin *mamma*, "breast." The primitive monotremes (the duck-bill platypus [*Ornithorhynchus anatinus*], and the spiny anteaters or echidnas [*Tachyglossus and Zaglossus*]), although classed as mammals, lack true mammae, but paired areas of tubular glands on the abdomens of the females secrete milk which is lapped up by the young. Except for the monotremes, which lay eggs, mammals bear live young.

Mammals are also characterized by having hair during at least some stage of their development. Hair, a multicellular derivative of the skin, serves mainly for insulation. The coat of hair, called the pelage, usually consists of two types of hairs. The underhairs are generally soft and thick and lie close to the skin. The guard hairs, longer and coarser, project beyond the underhairs to protect them from wear. (The quills of porcupines are specialized hairs.) Mammals shed or molt periodically, the worn hairs dropping out and new hairs growing in. Molting may occur seasonally, once or twice a year, or continuously, but the pelage is usually thicker in winter. Members of the order Cetacea (whales, dolphins, and porpoises) are almost hairless.

Hair also contributes to protective coloration, and the color of a mammal is determined by both the structure of the hair and the pigment of the skin and hair. The pigments normally occurring in the hairs are melanin and xanthophyll. The presence of melanin results in brown or black hairs; xanthophyll makes hair yellowish or reddish; and white hairs lack all pigment. The color of most small land mammals closely resembles the color of their environment; so as a rule mammals that live beneath a dense canopy of vegetation or in forests are darker than those living in open areas and deserts.

Besides possessing mammary glands and hair, mammals have a muscular diaphragm that separates the lungs from the abdominal cavity, and a four-chambered heart. Birds and crocodilians also have a four-chambered heart, but in mammals the aorta, the large artery carrying blood from the heart, turns to the left, whereas in birds it turns to the right and in reptiles to both left and right.

Mammals are "warm-blooded," or homothermous, permitting them to maintain a nearly constant body temperature in spite of normal changes in environmental temperatures. The mature red blood cells are usually nonnucleated and are generally round, or rarely oval, as in the family Camelidae (camels). Except for birds, which are also warm-blooded, other classes of animals—fishes, amphibians, and reptiles—are referred to as "cold-blooded," or poikilothermous, because their body temperatures vary with their surroundings.

Some mammals can pass into a lethargic state —"winter sleep," or hibernation—for weeks or even months, in which their body temperatures are only slightly higher than that of the environment, and other species go into a transitory torpor during the hot dry summer and fall months without much reduction in body temperature (estivation). Ground squirrels (*Citellus*), woodchucks, and jumping mice are true hibernators. Opossums, skunks, and raccoons become torpid for several weeks but often become active during mild periods in winter, and even bats awaken and fly about when disturbed. Bears are not true hibernators, because they retain their normal temperature during long periods of drowsy sleep. Chipmunks (*Tamias*) do become dormant but are active during warm winter weather, and tree squirrels may become torpid during periods of inclement weather.

Mammals also possess seven cervical vertebrae, except the manatees (*Trichechus*), and the two-toed sloths (*Choloepus*), each with six; and the anteaters (*Tamandua*), with eight. The lower jaw articulates directly with the cranium. Finally, in mammals the brain is greatly enlarged and more highly developed and complex than in any other group of animals.

**Teeth.** Except for the monotremes, anteaters, pangolins (*Manis*), and baleen whales, all adult mammals have teeth. The mammalian tooth is an ectodermal structure consisting of several distinct tissues. Its center is composed of a vascular pulp which is abundantly supplied with blood and nerves when the animal is young. The bulk of the tooth, surrounding the pulp, consists of dentin, a bonelike material composed mainly

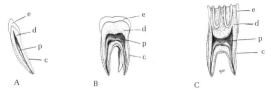

Lateral view of sections through mammalian teeth. *A*, typical incisor; *B*, low-crowned molar of a primate; *C*, high-crowned molar of a horse. The sections are *e*, enamel; *d*, dentine; *p*, pulp cavity; *c*, cement.

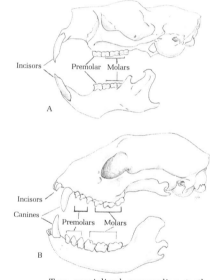

Two specialized mammalian teeth, reflecting two different kinds of diet. *A*. In the porcupine, a rodent, the prominent incisors are self-sharpening and are used mainly for gnawing plants. The canines are absent. The premolars and molars (cheek teeth) are separated from the incisors by a gap in the tooth row—a diastema. The cheek teeth are specialized for grinding plants. *B*. In the raccoon, a carnivore, the incisors are small, and the large, pointed canines are specialized for stabbing and tearing flesh. The premolars and molars are specialized for shearing flesh.

of calcium phosphate. The crown of the tooth is coated with hard, white, shiny enamel, and the root is coated with cementum.

Mammalian teeth are differentiated into incisors, canines, premolars, and molars. In general, the first set of teeth, except the molars, is deciduous and is replaced by the permanent set. These "milk teeth" may disappear before birth or a few days after, as in seals, whereas moles, for example, retain theirs throughout life. Animals having two successive sets of teeth are called "diphyodont." The teeth of mammals are thecodont, being set in alveoli, or sockets, in the jawbones. The bases of the teeth may become constricted and often divide into separate roots, each in a separate socket. Such teeth cease to grow after they mature and are called "rooted" teeth. Teeth whose bases remain widely open throughout life and continue to grow to compensate for the wearing away of their crowns are called "rootless" teeth. The incisors of rodents are of this type.

The mammalian dentition is usually heterodont; that is, the teeth vary both in structure and in function. Therefore the characters of the teeth are useful in classifying mammals. In some species four tooth types can be distinguished in the adult. In front are the incisors, unspecialized, chisellike teeth used for nipping and gnawing. Next in each series is a single canine which is generally simple, with a long, conical crown, used for stabbing and holding. Canines are well developed in the carnivores, and in walruses the upper canines form tremendous tusks. Following the canines is a series of premolars (called "bicuspids" by the dentist), employed for cutting, shearing, slicing, and grinding. In the herbivores these are greatly modified to accommodate a diet of roughage. Next are the molars, which are used mainly for crushing. The pre-

molars and molars together are called cheek teeth.

The number of teeth of each type on one side of the upper jaw and the corresponding number in one side of the lower jaw are expressed in the dental formula: I for incisors, C for canine, P for premolars, and M for molars. The total number of teeth is derived by adding the numbers in the dental formula, as in the following examples for moose, man, and pig. Zeros indicate that the teeth are absent.

$$I \frac{0-0}{3-3} \ C \frac{0-0}{1-1} \ P \frac{3-3}{3-3} \ M \frac{3-3}{3-3} = 32 \text{ total (moose)}$$

$$I \frac{2-2}{2-2} \ C \frac{1-1}{1-1} \ P \frac{2-2}{2-2} \ M \frac{3-3}{3-3} = 32 \text{ total (man)}$$

$$I \frac{3-3}{3-3} \ C \frac{1-1}{1-1} \ P \frac{4-4}{4-4} \ M \frac{3-3}{3-3} = 44 \text{ total (pig)}$$

The largest number of teeth for placental (eutherian) mammals is forty-four. This book gives dental formulas for adult specimens only.

The teeth of mammals closely reflect their eating habits, as is evident in complicated crown patterns of projecting cusps of molars. The tritubercular type of tooth is a primitive molar, low-crowned, with three simple cusps or tubercles, called protocone, paracone, and metacone, arranged in a triangle. In Recent mammals there are slight modifications from the primitive tooth, mainly the addition of accessory cusps. This type of tooth, found in some insectivores, bats, and carnivores, is used for cutting and crushing.

In most carnivores, cats, for example, the fourth upper premolar (P 4) and the first lower molar (M 1) are developed into shearing teeth (sectorial), a modification of the tritubercular.

Man and pigs, which are omnivorous, have low-crowned (bunodont) teeth for crushing soft foods. The bunodont is also derived from the tritubercular.

In deer, horses, and cows, the individual cusps of a bunodont tooth are converted into crescent-shaped cusps (selenodont) that wear down and expose enamel ridges on the flat surface. These selenodont molars are adapted for a herbivorous diet including highly abrasive foods such as grass. The selenodont molar is derived from the bunodont.

Lophodont molars have transverse ridges separated by low valleys. These teeth were present in the extinct mastodon, and modern ele-

Crown view of three major types of right upper molariform teeth. *A*, bunodont; *B*, lophodont; *C*, selenodont (outer edge of each tooth above, front edge to right). Stipple parts of lophodont and selenodont teeth are dentine.

phants possess high-crowned hypsolophodont, cross-crested molars. The lophodont tooth also is derived from the bunodont.

**Adaptation.** In the course of evolution, mammals have adapted to nearly all kinds of environments. Locomotion probably shows the most striking adaptation. *Aquatic* mammals like the whales and sirenians (sea cows) are descended from land-dwelling quadrupeds, but their limbs became modified for locomotion in the water when they reinvaded their original aquatic environment. The whales, dolphins, porpoises, and sirenians swim by oscillating their flukes or flippers in a horizontal plane. The earless seals (Phocidae) propel themselves by hind flippers pressed nearly together in a vertical plane, with the toes widely separated for maximum extension of the webs, which enables them to maintain a constant crusing speed.

*Terrestrial* mammals employ all four limbs in different types of locomotion. *Ambulatory* (walking or ambling) animals such as the opossum, shrews, bears, and man are *plantigrade* (walking on the soles of the feet) and are typically five-toed. The *cursorial* (running) animals are *digitigrade* (running on the toes) like foxes, cats, rabbits, and hares. Mammals like the deer, which run on the tips of their toes, are *unguligrade,* while others are *saltatorial* (jumping) like the jumping mice and kangaroos. *Fossorial* mammals live underground like the moles and pocket gophers and have remarkably large forefeet adapted for digging.

Mammals that live in trees, such as tree squirrels and monkeys, are referred to as *arboreal.* They usually have long toes with sharp claws, like squirrels, or grasping hands or feet, like monkeys. The flying squirrels are *glissant* or *volant,* for they launch themselves from treetops and are able to glide considerable distances from tree to tree or to the ground. Bats, however, are *aerial* animals capable of flying, because their forelimbs are remarkably modified by greatly elongated fingers, with a thin membrane linking the digits and connecting them with the hind limbs and the tail.

*Graviportal* (carrying great weight) locomotion is characteristic of animals like elephants. Their limbs are straight and pillarlike, and the digits are distributed in a circle around the edge of the flat-bottomed foot.

**Classification of Mammals.** The great Swedish naturalist Carolus Linnaeus (1707–78) classified the animal kingdom on the basis of diagnostic external characteristics. His *Systema naturae* went through twelve editions, beginning in 1735 as a brochure of twelve pages and ending in 1766 with 2,400 pages. The tenth edition, published in 1758, listed only eighty-six species of mammals. In the twelfth edition Linnaeus listed the mammalian groups under three principal divisions: mammals with claws, mammals with hoofs, and mammals without claws or hoofs, such as the whales. Linnaeus coined the word "mammal" and made the possession of mammae a character of the class Mammalia.

The scientific classification of animals is based upon a progression of subdivisions beginning with the phylum and ending with the species and subspecies. The basis of classification is similarity of character relationships such as embryology, morphology, ecology, and behavior.

Within the animal kingdom, animals possessing a dorsal structure or a notochord during some stage of life belong to the phylum Chordata. The phylum includes the subphylum Vertebrata, in which all animals possess pharyngeal gills or pouches and a notochord at some stage of development, and a hollow, fluid-filled dorsal nerve cord. The Vertebrata includes five classes: Pices, fishes; Amphibia, amphibians; Reptilia, reptiles; Aves, birds; and Mammalia, mammals.

The class Mammalia includes nineteen orders of Recent mammals and 4,060 Recent species belonging to 1,004 Recent genera (Anderson and Jones 1967). Classes are divided into orders, which are then divided into families, and each family is broken down into successively smaller groups in which the animals possess closer and closer character relationships. For example, the bionomial (a species name consisting of two terms) of the eastern meadow vole is *Microtus pennsylvanicus* (Ord), and that of the beach vole is *Microtus breweri* (Baird). The first word, *Microtus*, is the genus to which the species belongs. Generic characters are less marked than family characters and often relate to differences in the skull and teeth. *Pennsylvanicus* and *breweri* are species names, which usually relate to differences in size, markings, and other prominent external features. Subspecies are given trinominals (three-part names), as for the Block Island meadow vole, *Microtus pennsylvanicus provectus* Bangs. A subspecies or geographic race is distinguished by the minor morphological differences, such as size and shades of coloration, evident between populations of a species from two or more parts of its range.

The names Ord and Baird (following the species names in parentheses) refer to the persons who first named the eastern meadow vole and the beach vole, and the parentheses indicate that the generic names *Mus* (Ord) and *Arvicola* (Baird) originally used by these authors are no longer valid, although their specific names, *pennsylvanicus* and *breweri*, are still being used. If there are no parentheses around a founder's name nothing has changed, and the scientific name has remained as the author first used it.

Most formal biological names come from Latin or Greek, but some are latinized forms of modern words. Scientific names are uniform throughout the world and thus enable scientists to refer to animals and plants far more accurately than by using their popular names, which vary between languages and regions.

## DICHOTOMOUS KEYS

The keys help one correctly identify an unknown specimen through a series of numbered and lettered couplets or alternate choices. To identify a specimen, make the appropriate choice from each set of alternatives in the key to the orders, following the sequence until the order is identified. Then closely follow the descriptions in the series of alternatives in the key to that order until the species of the animal in question is reached, and turn to the pages where this mammal is illustrated and discussed. If the specimen is misidentified, work backward in the key until the mistake is found. With little experience and knowledge of mammals one can go directly to the characteristics of the order.

### KEY TO THE ORDERS (WHOLE SPECIMENS)

1*a*. Forelimbs modified into flippers . . . . . . . . . .  **2**
 *b*. Forelimbs not modified into flippers . . . . . .  **3**
2*a*. Hind limbs absent externally; body fusiform and tapering to flukes flattened horizontally; body essentially hairless; external nares on top of head; ears lack pinnae; strictly aquatic . . . . . . . . .
 . . . . . . . . . . . . . . . . . . . . . . . . . . Cetacea, p. 161
 *b*. Hind limbs markedly expanded to resemble fins; tail vestigial; body fusiform and covered with thick, short hair; ear pinnae reduced or absent; semiaquatic . . . . . . . . . . . . . . Pinnipedia, p. 249
3*a*. Forelimbs greatly modified for flight to serve as

wings and supporting a leathery membrane stretched between elongated fingers, also between hind limbs, including tail . . . . . . . . . . . . . .
 . . . . . . . . . . . . . . . . . . . . . . . . . . Chiroptera, p. 44
 *b*. Forelimbs not modified for flight; no leathery membrane between hind limbs . . . . . . . . . . .  **4**
4*a*. Toes modified into claws . . . . . . . . . . . . . . . .  **5**
 *b*. Toes modified into hoofs . . . . . . . . . . . . . . . .  **9**
5*a*. Innermost toe of hind foot thumblike and clawless and opposed to other toes for grasping; ears thin, leaflike, and naked; female with abdominal pouch . . . . . . . . . . . . . . . . . . Marsupialia, p. 16
 *b*. Innermost toe of hind foot not thumblike and clawed and not opposed to other toes for grasping, ears not thin or leathery, females without abdominal pouch . . . . . . . . . . . . . . . . . . . . . . . .  **6**
6*a*. Always five clawed toes on front foot; thumb never a mere knob with a nail; tail never flattened dorsoventrally or laterally . . . . . . . . . .  **7**
 *b*. Usually only four clawed toes on front foot; thumb sometimes present as small knob with nail; tail may be flattened dorsoventrally or laterally . . . . . . . . . . . . . . . . . . . . . . . . . . . . . .  **8**
7*a*. Total length 203 mm (8 in) or less; eyes small or not visible; fur plushlike . . . Insectivora, p. 22
 *b*. Total length longer than 203 mm (8 in); eyes large and visible; fur not plushlike . . . . . . . . . . . .
 . . . . . . . . . . . . . . . . . . . . . . . . . Carnivora, p. 196
8*a*. Ears longer than tail; tail a cottony tuft; hind foot with four clawed toes; entire bottom of foot densely furred . . . . . . . . . . Lagomorpha, p. 65
 *b*. Ears shorter than tail; tail not a cottony tuft; hind

foot with five clawed toes; entire bottom of feet not densely furred . . . . . . . . . . Rodentia, p. 83

9a. Toes ending in hoofs, two functional toes on each foot; antlers or horns sometimes present . . . . . . . . . . . . . . . . . . . . . . Artiodactyla, p. 259

## KEY TO SPECIES OF INSECTIVORA

1a. Forefeet slender, about same size as hind feet and not adapted for digging; ear conch (pinna) present; auditory bullae and zygomatic arch absent . . . . . . . . . . . . . . . . . . . . . . . . . . . . . . . **2**

b. Forefeet greatly enlarged, broader than hind feet, and adapted for digging; ear conch (pinna) absent; auditory bullae and zygomatic arch present . . . . . . . . . . . . . . . . . . . . . . . . . . . . . . **8**

2a. Tail short, less than one-fourth total length of animal . . . . . . . . . . . . . . . . . . . . . . . . . . . . . . . **3**

b. Tail long, at least one-half total length of animal . . . . . . . . . . . . . . . . . . . . . . . . . . . . . . . . **4**

3a. Coloration grayish; total length more than 90 mm (3.5 in); four or five unicuspid teeth visible in each side of upper jaw; total 32–34 teeth . . Short-tailed Shrew, *Blarina brevicauda*, p. 30

b. Coloration brownish; total length less than 90 mm (3.5 in); three unicuspid teeth visible in each side of upper jaw; total 30 teeth . . . . . . . . . . . . . . . . . . . . . Least Shrew, *Cryptotis parva*, p. 34

4a. Only three upper unicuspid teeth easily visible from the side, third and fifth upper unicuspids minute . . . . . . . . . . . . . . . . . . . . . . . . Thompson's Pygmy Shrew, *Microsorex thompsoni*, p. 28

b. Four upper unicuspid teeth easily visible from the side, fifth upper unicuspid minute . . . . . **5**

5a. Total length more than 135 mm (5.3 in); hind feet large and fringed with stiff hairs; third and fourth toes of hind feet thinly webbed for about half their length . . . . . . . . . . . . . . . . . . . . . . . . . . . . . . . . . . . . . . Water Shrew, *Sorex palustris*, p. 25

b. Total length 135 mm (5.3 in) or less; hind feet not conspicuously large and not fringed with stiff hairs; no webbing on any toes . . . . . . . . **6**

6a. Tail length more than 50 mm (2.0 in) . . . . . . . . . . . . . . . . Long-tailed Shrew, *Sorex dispar*, p. 28

b. Tail length less than 50 mm (2.0 in) . . . . . . . **7**

7a. Total length 110 mm (4.3 in) or more . . . . . . . . . . . . . . . . . . Smoky Shrew, *Sorex fumeus*, p. 26

b. Total length less than 110 mm (4.3 in) . . . . . . . . . . . . . . . . . . Masked Shrew, *Sorex cinereus*, p. 22

8a. Snout fringed with fleshy processes; tail more than 50 mm (2.0 in) . . . . . . . . . . . . . . . . . . . . . . . . . . . Star-nosed Mole, *Condylura cristata*, p. 39

b. Snout not fringed with fleshy processes; tail less than 50 mm (2.0 in) . . . . . . . . . . . . . . . . . . . . **9**

9a. Tail thick, blackish, and densely haired . . . . . . . . . Hairy-tailed Mole, *Parascalops breweri*, p. 36

b. Tail thin, flesh-colored, and scantily haired . . . . . Eastern Mole, *Scalopus aquaticus*, p. 37

## KEY TO SPECIES OF CHIROPTERA

1a. Interfemoral membrane partially or wholly furred on upper surface . . . . . . . . . . . . . . . . . **2**

b. Interfemoral membrane not furred on upper surface . . . . . . . . . . . . . . . . . . . . . . . . . . . . . . . **4**

2a. Fur reddish or yellowish orange . . . . . . . . . . . . . . . . . . . . . . . . . Red Bat, *Lasiurus borealis*, p. 58

b. Fur brownish or blackish, frosted with white . . . . . . . . . . . . . . . . . . . . . . . . . . . . . . . . . . . . . . . . . **3**

3a. Interfemoral membrane wholly furred; fur brownish, heavily frosted with white; individual hairs banded; total length more than 125 mm (4.9 in) . . . Hoary Bat, *Lasiurus cinereus*, p. 60

b. Interfemoral membrane partially furred; fur blackish, lightly frosted with white; individual hairs not banded; total length less than 120 mm (4.7 in) . . . . . . . . . . . . . . . . . . . . . . . . . . . . Silver-haired Bat, *Lasionycteris noctivagans*, p. 52

4a. Ears when laid forward extend considerably beyond nostrils; fur light reddish brown . . . . . . . . . . . . . . . . . . Keen's Myotis, *Myotis keenii*, p. 48

b. Ears when laid forward do not reach noticeably beyond nostrils . . . . . . . . . . . . . . . . . . . . . . . . **5**

5a. Total length more than 100 mm (3.9 in) . . . . . . . . . . . . . . . . Big Brown Bat, *Eptesicus fuscus*, p. 55

b. Total length less than 100 mm (3.9 in) . . . . . **6**

6a. Total length less than 85 mm (3.3 in); distinct black facial mask . . . . . . . . . . . . . . . . . . . . . . . . . . . . . . Small-footed Myotis, *Myotis leibii*, p. 51

b. Total length more than 85 mm (3.3 in); no black facial mask . . . . . . . . . . . . . . . . . . . . . . . . . . **7**

7a. Color of fur pale golden brown; nose, ears, and wing membranes reddish brown . . . . . . . . . . . . . Eastern Pipistrelle, *Pipistrellus subflavus*, p. 53

b. Color of fur dark brown; nose, ears, and wing membranes nearly black . . . . . . . . . . . . . . . . **8**

8a. Fur glossy, coloration yellowish to olive brown; calcar not keeled . . . . . . . . . . . . . . . . . . . . . . . . . . . Little Brown Myotis, *Myotis lucifugus*, p. 44

b. Fur fluffy, dull, coloration pinkish brown; calcar keeled . . . . . . . . . . . . . . . . . . . . . . . . . . . . . . . . . . . . . . . Indiana Myotis, *Myotis sodalis*, p. 49

## KEY TO SPECIES OF LAGOMORPHA

1a. Length of hind foot less than 115 mm (5.5 in); interparietal distinct, not fused with parietals . . . . . . . . . . . . . . . . . . . . . . . . . . . . . . . . . . . . . . . . **2**

b. Length of hind foot usually more than 115 mm (5.5 in); interparietal fused with parietals . . . **3**

2a. White spot on forehead in 50% of specimens, others with light brown spot; outside of ear grayish; leading edge of ear with thin, indistinct black stripe; inside of ear sparsely furred with thin brown stripe; general coloration of body grayish with little black; rump patch gray and distinct . . . . . . . . . . . . . . . . . . . . . . . . . . . . . . . . . . . Eastern Cottontail, *Sylvilagus floridanus*, p. 65

b. Black or dark brown spot on forehead; outside of ear brownish; leading edge of ear with broad, distinct black stripe; inside of ear densely furred with broad brown stripe; general coloration of body grizzled, washed with black; rump patch brown and indistinct .............. New England Cottontail, *Sylvilagus transitionalis*, p. 70

3a. Tail not black above; entire pelage except for tips of ears white in winter ......................
..... Snowshoe Hare, *Lepus americanus*, p. 73

b. Tail black above; pelage does not turn white in winter................................. **4**

4a. Upper sides of hind foot white or whitish; upper parts of body grayish or brownish... Black-tailed Jackrabbit, *Lepus californicus*, p. 77

b. Upper sides of hind feet without white; upper parts of body tawny .........................
....... European Hare, *Lepus europaeus*, p. 78

**KEY TO SPECIES OF RODENTIA**

1a. Tail large and flattened horizontally ...........
........... Beaver, *Castor canadensis*, p. 104

b. Tail not large and not flattened horizontally ........................................... **2**

2a. Tail compressed vertically ...................
......... Muskrat, *Ondatra zibethicus*, p. 130

b. Tail not compressed vertically ........... **3**

3a. Hind legs greatly elongated for jumping ... **4**

b. Hind legs not elongated for jumping ...... **5**

4a. Tail usually tipped with white; coloration of back bright orange ................. Woodland Jumping Mouse, *Napaeozapus insignis*, p. 146

b. Tail rarely tipped with white; coloration of back yellowish ......................... Meadow Jumping Mouse, *Zapus hudsonius*, p. 143

5a. Tail bushy ........................... **6**

b. Tail not bushy ......................... **12**

6a. Flying membrane present, connecting fore- and hind limbs ............................ **7**

b. Flying membrane absent ................. **8**

7a. Hairs on breast and belly white to bases .......
.................................. Southern Flying Squirrel, *Glaucomys volans*, p. 99

b. Hairs on breast and belly slate-colored near bases .......................... Northern Flying Squirrel, *Glaucomys sabrinus*, p. 102

8a. Lengthwise stripes present .............. **9**

b. Lengthwise stripes absent .............. **10**

9a. Stripes on side of face and on body ..........
.... Eastern Chipmunk, *Tamias striatus*, p. 84

b. Stripes absent on side of face; usually a black stripe on side of body in summer ............
Red Squirrel, *Tamiasciurus hudsonicus*, p. 96

10a. Tail fringed with white .....................
.... Gray Squirrel, *Sciurus carolinensis*, p. 91

b. Tail not fringed with white .............. **11**

11a. Hair on body and tail mixed with sharp quills ....... Porcupine, *Erethizon dorsatum*, p. 149

b. Hair on body and tail not mixed with sharp quills .... Woodchuck, *Marmota monax*, p. 87

12a. Tail well haired and not conspicuously scaly ... Eastern Woodrat, *Neotoma floridana*, p. 117

b. Tail scantily haired and noticeably scaly... **13**

13a. Tail as long as or longer than head and body ...
........................................ **14**

b. Tail shorter than head and body .......... **15**

14a. Total length of adults more than 300 mm (11.7 in) .......... Black Rat, *Rattus rattus*, p. 136

b. Total length of adults seldom exceeds 225 mm (8.6 in) ... House Mouse, *Mus musculus*, p. 140

15a. Total length of adults more than 300 mm (11.7 in) ...... Norway Rat, *Rattus norvegicus*, p. 137

b. Total length of adults less than 300 mm (11.7 in) ...................................... **16**

16a. Body and tail distinctly bicolored ......... **17**

b. Body and tail usually unicolored or not distinctly bicolored ........................ **18**

17a. Coloration on back grayish brown in adults .. Deer Mouse, *Peromyscus maniculatus*, p. 110

b. Coloration on back chestnut brown in adults ......................................White-footed Mouse, *Peromyscus leucopus*, p. 114

18a. Tail length nearly same as hind foot ....... **19**

b. Tail length twice as long as hind foot ..... **21**

19a. Coloration of back uniform brown; outer edges of upper incisors not grooved ...................
....... Pine Vole, *Microtus pinetorum*, p. 127

b. Coloration of back slate gray; outer edges of upper incisors with shallow lengthwise grooves ........................................ **20**

20a. A few hairs at base of ears with bright rusty tinge, distinctly brighter than rest of fur; females have eight mammae .............. Northern Bog Lemming, *Synaptomys borealis*, p. 135

b. No brighter hairs at base of ears, females have six mammae ........................ Southern Bog Lemming, *Synaptomys cooperi*, p. 134

21a. Broad reddish band on back ... Gapper's Red-backed Mouse, *Clethrionomys gapperi*, p. 119

b. No broad reddish band on back .......... **22**

22a. Coloration pale gray throughout ..............
........ Beach Vole, *Microtus breweri*, p. 124

b. Coloration grizzled; back brownish, belly yellowish ................................ **23**

23a. Coloration yellowish from nose to eyes........
.... Rock Vole, *Microtus chrotorrhinus*, p. 126

b. Coloration yellowish from nose to eyes absent .Meadow Vole, *Microtus pennsylvanicus*, p. 121

**KEY TO SPECIES OF CETACEA**

a. Nostril single; no long-fringed baleen or whale-bone in mouth; teeth in one or both jaws or bare gums ................... Odontoceti, p. 161

b. Nostrils double; long-fringed baleen or whale-bone suspended from upper jaw; teeth absent ............................ Mysticeti, p. 181

## KEY TO STRANDED SPECIES OF ODONTOCETI

1a. Throat with grooves ..................... **2**
b. Throat without grooves ................. **10**
2a. Mouth situated on lower side of head; anterior part of head far ahead of front end of lower jaw; lower jaw much narrower than upper; blowhole slit opening on left side of median line of skull; functional teeth in lower jaw eighteen or more ......................................... **3**
b. Mouth not situated on lower side of head; anterior part of head not ahead of front end of lower jaw; lower jaw somewhat ahead of front end of upper jaw; blowhole slit opening median; functional teeth in lower jaw never more than two ......................................... **4**
3a. Head one-third or more of total length of body; dorsal fin replaced by several low humps along posterior third of body; coloration bluish gray or dusky brown, paler below ................... ....... Sperm Whale, *Physeter catodon*, p. 166
b. Head about one-fifth of total length of body; dorsal fin slender and falcate posteriorly; general appearance sharklike; coloration blackish gray above, whitish below ...................... Pygmy Sperm Whale, *Kogia breviceps*, p. 168
4a. Beak abruptly set off from bulging forehead; coloration grayish black above, light grayish below ............................... Bottle-nosed whale, *Hyperoodon ampullatus*, p. 165
b. Beak merging gradually with forehead ... **5**
5a. Mouth large, reaching posteriorly nearly to level of eye ................................. **6**
b. Mouth small, reaching posteriorly less than half-way to level of eye ..................... **9**
6a. Functional teeth in lower jaw situated anterior to posterior margin of mandibular symphysis ......................................... **7**
b. Functional teeth in lower jaw situated posterior to mandibular symphysis ................ **8**
7a. Coloration blackish above, lower sides yellowish purple, flecked with black, and somewhat darker on median line of belly; functional teeth in lower jaw small, situated at anterior tip of mandible ......................................... True's Beaked Whale, *Mesopolodon mirus*, p. 163
8a. Coloration blackish blue above and usually grayish or whitish below; functional mandibular tooth small, height from alveolus less than depth of mandible at shallowest place .............. ............................ North Atlantic Beaked Whale, *Mesoplodon bidens*, p. 162
b. Coloration almost completely black, some irregular grayish blue or whitish spots on belly in life, or some reddish patches on underside between flippers and about anal opening in death; functional mandibular tooth large, height from alveolus much more than depth of mandible at shallowest place ................... Tropical Beaked Whale, *Mesoplodon densirostris*, p. 162
9a. Coloration variable, usually grayish throughout, darker above than below ............. Goose-beaked Whale, *Ziphius cavirostris*, p. 164
10a. Coloration entirely white or whitish gray; dorsal fin reduced to indistinct ridge ................. .. White Whale, *Delphinapterus leucas*, p. 169
b. Coloration not white or whitish gray; dorsal fin distinct ................................ **11**
11a. Beak distinct .......................... **12**
b. Beak indistinct ......................... **16**
12a. Black ring around eye ................... **13**
b. Black ring around eye absent ............. **15**
13a. Black band running from base of flipper toward chin; coloration blackish above, sides shading to grayish green mixed, with elongated elliptical bands of whitish on flanks; belly whitish; whitish band over forehead with narrow black band in center connecting black eye rings .......... Common Dolphin, *Delphinus delphis*, p. 172
b. No black band running from base of flipper toward chin ........................... **14**
14a. Beak tip white; coloration dusky above and flanks sometimes with small gray spots ......... ..................................... White-beaked Dolphin, *Lagenorhynchus albirostris*, p. 174
b. Beak tip dusky; coloration grayish above, yellowish on flanks, whitish below; frequently a narrow black band from base of flukes along side to level of dorsal fin ........................ ..................................... Atlantic White-sided Dolphin, *Lagenorhynchus acutus*, p. 175
15a. Coloration dark steel blue above; lower sides of flanks and below white; dark blue stripe from dorsal fin passing forward and ending abruptly in front; narrow stripe from across eye passing laterally to anal region; single or double stripe extending from eye to base of flipper .......... Striped Dolphin, *Stenella caeruleoalba*, p. 171
b. Coloration purplish gray to clear gray above; sides pale gray blending whitish below ....... Bottle-nosed Dolphin, *Tursiops truncatus*, p. 173
16a. Dorsal fin strongly falcate posteriorly ..... **17**
b. Dorsal fin not falcate posteriorly ........ **18**
17a. Forehead prominent; coloration grayish above, paler below ............................... .......... Grampus, *Grampus griseus*, p. 178
b. Forehead prominent; coloration largely black, sometimes narrow white stripe extending from throat to belly ..................... Common Pilot Whale, *Globicephala melaena*, p. 179
18a. Forehead not prominent; dorsal fin prominent

and high; coloration black above; white patch posterior to eye and below to dorsal fin; white patch below forms trident area posteriolaterally ............ Killer Whale, *Orcinus orca*, p. 176

b. Forehead not prominent; dorsal fin short and triangular; coloration blackish gray above; flanks grayish, fading to white below ............... Harbor Porpoise, *Phocoena phocoena*, p. 180

## KEY TO STRANDED SPECIES OF MYSTICETI

1a. Pleats or grooves present on chin, throat, and belly; dorsal fin present .................. **2**

b. Pleats or grooves absent; dorsal fin absent .. **6**

2a. Flippers very long, about one-third of total length of animal and scalloped on anterior border; deep grooves on throat usually encrusted with barnacles; dorsal fin small and triangular; coloration black above, whitish below ........ ............................................ Humpback Whale, *Megaptera novaeangliae*, p. 188

b. Flippers short, one-sixth to one-seventh of total length of animal and not scalloped on anterior border ................................. **3**

3a. Baleen black; coloration bluish, marbled with patches of gray on back and sides; yellowish diatom film below .......................... ... Blue Whale, *Balaenoptera musculus*, p. 187

b. Baleen of different color, but if black, then fringed with gray ........................ **4**

4a. Baleen yellowish; coloration grayish black above, white below; flippers above with wide transverse white band ................ Minke Whale, *Balaenoptera acutorostrata*, p. 185

b. Baleen not yellowish; flippers without white band .................................. **5**

5a. Baleen grayish blue, except on anterior third of right side of jaw, where it is white; coloration grayish black above, white below ............. ... Fin Whale, *Balaenoptera physalus*, p. 182

b. Baleen blackish, but fringed with white; coloration bluish gray above and white below; white below not extending to flippers and flukes ..... .... Sei Whale, *Balaenoptera borealis*, p. 184

6a. Coloration blackish throughout, sometimes with whitish spots on belly; mouth strongly arched with long black baleen ...................... ..... Right Whale, *Eubalaena glacialis*, p. 190

## KEY TO SPECIES OF CARNIVORA

1a. Tail very short and hidden in long heavy fur; body massive; coloration usually black with brown muzzle ............................ ....... Black Bear, *Ursus americanus*, p. 208

b. Tail conspicuous; body not massive ....... **2**

2a. Tail annulated with five or six black rings; black facial mask present ......................... ............. Raccoon, *Procyon lotor*, p. 212

b. Tail not annulated; black facial mask absent ........................................... **3**

3a. Toes webbed; tail thickened at base and tapering toward tip; coloration dark brownish black ..... ....... River Otter, *Lutra canadensis*, p. 234

b. Toes not webbed; tail not noticeably thickened at base .................................. **4**

4a. Tail with conspicuous black tip ........... **5**

b. Tail without conspicuous black tip ........ **6**

5a. Tail length less than 75 mm (3 in) ............ ............ Ermine, *Mustela erminea*, p. 221

b. Tail length more than 75 mm (3 in) ........... .... Long-tailed Weasel, *Mustela frenata*, p. 224

6a. Coloration black; back usually with two white stripes ................................... ..... Striped Skunk, *Mephitis mephitis*, p. 231

b. Coloration not black; back without white stripes ........................................... **7**

7a. Spot or patch present on chin or throat ... **8**

b. Spot or patch sometimes present on chin or throat .................................. **9**

8a. Coloration rich brown; orange patch on throat and chest .. Marten, *Martes americana*, p. 216

b. Coloration dark brown to blackish with frosted hairs on back; white spots on throat and chest ... .......... Fisher, *Martes pennanti*, p. 218

9a. Coloration dark rich brown; white spots sometimes present on chin and throat .............. ............... Mink, *Mustela vison*, p. 226

b. Spot or patch on chin and throat absent ... **10**

10a. Claws nonretractile, not entirely hidden in fur; muzzle long; tail bushy, usually with stiff mane ........................................... **11**

b. Claws retractile and hidden in fur; muzzle short; tail not bushy and without mane ........ **13**

11a. Pupils of eyes round; tip of tail black, sometimes with tuft of white ........................... ............... Coyote, *Canis latrans*, p. 196

b. Pupils elliptical; tail with or without mane; tip of tail either all black or all white ........... **12**

12a. Tail with mane; tip of tail black; legs and feet not blackish................................... .. Gray Fox, *Urocyon cinereoargenteus*, p. 205

b. Tail without mane; tip of tail white; legs and feet blackish ..... Red Fox, *Vulpes vulpes*, p. 201

13a. Tip of tail black all around .................. .............. Lynx, *Lynx canadensis*, p. 237

b. Tip of tail black on top only ................. ................. Bobcat, *Lynx rufus*, p. 240

## KEY TO SPECIES OF PINNIPEDIA

1a. Incisors 3/2 × 2; proboscis absent ........ **2**

b. Incisors 2/1 × 2; proboscis present ....... **4**

2a. Color of adults variable, with irregular small dark spots; closed nostril slits almost meet in broad V ... Harbor Seal, *Phoca vitulina*, p. 249

b. Color of adults not variable, with irregular large

blotches or elliptical bands; closed nostril slits separated by an inch or more . . . . . . . . . . . . **3**

3a. Body with blotches; snout long, distance between tip of nose and eye almost twice that between ear and eye . . . . . . . . . . . . . . . . . . . . . . . . . . . . . . . Gray Seal, *Halichoerus grypus*, p. 253

b. Males with dark elliptical band on side from shoulder nearly to tail; in females, the band is usually paler and may be interrupted . . . . . . . . . . . . . . Harp Seal, *Pagophilus groenlandicus*, p. 251

4a. Inflatable tubular proboscis and bright red nasal sac in adult males, absent in females and immature males; body dark with blotches . . . . . . . . . . . . . . Hooded Seal, *Cystophora cristata*, p. 255

**KEY TO SPECIES OF ARTIODACTYLA**

1a. Head with an elongated, mobile, flat, oval snout, with nostrils opening at terminal surface; antlers absent; legs short . . . . . . . . . . . . . . . . . . . . . . . . . . . . . . . . . European Wild Boar, *Sus scrofa*, p. 259

b. Head without an elongated, mobile, flat, oval snout, and nostrils not opening at terminal surface; antlers usually present, at least in males; legs long . . . . . . . . . . . . . . . . . . . . . . . . . . **2**

2a. Nose covered with hair all around nostrils . . . . . . . . . . . . . . . . . . . . . . . . . . . . . . . . . . . . . . . . . . . . . **4**

b. Nose not covered with hair all around nostrils . . . . . . . . . . . . . . . . . . . . . . . . . . . . . . . . . . . . . . **3**

3a. Coloration of adults reddish in summer, grayish in winter; antlers not palmate, if present . . . . . . . . . . . . . . . . . . . . . . . . . . . . . . . . . . . . . . . . . . . . Whitetailed Deer, *Odocoileus virginianus*, p. 262

b. Coloration of adults bright fawn or yellowish brown, dappled with white or yellowish white spots on back and flanks in summer; winter color dark grayish brown on upper half of body, paler below, and spots on body indistinct; antlers palmate, if present . . . . . . . . . . . . . . . . . . . . . . . . . . . . . . . . . . . . Fallow Deer, *Cervus dama*, p. 275

4a. Coloration dark brown, paler on legs in adults; upper lip long, thick, overhanging the lower; neck and tail short; dewlap present; antlers, if present, without brow plate extending down over nose between eyes . . . . . . . . . . . . . . . . . . . . . . . . . . . . . . . . . . . . Moose, *Alces alces*, p. 269

**REFERENCES**

Anderson, Sydney, and Jones, J. Knox, Jr. 1967. *Recent mammals of the world.* New York: Ronald Press.

Burt, William Henry. 1957. *Mammals of the Great Lakes Region.* Ann Arbor: University of Michigan Press.

Hall, E. Raymond, and Kelson, Keith R. 1959. *The mammals of North America.* 2 vols. New York: Ronald Press.

Kellogg, Charles E. 1941. Climate and soil. In *Yearbook of agriculture, climate, and man*, pp. 265–91. Washington, D.C.: U.S. Department of Agriculture.

Scheffer, Victor B., and Rice, Dale W. 1963. A list of the marine mammals of the world. Washington, D.C.: *U.S. Fish and Wildl. Serv. Spec. Rept. Fish.* 431:1–12.

Simpson, George Gaylord. 1945. The principles of classification and a classification of mammals. *Bull. Amer. Mus. Nat. Hist.* 85:1–350.

# New England

**Physiographic Features.** New England comprises the states of Maine, New Hampshire, Vermont, Massachusetts, Connecticut, and Rhode Island, an area of 67,384 square miles. The region lies between 41° and 47° north latitude and between 73° and 67° west longitude. It is bounded on the north by Canada, on the east by the Atlantic Ocean, on the south by Long Island Sound, and on the west by the State of New York.

New England has been modified by the retreat and advance of the continental ice sheets which began about a million years ago. The ice overrode all the mountains, including Mount Washington (6,290 ft) in New Hampshire and Mount Katahdin (5,268 ft) in Maine. At its maximum extension the ice sheet passed off the coasts of Maine, New Hampshire, and Massachusetts—the shoals of big submarine banks near Cape Cod—and marked the extreme southeastern limit at Nantucket Island, Massachusetts. In the Gulf of Maine, it surmounted the eminences of Mount Desert and all the rock-bound coast.

The ice left large U-shaped valleys, smooth-sided mountains, irregular hills of gravel and boulders, several thousand scoured lakes, peat bogs, and a heterogeneous mantle of glacial drift a few inches to several feet thick, ranging from stone-free silt loam to boulder fields where stones made agriculture impossible until they

Rocky coast at Cape Elizabeth, York County, Maine.

John Norton, Maine Department of Commerce and Industry, Augusta.

The Green Mountain range near Waterbury, Washington County, Vermont.

were cleared away. Some of these boulders were built into networks of stone walls.

**Original Forest.** The original forest of New England consisted mainly of mixtures of coniferous and broad-leaved trees. Immense tracts of white pine stretched across Massachusetts to the Berkshires and extended well up the rivers of Maine and to Lake Winnipesaukee in New Hampshire. The pines reached still farther north through the Connecticut Valley and its tributaries, and from the borders of Lake Champlain as far inland as the base of the Green Mountains in Vermont. The better upland soils supported hemlock, fir, and spruce, as well as white pine and mixtures of beech, birch, and sugar maple in the north and oak, hickory, and chestnut in the south. Pitch pine was perhaps more common than white pine in the Cape Cod region and in the Connecticut valley in Connecticut.

**Forest Zones.** New England is divided into six major natural forest zones, in each of which the forest has a distinctive character. The names used to designate the six zones are based upon several tree species that characterize the zone rather than a single species of high economic value. The following brief account of the zones is taken from Westveld et al. (1956).

The Spruce–Fir–Northern Hardwoods Zone contains pure stands of spruce-fir, which commonly occur on high slopes, old fields, and swamps. Red spruce predominates on the high slopes; white spruce is frequently encountered on old fields as an associate; black spruce, tamarack, and northern white cedar occur in swamps. Beech, sugar maple, and particularly yellow birch are commonly found on the lower slopes and well-drained flats. Red maple, aspen, paper birch, cedar, black ash, tamarack, and white pine occur frequently on the shallower, more slowly drained soils. Hemlock is more common at the lower elevations and in some sections, such as eastern Maine, it may be the predominant conifer.

The Northern Hardwoods–Hemlock–White Pine Zone contains beech, red and sugar maple, and yellow birch, which are the predominant forests. Hemlock, white pine, and occasionally red and white spruce and balsam fir occur in mixture with some white ash, black cherry,

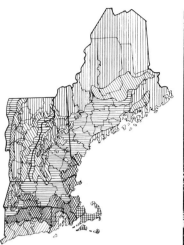

Spruce, fir, northern hardwoods
Northern hardwoods, hemlock, white pine
Transition hardwoods, white pine, hemlock
Central hardwoods, hemlock, white pine
Central hardwoods, hemlock
Pitch pine, oak

Natural forest vegetation zones of New England (from Westveld *et al.* 1955).

Mount Monadnock (3,166 ft.), at Jaffrey, Cheshire County, New Hampshire.

The Connecticut River at Hamburg Cove, Middlesex County, Connecticut.

Stone wall, Litchfield, Litchfield County, Connecticut.

sweet birch, paper birch, northern red oak, American elm, and basswood. White pine and hemlock and occasional spruce and fir are encountered in old-field associations. Red oak is common throughout this zone.

The Transition Hardwoods–White Pine–Hemlock Zone contains a variety of species that are characteristic of the zone adjacent to the north and south. A variety of oaks and hickories occupies hilltops and sandy terraces, along with white pine, some paper birch, quaking and bigtoothed aspen, and occasionally red pine. On the lower slopes and coves, white ash, red oak, and black birch are more abundant than to either north or south, but they are so widespread elsewhere that they can hardly be considered as indicators by themselves.

The Central Hardwood–Hemlock–White Pine Zone is characterized by black, red, and white oaks, chestnut, shagbark and bitternut hickories as the dominant species. Red maple, chestnut and scarlet oaks, and black birch are usually present. White birch is very local except in the northern fringe of the central hardwoods. White pine is abundant in some old fields and on sandy sites, but eastern red cedar tends to dominate old fields.

The Central Hardwoods–Hemlock Zone is characterized by hemlock, through over extensive areas it occurs only as scattered specimens. Various species of oaks and hickories are prominent hardwood species in this zone. Eastern red cedar is the characteristic tree of old fields and pastures.

The Pitch Pine–Oak Zone contains pitch pine and mixtures of white, black, and red oak compete for dominance. Repeated fires encourage the increase of pitch pine and scrub oak. Fires and repeated cuttings have left the majority of the tree stands in poor condition, with slow-growing, unhealthy trees.

**Climate.** The climate of New England is influenced by constant conflicts between dry arctic masses flowing out of the subpolar region to the west and the moisture-bearing tropical marine air from the south. Topography and configuration of the coast present varied influences with the moving air masses. In winter the great snowstorms usually occur with the northeasters, as a wedge of cold, dry air displaces the moist air.

Maine

Vermont

New Hampshire

Massachusetts

Connecticut     Rhode Island

List of counties of the states of New England.

MAINE: *1*, Aroostook; *2*, Somerset; *3*, Piscataquis; *4*, Penobscot; *5*, Washington; *6*, Hancock; *7*, Waldo; *8*, Knox; *9*, Lincoln; *10*, Kennebec; *11*, Franklin; *12*, Oxford; *13*, Androscoggin; *14*, Sagadahoc; *15*, Cumberland; *16*, York.

NEW HAMPSHIRE: *1*, Coos; *2*, Grafton; *3*, Carroll; *4*, Belknap; *5*, Sullivan; *6*, Merrimack; *7*, Strafford; *8*, Rockingham; *9*, Hillsborough; *10*, Cheshire.

VERMONT: *1*, Grand Isle; *2*, Franklin; *3*, Orleans; *4*, Essex; *5*, Chittenden; *6*, Lamoille; *7*, Caledonia; *8*, Washington; *9*, Addison; *10*, Orange; *11*, Rutland; *12*, Windsor; *13*, Bennington; *14*, Windham.

MASSACHUSETTS: *1*, Berkshire; *2*, Franklin; *3*, Hampshire; *4*, Hampden; *5*, Worcester; *6*, Middlesex; *7*, Essex; *8*, Suffolk; *9*, Norfolk; *10*, Bristol; *11*, Plymouth; *12*, Barnstable; *13*, Dukes; *14*, Nantucket.

CCONNECTICUT: *1*, Litchfield; *2*, Hartford; *3*, Tolland; *4*, Windham; *5*, Fairfield; *6*, New Haven; *7*, Middlesex; *8*, New London.

RHODE ISLAND: *1*, Providence; *2*, Kent; *3*, Washington; *4*, Bristol; *5*, Newport.

Mount Greylock (3,500 ft.), Berkshire County, Massachusetts.

Salt marsh–farmlands at Tiverton, Newport County, Rhode Island.

The northeasters are followed by the prevailing northwesterly winds, accompanied by clearing skies and temperatures often falling below zero in the north.

Winters are somewhat warmer near the coast than farther inland because the warmer waters of the Atlantic Ocean furnish heat to the lower layers of the atmosphere. Summers along the coast and in the mountains to the west are cooler than those of the inland lowlands. Precipitation is rather evenly distributed throughout the year in all the New England states.

## REFERENCES

Baldwin, John L. 1973. *Climates of the United States.* Washington, D.C.: U.S. Department of Commerce, National Oceanic and Atmospheric Administration.

Bromley, Stanley W. 1935. The original forest types of southern New England. *Ecol. Monogr.* 5(1):62–89.

Westveld, Marinus; Ashman, R. I.; Daldwin, H.I.; Holds, R. P.; Lutz, J.; Swain, Louis; and Standish, Myles. 1956. Natural forest vegetation zones of New England. *J. Forestry* 54(5):332–38.

# Order Marsupialia

The Marsupialia is one of the most remarkable of all mammalian orders because it has remained structurally unchanged for the past 50 million years. Its members are referred to as living fossils. In their method of reproduction marsupials are considered intermediate between the primitive egg-laying monotremes of Australia and its adjacent islands and the higher placental mammals, since their young are born live, but not fully developed.

In most families of marsupials the female has an external marsupium or pouch on her abdomen in which the undeveloped young are carried, nursed, and protected after birth. All marsupials possess fully developed epipubic bones. The order contains such mammals as the kangaroos, bandicoots, wallabies, and opossums. Marsupials may be arboreal, cursorial, fossorial, or semiaquatic.

Marsupials are found in the region of Australia, west to Celebes and the Moluccas; in North, Central, and South America; and they have been introduced on New Zealand (Van Deusen and Jones 1967). Only one family is represented in New England.

## FAMILY DIDELPHIDAE (OPOSSUMS)

Females of the family Didelphidae possess two uteri, which remain separate and do not fuse at the bottom as happens in true placental (eutherian) mammals. Opossums are small to medium in size; their limbs are short, and all four feet have five distinct digits, with the hallux, or great toe of the hind foot, opposable and clawless. The tail is long and partially naked and is prehensile. The marsupium is present in some genera and vestigial or absent in others. Some members of the family are omnivorous, others are carnivorous or insectivorous.

Representatives of the family are found from Argentina northward through Mexico, the eastern and central United States (also introduced in several areas in the western United States), and southeastern Canada (Ontario). In New England the family is represented by only one species.

## Virginia Opossum

*Didelphis virginiana* Kerr

*Didelphis virginiana* Kerr, 1792. *The animal kingdom*, p. 193.
Type Locality: Virginia
Dental Formula:
  I 5/4, C 1/1, P 3/3, M 4/4 × 2 = 50
Names: Possum and wooly shoat. The meaning of *Didelphis* = doubled womb, and *virginiana* = of Virginia.

**Description.** The adult opossum is about the size of a large domestic cat but has a heavier body and shorter legs. The head is cone-shaped with a pointed snout, beady eyes, and prominent thin, naked, leathery ears. The tail is long, tapering, scantily haired except at the base, scaly, and prehensile. The feet have five toes, each with a claw, except the hallux, which is clawless, thumblike, and opposable to the other toes of the hind foot. The position of the clawless toe and the fanlike characteristics of the front toes make the tracks of the opossum identifiable. Adult females have a prominent pouch lined with soft fur, containing from nine to seventeen mammae; the usual number is thirteen. The mammae are arranged in a horseshoe shape, with one or two in the center.

The skull is characterized by a small braincase, poorly developed auditory bullae, strong sagittal crest, and inflected angular process. The incisors are small and weak; the upper canines large and curved; premolars have compressed, pointed crowns; and molars have numerous sharp cusps.

The fur is coarse and rather long, and the whiskers are moderately long. The fur and skin coloration vary markedly from region to region.

In the northern latitudes opossums have relatively thick underfur, whitish at the base and occasionally tipped with black, which is overlain by a thin covering of pale guard hairs, giving the animal a gray, grizzled appearance ranging to grayish white. Opossums in southern latitudes have sparser underfur with a greater proportion of black hairs and thus are darker (McManus 1974). The head is pale yellow to white, except for a narrow black line down the center of the crown, with black ears usually edged with white. The belly appears more grizzled or darker than the back, where the white guard hairs are more abundant. The legs and feet are dark brown or black, with the toes partly white, and the tail is black at the base and usually flesh-colored for the distal (last) two-thirds of its length. The sexes are colored alike, with no apparent seasonal variations, and immatures are colored like adults. All-white, albino, and cinnamon-colored opossums occur infrequently. The molt is not marked, since individual hairs are usually lost a few at a time all over the body.

Males are larger than females. Measurements range from 610 to 813 mm (24–32 in); tail, 229 to

VIRGINIA OPOSSUM

330 mm (9–13 in); and hind foot, 51 to 76 mm (2–3 in). Weights vary from 1.8 to 5.4 kg (4–12 lb).

**Distribution.** Opossums are found throughout wooded areas of the eastern United States and southeastern Canada, south to Middle America (Gardner 1973). They have expanded their range northward and westward within the past century (Hamilton 1958) and have been introduced widely in areas of the western United States.

Opossums did not occur in New England before 1900 (Seton 1929). Earliest reports of the species in New England have been documented by Kirk (1921), Kennard (1925), and Seton (1929). The species occurs throughout Connecticut, Rhode Island, and Massachusetts, except the offshore islands, and is encountered in southern Vermont, New Hampshire, and extreme southern Maine.

**Ecology.** Opossums inhabit a wide variety of habitats ranging from dry to wet. They are found more frequently in the wetter areas of their distribution, particularly near streams and swamps. Since they are not adept at digging, for cover opossums use abandoned burrows and dens of other animals, cavities in trees, logs, or old stumps, rock piles, brush and wood piles, and trash piles, and hide underneath buildings and even in drainpipes. They seldom use the same den site on successive nights.

Nest building was described by Pray (1921), Smith (1941), and Layne (1951). In gathering nesting material the opossum grasps leaves in its mouth, passes them under the thorax with the forepaws, and places them on the tail, which is drawn forward between the hind legs. The tail encircles and carries the litter, freeing the legs for locomotion. This method is adapted for construction of nests in trees.

Man, dogs, foxes, bobcats, hawks, and the great horned owl prey on opossums, and hundreds are killed each year by automobiles. Opossums are parasitized by ticks, fleas, cestodes, nematodes, and trematodes. Mease (1929) and Roth and Knieriem (1958) reported tularemia and leptospira in opossums. Beamer, Mohr, and Barr (1960) stated that opossums are highly resistant to rabies.

**Behavior.** The opossum has a skull about the same size as a raccoon's, but the raccoon's brain is about five times larger. The small and primitive structure of the brain suggests the opossum's slow-witted behavior; yet the animal has expanded its range in North America to an amazing degree and spread widely in a short time.

Opossums are rather unsociable, and most encounters between adults are agonistic. The species is almost exclusively nocturnal: activity of captive opossums begins at or soon after dusk and continues until dawn. In spring and summer maximum activity occurs between 2300 and 0200 hr. In late autumn and winter activity is markedly depressed, particularly during cold temperatures, but hibernation does not occur (McManus 1971). Although Reynolds (1952) was able to keep groups of several females together in a large enclosure, McManus (1970) found that males could not be confined together without continual fighting, resulting in the death of the weakest. Newell and Yates (1969) observed that among captive opossums one male and several females slept in a group while the other males slept separately. Although the females did not fight among themselves, they threatened the subordinate males.

Opossums are adept climbers and take to trees to escape an enemy or to seek fruit. They climb hand over hand and use the tail for balance and support; but arboreal locomotion is typically slow and clumsy. The animals usually descend head first. On the ground they amble in a slow, awkward gait, the entire foot touching the ground and the tail swinging in pronounced lateral and vertical motions. The walking speed is from 0.3 to 1.0 miles per hour, averaging 0.7 mph, and the running speed is 4.6 mph (McManus, 1970).

While walking opossums often sniff the air and may stand on their hind feet for a second or two to look around. Their senses of smell and touch are well developed, but hearing is not keen and they seem myopic.

Opossums rarely turn on their pursuers but attempt to escape to the safety of trees or other protective cover. If they cannot reach cover, they either assume a defensive posture, accompanied by low hisses, growls, and screeches, or, more usually, feign death or "play possum." When feigning death, opossums become limp and motionless, usually with the mouth open and the angles of the lips drawn back to expose the front cheek teeth as if in a cynical grin, and the tongue protrudes slightly. Drooling may occur, and respiration becomes very shallow. This catatonic state may last from several minutes up to two to six hours (Wiedorn 1954), then death feigning may cease as suddenly as it began (McManus 1970).

Feigned death is a highly stereotyped behavior. Perhaps it evolved for survival, or perhaps it is physiological, caused by a nervous paralysis. Francq (1969) fround that opossums raised in cages could not be induced to feign death until they were 120 days old. The response can be induced by grasping the animal's body and shaking it roughly (Francq 1969) or by lifting the opossum by the tail and shaking it (McManus 1970). Death feigning may be a passive defense which compensates for the animal's slow run-

ning speed (James 1937); Francq (1967) suggested that this behavior increases its chance of escape when threatened by predators, because predators react differently to dead animals than to living ones. Opossums that are grasped commonly defecate, pass gas, and exude, from two papillae on either side of the anus, a foul-smelling greenish secretion that varies in consistency from a viscous liquid to nearly a paste (Wiedorn 1954; Francq 1969; McManus 1970). This secretion is probably a defense mechanism, since it is associated with other defense behaviors (McManus 1970).

Opossums are normally silent, but when angered they may emit a faint hissing or clicking sound with the teeth bared or give a low growl. They are generally clean, and caged opossums have been observed to lick their forepaws and face like a cat and to groom their fur after each meal; but like some other mammals, opossums drop their feces or scats anywhere, even in their nests, without making an effort to remove them (McManus 1970).

This species is at ease in the water and is a good though slow swimmer (Doutt 1954; Moore 1955). In swimming the animal uses two types of swimming patterns: the pacing stroke, in which both legs on one side move in unison, and a walking gait. In both strokes the tail performs sculling movements. When seeking refuge in water, opossums dive readily and can swim at least 20 feet underwater, keeping the eyes open. The maximum swimming speed is about 0.62 miles per hour, and distances of at least 328 feet can be covered with relative ease (Doutt 1954).

The opossum leaves its den at dusk and wanders about, never far from water, covering much ground in search of food. Its range depends on its food requirements and on the transient nature of the individual opossum. In Texas, Lay (1942) found that a sample of 29 individuals captured three or more times had an average home range of 11.5 acres, ranging from one-third of an acre to 58 acres. In Kansas, Fitch and Shirer (1970) determined the mean radius of activity from the nest site to be 2,034 feet. In Maryland, Llewellyn and Dale (1964) noted that ranges were typically elongated rather than circular, often following watercourses; the average home range length for a sample of 25 opossums was 0.6 mile. Verts (1963) found that movement was considerably greater in an intensively cultivated area than in partly wooded or wooded areas.

In autumn and winter the opossum devotes nearly twice as much time to feeding and nest construction or maintenance as at other times of the year (McManus 1971). The animal is not well adapted to subzero weather. Although it usually has a heavy layer of fat at the approach of winter, its fur has relatively poor insulative properties (Scholander et al. 1950). Its body temperature remains near 35°C throughout the year (McManus 1969). During extremely cold weather the animal usually remains in its den, but a hungry opossum may be seen in temperatures below zero. As a result, opossums suffer heavy frost damage to the ears and tail, and it is not uncommon to see one in the northern latitudes with the tips of its ears sloughed off and with a stubby tail. Layne (1951) reported a few instances of opossums biting off tail tips injured by frost.

The opossum laps water like a dog, and McManus (1970) reported that they drink between 7 and 17 ounces of water over a 24-hour period. Opossums are opportunists and eat almost anything, including carrion. Hamilton (1958) found from stomach analyses of 461 opossums taken in New York throughout the year that the bulk of their food consisted of animal material, chiefly insects and carrion, but that considerable plant material was eaten, particularly fruits and grains in season.

It has been recorded in folklore that the female is inseminated through her nostrils by the forked penis of the male and that the "fruits of conception" are blown into the pouch when newly born. This superstition stems from two observations: the bifurcated shape of the penis, and the fact that the undeveloped young are not seen in the pouch. This erroneous belief was enhanced by observations of the female licking out her pouch immediately before parturition.

Copulation occurs soon after the female comes into heat. The male straddles the female, clasping her hind legs with his hind feet and her nape with his jaws. The pair topple to one side and insertion of the penis occurs within 2 minutes. The male maintains his grasp for up to 20 minutes. Afterward the female resumes her aggressive attitude and resists any further advances of the male (McManus 1967).

The mating season extends from January or February to June or July (McManus 1974). In New York, peak periods of reproduction appear to be late January through late March and mid-May to early July (Hamilton 1958). In Connecticut, opossums produce two or three litters annually (Goodwin 1935). The estrous cycle averages 29.5 days (range 22 to 38), and the period of receptivity is not longer than 36 hours (Reynolds 1952). Receptivity apparently terminates with copulation (McManus 1967).

Parturition, which occurs approximately 13 days after copulation, has been described by Hartman (1920). When birth begins, the female sits with the cloacal opening directed forward and begins licking the area. Spasmodic contractions of her abdominal wall force the young out of the opening. Immediately after the mother frees them from the chorionic fluid and afterbirth the young crawl upward, with an overhand stroke as if swimming, into the pouch, where each attaches to a nipple. They must find this or perish. Hartman (1920) noted that the newborn young can be made to travel upward even away from the pouch, "for they are negatively geotrophic." While attached to the nipples the young can breathe and swallow at the same time, since the passage from the nasal chamber to the larynx is separated from the passage to the esophagus (Hartman 1920).

The female produces from five to thirteen young, averaging eight. As many as twenty-one young have been born in one litter (Hartman 1923). At birth the embryolike young are almost transparent and are as small as honeybees. Each weighs about 0.13 g (0.0046 oz) and measures approximately 13 mm (0.51 in) long. The hind limbs are like embryonic buds, and the undeveloped mouth serves only to grasp the nipple. The forelimbs, however, are muscular, with well-developed claws, which enables the young to make the perilous trip to the pouch.

The young remain attached to the nipples for some 50 to 65 days, during which they develop rapidly. Weaning occurs between 95 and 105 days after parturition, but the young may eat solid food before weaning. The nipples of the female lengthen and enlarge within the mouths of the young, making detachment unlikely (McManus 1974). The eyes open at approximately two months of age, and after this the young leave the pouch for short periods of time.

Some of them usually spend part of the time in the pouch, but most prefer to ride on the mother's back, clinging with their feet and tails to her fur. They stay with the mother for about one more month, returning to the pouch when necessary for milk or protection, before they begin to fend for themselves completely. Usually at least two weeks must elapse between the weaning of the first litter and the birth of the second, since lactation generally inhibits the recurrence of estrus and the greatly enlarged nipples must shrink to allow the newly born young to attach themselves (Hartman 1920).

Females become sexually mature at one year of age. Hartman (1923) reported that opossums live at least 7 years, and Petrides (1949) computed an average longevity of 1.33 years, with a 4.8-year turnover in the population. Llewellyn and Dale (1964) felt that Hartman's figure was far too high and that even Petrides's estimate was slightly high.

**Sex Determination.** The sex of live adults can be recognized by the pouch and mammae on the belly of females and by the pendulant scrotum located anterior to the forked penis in males. In cased pelts, sex can be determined by the presence of the pouch of the female, or by a black-blotched patch of heavily pigmented skin characteristic of the male's scrotum. Also, the chest of males is often stained yellowish, perhaps by skin glands whose function is unknown.

In pouch young, sex can usually be recognized after the second week by the presence of a rudimentary scrotum in the male or by minute mammae and outlines of a pouch in the female.

**Age Determination.** The age of immatures can be determined by the dental formula as described by Petrides (1949). At three months, $I$ 5/2, $C$ 1/1, $P$ 2/2, $M$ 0/0; at four months, $I$ 5/4, $C$ 1/1, $P$ 3/3, $M$ 1/2; at five to eight months, $I$ 5/4, $C$ 1/1, $P$ 3/3, $M$ 2/3; and at seven to eleven months, $I$ 5/4, $C$ 1/1, $P$ 3/3, $M$ 3/4.

**Specimens Examined.** VERMONT, total 2: Rutland County, Tinmouth 1 (UCT), Wells 1 (UVT).

MASSACHUSETTS, total 6: Hampden County, Westfield 1; Hampshire County, Hadley 1, Holyoke 2, Pelham 1; Worcester County, Southboro 1 (UMA).

CONNECTICUT, total 29: Fairfield County, Fairfield 4, Monroe 1, Stratford 1, Westport 1; Hartford County, Bristol 1, East Berlin 2, Newington 1, Southington 1; Litchfield County, Terryville 1; New Haven County, Branford 1, Milford 1, Orange 1; New London County, Canterbury 1, Lebanon 1; Tolland County, Coventry 1, Ellington 1, Mansfield 2, Storrs 2; Windham County, Pomfret 1, Scotland 1, Westford 2, Windham 1 (UCT).

RHODE ISLAND, total 8: Washington County, Kingston 1–5–1 (UCT–UME–URI), Matunuck 1 (URI).

## REFERENCES

Beamer, P. D.; Mohr, C. O.; and Barr, T. R. B. 1960. Resistance of the opossum to rabies virus. *Amer. J. Vet. Res.* 21(82):507–10.

Doutt, Kenneth J. 1954. The swimming of the opossum (*Didelphis marsupialis virginiana*). *J. Mamm.* 35(4): 581–83.

Fitch, Henry S., and Shirer, Hampton W. 1970. A radiotelemetric study of spatial relationships in the opossum. *Amer. Midland Nat.* 84(1):170–86.

Francq, Edward N. 1967. Feigned death in the opossum, *Didelphis marsupialis*. Ph.D. diss., Pennsylvania State University, College Park.

———. 1969. Behavioral aspects of feigned death in the opossum, *Didelphis marsupialis*. *Amer. Midland Nat.* 81(2): 556–68.

Gardner, A. L. 1973. The systematics of the genus *Didelphis* (Marsupialia: Didelphidae) in North and Middle America. *Spec. Pub. Mus., Texas Tech. Univ.* 4:1–81.

Goodwin, George Gilbert. 1935. The mammals of Connecticut. *Connecticut Geol. Nat. Hist. Surv. Bull.* 53:1–221.

Hamilton, William J., Jr. 1958. Life history and economic relations of the opossum (*Didelphis marsupialis virginiana*) in New York State. *Cornell Univ. Agric. Exp. Sta. Mem.* 354:1–48.

Hartman, Carl G. 1920. Studies in the development of the opossum (*Didelphis virginiana* L.). V. The phenomena of parturition. *Anat. Record* 19(5): 251–56.

———. 1923. Breeding habits, development and birth of the opossum. *Smithsonian Rept.* (for 1921), pp. 347–64.

James, W. T. 1937. An experimental study of the defense mechanism in the opossum, with emphasis on natural behavior and its relation to mode of life. *J. Genet. Psychol.* 51:95–100.

Kennard, Fred H. 1925. The Virginia opossum in

Massachusetts and New Hampshire. *J. Mamm.* 6(3):196.

Kirk, George L. 1921. Opossum in Vermont. *J. Mamm.* 2(2):109.

Lay, Daniel W. 1942. Ecology of the opossum in eastern Texas. *J. Mamm.* 23(2): 147–59.

Layne, James N. 1951. The use of the tail by an opossum. *J. Mamm.* 32(4):464–65.

Llewellyn, Leonard M., and Dale, Fred H. 1964. Notes on the ecology of the opossum in Maryland. *J. Mamm.* 45(1):113–22.

McManus, John L. 1967. Observations on sexual behavior of the opossum (*Didelphis marsupialis*). *J. Mamm.* 48(3):486–87.

———. 1969. Temperature regulation in the opossum, *Didelphis marsupialis virginiana*. *J. Mamm.* 50(3):550–58.

———. 1970. Behavior of captive opossums, *Didelphis marsupialis virginiana*. *Amer. Midland Nat.* 84(1):144–69.

———. 1971. Activity of captive *Didelphis marsupialis*. *J. Mamm.* 52(4):846–48.

———. 1974. *Didelphis virginiana. Mammalian Species* no. 40, pp. 1–6. Amer. Soc. Mammalogists.

Mease, J. A. 1929. Tularemia from opossums. *J. Amer. Med. Assoc.* 92:1042.

Moore, Joseph Curtis. 1955. Opossum taking refuge under water. *J. Mamm.* 36(4):559.

Newell, Terry G., and Yates, Allen T. 1969. The behavior of group reared opossums (*Didelphis virginiana*) in captivity. *Amer. Zool.* 9(3):572.

Petrides, George A. 1949. Sex and age determination in the opossum. *J. Mamm.* 30(4):364–78.

Pray, Leon L. 1921. Opossum carries leaves with its tail. *J. Mamm.* 2(2):109–10.

Reynolds, Harold C. 1952. Studies on reproduction in the opossum *Didelphis virginiana virginiana. Univ. Calif. Pub. Zool.* 52(3):223–84.

Roth, E. E., and Knieriem, B. B. 1958. The natural occurrence of *Leptospira pomona* in an opossum—a preliminary report. *J. Amer. Vet. Med. Assoc.* 132(3):97–98.

Scholander, P. F.; Walters, V.; Hock, R.; and Irving, L. 1950. Body insulation of some arctic and tropical mammals and birds. *Biol. Bull.* 99:225–36.

Seton, Ernest Thompson. 1929. *Lives of game animals.* Vol. 4, pt. 2. Garden City, New York: Doubleday, Doran and Co.

Smith, Luther. 1941. An observation on the nest-building behavior of the opossum. *J. Mamm.* 22(2):210–2.

Van Deusen, Hobart M., and Jones, J. Knox, Jr. 1967. Marsupials. In *Recent mammals of the world,* ed. S. Anderson and J. K. Jones, Jr., pp. 61–68. New York: Ronald Press.

Verts, B. J. 1963. Movements and populations of opossums in a cultivated area. *J. Wildl. Manage.* 27(1):127–29.

Wiedorn, W. S. 1954. A new experimental animal for psychiatric research: The opossum, *Didelphis virginiana. Science* 119:360–61.

# Order Insectivora

THE INSECTIVORA comprises a diverse group of mammals of doubtful relationships and many specializations. Insectivores are considered the most primitive of true placental mammals, and they are the smallest of all mammals. They are characterized by a long snout that extends well beyond the mouth, beady eyes, and a slender skull, usually without a zygomatic arch.

Insectivores are voracious and consume a great amount of food daily. They feed mainly on insects, although some species also eat vertebrate animals. Insectivores may be terrestrial, fossorial, or semiaquatic.

These mammals are found on all large land masses except Australia, Greenland, the southern half of South America, and the polar regions (Findley 1967). The order is represented in New England by two families.

## FAMILY SORICIDAE (SHREWS)

Shrews are elusive mouselike mammals with long pointed snouts, short legs, and tiny eyes often hidden in short, velvety fur. Their fur is well suited to subterranean life because the hairs are not displaced no matter what direction the shrew moves. Shrews are active throughout the year, and although most species are chiefly nocturnal, some are active day and night. They are extremely nervous and restless and may die of shock from a sudden fright. They have a high metabolic rate and must feed almost continuously or die of starvation.

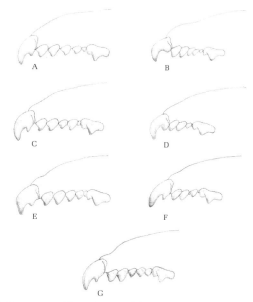

Left upper partial tooth rows of shrews. A, Sorex cinereus; B, Blarina brevicauda; C, Sorex palustris; D, Cryptotis parva; E, Sorex fumeus; F, Microsorex thompsoni; G, Sorex dispar.

The Soricidae contains the smallest mammal in the world, the musk or dwarf shrew, which weighs only about 2 g (0.08 oz).

The family is cosmopolitan in distribution, except for the polar regions, Australia, Greenland, and southern South America (Findley 1967). There are seven species in New England.

## Masked Shrew

Sorex cinereus cinereus Kerr

Sorex arcticus cinereus Kerr, 1792. The animal kingdom, p. 206.
Sorex cinereus cinereus, Jackson, 1925. J. Mamm. 6(1):56.
Type Locality: Fort Severn (Severn Settlement, now Severn), mouth of the Severn River, southwest side of Hudson Bay, Ontario, Canada
Dental Formula:
    I 3/1, C 1/1, P 3/1, M 3/3 × 2 = 32
Names: Common shrew, cinerous shrew, copper shrew, and shrewmouse. The meaning of Sorex = a shrew, and cinereus = ashy or the color of ashes.

**Description.** The masked shrew has a sharply pointed snout, beady eyes, and small ears nearly hidden in fine, soft, thick, velvety fur; the feet are delicate, each with five sharp, slender, weak claws; the tail is long, almost three-fourths the length of the head and body, and covered with hair. The masked shrew is distinguished from the smoky shrew by its smaller size and lighter color and from the least shrew by its much longer tail.

The skull is fragile, with a narrow rostrum. The incisors are sharp and pointed, project forward, and are notched posteriorly. There are five small unicuspidate teeth behind the upper incisors. The fifth unicuspid tooth is minute, the fourth is generally smaller then the third, and both of

these are smaller than the first and second. The tips of the teeth are dark chestnut.

The hairs everywhere are slate-colored at the base. The coloration in winter is dark brown to almost black on the upperparts, lighter brown or grayish on underparts; in summer, coloration is somewhat lighter and more brownish. The feet are whitish. The tail above is colored like the back and below is yellowish white. The immature pelage is very much like that of the adult. Color variants are rare, but albinos, white banded, and Himalayan white pattern masked shrews have been reported. Spring molt takes place from April to June, and the autumn molt is usually completed in October.

The sexes are equal in size. Measurements range from 85 to 110 mm (3.3–4.3 in); tail, 35 to 45 mm (1.4–1.8 in); and hind foot, 10 to 14 mm (0.39–0.55 in). Weights vary from 3.5 to 6.0 g (0.12–0.21 oz).

**Distribution.** The masked shrew is the most widely distributed shrew in North America. It occurs from Alaska through all of Canada, south in the Appalachian Mountains to North Carolina and Tennessee, in the northern half of the lower peninsula of Michigan, and through the mountains of Idaho, western Montana, and western Wyoming to northern New Mexico and to northeastern and central Washington.

**Ecology.** This species lives mainly in damp deciduous and coniferous woods underneath leaf litter, logs, and rock piles. It prefers the cover of damp swamp bogs of spruce and cedar, hemlock ravines, mossy banks, and brushy areas near streams that do not freeze or become stagnant. It is occasionally found in salt marshes but rarely if ever in dry fields or woods. At times it inhabits deserted buildings. Woolfenden (1959) discovered twenty masked shrews under the roof section of a dismantled shed. This species spends most of its time in underground runways that it constructs or in runways and tunnels that have been dug by mice and other small mammals. The nest, situated in a cavity under a log, stump, rock, or such, is a flattened sphere some 3 inches in diameter, made of leaves, grasses, and fine rootlets.

Mammalian predators kill many shrews, but few eat them except perhaps when very hungry. Owls, hawks, and herons, and shrikes, weasels, foxes, cats, and short-tailed shrews are known to prey on masked shrews. Many shrews die from floods, accidents, or being trapped in postholes, ditches, and springs. Many perish from extreme temperatures and from shock due to fright. Ticks, fleas, chiggers, and mites are known to parasitize masked shrews.

**Behavior.** Masked shrews are active throughout the year, day and night, but mainly at dusk. Doucet and Bider (1974) found that the greatest period of activity for this species was on cloudy nights. The senses of sight and smell are poor, but touch is well developed.

Masked shrews are restless, pugnacious, and voracious. They will fight with little provocation and will devour their own kind. Tuttle (1964) stated that frightened masked shrews "would usually disappear almost immediately into small burrows, but if surprised away from a burrow, they could vanish almost as rapidly by merely forcing their way directly down into the loose 'sod' below. They would soon reappear uttering their high-pitched calls. Repeatedly the shrews would answer each other's calls and follow them until they came together in one of their minute runways. They would then rear up on their hind legs, striking each other in the face until one or both fell over backwards, at which time they would run away in opposite directions and cease to call for several minutes."

Masked shrews are good swimmers but seldom enter the water. Although terrestrial, they climb low bushes, small branches, and fallen trees. They venture about the surface of the ground or snow, skipping from tree to tree and poking their delicate snouts into crannies in search of food, which they need in large quantities to sustain their high energy requirements. Blossom (1932) recorded that a captive masked shrew ate 3.3 times its own weight every 24 hours.

MASKED SHREW

As far as is known, this species does not store food. It consumes a great variety of insects, worms, centipedes, slugs, snails, mollusks, and spiders as well as vegetable matter such as moss and seeds, and the flesh of mice and other shrews. Horváth (1965) reports seeing a masked shrew prey on the young of a solitary vireo: the bird's nest was 5½ feet above the ground.

Little is known about the reproduction of this species. The breeding season may extend from March to September, and there may be as many as three litters in a single season. The gestation period is probably 18 days, and from two to ten young are produced per litter, the usual number being seven. At birth the young are blind, naked, and helpless and weigh about 0.10 g (0.0035 oz). They grow rapidly, attaining half the size of adults in 10 days, and are fully furred with short hair. The young shift for themselves when they are nearly one month old. The male remains with the female during the early life of the young.

Short (1961) reported that sexual maturity is attained when the animals are 20 to 26 weeks old, and mating occurs in the first year of life, indicating that masked shrews born in early spring probably produce young in their first fall. Connor (1960) suggested that mating may follow birth of the young.

**Specimens Examined.** MAINE, total 77: Aroostook County (no location) 6 (UCT), Madawaska 5 (HMCZ), Orient 1 (UCT); Franklin County, Mt. Tumbledown 7 (UME); Hancock County, Mt. Desert Island 1 (HMCZ); Kennebec County, Oakland 2 (UCT); Oxford County, Fryeburg 1 (AMNH), Upton 1 (HMCZ), Woodstock 1 (UMA); Penobscot County, Bradley 1 (UME), Brewer 1 (HMCZ), Eddington 1, Halden 1 (UME), Lakeville 1 (UCT), Lincoln 2 (UME), Orono 3–10 (HMCZ–UME), South Twin Lake 11 (AMNH); Piscataquis County (no location) 3 (UCT), Mt. Katahdin 1–1 (AMNH–DC); Sagadahoc County, East Harpswell 3; Somerset County, Enchanted Pond 5 (AMNH); Washington County (no location) 3 (UCT), Princeton 1 (UME), Vanceboro 4 (UCT).

NEW HAMPSHIRE, total 304: Carroll County, Albany 2, Bartlett 49, Jackson 95, Harts Location 65 (UCT), Ossipee 1 (USNM); Coos County, Colebrook 1 (UCT), College Grant (Errol) 1–7 (AMNH–DC), Fabyan 2 (USNM), Jefferson–Randolph 3 (UCT), Mt. Washington 8–1–6 (AMNH–UCT–USNM), Twin Mt. 3 (UCT); Grafton County, Franconia 1 (HMCZ), Hanover 3 (DC), Livermore 44 (UCT), Lyme 1, Mt. Moosilauke 5 (DC); Hillsborough County, Brookline 1 (UCT), Greenfield 1 (DC), Hollis 1;

Rockingham County, Hampton Falls 1 (HMCZ); Strafford County, Durham 1–1 (HMCZ–UNH).

VERMONT, total 103: Bennington County, Readsboro 4 (AMNH); Caledonia County, Burke 2, East Haven 1, Hardwick 10, Lyndon 26 (UCT); Chittenden County, Burlington 1 (USNM); Essex County, Brighton 5, Victory 5 (UCT); Lamoille County, Cambridge 1 (DC), Mt. Mansfield 4 (USNM); Orleans County, Westmore 1 (UVT); Orange County, Williamstown 1 (UCT); Rutland County, Chittenden 3 (AMNH), Clevendon 1 (UCT), East Wallingford 15 (DC), Mendon 1 (UCT), Mt. Pico 5 (AMNH), Sherburne 8, Wallingford 1 (UCT); Washington County, Duxbury 5 (UVT), Plainfield 1 (UCT); Windham County, Newfane 1 (AMNH); Windsor County, Pomfret 1 (DC).

MASSACHUSETTS, total 138: Barnstable County, Barnstable 1–2 (HMCZ–UCT), Chatham 1 (USNM), Falmouth 1, Provincetown 2, South Dennis 1, South Wellfleet 1 (HMCZ), West Falmouth 1 (USNM), Woods Hole 2–1 (AMNH–USNM), Yarmouth 1 (HMCZ); Berkshire County, Adams 6 (UCT), Monterey 1 (UMA), Mt. Greylock 2 (HMCZ), Peru 2 (UCT), Williamstown 1 (USNM); Dukes County, Martha's Vineyard 1 (USNM); Essex County, Andover 1 (UCT), Danvers 1, Hamilton 1 (HMCZ), Parker River National Wildlife Refuge 12 (USNM), Topsfield 1 (UCT); Franklin County, Leverett 1, Sunderland 1 (UMA), Warwick 2; Hampden County, Brimfield 3, Chester 1 (UCT), Feltonville 1 (HMCZ), Holland 1, Wales 12 (UCT); Hampshire County, Amherst 2 (UMA), Chesterfield 1 (UCT), Mt. Toby 4, Williamsburg 1 (UMA); Middlesex County, Ayer 1 (UMA), Bedford 3 (HMCZ), Cambridge 11–5 (HMCZ–USNM), Concord 1 (UMA), Lexington 1 (HMCZ), Wakefield 10, Wilmington 4, Woburn 1 (USNM); Nantucket County, Nantucket 7–3 (HMCZ–UCT); Norfolk County, Randolph 2 (HMCZ); Plymouth County, Marshfield 2, Middleboro 4 (USNM), Wareham 4 (HMCZ), West Bridgewater 1 (UMA); Worcester County, Harvard 2–2 (HMCZ–USNM).

CONNECTICUT, total 186: Fairfield County, Easton 1, Weston 4, Westport 3, Wilton 2 (UCT); Litchfield County, Bear Mt. 1 (AMNH), Kent 1–4 (AMNH–UCT), Macedonia Park 1 (AMNH), Norfolk 1, Salisbury 1 (UCT); New Haven County, Hamden 1, Short Beach 1 (AMNH), Southbury 5, Waterbury 5; New London County, Barn Island 3, Griswold 6, Groton 3, Hopeville 4, Lebanon 1, Mystic 1 (UCT), Stonington 1–3 (UCT–USNM), Voluntown 4, Waterford 1; Tolland County, Columbia 3, Coventry 5, Mansfield 34, Storrs 12, Tolland 2, Union 45; Windham County, Ashford 2, Brooklyn 10, Chaplin 1, Coventry 2 (UCT), Putnam 1 (AMNH), Scotland 1 (UCT), South Woodstock 2 (AMNH), Sterling 3, Strafford 1, Thompson 2, Windham 2 (UCT).

RHODE ISLAND, total 16: Newport County, Newport 1; Washington County (no locations) 15 (USNM).

WATER SHREW

# Water Shrew

*Sorex palustris albibarbis* (Cope)

*Neosorex albibarbis* Cope, 1862. *Proc. Acad. Nat. Sci. Philadelphia* 14:188.
*Sorex palustris albibarbis*, Rhoads, 1903. *The mammals of Pennsylvania and New Jersey*, p. 191.
Type Locality: Profile Lake, Franconia Mountains, Grafton County, New Hampshire
Dental Formula:
I 3/1, C 1/1, P 3/1, M 3/3 × 2 = 32
Names: White-lipped water shrew, big water shrew, Cope's water shrew, eastern marsh shrew, and black and white shrew. The meaning of *Sorex* = a shrew, and *palustris* = marshy, and *albibarbis* = white-bearded.

**Description.** The water shrew is the largest long-tailed shrew in New England. It is especially adapted for a semiaquatic life, having big hind feet with a fringe of short stiff hairs, the third and fourth hind toes being joined by a thin web at the base for slightly more than half their length, an adaptation for swimming. The eyes are beady, ears small and hidden in velvety fur, and snout long, pointed, and slightly decurved. Females have six mammae.

The rostrum is narrow and short, and the anterior part of the premaxillae is nearly as deep as the middle part. The third unicuspid is smaller than the fourth.

The sexes are colored alike and show slight seasonal color variation. In winter the upperparts are dark, fuscous black or gray-black with some of the hairs white-tipped, producing a frosted appearance. The underparts are a paler grayish white. The tail is distinctly bicolor; brownish-black above, paler below. The chin and lips are pale gray. The outer sides of the feet are dark, the inner sides whitish. The summer pelage is more brownish above and slightly paler below, with a less frosted appearance. The color of immatures is much like that of adults, and color variants apparently have not been recorded. The fall molt takes place during late August or early September; the spring molt seems to occur during May or early June.

The sexes are equal in size. Measurements range from 137 to 157 mm (5.3–6.1 in); tail, 62 to 89 mm (2.4–3.5 in); and hind foot 18 to 20 mm (0.70–0.78 in). Weights vary from 10 to 17 g (0.35–0.60 oz).

**Distribution.** The distribution of this subspecies is from Nova Scotia and southern Quebec to British Columbia, the New England states, and eastern and southwestern New York, and south to northeastern Pennsylvania.

**Ecology.** This species is found in wet areas along streams, ponds, and lakes. It prefers heavily wooded areas and is rarely found in marshes that are devoid of bushes and trees. It may be found in beaver lodges and muskrat houses in winter. Its nest is usually made of dry moss.

Weasels, mink, otters, hawks and owls, snakes, black bass, trout, pickerel, and other fishes prey on water shrews. Fleas, cestodes, and nematodes have been reported to parasitize them.

**Behavior.** Water shrews are secretive and elusive, seeking cover along the waterways. They are active during most of the winter; although mainly nocturnal, they are also active at dusk, on dull, cloudy days, and in shaded areas on sunny days. The sense of sight seems to be poorly developed, but smell and touch appear well developed. Water shrews swim well and can dive, turn, twist, and run along the bottom of the water, employing the four feet with the same motions used in running on the ground.

This species is said to run or skate on the surface of the water. In one instance Jackson (1961) observed a water shrew run more than 5 feet across the smooth surface of a pool. The animal's head and body were entirely out of the water, supported by surface tension, and at each step the animal took there appeared to be a little globule of air held by the hair fringe of the hind feet.

Water shrews feed mainly on aquatic insects, chiefly mayflies, caddis flies, stone flies, and other flies and beetles and their larvae. Snails, flatworms, small fish, and fish eggs are also eaten.

Little is known about the reproduction of the water shrew, but the species apparently has an extensive breeding season. Conaway (1952), who studied the subspecies *navigator*, collected pregnant females in March, suckling females the first week of June, half-grown young early in July, a female with five small embryos on 2 August, and a male with enlarged testes on 9 August. He reported that the gestation period is probably about 21 days, the number of embryos from four to eight, and longevity about 18 months.

**Specimens Examined.** NEW HAMPSHIRE, total 20: Coos County, Brenton Woods 2 (USNM), Carroll 1 (UCT), College Grant (Errol) 1 (DC), Jefferson 1 (UCT), Mt. Washington 8–1 (AMNH–DC); Grafton County, Franconia 1 (DC), Profile Lake 2 (USNM), Mt. Moosilauke 2 (DC); Strafford County, Durham 1 (UNH).

VERMONT, total 52: Bennington County (no location) 1 (UVT); Caledonia County, Lyndon 1 (UCT); Chittenden County, Duxbury 1 (UVT); Essex County, Brighton 1–3 (HMCZ–UCT), Island Pond 7, Maidstone 1; Lamoille County, Cambridge 1 (DC); Rutland County, Chittenden 2 (AMNH), East Wallingford 15 (DC), Mendon 3–3–1 (AMNH–UCT–USNM), Rutland 1–1–1 (AMNH–DC–UCT), Sherburne 1–3 (DC–UCT); Washington County, Plainfield 2 (UCT); Windsor County, Pomfret 2–1 (AMNH–DC).

MASSACHUSETTS, total 4: Berkshire County, Monterey 1; Franklin County, Montague 2, Shutesbury 1 (UMA).

CONNECTICUT, total 16: Litchfield County, Bear Mt. 1, Macedonia Park 3 (AMNH), Sharon 4–2 (AMNH–UVT), Salisbury 1; New London County, East Lyme 1; Tolland County, Bigelow Hollow 1, Vernon 1, Willington 1; Windham County, Hampton 1 (UCT), South Woodstock 1 (AMNH).

# Smoky Shrew

*Sorex fumeus*

There are two subspecies of smoky shrews recognized in New England:
*Sorex f. fumeus* Miller, 1895. *N. Amer. Fauna* 10:50.
Type Locality: Peterboro, Madison County, New York
*Sorex f. umbrosus* Jackson, 1917. *Proc. Biol. Soc. Washington* 30:149.
Type Locality: James River, Antigonish County, Nova Scotia (Hall and Kelson 1959)
Dental Formula:
I 3/1, C 1/1, P 3/1, M 3/3 × 2 = 32
Names: Northern smoky shrew, smoky mountain shrew, and gray shrew. The meaning of *Sorex* = shrew, and *fumeus* = smoky, referring to the general appearance of the animal.

**Description.** The smoky shrew resembles the masked shrew but is larger, stouter, and darker, with more prominent ears, shorter tail, bigger feet, and somewhat paler coloration. The sexes are colored alike and show seasonal color variation. In winter the upperparts are deep mouse gray or slate-colored, with some hairs whitish-tipped, and the underparts are slightly paler, almost silvery. In summer the color is somewhat darker and browner. The tail is bicolor in all seasons, fuscous above, yellowish below. The feet are pale buffy, the outer edges dusky. Apparently albino and melanistic specimens have not been recorded. Old animals may have small white patches on the thighs or some white-tipped hairs.

The skull is relatively broad and short, with a large, somewhat flattened cranium and a broad interorbital region. The molariform teeth are rather deeply emarginate (notched) posteriorly, and the unicuspids are broader than long, while the third unicuspid is larger than the fourth.

There are two molts each year. The summer pelage is attained from April to late June and the winter pelage is acquired by mid-September to early November. The molt is completed in several weeks and may be well marked or show little demarcation between the winter and summer pelage. Molting usually begins on the belly, extending over the shoulders and middle of the back, and ends about the ears and rump.

The sexes are equal in size. Measurements range from 95 to 129 mm (3.7–5.0 in); tail, 37 to 49 mm (1.4–2.0 in); and hind foot 11 to 14 mm (0.43–0.55 in). Weights vary from 6 to 10 g (0.21–0.35 oz).

**Distribution.** This species occurs from southern Ontario, through the east and north sides of the Saint Lawrence River, south from southern Maine, to most of New Hampshire and Vermont, much of Massachusetts, Connecticut, and Rhode Island, west to central Ohio and Kentucky, and south to northwestern Georgia. Jackson (1961) reported that a smoky shrew was collected in southeastern Wisconsin in 1853.

**Ecology.** Smoky shrews are essentially a northern and mountain species. They prefer cool, damp woods and bogs with deep, loose leaf

litter and friable soils, and areas near streams with moss-covered boulders and logs. The nest is 4 to 19 inches underground, beneath a stump or rotten log, and is roughly the size of a baseball—more compact and somewhat smaller and bulkier than that of the short-tailed shrew. It is lined with shredded grasses, leaves, and hair.

Short-tailed shrews, weasels, foxes, bobcats, owls, and hawks prey on smoky shrews. Fleas, mites, and nematodes are known to parasitize them.

**Behavior.** Little is known about the biology of this species. Smoky shrews are generally active at all hours throughout the year. Their weak feet are not adept at digging, and they use the runways of other small mammals. They appear to be social animals.

Captive smoky shrews are not prodigious feeders and do not eat the equivalent of their body weight daily. Insects are their main food, but they also eat spiders, salamanders, snails, mammals, and birds (Hamilton 1940).

In New York the breeding season occurs from late March into early August. The gestation period is about 20 days, and there may be as many as three litters each season, the last appearing in late August. From two to eight young are born per litter, the usual number being six. Smoky shrews are born naked and blind. After their season of reproduction, smoky shrews die of old age at 14 to 17 months of age (Hamilton 1940).

**Age Determination.** Immatures at least 8 to 11 months of age can be distinguished by their hairy tails, particularly at the tip. Adults have glabrous tails, often with numerous scars, and the tip of the tail is more or less blunt and rounded. They also show prominent tooth wear (Hamilton 1940).

**Specimens Examined.** MAINE, total 39: Aroostook County (no location) 12, Orient 1 (UCT); Franklin County, Dryden 8 (AMNH); Lincoln County, Boothbay Harbor 1 (HMCZ); Oxford County, Dixfield 1 (UCT); Penobscot County, Holden 5 (UME), Mud Pond 2 (AMNH); Piscataquis County, Greenville 2 (UCT), Mt. Katahdin 1 (DC), Squaw Brook 1; Washington County (no location) 3, Vanceboro 2 (UCT).

NEW HAMPSHIRE, total 112: Carroll County, Bartlett 8, Harts Location 17, Jackson 57 (UCT), Ossipee 4 (USNM), Washburn 1 (UCT); Cheshire County, Dublin 1 (HMCZ); Coos County, Carroll 1 (UCT), College Grant (Errol) 3–8 (AMNH–DC), Mt. Washington 1–2 (AMNH–DC); Grafton County, Franconia 1 (DC), Livermore 2 (UCT), Waterville 1 (USNM); Hillsborough County, Antrim 1 (HMCZ), Brookline 1 (UCT), Mt. Monadnock 1 (USNM), Peterboro 1; Merrimack County, Webster 1 (HMCZ).

VERMONT, total 96: Addison County, Ripton 1 (HMCZ); Caledonia County, Groton 1, Hardwick 1 (UCT); Essex County, Island Pond 1; Lamoille County, Cambridge 1 (DC), Mt. Mansfield 3 (USNM); Orange County, West Fairlee 1 (DC); Rutland County, Mendon 20–3–28 (AMNH–DC–UCT), Peru 1 (UVT), Rutland 1–3–6 (AMNH–DC–UCT), Sherburne 3–8 (DC–UCT); Windham County, Rockingham 1, Royalton 5, Saxton's River 2 (UCT); Windsor County, Norwich 2 (DC), Pomfret 1, West Bridgewater 1 (AMNH), Woodstock 2 (HMCZ).

MASSACHUSETTS, total 32: Berkshire County, Adams 1 (UCT), Mt. Washington 1–1 (AMNH–UCT), Peru 1 (UCT); Franklin County, Leverett 1, Montague 1, Quabbin Reservoir 1 (UMA), Whately 3; Hampden County, Brimfield 1, Chester 1 (UCT); Hampshire County, Amherst 12, Belchertown 1 (UMA), Chesterfield 6 (UCT), Pelham 1 (UMA).

CONNECTICUT, total 89: Fairfield County, Cos Cob 2

SMOKY SHREW

(UCT), Monroe 1–1 (UCT–USNM), Trumbull 1, Weston 2 (UCT), Wilton 1 (AMNH); Litchfield County, Barkhamsted 1 (UCT), Bear Mt. 9 (AMNH), Kent 3–1 (AMNH–UCT), Lime Rock 2 (UCT), Macedonia Park 7 (AMNH), Norwalle 1 (UCT), Sharon Mt. 8–1 (AMNH–UCT); Middlesex County, Clinton 1 (AMNH), Portland 1, Salisbury 1; New Haven County, Mt. Carmel 1, Southbury 1, Voluntown 2, Waterbury 4; New London County, Griswold 4, Stonington 1; Tolland County, Charter Marsh 3, Coventry 2, Mansfield 5, Storrs 6, Union 6, Willington 2; Windham County, Ashford 1, Brooklyn 4, Scotland 1 (UCT), South Woodstock 2 (AMNH).

# Long-tailed Shrew

*Sorex dispar dispar* Batchelder

*Sorex dispar* Batchelder, 1911. *Proc. Biol. Soc. Washington* 24:97. A renaming of *S. macrurus* Batchelder.
Type Locality: On a U.S.D.I. Geological Survey topographic map, 0.6 mile east of Saint Huberts, Essex County, New York, lat. 44°09′, long. 73°46′ (Martin 1966), *J. Mamm.* 47(1):130–31.
Dental Formula:
  I 3/1, C 1/1, P 3/1, M 3/3 × 2 = 32
Names: Gray long-tailed shrew, big-tailed shrew, and rock shrew. The meaning of *Sorex* = shrew, and *dispar* = unequal or different.

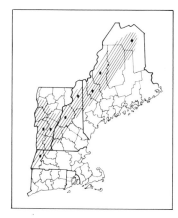

**Description.** The long-tailed shrew is similar to the smoky shrew but somewhat smaller, with a longer tail and a uniform slate-gray pelage in all seasons. The tail is fuscous black above, usually paler below.

The skull is long and depressed. The teeth are narrow, and the third and fourth unicuspids are about equal in size and are smaller than the first and second, while the fifth is quite large but smaller than the third.

The sexes are equal in size. Measurements range from 101 to 135 mm (3.9–5.3 in); tail, 50 to 60 mm (2.0–2.3 in); and hind foot, 12 to 15 mm (0.47–0.59 in). Weights vary from 4 to 6 g (0.14–0.21 oz).

**Distribution.** The long-tailed shrew is distributed in the mountains of Maine, south through western Massachusetts to eastern New York, central Pennsylvania, and western New Jersey, and south into North Carolina and Tennessee.

**Ecology.** The long-tailed shrew prefers cold, deep, damp coniferous forests in boreal pockets at altitudes as high as 6,000 feet. It is found in depressions of moist moss-covered logs, in crevices of large mossy rock piles, among shaded, wooded rock slides or talus, just beneath low, shaded cliffs, and at the edges of moist grassy clearings surrounded by swampy woods.

Predators and parasites of this species are apparently similar to those of other shrews.

**Behavior.** Very little is known of the biology of the long-tailed shrew, but it is probably similar to other shrews in the same habitat. In New York, Connor (1960) found that nine stomachs collected from June to November yielded the following food percentages: insects 74.9, centipedes 14.7, spiders 10.6, and plant material 1.1.

Little is known of reproduction in this species. Tate (1935) collected a female with two embryos from the Adirondacks in August, and Richmond and Rosland (1949) trapped a female with five embryos in Pennsylvania in May. Connor (1960) collected an adult male with testes 4 mm (0.16 in) long in central New York in October.

**Specimens Examined.** MAINE, total 3: Aroostook County, Beaver Brook 1; Franklin County, Tumbledown Mt. 1 (UME); Somerset County, Enchanted Pond 1 (AMNH).

NEW HAMPSHIRE, total 29: Coos County, College Grant (Errol) 2 (DC), Mt. Washington, 7–13–1–12–4–1 (AMNH–DC–HMCZ–UCT–UME–USNM); Grafton County, Mt. Moosilauke 4 (DC).

VERMONT, total 8: Rutland County, Mendon 3–2 (HMCZ–UCT), Sherburne 2 (UCT); Windsor County, Windsor 1 (HMCZ).

MASSACHUSETTS, total 1: Berkshire County, Adams 1 (HMCZ).

**Records and Reports.** MAINE: Piscataquis County, Baxter State Park 1 (Starrett 1954).

NEW HAMPSHIRE: Coos County, Mt. Washington 1–1 (Mather 1933; Lincoln 1935); Grafton County, Woodstock 1 (Preble 1937a).

VERMONT: Rutland County, Rutland 4 (Osgood 1935).

MASSACHUSETTS: Berkshire County, Mt. Greylock 1 (Jackson 1928).

# Thompson's Pygmy Shrew

*Microsorex thompsoni thompsoni* (Baird)

**Description.** Thompson's pygmy shrew is closely related to the long-tailed shrews but has a shorter tail. It is often confused externally with the masked shrew, but its skull is more flattened and narrower, with a short, broad rostrum, a very short and heavy lower jaw, and a relatively small infraorbital foramen. In lateral view the third unicuspid tooth is disklike and com-

LONG-TAILED SHREW

*Sorex thompsoni* Baird, 1858. Mammals, in *Repts. Expl. Surv. Railr. to Pacific* 8(1):34.
*Microsorex thompsoni thompsoni*, Long, 1972. Taxonomic revision of the mammalian genus *Microsorex* Coues. *Trans. Kansas Acad. Sci.* 74:181–96.
Type Locality: Burlington, Chittenden County, Vermont
Dental Formula:
  I 3/1, C 1/1, P 3/1, M 3/3 × 2 = 32
Names: The meaning of *Microsorex* = minute shrew, and *thompsoni* = for Thompson, a zoologist.

pressed anteroposteriorly, and the minute fifth unicuspid is sometimes indistinct. The third and fifth unicuspids usually are so small that only three teeth show in lateral view.

The sexes are colored alike and show little seasonal color variation. In summer the coloration is reddish brown or grayish brown above, paler on the sides, and whitish or grayish below. The winter coloration is olive brown above and smoky gray below, occasionally tinged with light buff. The tail is indistinctly bicolor, brown above, paler below, and darker toward the tip in all seasons. The coloration of immatures is very much like that of adults.

This species is the smallest mammal in New England and among the smallest mammals in the world. Measurements range from 82 to 98 mm (3.2–3.8 in); tail, 27 to 33 mm (1.1–1.3 in); hind foot, 9 to 10 mm (0.35–0.39 in). Weights vary from 2.2 to 3.6 g (0.08–0.13 oz).

**Distribution.** Thompson's pygmy shrew occurs from the Gaspé Penninsula to southern Wisconsin, south of the Great Lakes, and extends southward along the Allegheny-Appalachian Mountains into northern Georgia.

**Ecology.** This species is often found in wet or closely mingled wet and dry habitats and occasionally in dry habitats near water. It is encountered under old stumps and rotting logs, among the litter in sedges, ferns, clumps of aspens, beech-maple forests, and in heavy spruce and pine bordering water.

The enemies of this shrew are probably the same as those of other shrews.

**Behavior.** Little is known about the behavior of Thompson's pygmy shrew. Saunders (1929) observed a captive shrew that sat up on its hind legs like a kangaroo, emitted a combination of whispering and whistling musical sounds, discharged a powerful musk when excited, and was far tamer than the short-tailed shrew. The author reported that the speed of the animal was so great that at times it was difficult to follow its movements. Prince (1940) described the behavior of a captive pygmy shrew as follows: jumped 4½ inches high, was continually active, making many sudden stops and starts, and while running held its tail straight out from the body with a slight upward curve; could easily climb the sides of the cage and walk upside down on the

THOMPSON'S PYGMY SHREW

wire-mesh cover; sometimes hung down from the cover monkey fashion with its hind feet clinging to the cover while the body and forefeet dangled; was active during both the day and the night, but appeared more active at night; slept with its limbs drawn under its body and its head and tail curled alongside, much like a sleeping dog; and ate 107.5 g (3.8 oz) of meat, 22 grasshoppers, 20 houseflies, two crane flies, one beetle, and the liver of a meadow vole during a period of 10 days. The reproduction of this species is unknown.

**Specimens Examined.** MAINE, total 11: Aroostook County, Madawaska 3; Oxford County, Andover 1 (HMCZ), Norway 1 (USNM); Penobscot County, Brewer 2 (HMCZ), Holden 3 (UME); Piscataquis County, Mt. Katahdin 1 (AMNH).

NEW HAMPSHIRE, total 12: Carroll County, Harts Location 1, Jackson 4 (UCT); Coos County, Mt. Washington 2 (DC); Grafton County, Livermore 5 (UCT).

VERMONT, total 8: Caledonia County, Lyndon 5 (UCT); Chittenden County, Burlington 1 (USNM); Washington County, Plainfield 2 (UCT).

**Records and Reports.** MAINE: Hancock County, Mt. Desert Island (Manville 1942); Piscataquis County, Mt. Katahdin (Heinrich 1953).

NEW HAMPSHIRE: Coos County, First Connecticut Lake, Pinkam Notch (Preble 1937b, 1938).

VERMONT: Chittenden County, Burlington (Osgood, 1938b); Caledonia County, Lyndon (Miller 1964).

# Short-tailed Shrew

*Blarina brevicauda*

There are five recognized subspecies of short-tailed shrews in New England:

*Blarina b. aloga* Bangs, 1902. *Proc. New England Zool. Club* 3:76.
Type Locality: West Tisbury, Martha's Vineyard Island, Dukes County, Massachusetts

*Blarina b. compacta* Bangs, 1902. *Proc. New England Zool. Club* 3:77.
Type Locality: Nantucket Island, Nantucket County, Massachusetts

*Blarina b. hooperi* Bole and Moulthrop, 1942. *Sci. Publ. Cleveland Mus. Nat. Hist.* 5(6):110–12.
Type Locality: Lyndon, Caledonia County, Vermont

*Blarina b. pallida* R. W. Smith, 1940. *Amer. Midland Nat.* 24(1):223–24.
Type Locality: Wolfville, Kings County, Nova Scotia

*Blarina b. talpoides* (Gapper). *Sorex talpoides* Gapper, 1830. *Zool. J.* 5:202—*Blarina b. talpoides* Bangs, 1902. *Proc. New England Zool. Club* 3:75.
Type Locality: Between York and Lake Simco, Ontario, Canada (Hall and Kelson 1959)

Dental Formula:
I 3/1, C 1/1, P 3/1, M 3/3 × 2 = 32

Names: Large short-tailed shrew, mole shrew, and blarina. The meaning of *Blarina* is uncertain, probably a coined name, and *brevicauda* = short-tailed.

**Description.** The short-tailed shrew is larger and more robust than the other shrews of New England. It is recognized by its very short, scantily haired tail, which is less than half the length of the head and body. The eyes are beady, and the small ears are hidden in fur. The legs are short and the five-clawed feet rather broad and large. The paired scent glands on the flanks are more developed in males than in females.

The skull is much more angular than in other shrews, with higher ridges. The first and second upper unicuspids are large, the third and fourth are smaller and subequal, and the fifth is minute. The surfaces of the first two upper large molars have a distinct V-pattern. All the teeth are stained reddish brown.

The sexes are colored alike, the summer coloration being slightly lighter than the winter coloration. In winter adults are a dark slate color above and paler below. The tail is darker above than below. Immatures are somewhat darker and glossier than adults, whereas old individuals often have white hairs interspersed in their fur. Albinos and all-white specimens occur rarely.

Molt occurs for about six weeks in May or June and again in November. Molting usually begins on the back and shoulders and proceeds backward and forward without definite pattern.

The sexes are equal in size. Measurements range from 105 to 134 mm (4.1–5.2 in); tail, 17 to 30 mm (0.7–1.2 in); and hind foot, 12 to 19 mm (0.47–0.74 in). Weights vary from 12.5 to 23.5 g (0.44–0.82 oz).

**Distribution.** This species is one of the commonest shrews in the eastern United States and southeastern Canada. It occurs from Nova Scotia west into Manitoba, and south through eastern Colorado and southeastern Texas into Florida.

**Ecology.** The short-tailed shrew is one of the most abundant small mammals in New England and occurs in nearly all habitats. It frequents wooded areas, where it tunnels under leaves, moss, and loose loam in search of food, and is also found along the banks of small streams, under logs, and piles of brush, and in tall, rank grass.

Short-tailed shrews are encountered in the tunnels and runways of moles, voles, and other small mammals. They seek shelter in cavities in old logs and stumps, crevices of building foundations, and recesses of stone walls. They build two types of nests in the runways or under stumps, rocks, or logs; a small resting nest, and much larger, more elaborate mating nest. The mating nest is from 6 to 10 inches long and 4 to 6 inches in diameter outside, and 2½ to 4 inches inside. The resting nest is 6 to 7 inches long and about 3 inches in diameter outside, and 1½ to 3 inches inside. Both nests are made of shredded grass, leaves, and other plant fibers. They usu-

ally have three openings, one at the bottom for escape and one on each end of the burrow. Short-tailed shrews deposit their scats in unused sections of burrows.

Short-tailed shrews have many enemies, but few mammalian predators eat them, probably because of their musky odor. Coyotes, foxes, dogs, bobcats, cats, skunks, weasels, hawks, owls, shrikes, snakes, pike, black bass, and trout kill many shrews. Fleas, mites, acarinids, cestodes, nematodes and trematodes, and botfly larvae parasitize them. Many shrews die from floods, starvation, shock from sudden temperature changes, accidents, and fights with other shrews.

Some entomologists regard this species as one of the most important enemies of the European pine sawfly, particularly in the cocoon stage, when at least half of the sawflies present during pest outbreaks are eaten (Holling 1959).

**Behavior.** Short-tailed shrews are active throughout the year and may be seen at any time of the day and night, but they tend to be more active on cloudy, damp days. These animals are extremely nervous, aggressive, and ceaselessly active.

In captivity, short-tailed shrews live peacefully (Hamilton 1931b), some of them sleeping huddled together in piles (Rood 1958), which may help conserve body heat (Pearson 1947). In the wild, however, short-tailed shrews are unsociable, belligerent, and pugnacious.

Captive short-tails exhibit four action patterns —approach, attack, retreat, and combat—and five postures—tripedal, upright, sideways, back, and stance. Each of the five threat postures serves to intimidate the competitor in varying degrees (Olsen 1969).

These shrews travel under the leaf mold and logs during the day and venture above ground at night. In winter they burrow through the snow and occasionally are seen above the snow during inclement weather.

Short-tailed shrews are not fast but are incessant runners and when pressed can run at least three miles per hour, often changing the gait to a series of short leaps with the body somewhat arched and the tail elevated (Jackson 1961). Although generally terrestrial, short-tailed shrews may climb in search of food. Carter (1936) saw

one climb a tree to get suet nailed 6 feet above the ground. When necessary these shrews take to the water.

They expend a tremendous amount of energy during their activities. Doremus (1965) recorded the mean heart rate of anesthetized captive animals was 760 beats per minute, and in active shrews the average body temperature was 38.5°C and the average respiration rate 164 per minute.

Short-tailed shrews have poor eyesight and a moderately developed sense of smell, but highly developed senses of hearing and touch. They emit a continuous low, musical twitter like that of a sparrow when contented or feeding and a continuous shrill grating note or high-pitched chatter when alarmed or enraged. Hamilton (1931b) remarked that young shrews are vocal when 8 days old and are able to chatter at 20 days.

The short-tailed shrew secretes a venom from the submaxillary glands located between the bases of the lower incisors; this venom, mixed with saliva, seeps into the bite from a groove between the long incisors and overcomes the prey. As far as is known this is the only mammal with a poisonous bite, though some African and European shrews are also suspected (Pearson 1954). Pearson wrote: "A three-pound rabbit succumbed in less than five minutes to an intravenous injection of extract from only 10 milligrams of submaxillary tissue. One shrew provides more than seven times this much venom. There seems little doubt that the venom could kill a human being if injected intravenously." Fortunately the shrew lacks an effective injection mechanism. Maynard (1889), who was bitten on the fingers while holding a short-tailed shrew, felt a burning sensation 30 seconds later. Although the punctures were slight and barely drew blood, "the burning sensation, first observed, predominated in the immediate vicinity of the wounds, but now greatly intensified, accompanied by shooting pains, radiating in all directions from the punctures but more especially running along the arm, and in half an hour, they had reached as high as the elbow. All this time, the parts in the immediate vicinity of the wounds, were swelling, and around the punctures the flesh had become whitish.

"I bathed the wounds in alcohol and in a kind of liniment, but with little effect. The pain and swelling reached its maximum development in about an hour, but I could not use my left hand without suffering great pain for three days, nor did the swelling abate much before that time. At its greatest development, the swelling on the left hand caused that member to be nearly twice its ordinary thickness at the wound, but appeared to be confined to the immediate vicinity of the bites, and was not as prominent on the right hand; in fact, the first wound given was by far the most severe." Maynard remarked that the burning sensation and shooting pains lasted with lessening severity for a week. Pearson (1942) found that the poison slowed the heart, inhibited respiration, and lowered the blood pressure of the victim. Lawrence (1946) compared the poison of the short-tailed shrew with reptile poison and concluded: "All the evidence available on the effect of *Blarina* poison and of reptile venom, particularly that of the cobra, suggests very strongly that these salivary secretions resemble each other closely. Since snake venom is known to have a strong digestive action on protein, it is further suggested that *Blarina* saliva, in contrast to that of other mammals, may also have this property."

Short-tailed shrews have a voracious appetite. They feed mainly on insects, their larvae and pupae, on earthworms, and occasionally on crustaceans, centipedes, millipedes, snails, slugs, spiders, salamanders, snakes, songbirds, mice, voles, young hares and rabbits, and other shrews. Berries, fruits, acorns, nuts, and roots are eaten in season. They cache food in small chambers in their burrows. Maurer (1970) reported that a short-tailed shrew killed a meadow vole and after partly eating it built a nest over the carcass.

Buckner (1964) demonstrated that short-tailed shrews have a food assimilation efficiency of 78%. Martinsen (1969) reported that the animals can exist on much less food and can attain significantly lower basal metabolic rates than has been reported. He found that captives lived on 10% of their body weight per day on mealworms, and on about 20% of the body weight per day on cracked corn.

Little is known about the reproduction of this species in New England. In Pennsylvania, Christian (1969) found that the breeding season began in early January and was well advanced by mid-February, and that births probably occurred about the end of February. Pearson (1944) noted that these shrews can mate at least twenty or more times per day and that copulating pairs become locked on the average of 4.7 minutes, based on 315 timed matings. One mating lasted 25 minutes. When a pair is locked together, the female usually drags the male behind her tail-first.

The gestation period is 21 to 22 days, and between three and ten young are born per litter, although the usual is five or seven. Two or three litters may be produced annually. At birth the young are blind, helpless, and naked; they are approximately 25 mm (1 in) long and weigh slightly more than 1 g (0.04 oz). They grow rapidly, and at 14 days their ears open, they begin to crawl, and they are half-grown and covered with fur (Hamilton 1931b). The young leave the nest when 18 to 20 days old, and most are weaned at 25 days of age (Blus 1971). Longevity is 18 to 20 months, but some shrews live up to three years (Jackson 1961).

Blus and Johnson (1969) reported an unusual interspecific adoption of a nestling house mouse by a captive nursing short-tailed shrew. The mouse was observed in the shrew's nest three days after it was introduced into the cage, but it was found dead in the nest four days later. The writers found no scars on the mouse, so the reason for death was unknown.

**Specimens Examined.** MAINE, total 310: Androscogin County, Auburn 3, Lisbon 4; Aroostook County (no location) 54 (UCT), Madawaska 8; Cumberland County, Sebago Lake 1 (HMCZ); Franklin County, Carthage 1 (UCT), Dryden 10 (AMNH), Farmington 47 (UME), Mt. Blue 2, Weld 3 (UCT); Hancock County, Bar Harbor 1, Brooklin 6 (USMN), Castine 5 (AMNH), Mt. Desert Island 12–4 (HMCZ–UCT); Kennebec County, Gardner 1 (UCT), Oakland 3 (USNM); Oxford County, Dixfield 33 (UCT), Fryeburg 1, South Waterford 1 (AMNH), Upton 3 (UCT), Walkers Mill 7 (USNM); Penobscot County (no location) 1 (UCT), Brewer 1, Holden 4, Old Town 1, Orono 15 (UME), Penobscot River (East Branch) 21 (USNM); Piscataquis County (no location) 8, Frost Pond 4 (UCT), Greenville 1 (HMCZ), Mt. Katahdin 2–10 (UME–USNM); Sagadahoc County, East Harpswell 1 (AMNH), Small Point 1 (USNM); Somerset County, Mercer 1 (UCT), Seboonock 1 (UME), Taunton 1 (UCT); Waldo County, Belfast 1, Islesboro Island 3 (USNM); Washington County, Baring 1, Calais 11, Edmunds Unit 1, Grand

Lake 2, Vanceboro 7 (UCT); York County, Wells 1 (USNM).

NEW HAMPSHIRE, total 595: Belknap County, Alton 2, Gilford 2; Carroll County, Bartlett 61, Harts Location 73, Jackson 151 (UCT), Ossipee 18 (USNM), Snowville 1 (AMNH), Tuftonboro 2, Wolfeboro 5; Cheshire County, Dublin 1 (UCT); Coos County, Ammonoosuc Ravine 2 (UNH), Carroll 8 (UCT), College Grant (Errol) 10, Gorham 2 (DC), Mt. Washington 5–30–21–1 (AMNH–DC–UCT–USNM), Shelburne 1; Grafton County, Canaan 8, Dorchester 2, Etna 1, Franconia 10, Hanover 56 (DC), Lincoln 3, Livermore 35 (UCT), Lyme 5 (DC), Mt. Moosilauke 13–21–4 (AMNH–DC–HMCZ), Warner 3 (HMCZ); Hillsborough County, Antrim 3 (HMCZ), Brookline 4 (UCT), Hollis 1, Mt. Vernon 1 (UCT), Nashua 1 (HMCZ), Peterborough 7 (DC); Merrimack County, Pembroke 16 (UCT), Webster 5; Rockingham County, Exeter 5 (HMCZ), Northwood 3, Rye 1 (UCT); Strafford County, Durham 1–1–3 (HMCZ–UCT–UNH); Sullivan County, Grantham 1 (UCT), Sunapee Harbor 4 (USNM).

VERMONT, total 411: Addison County, Cornwall 1, Middlebury 7 (UVT), Starkboro 1 (UCT); Bennington County, Dorset 1–6 (UCT–USNM), Landgrove 2 (UVT), Peru 5–1–2 (DC–UCT–UVT), Readsboro 1 (AMNH); Caledonia County, Lyndon 15, Groton 1, Hardwick 40 (UCT); Chittenden County, Burlington 2–41 (USNM–UVT), Charlotte 1–1 (DC–UVT), Essex Junction 6, Jericho 1, Richmond 3, Shelburne 5, Westford 1, West Milton 9, Winooski 1 (UVT); Essex County, Island Pond 1, Maidstone 15 (DC), Victory 2 (UCT); Franklin County, Bakersfield 1, Georgia 9 (UVT), Swanton 2; Grand Isle County, Isle La Motte 1 (UCT); Lamoille County, Cambridge 4, Eden 3, Mt. Mansfield 10–1–17 (DC–HMCZ–USNM), Wolcott 1; Orange County, Randolph 3 (UVT), Thetford 5 (UCT), West Fairlee 1 (DC), Williamstown 5 (UCT); Rutland County, Chittenden 3 (AMNH), Mendon 4–18 (UCT–USNM), Pittsford 1, Mr. Tabor 3 (UCT), Rutland 3–40 (AMNH–UCT), Sherburne 1 (USNM), Wallingford 2, Williamstown 1 (UCT); Washington County, Barre 1, Duxbury 1 (UVT), Plainfield 9, Woodbury 2;

Windham County, Londonderry 6, Rockingham 16 (UCT), Whitingham 3; Windsor County, Hartland 2 (AMNH), Ludlow 1 (UCT), Norwich 24 (DC), Pomfret 2–20 (AMNH–DC), Royalton 4 (UCT), Woodstock 1–8 (DC–HMCZ).

MASSACHUSETTS, total 490: Barnstable County, Barnstable 3 (HMCZ), East Sandwich 1 (UCT), Monomoy Island 1 (AMNH), Orleans 1 (UCT), South Dennis 3 (HMCZ), Wellfleet 1–1 (HMCZ–UCT); Berkshire County, Adams 11, Ashford 1, Ashley Falls 19 (UCT), East Otis 2 (UMA), Hinsdale 5 (UCT), Lenox 2, Monterey 1 (UMA), Mt. Washington 1–10 (AMNH–UCT), Peru 15 (UCT), Pittsfield 1, Savoy 12 (UMA), Sheffield 1; Bristol County, Raynham 1 (UCT), Seekonk 1 (USNM); Dukes County, Martha's Vineyard 18–2–4–7 (HMCZ–UMA–UCT–USNM); Essex County, Andover 1–1 (UMA–USNM), Gloucester 1 (AMNH), Mandsleigh 7 (USNM); Franklin County, Ashfield 13 (UCT), Charlemont 7, Deerfield 2, Greenfield 4, Leverett 9 (UMA), Northfield 8 (UCT), Shutesbury 13, Sunderland 1 (UMA), Warwick 2, Whately 7; Hampden County, Brimfield 17, Chester 11, Holland 3 (UCT), Montgomery 6–1 (UCT–UMA), Springfield 2 (HMCZ), Wales 2 (UCT), Westfield 3 (UMA); Hampshire County, Amherst 51 (2 albinos) (UMA), Chesterfield 16 (UCT), Easthampton 1, Hadley 1 (UMA), Huntington 3 (UCT), Northampton 1 (UMA), Pelham 1, Ware 1 (UMA), West Chesterfield 1 (AMNH); Middlesex County, Bedford 6, Belmont 1 (HMCZ), Cambridge 33–15 (HMCZ–USNM), East Lexington 2 (HMCZ), Framingham 1, Hopkington 1 (UMA), Hudson 2, Lexington 1, Lincoln 1, Newton 2, Newtonville 1 (HMCZ), Waban 1 (AMNH), Weston 3 (HMCZ), Wilmington 18 (USNM), Winchester 2 (HMCZ); Nantucket County, Nantucket 8–8 (HMCZ–UCT), Seasconset 5 (UCT); Norfolk County, Braintree 1, Stoughton 1 (albino) (UMA), Quincy 1 (HMCZ), South Weymouth 4; Plymouth County, Bridgewater 1, Brocton 1 (UMA), Hanson 2 (UCT), Marshfield 1, Middleboro 11 (USNM), Plymouth 1 (UCT), Wareham 11 (HMCZ), Whitman 2 (UCT); Worcester County, Auburn 1 (UMA), Bolton 4

SHORT-TAILED SHREW

(AMNH), East Holden 1 (UMA), Harvard 5–3 (HMCZ-USNM), Lancaster 2 (HMCZ), Petersham 2 (UCT), Wes Brookfield 1, Worcester 3 (UMA).

CONNECTICUT, total 658: Fairfield County, Cos Cob 2–4 (UCT–USNM), Easton 4 (UCT), Fairfield 1 (AMNH), Monroe 1 (USNM), Norwalk 3 (UCT), Ridgefield 1 (AMNH), Shelton 1, Trumbull 1 (UCT), Westport 3–17 (AMNH–UCT), Weston 1, Wilton 3; Hartford County, Avon 1, East Granby 2 (UCT), East Hartford 8–2–2 (AMNH–UCT–USNM), Glastonbury 5, Newington 1 (UCT), Suffield 3, West Hartford 2 (AMNH), Wethersfield 1–2 (UCT–USNM), Windsor 17–3 (AMNH–UCT); Litchfield County, Barkhamsted 5 (UCT), Bear Mt. 6 (AMNH), Canaan 1, Cornwall 1 (UCT), Goshen 3 (UCT), Kent 2–5 (AMNH–UCT), Lime Rock 5, Litchfield 2 (UCT), Macedonia Park 17, Mt. Riga (AMNH), New Milford 2, Norfolk 2, Plymouth 1, Salisbury 4 (AMNH), Sharon 6–16 (AMNH–UCT), Thomaston 1 (UCT); Middlesex County, Clinton 13 (AMNH), East Haddam 1, Middlefield 1, Middletown 1; New Haven County, Ansonia 2 (UCT), Cheshire 4 (USNM), Guilford 1, Hamden 1, Meriden 1, Middlebury 5, Naugatuck 5 (UCT), New Haven 1–1 (AMNH–UCT), Oxford 1, Southbury 5, Walcott 4, Wallinford 3, Waterbury 25; New London County, Bozrah 3, Colchester 2, Franklin 1, Griswold 12, Hopeville 1, Lebanon 7, Ledyard 4 (UCT), Niantic 3–2 (AMNH–UCT), Norwich 1, Pachaug 1, Phoenixville 1, Preston 3 (UCT), Stonington 16–3 (UCT–USNM), Voluntown 6, (UCT), Waterford 6–7 (AMNH–UCT); Tolland County, Andover 2, Bolton 2, Columbia 4, Coventry 19, Hebron 2, Mansfield 85, Stafford 1, Storrs 46 (1 albino), Tolland 7, Union 84, Vernon 22, Willington 12 (1 albino); Windham County, Ashford 4, Brooklyn 6, Chaplin 1, Eastford 4, Scotland 1 (UCT), South Woodstock 12 (AMNH), Sterling 2, Warrenville 8, Westford 2, Willimatic 1, Windham 7, Woodstock 1 (UCT).

RHODE ISLAND, total 63: Kent County, Warwick 2 (URI); Newport County, Fort Adams 19 (USNM), Newport 2–19 (AMNH–USNM); Providence County, Chepacket 5 (USNM); Washington County (no location) 14 (USNM), Kingston 1–1 (UME–URI).

# Least Shrew

*Cryptotis parva* (Say)

*Sorex parvus* Say, 1823. In Long, *Account of an expedition from Pittsburgh to the Rocky Mountains,* 1:163.
*Cryptotis parva,* Miller, 1912. *U.S. Natl. Mus. Bull.* 79:24.
Type Locality: West bank of Missouri River, near Blair, formerly Engineer Cantonment, Washington County, Nebraska.
Dental Formula:
I 3/1, C 1/1, P 2/1, M 3/3 × 2 = 30
Names: Little short-tailed shrew, little shrew, field shrew, and small blarina. The meaning of *Cryptotis* = covered or hidden ears, and *parva* = small, relating to the tiny size of this shrew.

**Description.** The least shrew resembles the short-tailed shrew but is smaller and differs in certain skull and dental characters. The snout is long and pointed, extending forward of the mouth, the eyes are beady and black, and the ears are almost concealed in fine velvety fur. The tail is short and slender, less than half of the length of the head and body. Each foot has five toes.

The skull is compact, with no zygomatic arch. The anterior nares are wide, the rostrum is short and broad, and the braincase is low, laterally angular. There are 30 teeth instead of 32 as in other shrews. There are four unicuspidate teeth, never in two pairs; the fourth unicuspid is minute and always smaller than the third unicuspid.

The sexes are colored alike, with some seasonal color variation. In winter the back is dark brown and the underparts are ashy gray. The tail is bicolor, being like the back above and the color of the underparts below. The color in summer is somewhat paler. Immatures appear darker than adults, and color variants are rare.

There are two molts each year, in April or May and in September or October. The molt may be delayed for several weeks. Molting begins on the face and the rump and belly and progresses backward.

The sexes are equal in size. Measurements range from 70 to 90 mm (2.7–3.5 in); tail, 12 to 20 mm (0.47–0.78 in); and hind foot, 9 to 12 mm (0.35–0.47). Weights vary from 4.0 to 5.7 g (0.14–0.20 oz).

**Distribution.** This species is distributed from western coastal Connecticut to central New York, west to South Dakota, south to west and central Texas and the Gulf of Mexico, and east into Florida.

**Ecology.** Least shrews prefer open dry grassy fields—such as old pastures or meadows and along edges of woodlands. They use the runways and burrows of moles, voles, and other small mammals but will make their own runways in loose, soft soil. The nest is built in a shallow depression of a burrow, under a stump, log, rock slab, small boxes or cans, or other objects. It is a compact, roughly spherical structure about 2 to 5 inches in diameter, made of dry grass or leaves and finely shredded material. The nest is kept clean.

Owls, some hawks, foxes, cats, short-tailed

LEAST SHREW

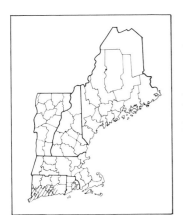

shrews, and snakes are known to prey on least shrews. Fleas have been reported to parasitize them.

**Behavior.** Least shrews are restless, nervous, and active throughout the year, at all hours of the day, but mainly at night. Their movements are quick and jerky and they dart about with the back arched. Least shrews are remarkably sociable and gregarious, often cooperating with others of their kind in building burrows and nests, which they share during the nesting and wintering seasons. Davis and Joeris (1945) found twelve least shrews in one nest, and McCarley (1959) found thirty-one of them in another. When food is scarce cannibalism may occur, as in other shrews.

Least shrews rely mainly on touch and to a lesser extent on smell. Sight and hearing are not well developed. These animals are normally quiet but utter sharp, high-pitched squeals when alarmed or disturbed. They frequently wash and comb their fur. They are not known to climb. Although they can swim, they avoid water.

Least shrews have a voracious appetite and consume 1.1 g (0.39 oz) of live food per gram of their own body weight daily (Barrett 1969). They feed on insects, earthworms, centipedes, millipedes, snails, mollusks, and salamanders, the remains of mice, rabbits, and other small mammals, and plant material in season. They are fond of frogs, which they catch by hamstringing them. They have a penchant for both raw and cooked meat, apples, pears, oats, cheese, grapes, bananas, and peanut butter. Least shrews have a reputation for entering beehives, where they eat the honeybees and their larvae and build their nests. They drink water freely, especially when the food is dry.

The reproductive behavior of the least shrew is poorly known. The breeding season extends from mid-March to early November in the northern latitudes and possibly throughout the year in the southern latitudes (Hamilton 1944). The gestation period is 15 days or slightly less. The litter size is three to six, usually four or five. Females are known to give birth, raise and wean a litter, then give birth to another litter all within 24 days (Walker et al. 1964). Mock (1970) found that captive females generally mated 1 to 4 days postpartum, and that both sexes matured sexually at about 40 days of age.

At birth the young are blind and naked; they are about 22 mm (0.86 in) long and weigh 0.32 g (0.112 oz). They grow rapidly and are fully grown and completely furred on the 9th day after birth. Their eyes open at approximately 14 days of age, and they are weaned at 21 days. Both parents care for the young, and the mother is very protective (Hamilton 1944). Longevity of a captive shrew was 21 months (Pfeiffer and Cass 1963).

**Records and Reports.** CONNECTICUT: Fairfield County, Darien 1 (Goodwin 1935); Middlesex County, Westbrook 1 (Goodwin 1942).

## FAMILY TALPIDAE (MOLES)

Moles are highly specialized for living underground, though some species show amphibious or semiaquatic behavior. These mammals possess greatly enlarged, modified, and powerful shoulders and front legs and have large handlike forefeet which are specialized for digging through the soil. The body form is cylindrical and stout, with a much reduced pelvic girdle and hind limbs which enable the animal to turn easily in a narrow tunnel. The snout is proboscislike and the ear conches vestigial, hidden in short, velvety fur that lacks definite direction.

Moles are active throughout the year. They are primarily insectivorous, but some species are omnivorous.

Members of the Talpidae are found in the temperate zone of North America and Eurasia, north to the 63d parallel (Findley 1967). There are three species in New England.

## Hairy-tailed Mole

*Parascalops breweri* (Bachman)

*Scalops breweri* Bachman, 1842. *Boston J. Nat. Hist.* 4:32.
*Parascalops breweri*, True, 1895. *Science* 1:101.
Type Locality: Eastern North America; type supposed by Bachman to have been taken on the island of Martha's Vineyard, Massachusetts, a locality where the animal probably does not occur
Dental Formula:
I 3/3, C 1/1, P 4/4, M 3/3 × 2 = 44
Names: Brewer's mole and Brewer's hairy-tailed mole. The meaning of *Parascalops* pertains to the large, rounded forefeet which act as a shield, and *breweri* = for Brewer, a zoologist.

**Description.** The hairy-tailed mole can be distinguished from the eastern mole by its hairy tail, constricted at the base, and from the star-nosed mole by the absence of fleshy nasal projections. Also, the soft, thick fur of the hairy-tailed mole is coarser than that of the eastern mole and star-nosed mole. The nostrils are lateral and crescentic. The large forefeet are almost circular in outline, and the toes are not webbed. Females have eight mammae. The skull is depressed, with the rostrum long and slender, and the auditory bullae are incomplete.

The sexes are colored alike, dark slate to black above and slightly paler below. Some specimens have a small white spot on the breast or abdomen. Apparently color variants have not been recorded.

Males average larger than females. Measurements range from 140 to 153 mm (5.5–6.0 in); tail, 23 to 36 mm (0.90–1.4 in); and hind foot, 18 to 20 mm (0.70–0.78 in). Weights vary from 40 to 63 g (1.4–2.2 oz).

**Distribution.** This species occurs from New Brunswick, southeastern Canada, into eastern Ontario, south through eastern Ohio and in the Appalachian mountains to western North Carolina.

**Ecology.** Hairy-tailed moles inhabit woods and meadows with loose, well-drained sandy loam soils. They are encountered frequently at high elevations, and occasionally near sea level. Though the geographic distribution of the hairy-tailed mole and eastern mole overlaps in some areas, they seldom are found together in the same habitat.

Hairy-tailed moles make irregular and complex subsurface tunnels, causing ridges less conspicuous than those made by the eastern mole. The winter tunnels may be from 10 to 20 inches below the surface of the ground. The nest is about 6 inches in circumference and lined with dried grasses and leaves. These moles keep their nests and tunnels clean of droppings.

Foxes, cats, weasels, owls, hawks, and snakes prey on hairy-tailed moles, and they harbor fleas, lice, and internal parasites.

**Behavior.** The behavior of this species is little known. Hairy-tailed moles are active day and night, and they may be seen above ground

HAIRY-TAILED MOLE

during the night in search of earthworms. Foote (1941) observed a hairy-tailed mole swimming some 50 yards from shore.

Hairy-tailed moles feed voraciously upon earthworms, insects, insect larvae and pupae, millipedes, centipedes, snails and slugs, and sow bugs. Hamilton (1943) reported that a captive hairy-tailed mole weighing 50 g (1.8 oz) consumed 66 g (2.3 oz) of earthworms and insect larvae in 24 hours.

Little is known about the reproductive behavior of this species. Breeding occurs in March and April, and after a gestation period of about a month four or five young are born in late April or May. At birth the young are blind and naked, but they grow rapidly and are probably weaned within a month and are able to shift for themselves. They probably become sexually mature and able to breed the following spring.

**Specimens Examined.** MAINE, total 19: Franklin County, Dryden 8 (AMNH), Farmington 8 (UME); Lincoln County, Sheepcot Township 1 (UCT); Oxford County, Brownfield 2 (UME).

NEW HAMPSHIRE, total 20: Carroll County, Bartlett 1 (UCT), Ossipee 2; Cheshire County, Dublin 1 (USNM); Coos County, Pittsburg 1 (AMNH); Grafton County, Hanover 5 (DC), Mt. Moosilauke 1 (AMNH); Merrimack County, Bow 1 (UNH), Pembroke 1; Rockingham County, Rockingham 1 (UCT); Strafford County, Durham 6 (UNH).

VERMONT, total 41: Bennington County, Sandgate 1 (USNM); Chittenden County, Burlington 1, Richmond 5; Essex County, Island Pond 3 (UVT), Orange County, South Newburg 1 (UVT), West Fairlee 3 (DC); Rutland County, Danby 1 (UMA), East Wallingford 10–1 (DC–UCT), Mendon 1–3 (AMNH–UCT), Rutland 1; Washington County, Plainsfield 1; Windham County, Saxton's River 2 (UCT), Whitingham 1 (AMNH); Windsor County, Norwich 4, Pomfret 2 (DC).

MASSACHUSETTS, total 22: Franklin County, Buckland 1, Quablin Reservoir 1, Sunderland 1; Hampshire County, Amherst 7, Easthampton 2 (UMA), Huntington 1 (UCT), Northampton 1, North Hadley 1 (UMA); Worcester County, Lunenburg 3 (USNM), Worcester 4 (UMA).

CONNECTICUT, total 9: Hartford County, Avon 1 (UCT); Litchfield County, Cornwall Bridge 1 (AMNH), Norfolk 1 (USNM), Northfield 1 (UCT), Sharon Mt. 4 (AMNH), West Winstead 1 (USNM).

# Eastern Mole

*Scalopus aquaticus aquaticus*
(Linnaeus)

**Description.** The eastern mole differs from the hairy-tailed mole by its short and naked tail and from the star-nosed mole by the absence of fleshy nasal projection. The eastern mole is robust and somewhat cylindrical in shape, with a short head. It has a distinctly pointed, movable snout, naked at the tip, crescentic nostrils that open upward, and minute eyes covered with thin skin. Its eyes and its small ears are concealed in fur. The scantily haired forefeet are greatly enlarged, broader than long, with heavy claws adapted for digging, and are normally held pointed outward with the soles vertical. The hind feet are small and weak, with one to three tubercles. Both forefeet and hind feet are webbed. The tail is short, thick, and round, slightly constricted at the base and tapering to the tip. The tail is nearly naked but is indistinctly annulated with scales. It probably serves as a tactile organ when the animal moves backward. The fur is soft, dense, and velvety, with guard hairs and underfur equally long. Females have six mammae.

The skull is conoidal and depressed, with a

EASTERN MOLE

[Sorex] *aquaticus* Linnaeus, 1758.
*Systema naturae,* 10th ed., 1:53.
*Scalops aquaticus aquaticus,* True,
1884. *Proc. U.S. Natl. Mus.* 7:606.
Type Locality: Eastern United States
(Philadelphia, Pennsylvania; fixed
by Jackson, *N. Amer. Fauna* 38:33).
Dental Formula:
I 3/2, C 1/0, P 3/3, M 3/3 × 2 = 36
Names: Common mole, ground mole,
common shrew-mole, garden mole,
and naked-tailed mole. The meaning of *Scalopus* = digging foot or
digger, and *aquaticus* = occurring
in water, a misnomer given to this
mole from its large, paddle-shaped
forefeet, which suggest an aquatic
environment.

broad braincase, complete auditory bullae, and long, weak zygomatic arches. The rostrum is relatively short, and the palate extends beyond the molars. The first large incisors curve backward and downward, while the second and third incisors are quite small.

The sexes are colored alike and show some seasonal variation. In winter the coloration is black to brownish black above and paler below, in summer the coloration is paler. Immatures are much grayer than adults. Color variants occur as all-white, oranges, and creams, and some individuals have yellow blotches on the belly. There are two molts annually: in early spring and in late summer or early fall. Molting begins on the chest and belly and on the rump near the base of the tail and progresses upward and forward. The molt line is sometimes sharply defined.

Males are slightly larger than females. Measurements range from 145 to 205 mm (5.7–8.0 in); tail, 15 to 30 mm (0.6–1.2 in); and hind foot, 18 to 22 mm (0.70–0.86 in). Weights vary from 50 to 120 g (1.8–4.2 oz).

**Distribution.** The eastern mole has the most extensive range of all the North American talpids. It occurs in the eastern United States from Massachusetts west to eastern Wyoming and Colorado, and south to central Texas and the Gulf of Mexico.

**Ecology.** Eastern moles are found in open woodlands, meadows and pastures, cultivated fields, gardens, and lawns that have well-drained, firm sandy or light loam soils where digging is easy and food is plentiful. They frequent bottomlands along shaded streams and avoid wet, loose soils and swamps because they do not support tunnels and earthworms. Eastern moles are absent in arid soils, particularly in prairies and the Rocky Mountain states.

Eastern moles construct series of subsurface burrows by bringing the forepaws alternately forward nearly to the snout, then thrusting outward and backward. This pushes the dirt aside or beneath the body, where it is kicked farther behind, while the head and body are rotated about 45 degrees to each side, forcing the loose soil upward and creating ridges. When a pile of dirt has accumulated the mole turns around in a half-somersault and pushes the dirt out of the burrow with one of its forepaws, with the head and shoulders turned to one side. The dirt is either packed in an unused burrow or pushed to the surface by the laterally extended forepaws. The subsurface burrows are normally dug at a rate of 10 to 20 feet per hour, but a mole may dig 100 to 150 feet per day during wet weather (Hisaw 1923). In dry and cold weather, moles dig deeper permanent burrows 10 or more inches below the surface. The ridges made by eastern moles are more conspicuous than those of hairy-tailed moles.

The nest is built in one of the permanent burrows, usually a foot or more beneath the surface and under stumps, roots of shrubs, or boulders. It is placed on the bottom of a flattened ellipsoidal tunnel enlargement about 5 inches in diameter and 8 inches long and is usually lined with rootlets and grass, and sometimes with leaves. Several passages may lead to the nest, one often entering from below.

Man, weasels, short-tailed shrews, foxes, dogs, cats, skunks, owls, and snakes prey on eastern moles. They are known to be parasitized by fleas, lice, cestodes, and nematodes.

**Behavior.** Eastern moles are active throughout the year, day and night, and are most active in the subsurface tunnels during a winter thaw or after a rain, when the soil is friable and earthworms abound. They spend most of their lives underground but may venture to the surface at night.

Eastern moles are solitary except for the breeding season. Eastern moles are not aquatic as the scientific name *aquaticus* suggests, although they are good swimmers. They are quiet animals, but they utter shrill squeals when irritated.

Eastern moles are voracious eaters that can consume about one-third of their body weight daily and are capable of eating two-thirds of their body weight in 18 hours (Hisaw 1923). Food consists chiefly of earthworms, beetles, grubs, centipedes, millipedes, wireworms, ants, spiders, slugs, insects, and mollusk eggs. Corn, tomatoes, potatoes, apples, grass, wheat, and oats are occasionally eaten.

The breeding begins in March and continues into April. The gestation period is 42 to 45 days, and a litter of two to five young are born in late

STAR-NOSED MOLE

April or in May. A single litter may be born annually. Newborns are pink, blind, and hairless and are about 50 mm (2.0 in) long. They grow rapidly and by about 10 days old have fine, velvety gray fur which remains for several weeks. The young leave the nest at nearly 4 weeks old and shift for themselves. They become sexually mature at one year old.

**Remarks.** Yates and Schmidly (1975) found that the eastern mole has a diploid chromosome number (2N) of 34 and fundamental number (FN) of 64.

**Specimens Examined.** MASSACHUSETTS, total 56: Barnstable County, Wellfleet 2, Yarmouth 1 (HMCZ); Dukes County, Martha's Vineyard 4–1 (HMCZ–USNM);

Hampden County, Holyoke 1, Springfield 2 (HMCZ); Hampshire County, Hadley 4, Westfield 3 (UMA); Nantucket County, Nantucket 6–1–2 (HMCZ–UCT–UMA); Plymouth County, Middleboro 1 (USNM), Wareham 28 (HMCZ).

CONNECTICUT, total 85: Fairfield County, Trumbull 1, Westport 2 (UCT), Wilton 1 (AMNH); Hartford County, East Hartford 3–1 (HMCZ–UCT), Glastonbury 1, Manchester 1, Simsbury 1; Litchfield County, Bantam Lake 1, Bethlehem 2 (UCT), Macedonia Park 1 (HMCZ), New Milford 1 (UCT), Pleasant Valley 1 (HMCZ), Watertown 1 (UCT); Middlesex County, Clinton 5 (AMNH), East Haddam 1 (UCT); New Haven County, Guilford 1; New London County, Bozrah 1, Fitchville 2 (UCT), Liberty Hill 6 (HMCZ), New London 1, Norwich 1, Stonington 1; Tolland County, Amston 1, Coventry 2, Mansfield 9, Storrs 5, Tolland 2, Willington 4; Windham County, Ashford 8, Brooklyn 3, Chaplin 6, Paskerville 1, Scotland 1, Warrenville 6 (UCT).

# Star-nosed Mole

*Condylura cristata* (Linnaeus)

[*Sorex*] *cristatus* Linnaeus, 1758.
  *Systema naturae*, 10th ed., 1:53.
*Condylura cristata* Desmarest, 1819. *J. Phys. Chem. Hist. Nat. Arts* 89:230.
Type Locality: Eastern Pennsylvania
Dental Formula:
  I 3/3, C 1/1, P 4/4, M 3/3 × 2 = 44
Names: Long-tailed mole, black mole, and swamp mole. The meaning of *Condylura* = pertaining to three processes of the tail, and *cristata* = crested, referring to the fleshy tentacles.

**Description.** The star-nosed moles are easily identified from all other mammals by the 22 fleshy pink projections or tentacles which form a wide, naked nasal disk. These sensitive feelers are symmetrically arranged, 11 on each side of a vertical line separating the halves of the snout, and resemble a crude star. In winter the long, scaly tail is greatly enlarged and covered with coarse black hairs. The fatty tissue in the swollen tail may be stored food for the breeding season. Females have eight mammae.

The skull is long and slender with a high, narrow braincase and tapers anteriorly to a long, slender rostrum. The zygomatic arch is short, slender, and directed obliquely downward an-

teriorly; the premaxillae are extended far beyond the nasals; the auditory bullae are incomplete; and the first upper incisors are large, incurved, and project anteriorly.

The coloration of both sexes is blackish brown or nearly blackish above and browner and paler below. The tail is somewhat lighter on the underside. The nasal disk is rose colored in live specimens. Immatures are paler and browner than adults. Apparently color variants have not been reported. Molting occurs in June and July and in September and October.

The sexes appear to be of equal size. Measurements range from 175 to 208 mm (6.8–8.1 in); tail, 70 to 85 mm (2.7–3.3 in); and hind foot, 25

to 30 mm (1.0–1.2 in). Weights vary from 40 to 78 g (1.4–2.7 oz).

**Distribution.** Star-nosed moles occur from southern Labrador across southeastern Canada to southwestern Manitoba, south through the lake states to Illinois, Indiana, and Ohio, and from the northeastern United States south in the Appalachian Mountains to western North Carolina.

**Ecology.** This mole prefers deep, mucky soils in low, wet meadows, bogs, marshes, and swamps. It is occasionally found near streams and ponds and sometimes in damp spots in dry fields. It seems to be more confined to the loose soil of lowlands and woods than is the hairy-tailed mole.

The burrow system is irregular and may branch around rocks and trees. It may be as deep as 2 feet in places and then abruptly rise to the subsurface or surface in dense grass or marshes, disappear again underground, and end at or below the water level of a stream. Mice, shrews, and other small mammals also use these burrows. The ridges are approximately 1¾ inches wide by 1¼ inches high, and the mounds may be 1 to 2 feet wide by 6 to 9 inches high depending on the condition of the soil. The mounds are higher and narrower in wet ground than in moist soils.

The nest is a flattened sphere, about 5 to 7 inches in diameter and 3 to 5 inches high. It is built in some natural elevation well above the high-water level, beneath a stump, log, or fallen tree, and is lined with dead grass, straw, and leaves.

Hawks, owls, dogs, foxes, weasels, mink, snakes, pickerel, pike, muskelunge, and black bass are known to prey on star-nosed moles. Fleas, mites, cestodes and nematodes are reported to parasitize them.

**Behavior.** Star-nosed moles are active throughout the year, both at day and at night. In winter they burrow in the snow and even travel on its surface. They are less fossorial than the eastern mole and hairy-tailed mole. Jackson (1961) reported that star-nosed moles dig at a rate of 7 or 8 feet per hour and can run on the surface at 4 or 5 miles per hour for short distances.

They are efficient swimmers, using their broad forepaws alternately like oars and sculling with the tail. Goodwin (1935) reported that they can stay underwater for 10 seconds, come up for air, and dive again.

Star-nosed moles feed on aquatic insects, earthworms, crustaceans, slugs, snails, isopods, and occasionally small minnows, and on plant material.

Star-nosed moles live in small colonies; they appear to pair in the fall and remain together until the young are born. Both sexes have been captured with swollen tails as early as late October and some individuals with swollen tails have been taken in June (Hamilton 1931a). The gestation period is about 45 days, and from three to seven young are born in May and June. Newborns are blind, helpless, naked, and pinkish and are about 50 mm (2.0 in) long and weigh 1.5 g (0.05 oz). They grow rapidly, and when they are nearly 7 days old short hairs appear over the back; about 2 days later hairs appear on the belly. They leave the nest and disperse at nearly 3 weeks old, when they are two-thirds grown, well furred, and weigh nearly 33 g (1.2 oz). By August and September they are about the weight of adults. They probably become sexually mature at one year (Hamilton 1931a).

**Specimens Examined.** MAINE, total 60: Aroostook County (no location) 2 (UCT), Madawaska 2 (HMCZ); Cumberland County (no location) 1 (UCT), Freeport 1 (USNM), Gray 1 (UCT); Franklin County (no location) 1, Carthage 3 (UCT), Dryden 5 (AMNH), Rangeley Island 1; Hancock County, Mt. Desert Island 1, Trenton 2 (HMCZ); Kennebec County (no location) 1 (UCT), Waterville 14 (HMCZ); Knox County, Oakland 1; Lincoln County, Hog Island 1 (USNM); Oxford County, Dixfield 1 (UCT), Norway 3, Upton 1 (HMCZ); Penobscot County, Holden 3, Lincoln 1, Orono 3 (UME), Penobscot River (East Branch) 2 (USNM); Piscataquis County, Mt. Katahdin 1 (AMNH); Sagadahoc County, Small Point 1 (USNM); Waldo County, Freedom 2 (UME); Washington County, Calais 3 (UCT), Eastport 1 (USNM), Milltown 1 (HMCZ).

NEW HAMPSHIRE, total 26: Carroll County, Ossipee 1 (USNM); Coos County, College Grant (Errol) 1–2 (AMNH–DC), Mt. Washington 1 (AMNH), Twin Mt. 1 (UCT); Grafton County, Etna 1, Franconia 1, Hanover 1 (DC), Lisbon 1 (HMCZ), New Hampton 1 (UNH), Wentworth 1 (AMNH); Hillsborough County, Amherst 1, Milford 1, South Lyndeboro 1; Merrimack County,

Pembroke 2 (UCT), Webster 1; Rockingham County, New Castle 1 (HMCZ), Northwood 1 (UCT); Strafford County, Dover 1, Durham 5 (UMA).

VERMONT, total 51: Addison County, Addison 1, Middlebury 1 (DC); Bennington County, Bennington 1 (UMA); Caledonia County, Hardwick 1 (UCT); Chittenden County, Burlington 1–3 (URI–UVT), Milton 2; Orange County, Randolph 1 (UVT), West Fairlee 1 (DC), Williamstown 2; Rutland County (no location) 5 (UCT), East Wallingford 8 (DC), Mendon 1–2–3 (AMNH–DC–UCT), Rutland 1–1 (AMNH–USNM); Washington County, Barre 1, Cabot 1 (UVT), Plainfield 6 (UCT), Waterbury 1 (UVT); Windham County, Saxton's River 1 (UCT); Windsor County, Norwich 3, Pomfret 2 (DC), Royalton 1 (UCT).

MASSACHUSETTS, total 99: Berkshire County, Williamstown, 1–1 (HMCZ–USNM); Bristol County, New Bedford 1 (USNM), Taunton 1; Essex County, Beverley 1, Lynn 1, Merrimac 1 (HMCZ), Newburyport 2 (USNM), Wenham 1 (HMCZ); Franklin County, Montague 1 (UMA), Warwick 1; Hampden County, Feltonville 6, Springfield 3 (HMCZ); Hampshire County, Amherst 3 (UMA), Chesterfield 1 (UCT), Hadley 1–1 (HMCZ–UMA), Leverett 6, Palmer 1 (UMA); Middlesex County, Ar-lington 1, Belmont 3 (HMCZ), Cambridge 13–10 (HMCZ–USNM), East Lexington 2, Newtonville 2, Sherborn 2, Waltham 8, Watertown 2 (HMCZ), Wilmington 5 (USNM), Woburn 3; Norfolk County, Brookline 4 (HMCZ); Plymouth County, Middleboro 2, North Abington 1 (USNM), Plymouth 1, Scituate 1 (HMCZ); Worcester County, Gardner 1 (USNM), Harvard 2–1 (HMCZ–USNM), Lunenburg 1 (USNM).

CONNECTICUT, total 62: Fairfield County, Darien 1 (AMNH), Monroe 1 (USNM), Norwalk 2 (UCT), Westport 1–2 (AMNH–UCT), Weston 1, Wilton 1; Hartford County, Bristol 1 (UCT), East Hartford 1–1 (HMCZ–USNM), Windsor Locks 1; Litchfield County, Kent 1 (UCT), Macedonia Park 2 (AMNH), Norfolk 1–1 (UCT–USNM), Sharon 3; Middlesex County, East Haddam 1 (AMNH); New Haven County, Hamden 1, Middlebury 1, Southbury 1 (UCT); New London County, Liberty Hill 1 (HMCZ), Norwich 1, Voluntown 1; Tolland County, Bolton 1, Coventry 4, Mansfield 10, Storrs 5, Tolland 1, Vernon 1, Willington 2, Wornwood Hill 1; Windham County, Ashford 2, Brooklyn 2, Scotland 2, Warrenville 2, Willimantic 1 (UCT).

RHODE ISLAND, total 1: Washington County (no location) 1 (UCT).

# REFERENCES

Barrett, Gary W. 1969. Bioenergetics of a captive least shrew, *Crytotis parva. J. Mamm.* 50(3):629–30.

Blossom, Philip M. 1932. A pair of long-tailed shrews (*Sorex cinereus cinereus*) in captivity. *J. Mamm.* 13(2):136–43.

Blus, Lawrence J. 1971. Reproduction and survival of short-tailed shrews (*Blarina brevicauda*) in captivity. *Lab. Animal Sci.* 21(6, part 1):884–91.

Blus, Lawrence J., and Johnson, David A. 1969. Adoption of a nestling house mouse by a female short-tailed shrew. *Amer. Midland Nat.* 81(2):583–84.

Buckner, Charles H. 1964. Metabolism, food capacity, and feeding behavior in four species of shrews. *Canadian J. Zool.* 42:259–79.

Burt, William H. 1948. *The mammals of Michigan.* Ann Arbor: University of Michigan Press.

Carter, T. D. 1936. The short-tailed shrew as a tree climber. *J. Mamm.* 17(3):285.

Christian, John J. 1969. Maturation and breeding of *Blarina brevicauda* in winter. *J. Mamm.* 50(2):272–76.

Conaway, Clinton H. 1952. Life history of the water shrew (*Sorex palustris navigator*). *Amer. Midland Nat.* 48(1):219–48.

Connor, Paul F. 1960. The small mammals of Otsego and Schoharie counties, New York. *New York State Mus. Sci. Bull.* 382:1–84.

Davis, William B., and Joeris, Leonard. 1945. Notes on the life-history of the little short-tailed shrew. *J. Mamm.* 26(2):136–38.

Doremus, Henry M. 1965. Heart rate, temperature and respiration rate of the short-tailed shrew in captivity. *J. Mamm.* 46(3):424–25.

Doucet, G. Jean, and Bider, J. Roger. 1974. The effects of weather on the activity of the masked shrew. *J. Mamm.* 55(2):348–63.

Findley, James S. 1967. Insectivores and dermopterans. In *Recent mammals of the world*, ed. S. Anderson and J. K. Jones, Jr., pp. 87–108. New York: Ronald Press.

Foote, Leonard E. 1941. A swimming hairy-tailed mole. *J. Mamm.* 22(4):452.

Getz, Lowell L. 1961. Factors influencing the local distribution of shrews. *Amer. Midland Nat.* 65(1):67–87.

Goodwin, George Gilbert. 1935. The mammals of Connecticut. *Connecticut Geol. Nat. Hist. Surv. Bull.* 53:1–221.

———. 1942. *Cryptotis parva* in Connecticut. *J. Mamm.* 23(3):336.

Hall, E. Raymond, and Kelson, Keith R. 1959. *The mammals of North America.* 2 vols. New York: Ronald Press.

Hamilton, William J., Jr. 1931a. Habits of the star-nosed mole (*Condylura cristata*). *J. Mamm.* 12(4):345–55.

——. 1931*b*. Habits of the short-tailed shrew (*Blarina brevicauda*) (Say). *Ohio J. Sci.* 31(2):97–106.

——. 1939. Activity of Brewer's mole (*Parascalops breweri*). *J. Mamm.* 20(3):307–10.

——. 1940. The biology of the smoky shrew (*Sorex fumeus fumeus* Miller). *Zoologica* 25(4):473–92.

——. 1943. *The mammals of eastern United States.* New York and London: Hafner Publishing Co.

——. 1944. The biology of the little short-tailed shrew (*Cryptotis parva*). *J. Mamm.* 25(1):1–7.

Heinrich, Gerd H. 1953. *Microsorex, Sorex palustris,* and *Microtus chrotorrhinus* from Mt. Katahdin, Maine. *J. Mamm.* 34(3):382.

Hisaw, Frederick Lee. 1923. Observations on the burrowing habits of moles (*Scalopus aquaticus machrinoides*). *J. Mamm.* 4(2):79–88.

Holling, C. S. 1959. The components of predation as revealed by a study of small-mammal predation of the European pine sawfly. *Canadian Entom.* 91(5):293–320.

Horváth, Otto. 1965. Arboreal predation on bird's nest by masked shrew. *J. Mamm.* 46(3):495.

Jackson, Hartley H. T. 1925. The *Sorex arcticus* and *Sorex arcticus cinereus* of Kerr. *J. Mamm.* 6(1):55–56.

——. 1928. A taxonomy review of the American long-tailed shrews (genera *Sorex* and *Microsorex*). *N. Amer. Fauna* 51:1–238.

——. 1961. *Mammals of Wisconsin.* Madison: University of Wisconsin Press.

Jarrell, Gordon H. 1965. A correction on the range of *Cryptotis parva* in New England. *J. Mamm.* 46(4):671.

Lawrence, Barbara. 1946. Brief comparison of the short-tailed shrew and reptile poisons. *J. Mamm.* 26(4):393–96.

Layne, James N., and Shoop, C. Robert. 1971. Records of the water shrew (*Sorex palustris*) and smoky shrew (*Sorex fumeus*) from Rhode Island. *J. Mamm.* 52(1):215.

Lincoln, Alexander, Jr. 1935. *Sorex dispar* in New Hampshire. *J. Mamm.* 16(3):223.

Long, Charles A. 1972. Taxonomic revision of the mammalian genus *Microsorex* Coues. *Trans. Kansas Acad. Sci.* 74:181–96.

——. 1974. *Microsorex hoyi* and *Microsorex thompsoni. Mammalian Species* 33:1–4. Amer. Soc. Mammalogists.

McCarley, W. H. 1959. An unusually large nest of (*Cryptotis parva*). *J. Mamm.* 40(2):243.

Manville, Richard H. 1942. Notes on the mammals of Mount Desert Island, Maine. *J. Mamm.* 23(4):391–98.

Martin, Robert L. 1966. Redescription of the type locality of *Sorex dispar. J. Mamm.* 47(1):130–31.

Martinsen, David L. 1969. Energetics and activity patterns of short-tailed shrews (*Blarina*) on restricted diets. *Ecology* 50(3):505–10.

Mather, Deane W. 1933. Gray long-tailed shrew from New Hampshire. *J. Mamm.* 14(1):70.

Mauer, Frank W., Jr. 1970. Observation of fighting between a vole and a shrew. *Amer. Midland Nat.* 84(2):549.

Maynard, Charles J. 1889. Singular effects produced by the bite of a short-tailed shrew, *Blarina brevicauda. Contrib. Sci.* 1(2):57–59.

Merriam, C. Hart. 1884. *The mammals of the Adirondack region, northeastern New York.* New York: Published by the author.

Miller, Donald H. 1964. Pigmy shrew in Vermont. *J. Mamm.* 45(4):651–52.

Miller, Gerrit S., Jr. 1895. The long-tailed shrews of the eastern United States. *N. Amer. Fauna* 10:35–56.

——. 1912. List of North American land-mammals in the United States National Museum, 1911. *U.S. Natl. Mus. Bull.* 79:1–455.

Mock, Orin Bailey. 1970. Reproduction of the least shrew (*Cryptotis parva*) in captivity. Ph.D. diss., University of Missouri, Columbia.

Olsen, Ronald W. 1969. Agonistic behavior of the short-tailed shrew, *Blarina brevicauda. J. Mamm.* 50(3):494–500.

Osgood, F. L. 1935. Four Vermont records of the big-tailed shrew. *J. Mamm.* 16(2):146.

Pearson, Oliver P. 1942. On the cause and nature of a poisonous action produced by the bite of a shrew (*Blarina brevicauda*). *J. Mamm.* 23(2):159–66.

——. 1944. Reproduction in the shrew (*Blarina brevicauda* Say). *Amer. J. Anat.* 75(1):39–93.

——. 1947. The rate of metabolism of some small mammals. *Ecology* 28:127–45.

——. 1954. Shrews. *Sci. Amer.* 191(2):66–68, 70.

Pfeiffer, Carl J., and Cass, George H. 1963. Note on the longevity and habits of captive (*Cryptotis parva*). *J. Mamm.* 44(3):427–28.

Preble, Norman A. 1937*a. Sorex dispar* in New Hampshire. *J. Mamm.* 18(1):95.

——. 1937*b.* Pigmy shrew in New Hampshire. *J. Mamm.* 18(3):95.

——. 1937*c.* The Nova Scotian smoky shrew in New Hampshire. *J. Mamm.* 18(4):513.

——. 1938. Additional records of the pigmy shrew in New Hampshire. *J. Mamm.* 19(3):371–72.

Prince, Leslie A. 1940. Notes on the habits of the pigmy shrew, *Microsorex hoyi* in captivity. *Canadian Field-Nat.* 54(7):97–100.

Rhoads, Samuel N. 1903. The mammals of Pennsylvania and New Jersey. Philadelphia: Privately published.

Richmond, Neil D., and Rosland, Harry R. 1949. Mammal survey of northwestern Pennsylvania. Pennsylvania Game Comm. P-R Proj. 20-R.

Rood, John P. 1958. Habits of the short-tailed shrew in captivity. *J. Mamm.* 39(4):499–507.

Saunders, P. B. 1929. *Microsorex hoyi* in captivity. *J. Mamm.* 10(1):78–79.

Short, Henry L. 1961. Fall breeding activity of a young shrew. *J. Mamm.* 42(1):95.

Starrett, Andrew. 1954. Longtail shrew, *Sorex dispar,* in Maine. *J. Mamm.* 35(4):583–84.

Tate, G. H. H. 1935. Observations on the big-tailed shrew *Sorex dispar* Batchelder. *J. Mamm.* 16(3):213–15.

Tuttle, Merlin D. 1964. Observation of (*Sorex cinereus*). *J. Mamm.* 45(1):148.

Walker, Ernest P.; Warnick, Florence; Lange, Kenneth I.; Uible, Howard E.; Hamlet, Sybil E.; Davis, Mary A.; and Wright, Patricia F. 1964. *Mammals of the world.* 2 vols. Baltimore: Johns Hopkins University Press.

Wetzel, Ralph M., and Shelder, Eugene. 1964. The water shrew in southern Connecticut. *J. Mamm.* 45(2):311.

Woolfenden, Glen E. 1959. An unusual concentration of (*Sorex cinereus*). *J. Mamm.* 40(3):437.

Yates, Terry L., and Schmidly, David J. 1975. Karyotype of the eastern mole (*Scalopus aquaticus*), with comments on the karyology of the family Talpidae. *J. Mamm.* 56(4):902–5.

# Order Chiroptera

The order Chiroptera (bats) comprises the only mammals that fly. The finger bones, or phalanges, are slender and greatly elongated and are covered by an elastic, leathery membrane which extends to the sides of the body, legs, and tail, forming the wing. The short-clawed pollex, or thumb, is free of the wing and is used for grasping and as a claw when the bat is walking. The knee is directed outward and backward, enabling the bat to hang upside down by its toes. From this position bats can launch themselves into flight by simply releasing their toehold and spreading the wing membranes. The calcar is a long cartilaginous spur that projects toward the tail from the median side of the heel and helps support the interfemoral membrane between the hind limbs and the tail. It is absent in some species. The pectoral girdle is better developed than the pelvic girdle, the ulna is reduced, and the sternum is usually keeled, or ridged.

The eyes are small, but the ears are relatively large and well developed. The ear conch has a prominent projection called the tragus. Henson (1970) wrote that the function of the tragus is to increase directional sensitivity of hearing in "the vertical plane." In flight, bats are guided by echoes received from their supersonic cries. These echolocations keep them from hitting obstacles and pinpoint flying prey.

Most bats are gregarious, but as a rule the sexes are segregated in summer. They are crepuscular or nocturnal and spend most of the day sleeping in dark places in attics and belfries, behind shutters, in the foliage of trees, or in caves. Some species migrate in the fall to a warmer climate, where most of them hibernate in caves. Some bats feed mainly on insects, others on nectar, fruits, fish, meat, and even blood.

Bats occur in the temperate and tropical zones of the world, except for a few small, remote islands (Koopman and Cockrum 1967). There is only one family in New England.

## FAMILY VESPERTILIONIDAE (COMMON BATS)

The vespertilionids, or common bats, are small to moderate in size. Their muzzles lack leaflike outgrowths, and their long tails reach to the back edge of the tail membrane but never much beyond. The incisors are small, and the molars have well-developed W-shaped cusps and ridges. Some genera show a tendency toward reduced cusps.

These bats are mainly insect eaters. They capture prey in flight, on the ground, and over water. They may roost singly, in pairs, in small groups, or in large colonies during the entire year, but some species congregate only in winter. Some species migrate between summer and winter roosts, and colonial forms usually return to the same roosting site each year. Some species are known to home successfully when transported away from their normal range.

The distribution of this family is the same as for the order, except that it is absent from some islands in the South Pacific. There are nine species of bats in New England.

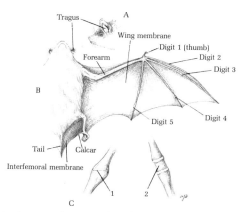

Major features of a vespertilionid bat. *A*, lateral view of head; *B*, ventral view of body; *C*, age indicators. *1*, digit joint of adult showing knobby joint and lacking cartilage. *2*, juvenile, showing clear cartilage between joints; the larger the area, the younger the animal. Baby bats cannot fly.

## Little Brown Myotis

*Myotis lucifugus* (Le Conte)

**Description.** The little brown myotis is distinguished from Keen's myotis by its shorter ears and shorter tail and from the Indiana myotis by its shorter tail and smaller, more delicate feet with longer hairs on the toes.

The little brown myotis has a hairy face, beady black eyes, broad, blunt nose, and rather short, narrow, naked ears which are rounded bluntly at the tips and barely reach the tip of the nose when laid forward. The slender tragus is almost

V[espertilio] lucifugus Le Conte, 1831. In McMurtrie, *The animal kingdom . . . by the Baron Cuvier,* 1 (app.):431.

*Myotis lucifugus,* Miller, 1897. *N. Amer. Fauna* 13:59.

Type Locality: Georgia; probably the Le Conte plantation, near Riceboro, Liberty County

Dental Formula:
  I 2/3, C 1/1, P 3/3, M 3/3 × 2 = 38

Names: Little brown bat, common bat, cave bat, Le Conte's bat, and blunt-nosed bat. The meaning of *Myotis* = mouse ear, and *lucifugus* = avoiding or fleeing light.

half as high as the ear; its tip is rounded and the inner edge is nearly straight, while the outer edge is almost convex, with a shallow notch at the base.

The wing membrane is attached to the foot near the base of the toes and is sparsely furred between the humerus and the knee. Long hairs extend beyond the sharp claws, and the interfemoral or tail membrane has only a few scattered hairs along the free edges. The fifth finger is a little shorter than the others, and the calcar is about 17 mm (0.66 in) long and not keeled. The tip of the long tail is usually free of the interfemoral membrane. Females have two mammae.

The skull is delicate and slender, rising gradually from the short, narrow rostrum. The braincase is broad and slightly flattened, usually with no sagittal crest. The palate is deeply emarginate in front, and the first two upper molars have a distinct W pattern on their surfaces, as in all New England bats. Frum (1946) and Findlay and Jones (1967) have reported a trend toward loss of the upper premolars in this species.

The fur is dense, fine, and glossy. The sexes are colored alike, a rich brown, almost bronze, usually with a dark spot on the shoulders. The ears and membranes are a glossy dark brown. Immatures are somewhat darker and grayer than adults, and pied or white-blotched and almost completely black specimens have been reported.

Females are somewhat larger than males. Measurements range from 80 to 95 mm (3.1–3.7 in); tail, 35 to 43 mm (1.4–1.7 in); hind foot, 8 to 11 mm (0.31–0.43 in); and wingspread, 220 to 270 mm (8.6–10.5 in). Weight is greatest just before hibernation and varies from 7 to 10 g (0.25–0.35 oz).

**Distribution.** This species occurs from Labrador to southern Alaska, south in the mountains of southern California, and in the Appalachians to Georgia and west into Arkansas. Stragglers have been collected in New Mexico, Texas, Mississippi, Alabama, and the coastal Carolinas. It is known to be one of the most abundant of all bats in the northern part of its range.

**Ecology.** The little brown myotis occurs almost everywhere. In spring and summer females form maternity colonies of hundreds in attics, barns, and various other retreats that are dark and hot during the daytime. Females can tolerate high temperatures; Davis, Hassell, and Rippy (1965) recorded a temperature of 131°F in an attic maternity colony. Solitary males roost in cooler, more exposed places, such as under loose bark, in hollow trees, behind loose shingles and sidings, under awnings, and in cracks and crevices of rocks. In late summer the maternity colonies break up, and both sexes roost in trees and other dark recesses. In winter the bats hibernate in caves and mines throughout the eastern part of their range. These bats frequently return year after year to the same nursery colony and hibernation cave.

Little brown myotis females end hibernation in Aeolus Cave in southern Vermont from early April to mid-May, and males do so between May 7 and June 7. The animals scatter to summer colonies as far distant as 172 miles, with the major exodus to the southeast. Some of the bats fly northward into Lake Champlain Valley, and a few scatter to other directions. Large numbers of bats again frequent the cave from mid-July through August, but after a one-night gathering or "swarming" at the cave the individual bats leave, then return in September and October to hibernate. The swarming seems to be associated with selecting a place for hibernation, which would be very important to the young bats. Mating has not been observed during swarming (Davis and Hitchcock 1965).

In Vermont, maternity colonies of 12 to 1,200 bats begin forming in attics as early as 22 April. The colonies break up gradually, from late July to mid-September. Adults enter hibernation before the juveniles (Davis and Hitchcock 1965).

Man, cats, mink, raccoons, hawks, owls, red-winged blackbirds, common grackles, rat snakes, leopard frogs, and black bass are known to capture unwary bats. Cave floodings drown thousands, and large numbers perish when caught in storms during migration. Some bats become impaled on barbed-wire fences or entangled with the burrs of burdock plants.

Fleas, wing mites, and the large blood-sucking bat bug, *Cimex pilosellus,* are found on bats. Although the bug does not infest man, it is responsible for the superstition that bats carry bedbugs. Cestodes and trematodes are known to parasitize the little brown myotis, and some bats become infested with rabies.

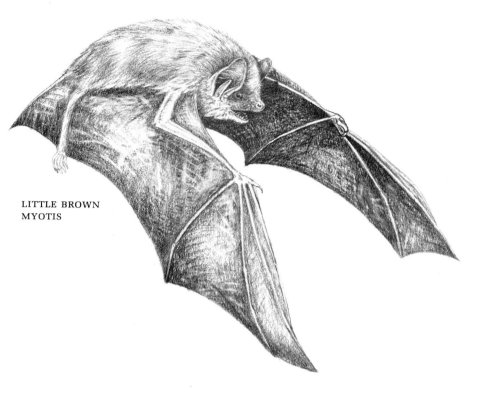

LITTLE BROWN
MYOTIS

**Behavior.** Little brown myotis take wing shortly after dusk and seek water, where they skim the surface for a drink and catch insects. They also forage over lawns and fields or among trees near their roost. When they return to the roost they usually make several passes at the entrance and may alight briefly several times before entering. The bats alight with the head up, bank slightly, then hook the claws of the thumbs and hind feet to the object to catch a hold and quickly turn head down and hang by the hind feet.

These bats fly at speeds from 12.5 to 21.7 miles per hour (Mueller 1965), and an average flight speed was recorded at 12.7 miles per hour for a distance of 67 feet in an open field (Patterson and Hardin 1969). In homing, the speed is approximately 4 miles per hour (Cope, Koontz, and Churchwell 1961). Blindfolded bats have returned home as rapidly as sighted bats when released 5 miles from their home colony (Mueller and Emlen, 1957). Homing orientation is potentially rapid, and homeward migration

takes a rather direct course (Stones and Branick 1969). Banded bats have returned to the home roost after being released more than 70 miles away (Hitchcock and Reynolds 1942), and some even returned to their home roost 17 and 22 days after they had been released 270 miles away (Schramm 1957).

Like other bats, little myotis move awkwardly on the ground, but they are fairly good swimmers. They do not hibernate continuously throughout the winter but awake spontaneously at intervals, fly about in the cave, lap moisture from the walls, and sometimes mate. They may shift to a spot where the temperature and humidity are more favorable.

These bats emit a wide range of sound waves from squeaks to supersonic cries. The sounds are made by the tiny larynx and mouth and, like radar, are reflected by unseen objects. As the bat takes flight it emits supersonic cries at about 30 kilocycles per second, and when it approaches an object this increases to 50 kilocycles; while passing the object the cries drop back to about 30 kilocycles per second. The bat judges its distance from the object by sensing the time delay between the outgoing pulse and the echo which is received by the inner ear. The bat is prevented from hearing its own cry by a minute ear muscle which contracts as the cry is emitted, preventing interference with reception of the echo. This echolocation guides bats in flight and helps them find the direction and distance of their insect prey. For an object 6 inches from the bat's mouth, the time delay for the echo is about a thousandth of a second, and a bat can detect a mosquito that quickly from the echo (McCue 1961).

Little brown myotis have voracious appetites. They feed chiefly on flying insects, particularly nocturnal moths, bugs, beetles, flies, and mosquitoes, which they catch on the wing. The bats regularly catch insects with the wing or tail membrane and transfer them to the mouth, all within half a second. Smaller insects are eaten immediately, whereas larger ones are held in the mouth and eaten when the bats alight.

Some noctuid moths have developed an amazing ability to hear the echolocation sounds of bats and zigzag in flight to avoid capture (Dunning and Roeder 1965). Little brown myotis catch flying vinegar flies at the rate of one each 7 seconds and other insects at about twice this rate (Griffin 1958).

The little brown myotis makes several food hunts during the night. Specimens shot within an hour after taking flight have full stomachs, weighing one-fifth their body weight. A colony of one hundred little brown myotis ate 19.2 kg (42 lb) of insects during a four-month period of summer activity (Gould 1955). The food transit time through the body for a fully active, free-flying little brown myotis is 35 minutes or less (Buchler 1975).

The breeding season occurs during the fall before the bats enter hibernation, though it is not unusual to find some active pairs breeding in winter. The viable sperm is stored in the uterus during the winter and fertilization occurs in the spring when the females emerge from hibernation. The gestation period lasts 50 to 60 days. There is usually only one litter per season, consisting of a single young, or occasionally two.

The dates of parturition vary with latitude and even within a colony. In southern Vermont parturition occurs between 7 June and 10 July, and old females give birth earlier than yearlings (Davis and Hitchcock 1965). Just before parturition the female becomes restless and irritable. As parturition begins she hangs head upward, the reverse of the normal resting position, and cups the tail and interfemoral membrane ventrally to form a cradle for the emerging offspring. It emerges breech first, the head being last to appear. Birth is completed in 15 to 30 minutes (Wimsatt 1960).

The newborn weighs from 1.5 to 1.9 g (0.53–0.67 oz), about one-fourth the weight of the mother. The wings, ears, legs, tail, and interfemoral membranes are blackish, but the head and body are flesh-colored and covered with fine, silky hair. The eyes open on the second day after birth. For the first 3 or 4 days after birth the young are usually carried by the female as she hunts (Griffin 1940b). One-day-old infants are able to emit "i call" sounds (Gould 1975). The young are weaned at 2 or 3 weeks of age. They grow rapidly and are able to fly when 3 to 4 weeks old. The colony disbands as soon as the young are able to catch their own food (Hitchcock 1974).

Sexual maturity is achieved at about 8 months in females, and males mature in their second summer. Longevity may be up to 25 years (Griffin and Hitchcock 1965), with males living longer than females (Hitchcock 1965).

**Age and Sex Determination.** Young bats are readily recognized by their darker pelage and the enlarged cartilaginous epiphyses in their fingers; the larger the cartilagenous area, the younger the bat. Closure of the epiphyses occurs as early as mid-August in some juveniles and is completed in nearly all by the time they enter hibernation.

Females are recognized by the vaginal opening, which is visible between the large urethral opening and anus, and by the pair of pectoral mammae. Males have a conspicuous penis.

**Specimens Examined.** MAINE, total 19: Cumberland County, West Gray 1 (HMCZ); Franklin County, Dryden 4 (AMNH); Hancock County, Mt. Desert Island 3 (HMCZ); Lincoln County, New Harbor 4 (USNM); Oxford County, South Waterford 1 (AMNH), Upton 1 (HMCZ); Penobscot County, Orono 2 (UME); Washington County, Eastport 3 (HMCZ).

NEW HAMPSHIRE, total 49: Belknap County, Laconia 1; Carroll County, Center Sandwich 2 (USNM); Coos County, Shelburne 3 (HMCZ), Mt. Washington 1–7 (DC–UNH); Grafton County, Enfield 6, Groton 4, Hanover 4 (DC), Mt. Mooselauke 1 (AMNH), Ruggles Mine 1 (DC), Squam Lake 3 (HMCZ); Hillsborough County, Antrim 1, Peterboro 5 (HMCZ); Merrimack County, Concord 1 (UNH); Sullivan County, George's Mill 9 (DC).

VERMONT, total 188: Addison County, Middlebury 1 (USNM); Bennington County, Dorset 9 (UCT), East Dorset 2–1–2–1–1 (AMNH–DC–HMCZ–UMA–USNM); Chittenden County, Burlington 5, Colchester 69; Franklin County, St. Albans 5 (UVT); Rutland County, Brandon 1–1–4–6 (AMNH–DC–UCT–USNM), Castleton 19 (UCT), Chittenden 13–10–4 (AMNH–DC–UCT), Hubbardston 4–1–9 (AMNH–DC–UCT), Mendon 1 (UCT), Proctor 3 (USNM); Washington County, Cabot 1 (UVT); Windham County, Putney 1 (AMNH), Saxton's River 1; Windsor County, Plymouth 7 (UCT), Plymouth Union 1–3–1 (AMNH–DC–HMCZ), Weathersfield 1 (HMCZ).

MASSACHUSETTS, total 37: Barnstable County, Centerville 1, Cotuit 1, Hatchville 6 (HMCZ), Woods Hole 1 (USNM); Berkshire County, Lanesboro 1–1 (AMNH–HMCZ), Monterey 4 (USNM); Essex County, Ipswich 1 (HMCZ); Franklin County, Sunderland 1 (AMNH); Hampden County, Chester 5–1–1 (UMA–USNM–UVT), Springfield 2; Middlesex County, Cambridge 1, Littleton 1; Plymouth County, Wareham 1; Worcester

County, East Templeton 1, Harvard 1 (HMCZ), Wind-hendon 2 (USNM), Worcester 3 (UMA).

CONNECTICUT, total 23: Fairfield County, Greenwich 1 (USNM), Monroe 1 (UCT), Stamford 1; Litchfield County, Sharon 1, Winchester 2 (AMNH); New Haven County, New Haven 1 (USNM); Tolland County, Mansfield 1, Storrs 2, Union 3; Windham County, Ashford 7, Hampton 1 (UCT), Woodstock 1–1 (AMNH–USNM).

**Records and Reports.** VERMONT: Bennington County, East Dorset 603; Orange County, Vershire 132; Rutland County, Chittenden 136; Windsor County, Plymouth 31 (Griffin 1940b).

MASSACHUSETTS: Hampden County, Chester 792 (Griffin 1940b).

CONNECTICUT: Litchfield County, Roxbury 413 (Griffin 1940b).

# Keen's Myotis

*Myotis keenii septentrionalis*
(Trouessart)

*Vespertilio gryphus* var. *septentrionalis* Trouessart, 1897. *Catalogus mammalium . . .*, fasc., 1, p. 131.
*Myotis keenii septentrionalis*, Miller and G. M. Allen, 1928. *U.S. Natl. Mus. Bull.* 144:105.
Type Locality: Halifax, Nova Scotia, Canada
Dental Formula:
I 2/3, C 1/1, P 3/3, M 3/3 × 2 = 38
Names: Keen's bat, Acadian bat, eastern long-eared bat, Say's bat, and Trouessart's bat. The meaning of *Myotis* = mouse ear, and *keenii* = for Rev. John Keen, who collected the first specimen of this species, and *septentrionalis* = belonging to the north or northern.

**Description.** Keen's myotis is similar in size and color to the little brown myotis, but it may be distinguished from that species by its longer tail and narrower and longer ears, which extend 4 to 5 mm (0.16–0.20 in) beyond the tip of the nose when laid forward. The tragus is long and pointed, and the calcar is not keeled. The coloration is brown but not glossy, and immatures are grayer than adults. Females have two mammae.

The skull is longer and narrower and the rostrum weaker than that of the little brown myotis. The skull is relatively slender and lightly built, and a sagittal crest is sometimes present. The length of the upper tooth row slightly exceeds the greatest palatal breadth including the molars. The interorbital constriction is less than 4 mm (0.16 in), but it is more than 4 mm in the little brown myotis.

The sexes are equal in size. Measurements range from 75 to 95 mm (3.0–3.7 in); tail, 35 to 40 mm (1.4–1.6 in); hind foot, 8 to 10 mm (0.31–0.39 in); and wingspread, 230 to 275 mm (9.0–10.7 in). Weights vary from 7 to 9 g (0.25–0.32 oz).

**Distribution.** This subspecies occurs in eastern North America, from Newfoundland, Nova Scotia, across southern Canada to west central Saskatchewan, south to northern Florida, and west to Wyoming.

**Ecology and Behavior.** Little is known of the ecology and behavior of Keen's myotis. This species roosts singly or in small colonies in caves, under loose bark on trees, behind window shutters, in cliff crevices, and around entrances of caves in summer. It forms nursery colonies in attics, roofs of barns, and cavities of trees. In winter it is found hibernating in mines and caves with the little brown myotis, big brown bat, and eastern pipistrelle.

A Keen's myotis was discovered hibernating in a dry well in Massachusetts in December, and some specimens were reported hibernating in cellars (Griffin 1945).

A blinded Keen's myotis was recovered after it flew a homing distance of 32 miles in 2.5 hours at a speed of 12.8 miles per hour, assuming a direct course (Stones and Branick 1969).

This species is probably more common than the number of collected specimens indicates. Barbour and Davis (1969) collected 43 Keen's myotis at the entrance of Aeolus Cave, Vermont, during August. Mumford and Cope (1964) found a nursery colony of some 30 Keen's myotis in a dead elm tree on 8 July. Small nurseries have been reported in roofs of barns.

Little is known regarding the feeding behavior

KEEN'S MYOTIS

of Keen's myotis, but it is probably very similar to that of the little brown myotis.

Essentially nothing is known of reproduction in this species. Hamilton (1943) suggested that birth of the single young occurs in July in New York and that parturition may occur later in this species than in most other bats in the United States. He found large embryos in late June and early July. Hall, Cloutier, and Griffin (1957) established the longevity of a Keen's myotis in New England to be 18.5 years based on tooth wear.

**Specimens Examined.** MAINE, total 9: Knox County, St. George 1; Oxford County, Norway 2 (HMCZ), South Waterford 1 (AMNH); Washington County, Eastport 3–2 (HMCZ–USNM).

NEW HAMPSHIRE, total 4: Grafton County, Bethlehem 2 (HMCZ), Lyman 1 (DC); Hillsborough County, Peterboro 1 (HMCZ).

VERMONT, total, 36: Bennington County, Dorset 10 (UCT), East Dorset 1 (UMA); Rutland County, Brandon 1 (UCT), Chittenden 2–2–1–1 (DC–UCT–USNM–UVT), Mendon 2 (UCT), Proctor 2 (USNM), West Haven 1; Windham County, Saxton's River 1 (UCT); Windsor County, Hartland 1 (AMNH), Plymouth 4 (UCT), Plymouth Union 3–2 (AMNH–DC), Weathersfield 2 (HMCZ).

MASSACHUSETTS, total 63: Barnstable County, Barnstable 2, Falmouth 6; Berkshire County, Lanesboro 1 (HMCZ); Hampden County, Chester 3–29 (HMCZ–UMA), Feltonville 4, Springfield 1; Hampshire County, Ware 1; Middlesex County, Shirley 1 (HMCZ), Wilmington 1 (USNM); Norfolk County, Brookline 1; Worcester County, East Templeton 2, Harvard 1 (HMCZ).

CONNECTICUT, total 16: Fairfield County, Westport 1 (UCT); Hartford County, Canton 1 (HMCZ); Litchfield County, Mt. Riga 1 (AMNH), Salisbury 1 (UCT); Middlesex County, Cobalt 1–1 (AMNH–USNM); New Haven County, New Haven 1 (USNM), Mt. Carmel 1 (HMCZ); New London County, Lebanon 1; Tolland County, Mansfield 2, Storrs 1, Tolland 5 (UCT).

**Records and Reports.** VERMONT: Bennington County, East Dorset 8; Orange County, Vershire 66; Rutland County, Chittenden 2; Windsor County, Plymouth 21 (Griffin 1940b).

MASSACHUSETTS: Hampden County, Chester 304 (Griffin 1940b).

CONNECTICUT: Litchfield County, Roxbury 123 (Griffin 1940b).

# Indiana Myotis

*Myotis sodalis* Miller and G. M. Allen

*Myotis sodalis* Miller and G. M. Allen, 1928. *U.S. Natl. Mus. Bull.* 114:130.
Type Locality: Wyandotte Cave, Crawford County, Indiana
Dental Formula:
I 2/3, C 1/1, P 3/3, M 3/3 × 2 = 38
Names: Indiana bat, social bat, pink bat, companion bat, and Wyandotte cave bat. The meaning of *Myotis* = mouse ear, and *sodalis* = a companion or comrade, referring to this mammal's habit of hibernating in large groups.

**Description.** The Indiana myotis resembles the little brown myotis, but has a slightly longer tail and smaller and more delicate hind feet; it also differs in coloration, the fur being dull grayish chestnut rather than bronze, with the base portion of the hairs of the back dull fuscous black. The ears reach the tip of the nose when laid forward, and the tragus is short and blunt and curves slightly forward. The calcar has a slight keel. Females have two mammae.

The skull is delicate, with a narrow braincase and a slight sagittal crest. The interorbital constriction of the Indiana myotis is always less than 4 mm (0.16 in), whereas in the little brown myotis it is 4 mm or more.

Adults apparently molt once a year during a rather short period in mid-June. Color variants are rare, but a few white-spotted or blotched individuals occur in some populations.

The sexes are equal in size. Measurements range from 75 to 95 mm (2.9–3.7 in); tail, 30 to 45 mm (1.2–1.8 in); hind foot, 8 to 10 mm (0.31–0.39 in); and wingspread, 240 to 265 mm (9.4–10.3 in). Weights vary from 5 to 8 g (0.18–0.28 oz).

**Distribution.** The Indiana myotis is found in the eastern United States from central New England west to Wisconsin, Missouri, and Arkansas, and south into northern Florida.

**Ecology.** Indiana myotis have a wide distribution, but have a very high degree of aggregation in winter, when 97% of the total estimated population hibernates in certain large caves in Kentucky, Missouri, southern Indiana, and southern Illinois. They often hibernate in large colonies with the little brown myotis, big brown bat, and eastern pipistrelle (Hall 1962).

Indiana myotis inhabit caves in summer and apparently do not use buildings as summer roosts. They cannot stand temperatures as high as those tolerated by little brown myotis. Adult females spend the summer singly or in small groups and bear their young in hollow trees or beneath loose bark.

Man, cats, hawks, owls, and snakes prey on this species. Some myotis become impaled on barbed-wire fences, and cave floodings may kill thousands of Indiana myotis. Mites and trematodes are known to parasitize this bat.

**Behavior.** Indiana myotis enter the hibernating caves about mid-September, and the major build-up occurs in late autumn. The hibernating sites are usually near the cave entrance, where the ambient temperature reaches 37° to 43°F in midwinter (Hall 1962; Henshaw and Folk 1966). During hibernation individual bats wake about every 8 to 10 days and form squeaking clusters at sites deeper in the cave where the ambient temperatures are higher than 54° to 57°F (Hardin 1967). Each bat is active for approximately one hour at each arousal period (Hardin and Hassell 1970).

During cold weather Indiana myotis hibernate in tightly packed clusters. The wings are always folded tightly against the body, not angled as in the small-footed myotis and the gray myotis. Both sexes seem to enter hibernation at the same rate, and the sex ratio is equal in the winter colonies. In spring most females leave the caves earlier than the males, probably either because they burn up fat more rapidly or because their departure is associated with ovulation and the onset of pregnancy. In winter Indiana myotis fly between caves where several hibernating populations are close together, and if they are disturbed some may fly to neighboring caves (Hall 1962).

Indiana myotis have an exceptional ability to

home to a particular area at a certain time of the year. Barbour and Davis (1969) reported that blinded Indiana bats home from distances of at least 40 miles, but Barbour, Davis, and Hassell (1966) recorded no homing in blinded Indiana bats released 125 to 200 miles from their cave, whereas 67% of the sighted controls returned from 200 miles. Hassell and Harvey (1965) suggested that this species is able to orient and navigate over many miles of rugged, unfamiliar country. These authors recovered nearly a third of the bats they released in North Carolina across the Smoky Mountains, outside their normal range, 200 miles from Bat Cave, Kentucky. Patterson and Hardin (1969) demonstrated that the average flight speed was 12.5 miles per hour for a distance of 67 feet in an open field and found that females flew faster than males.

Little is known about the feeding behavior of the Indiana myotis, but it probably is quite similar to that of the little brown myotis.

Little is known about the reproduction of this species. In Kentucky most breeding occurs over a period of about 10 days in early October. At night the bats scatter in pairs over the ceiling and copulate by the hundreds. Occasionally they breed in winter, and some bats breed in late April as they leave hibernation (Hall 1962).

The Indiana bat appears to produce a single young in late June (Mumford and Calvert 1960). The epiphyseal cartilage of the wings is closed by the beginning of hibernation in the fall in the southern parts of the animal's range, but in the northern portion, from Pennsylvania to Vermont, the cartilage may still be opened when the bats enter hibernation. Thus the bats can be distinguished throughout the winter as the young of that year (Hall 1962). Longevity is 12 years or more (Paradiso and Greenhall 1967).

**Remarks.** The U.S. Department of the Interior has placed the Indiana myotis on its list of endangered species because the population has decreased so significantly in recent years that it is threatened with extinction. It is nearly extinct over most of its former range in the northeastern states and has disappeared from its major winter caves of Illinois, Indiana, and West Virginia. The species may be extinct in New England. The Indiana myotis will probably disappear as a species in the near future unless adequate pro-

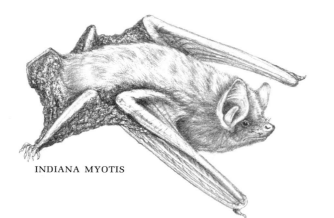

INDIANA MYOTIS

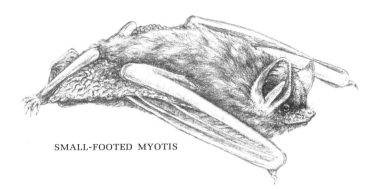

SMALL-FOOTED MYOTIS

**Specimens Examined.** NEW HAMPSHIRE, total 6: Coos County, Mt. Washington 3; Grafton County, Enfield 2; Sullivan County 1 (DC).

VERMONT, total 140: Bennington County, Dorset 11–1 (AMNH–UCT), Manchester 2 (DC); Orange County, Vershire 1 (UCT); Rutland County, Brandon 14–2–12–13 (AMNH–HMCZ–USNM–UVT), Chittenden 23–11–7 (AMNH–HMCZ–UCT), Proctor 40 (USNM); Windsor County, Plymouth 3 (UCT).

MASSACHUSETTS, total 4: Hampden County, Chester 4 (UMA).

**Records and Reports.** VERMONT: Bennington County, East Dorset 232; Orange County, Vershire 13; Rutland County, Chittenden 241; Windsor County, Plymouth 73 (Griffin 1940b).

MASSACHUSETTS: Hampden County, Chester 60 (Griffin 1940b).

CONNECTICUT: Litchfield County, Roxbury 224 (Griffin 1940b).

tection is given the hibernation caves. Fortunately the species is protected at least in the Bat Cave in Mammoth Cave National Park, Kentucky.

# Small-footed Myotis

*Myotis leibii leibii* (Audubon and Bachman)

*Vespertilio leibii* Audubon and Bachman, 1842. *J. Acad. Nat. Sci. Philadelphia* 1(8):284.
*Myotis leibii leibii,* Glass and Baker, 1965. *Bull. Zool. Nomenclature* 22(33):204–5.
Type Locality: Erie County, Ohio
Dental Formula:
  I 2/3, C 1/1, P 3/3, M 3/3 × 2 = 38
Names: Least myotis, least brown myotis, and Leib's myotis. This species is the smallest bat in northeastern North America. The meaning of *Myotis* = mouse ear, and *leibii* = for Dr. George Leib, who collected the first specimen.

**Description.** The small-footed myotis resembles the little brown myotis but differs in its golden-tinted, almost yellowish, fur and its shorter forearms. It is distinguished from the eastern pipistrelle by its lighter color, especially the light pinkish forearms, and its lack of a keeled sternum. It is recognized by the black facial mask, black ears, long-keeled calcar, and absence of a dark shoulder patch. When laid forward, the ears extend slightly beyond the nose. Females have two mammae.

The skull is much flatter than that of the little brown myotis, and the braincase is narrower. A sagittal crest may be present.

The sexes are equal in size. Measurements range from 72 to 84 mm (2.8–3.3 in); tail, 30 to 39 mm (1.2–1.5 in); hind foot, 6 to 8 mm (0.23–0.31 in); and wingspread, 212 to 248 mm (8.3–9.7 in). Weights vary from 5 to 8 g (0.18–0.28 oz).

**Distribution.** The small-footed myotis occurs from Ontario and southwestern Quebec south into the mountains of northern Georgia, and west to Arkansas, Missouri, southern Iowa, and eastern Kansas. It is one of the least common bats throughout the eastern part of its range.

**Ecology.** Little is known regarding the ecology of the small-footed myotis. Buildings seem to provide suitable places for shelter in summer. A

colony of about a dozen bats was found behind a sliding door on a shed in Ontario (Hitchcock 1955), and several were found beneath rock slabs in quarries in Tennessee (Tuttle 1964). In winter the species is found in caves and mines.

The small-footed myotis seems to be restricted to caves in the foothills of mountains rising to 2,000 feet, with hemlock, spruce, and white cedar predominating among the conifers. The characteristics of the particular cave itself determine where in the cave this species will hibernate (Hitchcock 1949).

**Behavior.** This hardy bat seldom enters the caves before mid-November and leaves by March, or possibly earlier in Vermont (Mohr 1936). It hibernates in narrow cracks in the wall, floor, or roof, both singly and in groups of fifty or more. Some individuals hibernate horizontally. Martin, Pawluk, and Clancy (1966) felt that the small-footed bat selects floor crevices because they provide a stable cooler temperature and greater protection from disturbance. Hitchcock (1955) recovered two small-footed myotis 10 and 12 miles away from a cave in which they had been banded during the winter. Little is known about the feeding behavior of this species.

Little is also known about reproduction in the small-footed myotis, but a single young is apparently born annually. Maternity colonies of 12 to 20 have been found in buildings (Hitchcock 1955; Koford and Koford, 1948). A pregnant female was caught on 12 July (Quay 1948), and embryos have been found on 17 and 24 June (Jones and Genoways 1966). Hitchcock (1965) reported a 12-year-old specimen.

**Specimens Examined.** VERMONT, total 19: Essex County, Island Pond 1 (DC); Orange County, Vershire 5 (HMCZ); Rutland County, Brandon 5–3 (DC–UCT), Chittenden 5 (UCT).

MASSACHUSETTS, total 1: Hampden County, Chester 1 (HMCZ).

CONNECTICUT, total 6: Litchfield County, Roxbury, 1–5 (AMNH–HMCZ).

**Records and Reports.** VERMONT: Orange County, Vershire 8; Rutland County, Chittenden 1; Windsor County, Plymouth 2 (Griffin 1940b).

# Silver-haired Bat

*Lasionycteris noctivagans* (Le Conte)

V[espertilio] noctivagans Le Conte, 1831. In McMurtrie, *The animal kingdom . . . by the Baron Cuvier* 1 (app.):431.
*Lasionycteris noctivagans*, H. Allen, 1894. *U.S. Natl. Mus. Bull.* 43:105.
Type Locality: Eastern United States
Dental Formula:
I 2/3, C 1/1, P 2/3, M 3/3 × 2 = 36
Names: Silvery bat, silvery black bat, and black bat. The meaning of *Lasionycteris* = hairy bat, referring to the well-furred basal half of the upper surface of the interfemoral membrane, and *noctivagans* = night wandering.

**Description.** This medium-sized bat is smaller than the big brown bat but larger than any of the myotis. It has short, naked, rounded ears which are nearly as broad as long; the tragus is broad, less than half the length of the ear, straight and bluntly tipped. The interfemoral membrane is well furred on the basal half of the upper surface, terminating in a curved line from between the ankles; the wing membrane is attached to the base of the toes, and the calcar is not keeled. Females have two mammae.

The skull is flattened, with the rostrum very broad and concave on each side between the lacrimal regions and the nares, and the sagittal crest is absent.

The fur is long, soft, and lax. The sexes are colored alike, and there is no noticeable seasonal variation. The coloration above and on the sides is blackish brown, strongly tipped with silvery white, which produces a bright frosted appearance. The belly is slightly duller than the back and sides, and the head, neck, and throat are black. Immatures are brighter than adults because the silvery white tips of the hairs are much more pronounced. Color phases have apparently not been recorded.

SILVER-HAIRED BAT

The sexes are equal in size. Measurements range from 95 to 116 mm (3.7–4.5 in); tail, 35 to 50 mm (1.4–2.0 in); hind foot, 7 to 12 mm (0.27–0.47 in); and wingspread, 270 to 310 mm (10.5–12.1 in). Weights vary from 7 to 10 g (0.25–0.35 oz).

**Distribution.** The silver-haired bat occurs throughout North America from Alaska across southern Canada, and south through all the states except perhaps Florida. Anderson (1972) speculated that it might occur in Chihuahua, Mexico.

**Ecology.** This species inhabits woodland areas that border lakes and streams. Although it occurs in most areas of North America, it is erratically distributed, being common in some localities and rare in others. This bat roosts in dense foliage of trees and may also be encountered behind loose bark or in a hollow of a tree, a woodpecker hole, or an old crow's nest. During migration it may be found in buildings, abandoned quarries, old fenceposts, piles of slabs, lumber, or railroad ties, or hulls of ships. It occasionally enters caves and mines.

Man, owls, hawks, and striped skunks prey on this bat, and fleas and cestodes are known to parasitize it.

**Behavior.** Silver-haired bats take wing longer before sunset than most bats. They are slow and erratic flyers, making frequent glides while darting about among trees over woodland streams and ponds at heights of 20 feet, and sometimes repeatedly flying the same circuitous hunting path during the same evening. Sometimes they fly so close to the ground that they can be readily captured in nets.

This species is sociable and may be encountered in large groups. Although a colony of some 300 bats was reported behind a closed window shutter (Snyder 1902) and a colony of 10,000 was described as inhabiting a house in Maryland in 1860 (Goodwin 1935; Doutt, Heppenstall, and Guilday 1966), Barbour and Davis (1969) doubted the accuracy of these accounts since they could not find any reports of colonial behavior of the silver-haired bat within the previous sixty years.

Little is known on the migration of silver-haired bats, but they migrate leisurely southward and occasionally make long oceanic journeys while flying beneath heavy storm clouds several hundred feet above choppy seas (Hamilton 1943). They have been reported at a lighthouse on Mount Desert Rock, Maine, 15 miles from the nearest island and 30 miles from the mainland. Bats visited this island every spring and fall but did not live there permanently (Merriam 1888). A male silver-haired bat was collected on a research ship about 95 miles SSE of Montauk Point, Long Island, on 19 August (Mackiewicz and Backus 1956), and Davis and Harding (1966) report that a banded silver-haired male flew 107 miles in 14 days. A specimen was netted when the ambient temperature was 20°F (Jones 1965), and Stones and Holyoak

(1970) describe a semitorpid silver-haired bat found under the loose bark of an elm on 29 May in Michigan. The species does well in captivity and will eat insects, bits of raw meat, and bananas (Barbour and Davis 1969).

Silver-haired bats in the wild seem to feed entirely on nocturnal insects, particularly those that fly high in the woodlands or over the borders of woodland streams and ponds.

Little is known about the reproduction of this species. One litter, usually comprising two young, apparently is born in late June or early July. Newborns are black, wrinkled, and nearly hairless. They grow rapidly and cling to the mother until they are about 3 weeks old and are able to fly (Jackson 1961).

**Specimens Examined.** New Hampshire, total 14: Hillsborough County, Peterboro 14 (HMCZ).

Vermont, total 14: Rutland County, Chittenden 4 (AMNH), East Wallingford 7 (DC), Fair Haven 1 (UVT), Rutland 1–1 (DC–UCT).

Massachusetts, total 9: Dukes County 1, Penikese Island 1; Essex County, Gloucester 1; Hampden County, Feltonville 1 (HMCZ); Hampshire County, North Hadley 1 (UMA); Middlesex County, Tynsboro 2 (HMCZ); Plymouth County, Mattopoisett 1 (UCT); Worcester County, Harvard 1 (HMCZ).

Connecticut, total 1: Litchfield County (no location) 1 (AMNH).

# Eastern Pipistrelle

*Pipistrellus subflavus obscurus*
(F. Cuvier)

V[espertilio] subflavus F. Cuvier,
1832. Nouv. Ann. Mus. Hist. Nat.
Paris 1:17.
Pipistrellus subflavus obscurus, Miller, 1897. N. Amer. Fauna 13:93.
Type Locality: Lake George, Warren County, New York
Dental Formula:
I 2/3, C 1/1, P 2/2, M 3/3 × 2 = 34
Names: Pipistrel, pipistrelle, eastern bat, Georgian bat, pygmy bat, and butterfly bat. The meaning of *Pipistrellus* = a little bat, and *subflavus* = yellowish, referring to its yellowish brown fur.

**Description.** The eastern pipistrelle is one of the smallest bats in New England and the only bat having 34 teeth. It is distinguished by its yellowish brown tricolor fur: leaden gray at the base of the hairs, followed by a band of yellowish brown, and dark brown at the tip. The wing and interfemoral membranes are blackish, the ears tan. The coloration of the body is dark brown above, yellowish brown below. Immatures are somewhat paler than adults. Nearly all-white and melanistic specimens occur rarely. The pattern of molting is unknown.

The ears are thin and naked, slightly longer than broad, tapering to a narrow rounded tip, and when laid forward reach just beyond the tip of the snout. The tragus is short, nearly straight, bluntly rounded, nearly half the length of the ear, and not sharply pointed as in the myotis

species. The thin and delicate wing membrane is attached to the base of the toes; the thumb is large, about one-fifth the length of the forearm. The interfemoral membrane is lightly haired at the basal third of the upper surface. The calcar is not keeled. Females have two mammae.

The braincase is relatively inflated, with the rostrum broad, its profile rising at a distinct angle to the forehead.

The sexes are equal in size. Measurements range from 75 to 90 mm (2.9–3.5 in); tail, 36 to 45 mm (1.4–1.8 in); hind foot, 8 to 10 mm (0.31–0.39 in); and wingspread, 208 to 258 mm (8.1–10.1 in). Weights vary from 4 to 7 g (0.14–0.25 oz).

**Distribution.** This species occurs in eastern North America from southern Quebec and Nova

EASTERN
PIPISTRELLE

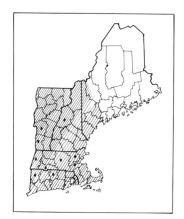

found in deep woods or in open fields unless there are large trees nearby. Its average speed in an open field was clocked as 11.7 miles per hour for a distance of 67 feet (Patterson and Hardin 1969). Griffin (1940a) reported successful homing from about 85 miles.

The eastern pipistrelle inhabits more caves in eastern North America than any other species of bat. Fat pipistrelles normally enter hibernating caves in mid-October and remain until late April, the females leaving before the males (Barbour and Davis 1969). These bats hibernate the most profoundly as well as the longest and may sleep in one position for a week or more, though periodically an individual will awaken, fly about, and return to the same spot it left (Rysgaard 1942). Eastern pipistrelles are unsociable and usually hang alone, or sometimes with a few others. They seldom shift from a roosting spot, and a pipistrelle may occupy the same spot in a cave during consecutive winters. Eastern pipistrelles select a hibernating site where the ambient temperature is about 52–55°F. Droplets of moisture may condense on their fur, and in a beam of light they appear almost pure white (Hall 1962).

This bat feeds on flies, grain moths, and smaller bugs and beetles. Hamilton (1943) suggested that pipistrelles feed in early evening and probably again toward midnight or early morning. He found pipistrelles with full stomachs 15 to 20 minutes after they first appeared in the evening.

Little is known about the reproduction of this species. Breeding occurs in November and a litter usually of two young is produced in mid-June to mid-July. At birth the young weigh 1.9 g (0.67 oz). They are able to fly by the time they are about one month old, and shortly thereafter shift for themselves (Lane 1946). Longevity is about 10 years in females and 15 years in males (Davis 1966).

Scotia west to Minnesota, south to Oklahoma and to eastern Mexico, Guatemala, and Honduras. Barbour and Davis (1969) stated that it is apparently absent from northern New England and southern Florida and is unknown in northern Indiana, northeastern Illinois, eastern Wisconsin, and most of Michigan.

**Ecology.** In summer this wide-ranging species is usually found in small clusters in more or less open woods near water, in crevices of cliffs and rocks, and less frequently in buildings and caves. In spring and autumn individuals are occasionally found hanging singly in abandoned buildings and beneath porch roofs or in lofts, often in daylight. It hibernates in caves, mines, and rock crevices.

Little is known of the enemies of the eastern pipistrelle, but the hoary bat has been known to kill one (Bishop 1947), and a pipistrelle has been recovered from the stomach of a frog (Barbour and Davis 1969).

**Behavior.** This dainty little bat is a weak, erratic flier and may be mistaken for a large fluttering moth. It takes wing in early evening and makes short elliptical flights at treetop level, covering so small an area that it is constantly in view. It often forages over ponds and streams and is not

**Specimens Examined.** NEW HAMPSHIRE, total 1: Grafton County, Enfield 1 (DC).

VERMONT, total 52: Bennington County, Dorset 3 (UCT), East Dorset 2 (USNM); Rutland County, Brandon 3–1–5–6 (DC–HMCZ–UCT–USNM), Chittenden 6–6–9 (AMNH–DC–UCT), Plymouth Union 1 (HMCZ), Proctor 10 (USNM).

MASSACHUSETTS, total 32: Berkshire County, Lanes-

BIG BROWN BAT

boro 3 (HMCZ), Sheffield 5–3 (HMCZ–UCT); Hampden County, Chester 6–9 (HMCZ–UMA); Hampshire County, Amherst 1 (UMA); Middlesex County, Concord 2, Woburn 1; Norfolk County, Braintree 1; Worcester County, East Templeton 1 (HMCZ).

CONNECTICUT, total 7: Litchfield County, Lakeville 2 (UCT), Roxbury 2 (HMCZ); Tolland County, Storrs 3 (UCT).

**Records and Reports.** VERMONT: Bennington County, East Dorset 12; Orange County, Vershire 3; Rutland County, Chittenden 5; Windsor County, Plymouth 2 (Griffin 1940b).

MASSACHUSETTS: Hampden County, Chester 31 (Griffin 1940b).

CONNECTICUT: Litchfield County, Roxbury 4 (Griffin 1940b).

## Big Brown Bat

*Eptesicus fuscus* (Palisot de Beauvois)

**Description.** The big brown bat is exceeded in size in New England only by the hoary bat. It is recognized by its uniformly dark brown pelage, varying in individuals from dark brown to cinnamon. Each hair is bicolor—the basal half is blackish and the outer half brown. The face, ears, and membranes are blackish. The sexes are colored alike and show no seasonal variation. Immatures are paler and grayer than adults, and albinos, pied-pattern, and white-blotched specimens have been recorded. Females have two mammae.

The ears are short, broad, somewhat rounded, and naked except for some fur at the base and reach barely to the nose when laid forward; the tragus is straight, rounded bluntly at the tip, curved slightly forward, and nearly half as long as the ear. The wing and interfemoral membranes are naked; and the calcar is keeled. The large skull is heavier than that of any New England bat except the red bat, which is noticeably wider. Both the big brown bat and the red bat have 32 teeth, but the big brown bat has two upper incisors and one upper premolar on each side, whereas the red bat has one upper incisor and two premolars on each side.

Vespertilio fuscus Palisot de
Beauvois, 1796. *Catalogue
raisonné du muséum de Mr. C. W.
Peale, Philadelphia*, p. 18. (English
edition by Peale and Palisot de
Beauvois, p. 14.)
*Eptesicus fuscus*, Méhely, 1900.
*Magyarorszyág denevéreinek
monographiája* [Monographia
Chiropterorum Hungariae], pp.
206, 338.
Type Locality: Philadelphia, Penn-
sylvania
Dental Formula:
I 2/3, C 1/1, P 1/2, M 3/3 × 2 = 32
Names: Brown bat, house bat, barn
bat, dusky bat, serotine bat, and
Carolina bat. The meaning of *Ep-
tesicus* = house flier or flying, and
*fuscus* = brown, referring to the
brown color of this species.

The sexes are equal in size. Measurements range from 105 to 123 mm (4.1–4.8 in); tail, 40 to 50 mm (1.6–2.0 in); hind foot, 10 to 13 mm (0.39–0.51 in); and wingspread, 310 to 330 mm (12.1–12.9 in). Weights vary from 12 to 16 g (0.42–0.56 oz).

**Distribution.** This species occurs from Alaska and southern Canada south to northern South America, including the Caribbean Islands. It is abundant throughout most of its range, except in the far northern states and the deep South. It has yet not been reported from southern Florida and much of central Texas.

**Ecology.** The big brown bat is probably the bat most familiar to man. In summer it commonly roosts in attics, belfries, and barns, behind awnings, doors, and shutters, in hollow trees, beneath bridges, or in similar shelters, but seldom in caves. In winter it hibernates in buildings, caves, mines, storm sewers and other retreats. Hitchcock (personal communication) stated that in New England the big brown bat regularly hibernates in buildings—the only species of bats to do this.

This hardy species can endure subfreezing temperatures. Nero (1959) found a big brown bat alive in a building in Prince Albert, Saskatchewan, on 1 February. Goehring (1971) reported that a big brown bat was found alive outside a building in Minnesota when the air temperature was −4°F and stated that the air temperature had been −11°F four hours earlier.

Man, great horned owls, barn owls, screech owls, sparrow hawks, common grackles, cats, and snakes prey on big brown bats. Severe cold, storms, and accidents kill many, and blood-sucking bugs, mites, fleas, and trematodes have been reported to parasitize them. Hitchcock (personal communication) reported that in New England rabies occurs in this bat more commonly than in the little brown myotis.

**Behavior.** Big brown bats are not as tolerant of high temperatures as are little brown myotis and will abandon an attic roost for cooler places. During extremely hot weather they may be heard squeaking in partitions of houses, and occasionally young bats may crawl out into

rooms from crevices of fireplaces, or both young bats and adults may appear in basements of houses in which the space between the inside and outside walls is continuous from attic to basement.

Big brown bats do not form large colonies as some other species do. Rysgaard (1942) reported a colony of about 400 big brown bats in Kentucky, Barbour and Davis (1969) found populations ranging from 20 to 300 bats in Kentucky, and Mills, Barrett, and Farrell (1975) reported nursery colonies in Ohio ranging from 8 to 700 individuals, with a mean of 154.

Big brown bats take wing at dusk, although they may sometimes be seen flying at midday, and generally use the same feeding grounds each night. They fly in a nearly straight course at a height of 20 to 30 feet. Patterson and Hardin (1969) recorded the average speed of a big brown bat as 10.4 miles per hour. In flight, big brown bats can be recognized by their large size and slow wing beats; they emit an audible chatter as they fly.

Big brown bats are vicious when first captured and can inflict a painful bite, but they adapt remarkably well to captivity (Barbour and Davis 1969).

Big brown bats do not migrate far from their natal colony. Rysgaard (1942) banded a big brown bat in one cave and recaptured it later the same winter in another cave 400 yards away, and Barbour and Davis (1969) often found the same individuals using the same cave as both a summer night roost and a hibernating site. Beer (1955) and Davis, Barbour, and Hassell (1968) recovered banded big brown bats within 30 miles of their banding site. Mumford (1958) reported a big brown bat movement of 142 miles, and Mills, Barrett, and Farrell (1975) reported two exceptional movements of 155 and 180 miles.

Big brown bats have a remarkable homing instinct. Cope, Koontz, and Churchwell (1961) released banded big brown bats at various distances from the home roost. Nearly all the bats that were released 20 miles away returned the same night, bats released 40 miles away returned the second night, and those released 100 miles away returned the third night. Of bats released 250 miles from home, some returned during the fourth night, and nearly all returned on the fifth night. Smith and Goodpaster (1958) released 51 banded adult females and 104 juveniles 450 miles north of a maternity colony on 20 July 1957. A month later three adults had returned, and four more returned within the following two months. None of the juveniles returned.

Big brown bats are one of the last bats to enter hibernation and may be seen flying about in late November and early March. When they hibernate in caves and mines they use sites near the entrance where the ambient temperature range is about 32°–64°F and the humidity below 100%, so that water does not condense on their fur (Barbour and Davis 1969). Rysgaard (1942) and Davis and Hitchcock (1964) found that wintering male big brown bats commonly formed tight clusters, with females hanging as singles in the warmer sites.

The food of the big brown bat consists mostly of insects, mainly Coleoptera and Hymenoptera (Hamilton 1933). Captive big brown bats are known to kill and eat smaller species of bats (Barbour and Davis 1969). Gould (1959) reported that a big brown bat ate 2.7 g (0.10 oz) of food per hour.

The breeding season occurs in autumn and winter, and ovulation and fertilization from stored sperm occur about the first week in April. The gestation period is about two months, and birth occurs about the first of June. Two young are usually born per litter, though occasionally there may be only one. At birth the young are naked and blind and weigh about 4 g (0.14 oz). They grow rapidly and gain 0.5 g (0.02 oz) per day. As the young grow they are left alone for longer periods while the mother forages farther away and is gone from the nursery for longer periods of time. The young are weaned at three weeks of age and by early July are able to fly. In New England many nurseries disintegrate as soon as the young are weaned. As the young mature the males enter the nurseries and become as numerous as the adult females.

Davis, Barbour, and Hassell (1968) reported that the newborn young cling so tightly to the female's teats that they cannot be induced to release their grip. Barbour and Davis (1969) found that females carried their young only when moving from one roost to another, but Hall (1946) suggested that females may carry their young while feeding.

Paradiso and Greenhall (1967) reported that big brown bats live 10 years or more. Hitchcock (1965) recaptured a 19-year-old male and a 12-year-old female, and Goehring (1972) reported retrieving two banded males 17 years old.

**Specimens Examined.** MAINE, total 18: Franklin County, Dryden 1 (AMNH); Hancock County, Mt. Desert Island 12 (HMCZ); Kennebec County, Waterville 1; Oxford County, South Waterford 1 (AMNH); Penobscot County, Brewer 1, Orono 1 (DC); Washington County, Eastport 1 (USNM).

NEW HAMPSHIRE, total 25: Grafton County, Birch Island 1 (UCT), Hanover 8, Ruggles Mine 1 (DC); Hillsborough County, Peterboro 9 (HMCZ); Merrimack County, Concord 1; Strafford County, Durham 4 (UNH); Sullivan County, Charlestown 1 (USNM).

VERMONT, total 17: Bennington County, East Dorset 1; Orange County, Thetford Center 2 (DC), Vershire 3; Rutland County, Brandon 1 (UCT), Chittenden 2 (DC), Mendon 2 (UCT), Rutland 2–3 (DC–UCT), Sudbury 1 (UCT).

MASSACHUSETTS, total 78: Essex County, Salem 1, West Andover 1 (HMCZ); Hampden County, Chester 3, Holyoke 1; Hampshire County, Amherst 15, Belchertown 1, Pelham 1 (UMA); Middlesex County, Bedford 1, Cambridge 20 (HMCZ), Wilmington 7 (USNM); Norfolk County, Brookline 2 (HMCZ), Stoughton 1 (USNM); Plymouth County (no location) 1, Halifax 2, Mashpee 1 (UCT); Suffolk County, Boston 15 (HMCZ); Worcester County, Dudley 1 (UCT), Worcester 4 (UMA).

CONNECTICUT, total 138: Fairfield County, Bridgeport 1, Cos Cob 1, Greenwich 1, Monroe 1, Stevenson 2, Westport 8; Hartford County, Bristol 1, Glastonbury 1 (UCT); Litchfield County, Macedonia Park 1, Round Hill 2 (AMNH), Roxbury 5 (UCT), Sharon Mt. 7 (AMNH); Middlesex County, Deep River 1; New Haven County, Hampden 2 (UCT), New Haven 1–1 (AMNH–HMCZ), Southbury 47; New London County, Lebanon 2, Niantic 1; Tolland County, Columbia 1, Coventry 2, Mansfield 7, Stafford 1, Storrs, 15, Tolland 5, Union 1, Willington 1; Windham County, Ashford 16, Chaplin 1, Hampton 1, Willimantic 1 (UCT).

RHODE ISLAND, total 1: Providence County, Chepachet 1 (USNM).

**Records and Reports.** VERMONT: Orange County, Vershire 68 (Griffin 1940b).

CONNECTICUT: Litchfield County, Roxbury 10 (Griffin 1940b).

# Red Bat

*Lasiurus borealis* (Müller)

*Vespertilio borealis* Müller, 1776. *Des Ritters Carl von Linne . . . vollständiges Natursystem nach der zwölften lateinischen Ausgabe. . . . Suppl.* (*Mammalia*), p. 20.
*Lasiurus borealis*, Miller, 1897. N. Amer. Fauna, 13:105–15.
Type Locality: New York
Dental Formula:
I 1/3, C 1/1, P 2/2, M 3/3 × 2 = 32
Names: Northern bat, leaf bat, tree bat, and New York bat. The meaning of *Lasiurus* = hairy tail, referring to the furred upper surface of the interfemoral membrane that is characteristic of the genus, and *borealis* = northern or of the north.

**Description.** The medium-sized red bat is easily distinguished by its bright rusty color, short rounded ears, long pointed wings, interfemoral membrane thickly furred on the upper surface, and keeled calcar. The ears reach about halfway from the angle of the mouth to the nostril when laid forward, and the tragus has a slight forward bend at the top and is less than one-half the length of the ear. The bat extends its long tail straight out in flight. Females have four mammae.

The skull is short and deep and broad, with flaring zygomatic arches; the braincase is high and rounded, and the nares openings are large. The anterior upper premolar is peglike, or occasionally lacking, and displaced inwardly out of the normal tooth row.

Males tend to be redder and less frosted than females. The underparts are paler than the back and lack much of the white tipping. The wing membrane is brownish black. Apparently color variants have not been recorded, and there is no seasonal molt or color phase so far as is known.

Females are somewhat larger than males. Measurements range from 94 to 122 mm (3.7–4.8 in); tail, 42 to 54 mm (1.6–2.1 in); hind foot, 8 to 10 mm (0.31–0.39 in); and wingspread, 290 to 330 mm (11.3–12.9 in). Weights vary from 8 to 14 g (0.28–0.49 oz).

**Distribution.** The red bat ranges widely in North America, from Nova Scotia and New Brunswick west across southern Canada to Alberta and south to Texas and Florida. It is also found from British Columbia south to Panama and beyond but is absent from the Rocky Mountains and the central plateau of Mexico.

**Ecology.** Except for females with young, red bats roost individually in trees, usually on the south side and sometimes only a few feet above ground, concealing themselves in the foliage. They prefer deciduous trees on forest edges and wide hedgerows of trees and tall shrubs bordering orchards and city parks. They particularly prefer American elms (McClure 1942; Constantine 1966).

Red bats generally choose roosting sites that provide protection from wind and sun and hide them from view except from below, allowing them to drop downward freely to begin flight (Constantine 1966). They are rarely found in caves and buildings. Roost sites are often used by different individuals on different days (Downes 1964; Constantine 1966). Infrequently some individuals have been found on the undersides of sunflower leaves in August (Downes 1964) or in woodpecker holes in dense wooded areas in March (Fassler 1975).

Blue jays are said to prey persistently upon young red bats (Elwell 1962). Sharp-shinned hawks and other hawks, owls, cats, opossums, skunks, and tree-climbing snakes occasionally capture red bats. Some become impaled on barbed-wire fences or entangled in burrs of burdocks, and many are killed in sudden storms. Wilson (1965) reported dead red bats in insect light traps, and Koestner (1942) collected fifteen red bats caught on the surface of a heavily oiled road, which they probably mistook for water.

Fleas, the bat bug, *Cimex philosellus*, cestodes, and trematodes are known to parasitize the red bat, and rabies has been frequently confirmed in this species.

**Behavior.** Red bats take wing at dusk, usually flying high in slow, erratic flight, then descending and feeding from the treetops to a few feet above ground and along watercourses, sometimes flying swiftly in straight lines or wide arches. They may be seen flying on cloudy days and infrequently on sunny days.

Two or more red bats may be seen flying over the same general area, but more often one bat flies over an area which it forages regularly. Red bats may shift their hunting territory as the night progresses, depending somewhat on weather conditions (Burt 1948). Occasionally red bats encroach upon one another's feeding territory without conflict (Barbour and Davis 1969). These bats hunt 600 to 1,000 yards from their roosting site (Jackson 1961).

These bats fly at nearly 40 miles per hour in a straightaway flight (Jackson 1961) and average a flight speed of 7.9 miles per hour for a distance of 30 feet (Patterson and Hardin 1969).

Red bats are considered to be migratory, though migration patterns are unknown (Hamilton 1943; Barbour and Davis 1969). Apparently the sexes are segregated during migration,

RED BAT

according to late fall and winter records (Baker and Ward 1967; Davis and Lidicker 1956; Mackiewicz and Backus 1956; Terres 1956).

Red bats are strong fliers and cover great distances over water. A red bat was reported 65 miles off the southern New England coast, and another lone bat was sighted about 90 miles SSE of Montauk Point, Long Island (Mackiewicz and Backus 1956). Two male red bats were recovered among a few hundred dead migratory birds that were killed when they collided with the Empire State Building one mid-October night (Terres 1956). Red bats seem to be adapted for survival at cold temperatures, since the densely furred interfemoral membrane may act like a blanket during hibernation. It is not uncommon for red bats to winter in areas where freezing weather is frequently encountered. McKeever (1951) recorded two red bats taken at elevations of 2,400 and 2,500 feet in West Virginia on 4 January, and Swanson and Evans (1936) recorded a specimen taken from a building in Minnesota on 9 April. Davis and Lidicker (1956) found that red bats awake from hibernation only when ambient temperature reaches 68°F. Barbour and Davis (1969) stated that red bats did not awaken unless flying insects were available, since arousal requires considerable energy.

Red bats swim fairly well by flapping the wings, and with some effort can rise from the surface of the water if their fur is not too wet. They frequently give sharp chirps when flying.

Little is known about the food of red bats, but they are known to eat flying moths, flies, bugs, beetles, crickets, and cicadas, which they take from the foliage and from the ground. They occasionally capture moths clinging to light poles.

Red bats breed from August to October, somewhat earlier than most bats. Copulation begins while the bats are in flight and ends on the ground. Fertilization occurs in spring after the bats emerge from hibernation, and the young are born from late May to early July after a gestation period of 80 to 90 days (Layne 1958).

Red bats produce larger litters than other species of bats, varying from one to five young, but usually including two or three. At birth the young are blind, naked, and helpless. They grow rapidly and are able to fly at between 3 and 4 weeks of age. They are weaned when 5 to 6 weeks old (Barbour and Davis 1969). Longevity is about 12 years (Jackson 1961).

There are reports of foraging female red bats carrying young whose combined weight exceeded their mother's (Poole 1932; Hamilton

1943; Burt 1948; Packard 1956; Schwartz and Schwartz 1959; Jackson 1961; Stains 1965; Doutt, Heppenstall, and Guilday 1966; Peterson 1966, Hamilton and Stalling 1972).

**Specimens Examined.** MAINE, total 5: Franklin County, Dryden 3 (UME); Hancock County, Mt. Desert Island 1 (HMCZ); Penobscot County, Lincoln 1 (UME).

NEW HAMPSHIRE, total 1: Hillsborough County, Milford 1 (HMCZ).

VERMONT, total 9: Chittenden County, Burlington 1 (UVT); Rutland County, East Wallingford 1, Rutland 6 (DC); Sherbourne 1 (HMCZ).

MASSACHUSETTS, total 51: Dukes County, Martha's Vineyard 1–1 (AMNH–HMCZ); Essex County, Annis-quam 2, Salem 2, Wenham 1; Middlesex County, Cambridge 2, Concord 1, Malden 10, Newton 14, Wayland 1, West Medford 3, Woburn 3 (HMCZ); Nantucket County, Nantucket 1 (USNM); Norfolk County, Sharon 1; Plymouth County, Marshfield 1; Suffolk County, Boston 1, Roxbury 2 (HMCZ); Worcester County, Worcester 4 (UMA).

CONNECTICUT, total 14: Fairfield County, Stamford 2 (AMNH), Westport 1 (UCT); Hartford County, Farmington 1 (USNM), Glastonbury 1 (UCT), Wethersfield 1 (USNM); Middlesex County, Clinton 1, Saybrook 1 (AMNH); New Haven County, Branford 1, New Haven 1 (USNM), Riverside 1 (AMNH); Tolland County, Mansfield 2, Storrs 1 (UCT).

RHODE ISLAND, total 2: Newport County, Fort Adams 1; Providence County, Chepachet 1 (USNM).

# Hoary Bat

*Lasiurus cinereus* (Palisot de Beauvois)

*Vespertilio cinereus* (misspelled linereus) Palisot de Beauvois, 1796. *Catalogue raisonné du muséum de Mr. C. W. Peale, Philadelphia*, p. 18. (English edition by Peale and Palisot de Beauvois, p. 15).
*Lasiurus cinereus*, H. Allen. 1864. *Smithsonian Misc. Coll.* 7(Pub. 165):21.
Type Locality: Philadelphia, Pennsylvania
Dental Formula:
  I 1/3, C 1/1, P 2/2, M 3/3 × 2 = 32
Names: Great northern bat, frosted bat, gray bat, and twilight bat. The meaning of *Lasiurus* = hairy tail, and *cinereus* = ashy, referring to the grayish color of this bat.

**Description.** The hoary bat is the largest and the most strikingly colored bat in New England. Its pelage is dark brown or umber heavily tinged with white, producing a hoary or frosty appearance, especially on the back. When the fur is separated the hairs are seen to be brownish black basally, followed by a broad band of yellowish brown, then chocolate brown, and finally white at the tip. The cheek and throat are yellowish and mixed with less white, the ears are tan rimmed with black or dark brown, and the membranes are brownish black. The dorsal surface of the wing membrane is yellowish brown along the forearm and partway down the fingers. Some individuals are much darker than others, but the sexes are colored alike and there seem to be no seasonal color variations. Apparently color variants and molt have not been recorded.

The head is broad and robust, with a short, wide rostrum and large braincase; the zygomatic arches are widely spread and the nares are wide. The ears are large and rounded; the tragus is triangular and broad at the base, curving forward, and is less than half the length of the ear. The anterior edge of the underside of the wing membrane is furred for about half its length, the interfemoral membrane is thickly furred on the upper surface, and the hind feet are well furred above. The calcar is keeled and the tail makes up about 40% of the total length of the animal. Females have four mammae.

Females are larger than males. Measurements range from 130 to 150 mm (5.1–5.9 in); tail, 52 to 62 mm (2.0–2.4 in); hind foot, 10 to 14 mm (0.39–0.55 in); and wingspread, 375 to 420 mm (14.6–16.4 in). Weights vary from 25 to 45 g (0.88–1.58 oz).

**Distribution.** This hardy bat is the most widespread of North American bats, ranging from the tundra of Southampton Island southward in boreal Canada from the Atlantic to the Pacific coast, south into South America to Chile, and to the Dominican Republic and Bermuda. It is rare throughout most of the eastern United States.

**Ecology.** Hoary bats roost in trees. They prefer coniferous forests but are also found in deciduous woods, at the forest edge or in wide hedgerows, and in farmyards and city parks.

These bats choose leafy sites well covered from above but open below, and 10 to 15 feet above the ground (Constantine 1966). They sometimes wander far into caves in late summer and many never find their way out (Myers 1960). A hoary bat was found in a woodpecker hole high in a tree in British Columbia (Cowan and Guiguet 1965), and another was found in a squirrel's nest (Neill 1952).

Man and occasionally hawks and owls prey on hoary bats. Some individuals may die during long migrations. Not much is known about the parasites of this bat, but it is known to carry rabies.

**Behavior.** Hoary bats generally take wing late in the evening, but may sometimes be seen flying before dark, especially on cooler nights. They are strong and swift flyers. Jackson (1961) estimated that they can achieve speeds of up to 60 miles per hour for a mile or more.

These bats usually migrate to warmer climates in winter and migrate along with red bats and silver-haired bats. Hoary bats regularly emit audible chattering sounds during migration, although they may fly in silence. They generally do poorly in captivity, and Bogan (1972) found that captives are reluctant to drink or eat.

Hoary bats prey chiefly on insects but occasionally capture pipistrelles (Bishop 1947; Orr 1950). A snakeskin and some grass were found in the alimentary canal of a hoary bat that was collected in Indiana in the winter (Whitaker 1967).

Little is known about the reproduction of this species except that the young are born from mid-May into early July (Gottschang 1966; Provost and Kirkpatrick 1952). It is believed that fertilization takes place in early spring and that the gestation period is about 90 days. The litter size is usually two (Jackson 1961). This species does not form maternity colonies, but the sexes remain segregated during parturition and for most of the summer, except during migration and breeding.

The female gives birth while hanging in her leafy retreat. The newborns are blind and naked except that the top of the head, shoulders, uropalagium, and feet are covered with fine silvery gray hair. The skin on the body is a darker brown than the wing membrane (Munyer 1967). At birth the young weigh 5 g (0.18 oz) and by about 3 weeks old weigh 9 to 10 g (0.32–0.35 oz). The eyes open about 10 to 12 days after birth. The young grow rapidly and within a month are generally able to shift for themselves (Jackson 1961). Newborns cling tightly to the mother through the day but are left clinging to a leaf or twig while she forages at night (Barbour and Davis 1969). Sometimes a mother with her heavy, clinging young falls to the ground and becomes helpless.

**Specimens Examined.** MAINE, total 3: Penobscot County, Lincoln 2 (UME); Somerset County, Penobscot Lake 1 (AMNH).

NEW HAMPSHIRE, total 1: Strafford County, Durham 1 (UNH).

MASSACHUSETTS, total 1: Nantucket County, Monomoy Island 1 (USNM).

CONNECTICUT, total 2; Fairfield County, Westport 1 (UCT); Litchfield County, Sharon Mt. 1 (AMNH).

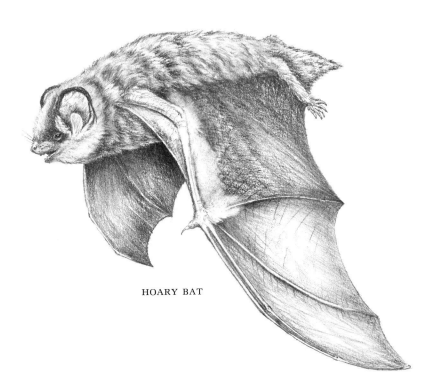

HOARY BAT

# REFERENCES

Allen, H. 1894. A monograph of the bats of North America. *U.S. Natl. Mus. Bull.* 43:1–198.

Anderson, Sydney. 1972. Mammals of Chihuahua: Taxonomy and distribution. *Bull. Amer. Mus. Nat. Hist.* 148(2):153–410.

Baker, Robert, J., and Ward, Claud M. 1967. Distribution of bats in southeastern Arkansas. *J. Mamm.* 49(1):130–32.

Barbour, Roger W., and Davis, Wayne H. 1969. *Bats of America.* Lexington: University of Kentucky Press.

Barbour, Roger W.; Davis, Wayne H.; and Hassell, Marion D. 1966. The need of vision in homing by *Myotis sodalis. J. Mamm.* 47(2):356–57.

Beer, James R. 1955. Survival and movements of banded big brown bats. *J. Mamm.* 36(2):242–48.

Bishop, Sherman C. 1947. Curious behavior of a hoary bat. *J. Mamm.* 28(3):293–94.

Bogan, Michael A. 1972. Observations on parturition and development in the hoary bat, *Lasiurus cinereus. J. Mamm.* 53(3):611–14.

Buchler, E. R. 1975. Food transit time in *Myotis lucifugus* Chiroptera: (Vespertilionidae). *J. Mamm.* 56(1):252–55.

Burt, William H. 1948. *The mammals of Michigan.* Ann Arbor: University of Michigan Press.

Constantine, Denny G. 1966. Ecological observations on lasiurine bats in Iowa. *J. Mamm.* 47(1):34–41.

Cope, J. B.; Koontz, K.; and Churchwell, E. 1961. Notes on homing of two species of bats, *Myotis lucifugus* and *Eptesicus fuscus. Proc. Indiana Acad. Sci.* 70:270–74.

Cowan, I. McTaggart, and Guiguet, C. J. 1965. *The mammals of British Columbia.* Publication no. 11. British Columbia Provincial Museum.

Davis, Wayne H. 1966. Population dynamics of the bat *Pipistrellus subflavus. J. Mamm.* 47(3):383–96.

Davis, Wayne H., and Hardin, James W. 1967. Homing in *Lasionycteris noctivagans. J. Mamm.* 48(2):323.

Davis, Wayne H., and Hitchcock, Harold B. 1964. Notes on sex ratios of hibernating bats. *J. Mamm.* 45(3):475–76.

———. 1965. Biology and migration of the bat, *Myotis lucifugus,* in New England. *J. Mamm.* 46(2):296–313.

Davis, Wayne H., and Lidicker, William Z., Jr. 1956. *Winter range of the red bat, Lasiurus borealis. J. Mamm.* 37(2):280–81.

Davis, Wayne H., and Mumford, Russell E. 1962. Ecological notes on the bat *Pipistrellus subflavus. Amer. Midland Nat.* 68(1):294–98.

Davis, Wayne H.; Barbour, Roger W.; and Hassell, Marion D. 1968. Colonial behavior of *Eptesicus fuscus. J. Mamm.* 49(1):44–50.

Davis, Wayne H.; Hassell, Marion D.; and Rippy, Charles L. 1965. Maternity colonies of the bat, *Myotis l. lucifugus,* in Kentucky. *Amer. Midland Nat.* 73(1):161–65.

Doutt, J. Kenneth; Heppenstall, Caroline A.; and Guilday, John E. 1966. *Mammals of Pennsylvania.* Harrisburg: Pennsylvania Game Commission.

Downes, William L., Jr. 1964. Unusual roosting behavior in red bats. *J. Mamm.* 45(1):143–44.

Dunning, D. C., and Roeder, K. D. 1965. Moth sounds and insect-catching behavior of bats. *Science* 147:173–74.

Easterla, David A., and Watkins, Larry C. 1967. Silver-haired bats in southwestern Iowa. *J. Mamm.* 48(2):327.

Elwell, Adela S. 1962. Blue jay preys on young bats. *J. Mamm.* 43(3):434.

Fassler, David J. 1975. Red bat hibernating in a woodpecker hole. *Amer. Midland Nat.* 93(1):254.

Findley, James S., and Jones, Clyde. 1964. Seasonal distribution of the hoary bat. *J. Mamm.* 45(3):461–70.

———. 1967. Taxonomic relationships of bats of the species *Myotis fortidens, M. lucifugus* and *M. occultus. J. Mamm.* 48(3):429–44.

Frum, W. Gene. 1946. Abnormality in dentition of (*Myotis lucifugus). J. Mamm.* 27(2):176.

Glass, Bryon P., and Baker, Robert J. 1965. *Vespertilio subulatus* Say, 1823: Proposed suppression under the plenary powers (Mammalia, Chiroptera). *Z. N. (S) 1701 Bull. Zool. Nomenclature* 22(33):204–5.

Goehring, Harry H. 1971. Big brown bat survives sub-zero temperatures. *J. Mamm.* 52(4):832–33.

———. 1972. Twenty-year study of *Eptesicus fuscus* in Minnesota. *J. Mamm.* 53(1):201–7.

Goodwin, George Gilbert. 1935. The mammals of Connecticut. *Connecticut Geol. Nat. Hist. Surv. Bull.* 53:1–221.

Gottschang, J. L. 1966. Occurrence of the hoary bat, *Lasiurus cinereus,* in Ohio. *Ohio J. Sci.* 66:527–29.

Gould, Edwin. 1955. The feeding efficiency of insectivorous bats. *J. Mamm.* 36(3):399–407.

———. 1959. Further studies on the feeding efficiency of bats. *J. Mamm.* 40(1):149–50.

———. 1975. Neonatal vocalizations in bats of eight genera. *J. Mamm.* 56(1):15–29.

Griffin, Donald R. 1940a. Migrations of New England bats. *Bull. Mus. Comp. Zool. Harvard Coll.* 86(6):217–46.

———. 1940b. Notes on the life histories of New England bats. *J. Mamm.* 21(2):181–87.

———. 1945. Travel of banded cave bats. *J. Mamm.* 26(1):15–23.

———. 1946. Mystery mammals of the twilight. *Natl. Geogr. Mag.* 90(7):117–34.

———. 1958. *Listening in the dark.* New Haven: Yale University Press.

Griffin, Donald R., and Hitchcock, Harold B. 1965. Probable 24-year longevity for *Myotis lucifugus*. *J. Mamm.* 46(2):332.

Hall, E. Raymond. 1946. *Mammals of Nevada.* Berkeley: University of California Press.

Hall, John S. 1962. *A life history and toxonomic study of the Indiana bat,* Myotis sodalis. Scientific Publication no. 12. Reading: Pennsylvania Public Museum and Art Gallery.

Hall, John S.; Cloutier, Roger J.; and Griffin, Donald R. 1957. Longevity records and notes on tooth wear of bats. *J. Mamm.* 38(3):407–8.

Hamilton, Robert B., and Stalling, Dick T. 1972. *Lasiurus borealis* with five young. *J. Mamm.* 53(1):190.

Hamilton, William J., Jr. 1933. The insect food of the big brown bat. *J. Mamm.* 14(2):155–56.

———. 1943. *The mammals of eastern United States.* New York: Hafner Publishing Co.

Hardin, James W. 1967. Waking periods and movements of *Myotis sodalis* during the hibernation season. M.S. thesis, University of Kentucky.

Hardin, James W., and Hassell, Marion D. 1970. Observation on waking periods and movements of *Myotis sodalis* during hibernation. *J. Mamm.* 51(4):829–31.

Hassell, Marion D., and Harvey, Michael J. 1965. Differential homing in *Myotis sodalis*. *Amer. Midland Nat.* 74(2):501–3.

Henshaw, R. E., and Folk, G. E., Jr. 1966. Relation of thermoregulation to seasonal changing microclimate in two species of bats, *Myotis lucifugus* and *M. sodalis*. *Physiol. Zool.* 39:223–36.

Henson, O. W., Jr. 1970. The ear and audition. In *Biology of bats* ed. William A. Wimsatt, pp. 181–263. New York: Academic Press.

Hitchcock, Harold B. 1949. Hibernation of bats in southeastern Ontario and adjacent Quebec. *Canadian Field-Nat.* 63(2):47–59.

———. 1955. A summer colony of the least bat, *Myotis subulatus leibii* (Audubon and Bachman). *Canadian Field-Nat.* 69(2):31.

———. 1965. Twenty-three years of bat banding in Ontario and Quebec. *Canadian Field-Nat.* 79(1):4–14.

———. 1974. Reflections of a batman. *Massachusetts Wildlife* 25(1):8–13.

Hitchcock, Harold B., and Reynolds, Keith. 1942. Homing experiments with the little brown bat (*Myotis lucifugus lucifugus*) Le Conte. *J. Mamm.* 23(3):358–67.

Jackson, Hartley H. T. 1961. *Mammals of Wisconsin.* Madison: University of Wisconsin Press.

Jones, Clyde. 1965. Ecological distribution and activity periods of bats of the Mogollon Mountains area in New Mexico and adjacent Arizona. *Tulane Studies in Zool.* 12:93–100.

Jones, J. Knox, Jr., and Genoways, Hugh H. 1966. Records of bats from western North Dakota. *Trans. Kansas Acad. Sci.* 69(1):88–90.

———. 1967. Annotated checklist of bats from South Dakota. *Trans. Kansas Acad. Sci.* 70(2):194–96.

Koestner, E. J. 1942. A method of collecting bats. *J. Tennessee Acad. Sci.* 17:301.

Koford, Carl B., and Koford, Mary R. 1948. Breeding colonies of bats, *Pipistrellus hesperus* and *Myotis subulatus melanorhinus*. *J. Mamm.* 29(4):417–18.

Koopman, Karl F., and Cockrum, E. Lendell. 1967. Bats. In *Recent mammals of the world*, ed. S. Anderson and J. K. Jones, Jr., pp. 109–50. New York: Ronald Press.

Krutzsch, Philip H. 1961. A summer colony of male little brown bats. *J. Mamm.* 42(4):529–30.

Lane, H. K. 1946. Notes on *Pipistrellus subflavus subflavus* (F. Cuvier) during the season of parturition. *Proc. Pennsylvania Acad. Sci.* 20:57–61.

Layne, James N. 1958. Notes on mammals in southern Illinois. *Amer. Midland Nat.* 60(1):219–54.

Mackiewicz, John, and Backus, Richard H. 1956. Oceanic records of (*Lasionycteris noctivagans*) and (*Lasiurus borealis*). *J. Mamm.* 37(3):442–43.

Manville, Richard H. 1963. Accidental mortality in bats. *Mammalia* 27(3):361–66.

Martin, Robert L.; Pawluk, John T.; and Clancy, Thomas B. 1966. Observations on hibernation of *Myotis subulatus*. *J. Mamm.* 47(2):348–49.

McClure, H. Elliott. 1942. Summer activities of bats (genus *Lasiurus*) in Iowa. *J. Mamm.* 23(4):430–43.

McCue, J. J. G. 1961. How bats hunt with sound. *Natl. Geogr. Mag.* 119(4):571–78.

McKeever, S. 1951. *A survey of the mammals of West Virginia.* Charleston: Conservation Commission of West Virginia.

Merriam, C. Hart. 1888. Do any Canadian bats migrate? Evidence in the affirmative. *Trans. Royal Soc. Canada.* 4:85–87.

Miller, Gerrit S., Jr. 1897. Revision of the North American bats of the family Vespertilionidae. *N. Amer. Fauna* 13:1–140.

Miller, Gerrit S., Jr., and Allen, G. M. 1928. The American bats of the genera *Myotis* and *Pizonyx*. *U.S. Natl. Mus. Bull.* 144:1–218.

Mills, Richard S.; Barrett, Gary W.; and Farrell, Michael P. 1975. Population dynamics of the big brown bat (*Eptesicus fuscus*) in southwestern Ohio. *J. Mamm.* 56(3):591–604.

Mohr, Charles E. 1936. Notes on the least bat, *Myotis subulatus leibii*. *Proc. Pennsylvania Acad. Sci.* 10:62–65.

Mueller, H. C. 1965. Homing and distance-orientation in bats. *A. Tierpsychol.* 23:403–21.

Mueller, H. C., and Emlen, J. T., Jr. 1957. Homing in bats. *Science* 126:307–8.

Mumford, Russell E. 1958. Population turnover in wintering bats in Indiana. *J. Mamm.* 39(2):253–61.

Mumford, Russell E., and Calvert, Larry L. 1960. *Myotis sodalis* evidently breeding in Indiana. *J. Mamm.* 41(4):512.

Mumford, Russell E., and Cope, James B. 1954. Distribution and status of the Chiroptera of Indiana. *Amer. Midland Nat.* 72(2):473–89.

——. 1958. Summer records of (*Myotis sodalis*) in Indiana. *J. Mamm.* 39(4):586–87.

Munyer, Edward A. 1967. A parturition date for the hoary bat, *Lasiurus c. cinereus*, in Illinois and notes on the newborn young. *Trans. Illinois State Acad. Sci.* 60(1):95–97.

Myers, Richard F. 1960. *Lasiurus* from Missouri caves. *J. Mamm.* 41(1):114–17.

Neill, Wilfred T. 1952. Hoary bat in a squirrel's nest. *J. Mamm.* 33(1):113.

Nero, R. W. 1959. Winter records of bats in Saskatchewan. *Blue Jay* 17:78.

Orr, Robert T. 1950. Unusual behavior and ocurrence of a hoary bat. *J. Mamm.* 31(4):456–57.

Packard, Ross L. 1956. An observation on quadruplets in the red bat. *J. Mamm.* 37(2):279–80.

Paradiso, John L., and Greenhall, Arthur M. 1967. Longevity records for American bats. *Amer. Midland Nat.* 78(1):251–52.

Patterson, Ann P., and Hardin, James W. 1969. Flight speeds of five species of vespertilionid bats. *J. Mamm.* 50(1):152–53.

Peterson, Randolph L. 1966. *The mammals of eastern Canada.* Toronto: Oxford University Press.

Poole, Earl L. 1932. Breeding of the hoary bat in Pennsylvania. *J. Mamm.* 13(4):365–66.

Provost, Ernest E., and Kirkpatrick, Charles M. 1952. Observations on the hoary bat in Indiana and Illinois. *J. Mamm.* 33(1):110–13.

Quay, W. R. 1948. Notes on some bats from Nebraska and Wyoming. *J. Mamm.* 29(2):181–82.

Rysgaard, G. N. 1942. A study of the cave bats of Minnesota with special reference to the large brown bat, *Eptesicus fuscus fuscus* (Beauvois). *Amer. Midland Nat.* 28(1):245–67.

Schramm, Peter. 1957. A new homing record for the little brown bat. *J. Mamm.* 38(4):514–15.

Schwartz, Charles W., and Schwartz, Elizabeth R. 1959. *The wild mammals of Missouri.* Columbia: University of Missouri Press.

Smith, E., and Goodpaster, W. 1958. Homing in nonmigratory bats. *Science* 127:644.

Snyder, W. E. 1902. A list, with brief notes, of the mammals of Dodge County, Wisconsin. *Bull. Wisconsin Nat. Hist. Soc.* 2:113–26.

Stains, Howard J. 1965. Female red bat carrying four young. *J. Mamm.* 46(2):333.

Stones, Robert C., and Branick, Leo P. 1969. Use of hearing in homing by two species of *Myotis* bats. *J. Mamm.* 50(1):157–60.

Stones, Robert C., and Holyoak, Garth W. 1970. Spring occurrence of a silver-haired bat in upper Michigan. *J. Mamm.* 51(4):811–12.

Swanson, Gustav, and Evans, Charles. 1936. The hibernation of certain bats in southern Minnesota. *J. Mamm.* 17(1):39–43.

Terres, John K. 1956. Migration records of the red bat, *Lasiurus borealis. J. Mamm.* 37(3):442.

Tuttle, Merlin D. 1964. *Myotis subulatus* in Tennessee. *J. Mamm.* 45(1):148–49.

Whitaker, John O., Jr. 1967. Hoary bat apparently hibernating in Indiana. *J. Mamm.* 48(4):663.

Wilson, Nixon. 1965. Red bats attracted to insect light traps. *J. Mamm.* 46(4):704–5.

Wimsatt, William A. 1960. An analysis of parturition in Chiroptera, including new observations on *Myotis l. lucifugus. J. Mamm.* 41(2):183–200.

# Order Lagomorpha

The order Lagomorpha includes the pikas, hares, and rabbits. Unlike rodents, with which they were formerly classed, lagomorphs have a second pair of small upper incisors directly behind the larger ones. These smaller incisors lack a cutting edge and are nearly circular in outline.

Lagomorphs have fenestrated skulls, and the incisors and molariform teeth are separated by a wide diastema, or gap, in place of the canines. There is a longitudinal groove on the anterior face of the first upper incisors, and the cheek teeth are rootless and high crowned. The rows of the maxillary cheek teeth are farther apart than the rows of the mandibular cheek teeth, so that only one pair of rows touches at the same time, which causes lateral jaw movements.

Lagomorphs are terrestrial, and their locomotion is digitigrade during running but plantigrade in hopping. They are herbivorous, and some species exhibit refection, a form of coprophagy, reingesting their soft, moist fecal pellets directly from the anus and swallowing them without chewing. Refection apparently allows for bacterial digestion of plant material, much like cud-chewing in ruminants.

Lagomorphs have become nearly cosmopolitan by introduction, except for Antarctica and some remote oceanic islands. In New England the order is represented by one family, having two genera and three species that have been introduced and are now established.

## FAMILY LEPORIDAE (HARES AND RABBITS)

Hares are not rabbits and rabbits are not hares, but the species are often confused because of the common practice of using their names as synonyms. Newborn hares are precocial, being born well furred and with their eyes open, and can run shortly after birth, whereas rabbits are altricial, born naked and helpless, with closed eyes. Hares are born not in a nest but in a sheltered spot; rabbits are born in a nest well concealed in a shallow depression. Hares generally prefer rock crevices or coves for shelter, while rabbits use burrows or take shelter under tree roots or logs or in dense bushes. Hares are larger than rabbits, with disproportionately long ears and limbs, and are excellent runners. Rabbits are short-legged and are not as strong runners as hares.

All leporids have long ears that are very sensitive to sound and large eyes situated on the sides of the head, which enable them to detect danger from any quarter. Their powerful hind limbs are much longer than their forelimbs, and their feet have four or five digits and furred soles. Leporids have cleft lips.

Leporids are crepuscular, but possibly more nocturnal than diurnal. They are strictly vegetarians, and some species exhibit refection. The leporids are among the most widely introduced mammals. There are two native species in New England—the New England cottontail rabbit and the snowshoe hare.

## Eastern Cottontail

*Sylvilagus floridanus* (J. A. Allen)

*Lepus sylvaticus floridanus* J. A. Allen, 1890. *Bull. Amer. Mus. Nat. Hist.* 3:160.
*Sylvilagus floridanus,* Lyon, 1904. *Smithsonian Misc. Coll.* 45:322.
Type Locality: Sebastian River, Brevard County, Florida
Dental Formula:
  I 2/1, C 0/0, P 3/2, M 3/3 × 2 = 28
Names: Cotton hare, gray rabbit, brush rabbit, cottontail, and bunny. The meaning of *Sylvilagus* = forest hare, and *floridanus* = of Florida.

**Description.** The cottontail rabbit is smaller and stockier than the snowshoe hare, and its hind legs are not as specialized. The soles of the feet are densely furred, with five toes on the front feet and four on the hind feet. The tail is short, thick, and fluffy.

The skull has a moderate-sized auditory bulla and the posterior extension of the supraorbital process is transversely thick. Cranial characters of this species differ markedly from those of the New England cottontail rabbit and are described more fully in the account of that species.

The pelage is long and coarse. Both sexes are colored alike and show slight seasonal variation. The coloration of the upperparts varies from reddish brown mixed with black to grayish brown, except for the cinnamon-rusty nape. The face and flanks are grayish and the chest is brownish, while the underparts are whitish. The legs are a rich cinnamon-rufous color. The outside of the ears is dull grayish buffy bordered by a thin, indistinct black stripe at the front and about the tip, and the inner surface of the ears is sparsely furred with short, very light buffy or grayish hairs. The hind feet are pale rusty buff to whitish above, and the fluffy tail is brownish above, whitish below. There is a white ring around the eyes. A white spot or blaze occurs on the forehead of 50% of eastern cottontails, and others have a light brown spot. This white blaze

is absent in the New England cottontail. The iris of the eye is deep brown and in bright light reflects a brilliant red color. Some individuals are darker than others because they have considerably more black hairs. Albino, all-white, and melanistic cottontails are rare.

Adult cottontails undergo a continuous molt from February to November, with two distinct phases. The spring molt occurs from March to August and the fall molt from September to November. In the spring molt the fur is replaced in spots or patches, whereas in the fall molt it is replaced in larger sheet areas. The spotted molt is characteristic of the breeding season and the sheet molt of the nonbreeding season (Dalke 1942). Cottontails do not turn white in winter.

Females are slightly larger than males. Measurements range from 380 to 461 mm (14.8–18.0 in); tail, 30 to 70 mm (1.2–2.7 in); and hind foot, 77 to 106 mm (3.0–4.1 in). Weights vary from 825 to 1,350 g (28.9–47.3 oz).

**Distribution.** The eastern cottontail occurs throughout the eastern United States and extreme southern Canada and south through eastern Mexico, parts of Yucatan, and Central America, with a disjunct population in Texas, New Mexico, and Arizona.

In New England the species occurs in western Vermont from the Massachusetts border to the Canadian border, especially in valleys where some farmland and brushland remain. In Vermont the species is common in Rutland and Bennington counties and less common in Addison, Chittenden, and Grand Isle counties. The first specimens recorded from the Connecticut River Valley extend to Brattleboro, Windham County, Vermont, and to Hindsdale, Cheshire County, New Hampshire. These animals have probably moved from Massachusetts. In New Hampshire the first specimen was taken in the coastal town of Hampton Falls, Rockingham County, in September 1971. Since then, fifteen more have been taken from all the coastal towns of Rockingham County and from the inland towns adjacent to the Massachusetts border east of the Merrimack River in Hillsborough County (Jackson 1973). The origin of these specimens is unknown, although they are likely the result of extensive stocking in Massachusetts in the 1930s. The population in coastal towns of New Hampshire may be the result of stocking by the New Hampshire Fish and Game Department in 1937.

The eastern cottontail occurs throughout Connecticut and Rhode Island. It is found in

EASTERN COTTONTAIL

Massachusetts, except in a narrow belt in the central northern portion of the state along the New Hampshire border, from the Connecticut River east to Worcester, Middlesex, and Essex counties (Johnston 1972).

**Introduction into New England.** The eastern cottontail was introduced into New England before 1900, probably on Nantucket Island (Bangs 1895; Nelson 1909). Introduction of eastern cottontails elsewhere in New England has been summarized by Dice (1927), Johnston (1972), and Jackson (1973). The eastern cottontail has been introduced into New England from Kansas, Missouri, Minnesota, and West Virginia, and this stocking may have included any of the following subspecies of *Sylvilagus floridanus*: *alacer, ilanensis, mallurus, mearnsi*, or *similis*, plus the desert cottontail.

**Ecology.** Eastern cottontails are found in fields, meadows, farmlands, dense high grass, and wood thickets, along fencerows, stone walls overgrown with brush, and forest edges as well as along edges of sedge swamps and marshes, but seldom in dense woods. They are encountered in thickets and tall grass in small towns, and sometimes even in large cities.

Cottontails spend most of the day hiding or resting in "forms" concealed in dense grass, weed patches, hayfields, along roadsides, in scrubby woods and thickets. The rabbits make a form by scratching out or trampling a shallow oval hollow in the ground. The size of a form depends on the size of the cottontail. It is usually about 6 to 8 inches wide by 9 to 11 inches long. The animals use the forms during moderately inclement weather, but seek shelter in dens, brushpiles, or stone walls, under shocks of grain, or in burrows during storms and on very cold days.

A pregnant female digs a brood nest in the ground, lining and covering it with grass and with fur plucked from her body. A nest may be made with alternate layers of grass, fur, and droppings (Madson 1959). The nest is usually concealed in grass, weeds, hayfields, thickets, orchards, or scrubby woods in well-drained areas. Peculiar nesting sites have been found in a manure pile (Beule 1940); in the crotch of a leaning tree, 9.5 feet above ground, where a nest

contained five young cottontails (Shadle, Austin, and Meyer 1940); in a rockpile at the edge of a field (Beule and Studholme 1942); and in the side of a two-foot pile of topsoil (Kirkpatrick 1960).

Although cottontails do not dig burrows as European or domestic rabbits do, they take advantage of abandoned woodchuck burrows, crevices of rocks, and cavities in stumps, particularly during winter. Although they generally prefer dry areas, cottontails sometimes remain in cool swamps and wet marshes during hot summers.

Man, dogs, foxes, coyotes, bobcats, cats, weasels, raccoons, mink, great horned owls, barred owls, large hawks, crows, and snakes prey upon cottontails, and many are killed by automobiles, farm machinery, fires, floods, storms, and severe cold weather. Cottontails are hosts to fleas, lice, ticks, cestodes, nematodes, and trematodes and to maggots of the gray flesh fly *Wohlfahrtia vigil* and botfly larvae, *Cuterebra* spp. They are victims of tularemia, or "rabbit fever," Shope's fibroma, torticollis, and strep-tothricosis cutaneous.

**Behavior.** Cottontails are active mainly at night and early in the morning. Although they spend most of the daylight hours concealed in a form, under a brushpile, or in a thicket or burrow, they enjoy sunbathing. They are active throughout the year except during storms, heavy snowfall, or sudden drops in temperature.

Cottontails are more easily trapped on rainy or foggy nights than on clear nights (Carson and Cantner 1963). They are also more vulnerable during cold weather (Huber 1962; Bailey 1969), presumably because they seek shelter from inclement weather in box traps.

Cottontails are solitary except for females with young. They are wary but are nervous and flush readily from a hiding place. Although they are timid, they can scratch deeply if captured and strike quick hard blows with the hind feet. Their sense of hearing is keen, and the ears move in various directions to catch sound. Sight is excellent, and their large eyes let them see at many angles at once, except for directly ahead.

Cottontail rabbits are comparatively silent, though sometimes a female with young utters low grunts. When terrified or caught, cottontails may emit a piercing, shrill cry. When playing, breeding, or fighting, they make low purring, growling, or grunting sounds. Most cottontails thump the ground with their hind feet, probably as a form of communication. These animals can swim but do not seem to take readily to water. Cottontails normally move along slowly in short hops and jumps, often traveling along regular trails or runways in vegetation or snow.

Frightened or fleeing cottontails may cover between 10 and 15 feet on the first several leaps, then hop shorter distances, often zigzagging to baffle a pursuing predator. They can attain speeds of up to 18 miles per hour for short distances (Jackson 1961). They often double back to their starting point by a circuitous route and sometimes will squat and "freeze" for several minutes to avoid detection.

The home range of cottontails varies from a half-acre or less to 40 acres or more, depending on the habitat, the season, and the age and sex of the animal. In Connecticut, Dalke and Sime (1938) found the home range of eastern cottontails to be 8.3 square acres for adult males and 3 acres for adult females. In Massachusetts, McDonough (1960) reported that the average distance cottontails traveled during all seasons was 450 feet and that the average home range size was 1.4 square acres for adult males and 1.2 square acres for adult females.

Hill (1967) suggested homing behavior for cottontails and recaptured a tagged female that had returned home after traveling a distance of 2⅓ miles in 331 days. Nugent (1968) reported that a female cottontail traveled a straight-line distance of 3,460 feet through two swamps within 24 hours. Chapman and Trethewey (1972) found that juvenile males traveled farther than adult males, males traveled farther than females, and juvenile and adult females traveled approximately the same distance.

Cottontails eat almost any kind of green vegetation, including bark, twigs, and buds of woody plants. They do not dig for bulbs or roots. In winter they feed mainly on tender bark, twigs, and buds of shrubs, on vines, and occasionally on dried herbaceous stems and leaves. Cottontails will dig deep into the snow to get to frozen apples when their natural food supply fails.

In Connecticut the most important winter food items are gray birch, red maple, and smooth sumac. The preferred summer foods are clover, alfalfa, timothy, bluegrass, quackgrass, crabgrass, redtop, ragweed, goldenrod, plantain, chickweed, and dandelion. Peas, beans, wheat, rye, oats, lettuce, cabbage, and cornstalks are also eaten. Food items taken during the fall are intermediate between the summer herbaceous plants and the woody winter foods (Dalke and Sime 1941).

The breeding season may extend from mid-February through September or later, but severe winter weather and deep snow may postpone its onset. The gestation period is about 30 days, and the litter size varies from three to eight, averaging five young. The first litter may arrive in early March, but more often comes in April or May.

Breeding is promiscuous, and there appears to be no difference in the breeding behavior of the eastern cottontail and the New England cottontail (Dalke 1942). Dalke observed that fighting often resulted when cottontails of either sex were introduced into pens with others of the opposite sex. However, breeding usually occurred within a week. Before copulation the male would chase the female around the pen, and both would leap into the air. Sometimes the female would stop when being chased by the male and then jump over him. In other instances after running around the pen the rabbits would stop and face each other, then one rabbit would leap straight in the air and the other would run underneath. Casteel (1966) noted that the act of copulation lasted 1 to 4 seconds, with 1 to 4 pelvic thrusts by the male. He reported that collisions between penned male rabbits were common as they dashed around the pen in search of a female. Dalke (1942) remarked that in every instance, from 24 to 48 hours after copulation the female rabbit would chase the male and bite fur from his sides and back. Marsden and Conaway (1963) reported that breeding activity may take place any hour of the day or night. Casteel (1966) noted that the female may breed again only a short time after she gives birth and leaves the brood nest.

At birth the young are blind and helpless and may be naked except for sparse covering of light-colored guard hairs. The incisors barely penetrate the gums. Newborns are about 100 mm (3.9 in) long and weigh about 25 g (0.9 oz). They grow rapidly, and their eyes and ears open at about 7 days of age. They leave the nest about 16 days after birth, though they may return to it occasionally for the next few days. They are weaned when 4 or 5 weeks old and become independent.

The female, or doe, cares for the young, but they get little attention except during nursing. The doe may remain in a form within calling distance of her young or may drift away during feeding. Protection of the young varies with individual does. A doe may defend her young, or she may move them to another nest when danger is imminent.

Most females breed in the spring following their birth, though some of them may breed in their first year. Longevity is less than 2 years in the wild (Lord 1961a) and about 10 years in captivity (Lord 1961b).

**Sex Determination.** In adults the external genitalia of the two sexes are similar in appearance, and it requires careful examination to distinguish the male from the female. The clitoris of the female resembles the penis. For examination, these organs can be erected by applying downward pressure with the thumb placed in front of the genital region and the forefinger behind. The penis is cylindrical and has an opening in the tip, while the clitoris is flattened and has no opening (Petrides 1951).

It is sometimes difficult to distinguish the sexes of young rabbits and rabbits during the nonbreeding season. During the breeding season, males have large testes in the scrotum. The testes move into the body cavity during the nonbreeding season, and sometimes as a result of shock during the breeding season (Schwartz and Schwartz 1959). Females have eight mammae—two pectoral, four abdominal, and two inguinal.

**Age Determination.** The length of the hind foot as an age criterion in nestlings has been used for Pennsylvania cottontails (Beule and Studholme 1942), and the presence or absence of epiphyseal cartilage on the proximal end of the humerus as determined from X-ray photographs was used by Thomsen and Mortensen (1946) and refined by Hale (1949). This cartilage is present in rabbits less than nine months of age, and after that age the cartilage closes. Eastern cottontails with

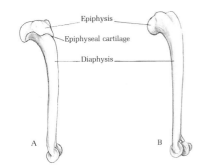

Lateral view of humerus of cottontail rabbit. *A*, presence of epiphyseal cartilage in immature; *B*, absence of epiphyseal cartilage in adult.

ossified epiphyses have a completely smooth bony surface in the epiphyseal region with no external evidence of cartilage.

Lord (1959) determined the age of cottontails by the weight of the lenses of the eyes of ninety-two known-age animals up to the age of 30 months. Rongstad (1966) found that lens weight varied too much to do more than separate young-of-the-year from adults, based on 188 known-age, pen-reared cottontails between 2 and 297 days old. Pelton (1969) reported that the epiphyseal groove in the humerus closes when the lens weight is 196 to 200 mg, or at approximately 10 months of age, but with considerable variation. In eastern cottontails from Georgia the groove was present in some rabbits at a lens weight of 232 to 236 mg, or 20 months old, and absent at a lens weight of 180 to 184 mg, or 8 months old. Bothma, Teer, and Gates (1972) found that closure of the epiphyseal plate was the most useful criterion of age up to about 9 months in Texas eastern cottontails.

**Specimens Examined.** VERMONT, total 25: Addison County, Middlebury 1 (AMNH); Chittenden County, Shelburne 1, South Burlington 4, Westford 1, Williston 10 (UVT); Rutland County, Mendon 2, Rutland 1–4 (DC–UCT), Sherburn 1 (AMNH).

MASSACHUSETTS, total 21: Bristol County, Dartmouth 1; Essex County, Andover 1 (UCT); Franklin County, Shutesbury 1; Hampden County, Mt. Tom 1 (UMA), Springfield 2 (USNM); Hampshire County, Amherst 5, Hadley 2 (UMA); Nantucket County, Nantucket Island 5 (USNM); Plymouth County, West Bridgewater 1 (UCT); Worcester County, Spencer 1–1 (UCT–UMA).

CONNECTICUT, total 134: Fairfield County,

Greenwich 1 (USNM), Monroe 3, Stamford 1, Trumbull 1, Weston 1; Hartford County, Avon 1 (UCT), East Hartford 1, Granby 1, Windsor 4 (USNM); Litchfield County, Barkhamsted 1 (UCT), Bear Mt. 1 (AMNH), Bethlehem 1, Canaan 1 (UCT), Litchfield 1, Salisbury 1 (USNM), Sharon 1 (AMNH), Watertown 1; Middlesex County, Old Saybrook 1; New Haven County, Cheshire 1, Derby 1, Meriden 2, New Haven 1 (UCT), North Branford 1 (USNM); New London County, Baltic 10, Colchester 2, Hanover 1 (UCT), Lebanon 1 (USNM), Norwich 2 (UCT), Old Lyme 1 (USNM), Sprague 1, Stonington 1; Tolland County, Coventry 4 (UCT), Ellington 6 (USNM), Mansfield 17(UCT), Rockville 1, Somers 2 (USNM), Storrs 16–3 (UCT–USNM), Vernon 4, Willington 10; Windham County, Ashford 15, Eastford 1, Warrensville 1, Willimantic 6, Windham 2 (UCT).

RHODE ISLAND, total 4: Washington County, Kingston 4 (DC).

# New England Cottontail

*Sylvilagus transitionalis* (Bangs)

*Lepus sylvaticus transitionalis*
    Bangs, 1895. *Proc. Boston Soc. Nat. Hist.* 26:405–7.
*Sylvilagus transitionalis*, Nelson, 1909. *N. Amer. Fauna* 29:195.
Type Locality: Liberty Hill, New London County, Connecticut
Dental Formula:
    I 2/1, C 0/0, P 3/2, M 3/3 × 2 = 28
Names: Wood rabbit and mountain rabbit. The meaning of *Sylvilagus* = forest hare, and *transitionalis* = the transition (brush) between grass and forest.

**Description.** The New England cottontail resembles the eastern cottontail but differs in pelage and in certain distinctive cranial characters. The pelage varies from individual to individual, and no single character can be used to separate the two species. Their coloration is similar, except in the New England cottontail the back is darker, an ochraceous buff overlaid with a wash of black-tipped guard hairs. The rump is slightly duller than the back, brown and indistinct, and usually not grayish as in the eastern cottontail. The short, rounded ears are heavily furred, with long whitish hairs on the anterior border near the base. The outer edge of the ear has a broad black stripe which does not blend gradually into the browner color of the ear as in the eastern cottontail. There is usually a black spot between the ears and never a white spot or blaze on the forehead. The coloration of the ears and the presence of the black spot or blaze on the forehead are the most reliable pelage characters separating the New England cottontail from the eastern cottontail. Color variants are rare, but Nelson (1909) reported an albino specimen. Females have eight mammae, two pectoral, four abdominal, and two inguinal.

The best way to separate the New England cottontail from the eastern cottontail is to examine the skull characters. In *transitionalis* the frontal-nasal suture is jagged, whereas in *floridanus* it is smooth or regular. The anterior supraorbital process in *transitionalis* decreases in width anteriorly and fuses with the skull with no apparent anterior process or notch, whereas in *floridanus* the notch is rather distinct. The posterior supraorbital process in *transitionalis* is narrow and tapers to a point which usually does not fuse to the frontal, whereas in *floridanus* it is broad and fused at the frontal at its tip or throughout its entire length. The tympanic process is lacking in adult *transitionalis* more often than in juveniles and is present more often in adult *floridanus* than in juveniles.

The three most reliable skull characters for interspecific identification are (1) the nasal-frontal suture, (2) the posterior supraorbital process; and (3) the tympanic process. The least reliable character is the anterior supraorbital process. The presence of the tympanic process is 96% reliable for interspecific identification of adults and about 78% reliable for juveniles (Johnston 1972).

Adult females are slightly larger than males. and autumn or during the nonbreeding season. Specimens taken in Connecticut from the third week in December to mid-February are nearly 100% prime fur (Dalke 1937).

Adult females are slightly larger than males. Measurements range from 363 to 483 mm (14.2–18.8 in); tail, 31 to 49 mm (1.2–1.9 in); and hind foot, 90 to 102 mm (3.5–4.0 in). Weights vary from 750 to 1,347 g (26.3–47.1 oz).

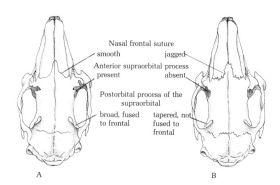

Skull characters separating *A*, Eastern cottontail rabbit, from *B*, New England cottontail rabbit.

**Distribution.** The species ranges from southern Maine, New Hampshire, Vermont, Rhode Island, and Massachusetts—except for Nantucket and Martha's Vineyard—southwest through New York, New Jersey, and eastern Pennsylvania, and along the southern Appalachian Mountains to Alabama.

The current distribution of the New England cottontail in Vermont, New Hampshire, and Maine is uncertain, and when found the animals appear few in number and marginal in habitat. In Vermont specimens are rare, and most occur in the southern third of the state. In New Hampshire, the animal is rare and restricted to the southern half of the state, with the largest concentration apparently in the Merrimack River Valley north of Concord, Merrimack County. In Maine the species is restricted to the coastal towns of the southwestern counties of York, Cumberland, Androscoggin, and Sagadahoc (Jackson 1973).

The species appears to be on the decline, and its distribution is spotty in Massachusetts, Connecticut, and Rhode Island. The type locality of the New England cottontail, Liberty Hill, New London County, Connecticut, no longer has a population (Johnston 1972).

**Ecology.** The New England cottontail occurs from sea level in New England to above 4,000 feet in the southern Appalachian Mountains (Chapman 1975). The species chiefly inhabits dense bushy, woodlands and pine-spruce areas (Llewellyn and Handley 1945; Fay and Chandler 1955; Linkkila 1971; Chapman and Morgan 1973). However, Eabry (1968) found considerable overlap in home ranges of individual New England cottontails and eastern cottontails. Johnston (1972) found the New England cottontail in different habitats varying from beach and salt marsh, to oak forest with a heavy understory, to open upland fields with little cover.

Enemies of this species are the same as those of the eastern cottontail.

**Behavior.** The behavior of the New England cottontail is much the same as that of the eastern cottontail, with a few interspecific differences. Chandler (1952) and Pringle (1960) have pointed out that the New England cottontail struggles vigorously and squeals more frequently when handled, whereas the eastern cottontail tends to freeze and remain docile. Olmstead (1970) suggested that these behavioral patterns may both have survival advantages in the differing habitats of the two species. For example, since the eastern cottontail is apt to be in the open much of the time, it would be a selective advantage for the animal to freeze when threatened by a predator. On the other hand, the New England cottontail, which is not likely to be found in the open, would not need this strong behavioral potential to freeze and in fact might escape more efficiently by dodging through the dense brush.

The New England cottontail deposits more fecal pellets in box traps than does the eastern

NEW ENGLAND COTTONTAIL

cottontail, according to Fay (1951), McDonough (1952), and Olmstead (1970); Olmstead (1970) reported that the New England cottontail ate more bait while in a trap than did the eastern cottontail. Eabry (1968) observed refection of fecal pellets in the New England cottontail. Pringle (1960) noted that the New England cottontail was less high-strung and wary and appeared to adapt to captivity much better than the eastern cottontail. Chandler (1952) reported that the New England cottontail made no attempt to run when released from a trap, whereas the eastern cottontail made a beeline for cover, and he also reported that the New England cottontail was less wary of traps. Nugent (1968) found that male New England cottontails were more likely to be captured a second time than were females, and that the likelihood of recapturing a New England cottontail was more than twice that of recapturing an eastern cottontail.

During autumn the New England cottontail has a home range of 0.5 to 1.8 square acres as determined by mark and recapture techniques. McDonough (1960) found that in the autumn season adults traveled an average distance of 450 feet and juveniles 390 feet between captures. In January and February, rabbits shift areas, apparently with the start of the breeding season. Dalke (1937) recaptured both males and females up to one-third of a mile from their December ranges. Nugent (1968) reported that both adult male and immature New England cottontails traveled over a greater distance than did adult females.

The summer food habits of the New England cottontail are essentially the same as those of the eastern cottontail. Pringle (1960) found that in autumn and winter the New England cottontail preferred seedlings, bark, twigs, and buds of maple and oak trees over field plants, whereas the eastern cottontail favored alder, apple, and aspen. Eabry (1968) reported that these food items of the eastern cottontail served as emergency food for the New England cottontail, and Nottage (1972) suggested that the eastern cottontail is adapted to a wider variety of foods than is the New England cottontail.

Little is known about reproduction in the New England cottontail. Courtship appears to be similar to that of the eastern cottontail. In Connecticut, enlargement of the testes begins during the last week in December, and McDonough (1960) noted that in Massachusetts New England cottontails began their breeding activity later than eastern cottontails. The litter size varies from three to eight young, averaging five, and from two to three litters per year are common (Dalke 1937). Dalke (1942) reported that the lactation period is about 16 days.

**Remarks.** The New England cottontail qualifies as a valid species because it has a diploid chromosome number of 52 (Holden and Eabry 1970) rather than the 42 of the eastern cottontail (Palmer and Armstrong 1967).

There is considerable interest in hybridization between the New England cottontail and the eastern cottontail. Dalke (1942) reported that one hybrid litter of three was born in captivity—the father was an eastern cottontail and the mother a New England cottontail. He noted that fighting was the main cause of failure to hybridize in captivity. Fay and Chandler (1955) believe that hybridization occurs in the wild, and Chapman and Morgan (1973) captured an adult female hybrid in western Maryland.

**Specimens Examined.** MAINE, total 13: Androscoggin County, Auburn 1 (HMCZ); Cumberland County, Gorham 3–3 (HMCZ–UME), Scarborough 1 (HMCZ), Standish 1–1 (HMCZ–UME); York County, Berwick 1 (UME), Buxton 1–1 (HMCZ–UME).

NEW HAMPSHIRE, total 30: Cheshire County, Stoddard 1 (HMCZ), Swanzey 2 (UNH), Winchester 1 (HMCZ); Grafton County, Canaan 1, Hanover 3; Hillsborough County, Peterborough 1 (DC); Merrimack County, Allentown 9, Concord 1, Webster 2 (HMCZ); Rockingham County, North Hampton 1 (UMA); Strafford County, Durham 7 (UNH); Sullivan County, Charlestown 1 (USNM).

VERMONT, total 9: Bennington County, Arlington 1 (HMCZ); Chittenden County, Cedar Beach 1 (DC); Rutland County, Mendon 1 (AMNH), Rutland 2–1 (DC–UCT), Sherburne 1 (AMNH); Windsor County, Pomfret 1, Woodstock 1 (DC).

MASSACHUSETTS, total 44: Barnstable County, Barnstable 1; Dukes County, Martha's Vineyard 5; Essex County, Andover 1 (HMCZ); Franklin County, Orange 1, Pelham 1 (UMA); Hampden County, Springfield 3 (HMCZ); Hampshire County, Easthampton 2; Middlesex County, Cambridge 1, Wilmington 13; Nantucket County, Nantucket Island 1; Norfolk County, Stoughton 1 (USNM); Plymouth County, Hanover 1 (HMCZ), Middleboro 1 (USNM),

Plymouth 1, Wareham 7 (HMCZ); Suffolk County, Boston 1 (USNM); Worcester County, Harvard 1–1 (HMCZ–USNM), Princeton 1 (HMCZ).

CONNECTICUT, total 81: Fairfield County, Stamford 1 (USNM), Trumbull 1 (UCT); Hartford County, Plainfield 3 (USNM); Litchfield County, Macedonia Park 1, Sharon Mt. 5 (AMNH); Middlesex County, East Haddam 1, Killingsworth 1 (USNM), Westbrook 1 (AMNH); New London County, Baltic 1 (UCT), East Lyme 1 (USNM), Franklin 3–1 (UCT–USNM), Hanover 1 (UCT), Lebanon 1 (USNM), Liberty Hill 14 (HMCZ), Montville 1 (UCT), Norwich 1, Salem 1 (USNM), Sprague 8 (UCT), Stonington 1, Waterford 1 (USNM); Tolland County, Coventry 1 (UCT), Mansfield 2–1 (UCT–USNM), Storrs 1–1 (UCT–USNM), Vernon 1, Willington 6; Windham County, Ashford 2 (UCT), Chaplin 1, Killingly 1 (USNM), Sterling 5, Warrensville 6 (UCT), Windham 1–2 (USNM–UVT), Woodstock 1 (USNM).

RHODE ISLAND, total 18: Washington County, Exeter 8, Lake Wooden 9 (USNM), South Kingston 1 (UCT).

# Snowshoe Hare

*Lepus americanus*

There are two subspecies recognized in New England:

*Lepus a. struthopus* Bangs, 1898.
  *Proc. Biol. Soc. Washington* 12:81.
Type Locality: Digby, Nova Scotia, Canada

*Lepus a. virginianus* Harlan. *Lepus virginianus* Harlan, 1825. *Fauna Americana;* . . ., p. 196.
Type Locality: Blue Mountains, northeast of Harrisburg, Pennsylvania (Hall and Kelson 1959)

Dental Formula:
I 2/1, C 0/0, P 3/2, M 3/3 × 2 = 28

Names: Varying hare, snowshoe rabbit, gray rabbit, white rabbit, big brown rabbit, brown jackrabbit, bush rabbit, and swamp jackrabbit. The meaning of *Lepus* = a hare, and *americanus* = of America.

**Description.** This medium-sized hare has moderately large ears, a short tail, and large, well-developed hind legs. There are five toes on the front feet and four on the hind feet. The name "snowshoe" comes from the large hind feet, whose long phalanges can be widely separated into a shape like a snowshoe. The soles of the feet are well furred in winter and help keep the animal from sinking in soft snow, which helps it escape predators. Females have six mammae, two pectoral and four abdominal.

The interparietals of the skull in adults are fused with the surrounding bones, and the supraorbitals are subtriangular and broadly wing-like.

The pelage is thick and soft. The sexes are colored alike and show marked seasonal variation. In summer the coloration is ochraceous or yellowish brown washed with black and is darkest along the back and rump. The top of the head is dark, dusky brown with a tuft of white hairs forming a blaze. The anterior half of the ear is like the top of the head and merges to darker brown or blackish at the tip, while the posterior half of the ear is grayish or whitish, becoming blackish at the tip. The inside of the ear is grayish in front and white along the posterior margin. The nape is grayish brown; chin, throat, and abdomen usually are bright buffy to roan color. The underparts of the hind feet are brownish to buffy white. During autumn the guard hairs are replaced by new white hairs, which causes the animal to appear all white even though the underfur remains dark. In winter the coloration is white except that the tips of the ears are rimmed with black. Melanistics and albinos occasionally occur.

Snowshoes molt in autumn and spring. In Maine the autumn molt begins in September or early October and ends in December. The spring molt begins in March and ends during May or early June. Both autumn and spring molts follow a definite pattern. The autumn molt starts on the ears, feet, and legs and ends along the back, whereas the spring molt starts on the forehead, muzzle, and body and ends on the ears and feet (Severaid 1942).

The snowshoe hare is also called the "varying hare" because of its protective coloration, which normally closely approaches the prevailing tone of its environment. In summer the brown hare is inconspicuous in its woodland haunts, and in winter the pure white animal is difficult to see on snow.

Females are slightly larger than males. Measurements range from 470 to 520 mm (18.3–20.3 in); tail, 36 to 50 mm (1.4–2.0 in); and hind foot, 135 to 147 mm (5.3–5.7 in). Weights vary from 1.4 to 2.0 kg (3.1–4.4 lb).

**Distribution.** This species occurs from Alaska to Newfoundland and the northern borders of the United States, south in the Appalachian Mountains to eastern Tennessee and western North Carolina, in the Rocky Mountains to New Mexico, and the Sierra Nevada in California.

**Ecology.** Snowshoe hares inhabit practically all types of forests with brushy understory, from near sea level to elevations over 4,000 feet. They prefer mixtures of hardwood-coniferous forests, coniferous swamps, young fir thickets, cut-over areas, poplar-birch second growth, spruce-pine forests, and old burns. They seldom are found in open places.

For shelter the hares select a knoll or hummock with high, dry vegetation, or the protected side of a ledge or large rock, or they may take

SNOWSHOE HARE

cover in clumps of small trees, among exposed tree roots, in hollow logs, or under logs or fallen trees. The shelter spot, or form, is used regularly throughout the year by the same animal.

Snowshoe hares seldom dig and do not frequently enter abandoned holes or dens of other animals as cottontail rabbits do. Occasionally they use hollow logs or scratch out hollows in recesses under roots, probably to make nesting places. They normally make definite beaten runways, especially in swamps, which they use throughout the year.

Man, foxes, Canada lynx, bobcats, martens, fishers, weasels, great horned owls, snowy owls, goshawks, and red-tailed hawks prey on snowshoe hares, and automobiles, fires, and storms take their toll. Ticks, fleas, mites, cestodes, nematodes, trematodes, and botflies, *Cuterebra* spp., are parasites of snowshoe hares. They are vulnerable to coccidiosis, tularemia, and lungworm infections.

**Behavior.** Snowshoe hares are active throughout the year and are primarily crepuscular and nocturnal. Occasionally they may be seen during the daytime basking in the sun or wallowing in dusty places. Snowshoe hares are not gregarious except during the breeding season, when several of them congregate and fight and chase one another.

These animals often sit quietly, the body drawn compactly together with the feet tucked out of sight and the ears pressed tightly against the back. They frequently sit on their haunches much like a dog or lie quietly on the side or belly with legs and feet outstretched like domestic rabbits. They may stand erect on their hind feet with the front legs limp, and often reach high twigs and buds in this manner. Snowshoe hares exhibit particular agility while turning in mid-air. In hopping the front feet are held close together, the hind feet spread far apart. When danger approaches they can burst almost instantly out of a relaxed position into a full run, using their long powerful hind legs. Terres (1941) recorded that a snowshoe hare ran 31 miles per hour on ice.

Snowshoe hares dislike water but will swim to get to the opposite bank of a stream and do not hesitate to leap into the water if necessary. These hares drink water freely and eat snow.

Both sexes frequently thump with their hind feet, at any time of the year. They make grunting noises resembling a snort, utter piercing and plaintive screams almost like a bleat when frightened or handled, and make a low clicking or chirping sound like a human saying "tsk, tsk." They are curious and often will investigate a strange object. Their senses of hearing and sight are excellent.

Snowshoe hares undergo great regional cyclic population fluctuations, with peaks of 9 to 11 years in Maine (Severaid 1942). The reasons for the complex cyclic mortality in this species are not fully understood. Theories that have been proposed include epizootic diseases (MacFarlane 1905), overcrowding (Hewitt 1921), the effect of increase and decrease of sunspots on reproductive rates (Wing 1935), adverse weather and mass migration (Cox 1936), and "shock disease" as the mechanism which causes mortality in the hare (Green and Larson 1938).

This mysterious "shock disease" is peculiar to snowshoe hares. It is characterized by an idiopathic degeneration of the liver and hypoglycemia, a condition in which the sugar content of the bloodstream becomes less than is required for life to continue. At times the disease may affect almost an entire hare population. The animals appear normal, but suddenly they go into coma or spasms for a few minutes or hours before dying. No causative bacterial or parasitic infection has been noted in any of the hares necropsied (Green, Larson, and Bell 1939). Green and Evans (1940a) believed that the factor which causes the die-off depends on a high population of hares for its initial spread, persists for several years, and despite a progressive decline in numbers of hares continues to deplete a population until only 10% survive. Green and Evans (1940b) found that mortality among adult hares was maintained at a relatively constant rate irrespective of the cyclic changes in a population and the same authors (1940c) concluded that the principal factor in cyclic changes was the changing mortality among immatures during the first nine months of life, since neither the adult mortality nor the rate of reproduction appeared to have varied enough to be of major importance in the cycles.

Dodds (1965) noted that productivity decreased sharply one year before the declines in reproductive rate and population density, and this decline in productivity could not be fully accounted for by an indicated drop in reproductive rate. Meslow and Keith (1968) observed that the recovery from the low population was a function of doubled adult survival and reproductive rate and a marked increase in juvenile survival, and the same authors (1971) found that cold temperatures and a deep accumulation of snow were correlated with poorer adult survival.

The home range of adult males is about 25 square acres, and that of females about 19 square acres (Adams 1959). The daily movements of males and females cover about 4 square acres at all seasons, and males move about more during the breeding season. Climatic and physical factors may dampen or activate hare movements within the home range (Bider 1961).

Snowshoe hares feed on a variety of plants. Their summer food consists mainly of clover, grasses, dandelion, berries, ferns, and occasionally succulent parts of wooded vegetation. Winter foods are twigs, buds, tender bark of shrubs and small evergreen trees, and seedlings of alders, aspens, poplar, spruces, eastern hemlock, balsam fir, American larch, northern white cedar, paper birch, willows, and white pine. These animals eat meat used as bait in traps. Like cottontail rabbits, they practice refection.

In Maine, snowshoe hares begin their courtship in late February, but the majority of them do not breed until March or early April. They produce young from April to August and go through a nonbreeding period from September to March. The gestation period is approximately 37 days. Females breed again on the day of parturition. They may sometimes produce four litters a year, but they average three. The number of young (called leverets) per litter is from one to six, averaging three (Severaid 1945). In New Hampshire, Siegler (1954) reported that a female taken on 28 November contained two well-developed fetuses, each approximately 115 mm (4.5 in) long.

Severaid (1942) found no appreciable difference in the average size of the first litter of the season between yearling females and females over one year old. The number of young per litter in the spring is very small compared with summer litters (Rowan and Keith 1956). Green and Evans (1940c) and Severaid (1942) found litter size to be independent of the hare population density. Meslow and Keith (1971) found that between-year differences in litter size were significantly correlated with temperature and snow depth—the colder the temperatures in the 250 days preceding mid-February and the deeper the winter's accumulation of snow, the larger the litters were the next spring.

Snowshoe hares are promiscuous. During the breeding season the males often fight fiercely and sometimes kill each other. Courtship behavior consists of the buck vigorously chasing the doe. Often, as he catches up with her, she jumps high in the air, occasionally swinging around and dropping in another direction. In the meantime the buck passes under and beyond her, then turns and follows. The chase is often preceded by the two animals touching noses and simultaneous crouching opposite each other. At the height of the courtship the doe gives up the chase, and often copulation takes place at this time. Upon ejaculation the buck jumps into the air and slightly backward, utters a gutteral hiss, and occasionally makes a complete about-face before landing. The doe utters the same gutteral hiss at this time and darts rapidly forward. The buck usually renews the chase immediately. A pair of snowshoe hares may copulate several times within 10 minutes (Severaid 1942). Like the European hare, both sexes of the snowshoe hare commonly urinate on their partner during courtship when one runs underneath the other, which has leaped into the air (Bourlière 1954).

Bookhout (1964) reported prenatal development of snowshoe fetuses at definite periods during the gestation period; these include the appearance of digits in 15 days; toenails in 19–20 days; fat storage in 21 days; whiskers in 23 days; and a distinctly furred appearance in 29 days.

At birth the leverets are fully furred, their eyes are open, and they weigh about 70 g (2.5 oz). The pelage is very fine and dense, brown interspersed with gray or black-tipped hairs. The lower legs and feet are generally lighter than the rest of the body.

The precocial young are able to walk and hop when one day old. They often emit a guttural sound very much like a growl when disturbed, and frequently will attempt to fight off an intruder. They grow rapidly and are weaned at about 4 weeks of age, although they may nurse for 6 weeks or more if the doe is not pregnant (Severaid 1942).

The doe nurses her young once each day for only 5 to 10 minutes. The time of nursing is remarkably constant and appears to be related to a certain light intensity during the evening twilight period. The home range of leverets varies from 1.6 to 1.9 square acres during their 2d to 6th weeks, then increases rapidly to about the size of the doe's home range—approximately 20 square acres (Rongstad and Tester 1971).

The doe does not build a natal nest, though infrequently she may give birth to her young in a selected spot. Snowshoe hares become sexually mature the spring following their birth, but

Keith and Meslow (1967) estimated that a juvenile female which had recently produced a litter must have conceived when barely 2 months old. Green and Evans (1940a) estimated longevity at about 5 years.

**Age Determination.** The epiphyseal cartilage in the humerus disappears by age 8 months (Dodds 1965), and lens weights of yearling snowshoe hares in August exceed 160 mg (Keith and Meslow 1967).

**Specimens Examined.** MAINE, total 86: Franklin County, Farmington 18 (UME); Hancock County, Bucksport 2, Mt. Desert Island 1 (HMCZ); Kennebec County, Waterville 1 (melanistic) (UME); Oxford County, Bethel 1 (USNM), Norway 13, Upton 27 (HMCZ); Penobscot County, Orono 9, Stetson 1 (UME); Piscataquis County, Greenville 8 (HMCZ), Sandy Stream Pond 2, Sebec Lake 1; Somerset County, Jackman 1 (USNM), Little Beach 1 (UCT).

NEW HAMPSHIRE, total 7: Grafton County, Sugar Hill 1; Merrimack County, Webster 2 (HMCZ); Strafford County, Barrington 1, Durham 1 (UNH); Sullivan County, Springfield 2 (DC).

VERMONT, total 48: Bennington County, Landgrove 1 (UVT), Woodford 2 (AMNH); Caledonia County, Hardwick 1; Chittenden County, Huntington 1, Milton 1 (DC), Williston 4; Grand Isle County, Isle La Motte 1 (UVT); Lamoille County, Mt. Mansfield 2–1 (HMCZ–UVT); Orange County, Newbury 1, Randolph 1, Thetford Center 1 (UVT), Topsham 1 (USNM); Orleans County, Barton 1 (UVT); Rutland County, East Wallingford 1–2 (UCT–UVT), Rutland 14 (AMNH); Washington County, Marshfield 2, Waterbury 1 (UVT); Windsor County, Hartland 8 (AMNH), Pomfret 1 (DC).

MASSACHUSETTS, total 30: Berkshire County, Beckett 6 (UMA), Stockbridge 1 (UNH), Windsor 1 (UCT); Franklin County, Monroe 1 (UMA); Hampden County, Feltonville 2, Springfield 1 (HMCZ); Hampshire County, Amherst 1 (UMA), Pelham 1–2 (UMA–UNH); Middlesex County, Concord 1 (HMCZ), Nantucket County, Nantucket 3 (USNM); Plymouth County, Hanover 1, Middleboro 1 (HMCZ); Worcester County, Barre 1 (UNH), Lunenburg 7 (USNM).

CONNECTICUT, total 4: Litchfield County, Salisbury 1; Tolland County, Mansfield 1, Union 1; Windham County, Westford 1 (UCT).

RHODE ISLAND, total 12: Newport County, Fort Adams 1; Washington County (no location) 10, Exeter 1 (USNM).

# Black-tailed Jackrabbit

*Lepus californicus melanotis*
Mearns

*Lepus melanotis* Mearns, 1890. *Bull. Amer. Mus. Nat. Hist.* 2:297.
*Lepus californicus melanotis* Nelson, 1909. *N. Amer. Fauna* 29:146.
Type Locality: Independence, Montgomery County, Kansas
Dental Formula:
  I 2/1, C 0/0, P 3/2, M 3/3 × 2 = 38
Names: Western jackrabbit, Great Plains jackrabbit, and jackrabbit. The word "jack" is probably applied to this hare because its ears resemble jackass ears. The meaning of *Lepus* = a hare, *californicus* = of California, and *melanotis* = black ear, referring to the black terminal ear patch.

**Description.** The black-tailed jackrabbit is a large hare with enormous ears, short black tail, powerful hind legs and large feet, and smaller front legs and feet. The upper lip is divided. The eyes are large and luminous. The front foot has five toes, the hind foot four, and the soles of the feet are densely haired. Scent glands on either side of the anus give off a musky odor which probably serves as an identification to other hares. Females have six mammae, two pectoral and four abdominal. Characters of the skull are typical *Lepus*.

The pelage is soft. The sexes are colored alike and show no seasonal change in color. The coloration above is grayish to ochraceous buffy with a wavy dark wash of black. The rump has a large, conspicuous whitish patch and a well-developed broad black band runs from the middle of the rump to the top of the tail. There is a whitish ring around the eye and sometimes a small white spot on the forehead. The tail is black above, white below. The chin and underparts are whitish. The insides of the ears are buffy, the backs whitish and usually tipped with black. Immatures are somewhat more yellowish brown than adults. Molt occurs in spring and fall.

Females are larger than males. Measurements range from 515 to 585 mm (20.1–22.8 in); tail, 50 to 115 mm (2.0–4.5 in); and hind foot, 115 to 145 mm (4.5–5.7 in). Weights vary from 2.3 to 3.3 kg (5.1–7.3 lb) (Tiemeier 1965).

**Distribution.** The black-tailed jackrabbit occurs in the Great Plains from southwestern South Dakota and southwestern Wyoming through Colorado and New Mexico east of the Rocky Mountains, all of Kansas, Nebraska, and Oklahoma, western Missouri and Arkansas, and south to north central Texas.

The species was introduced on Nantucket Island, Massachusetts, from Kansas in 1925 by the Nantucket Sportsmen's Club. The animal was released as a substitute for the red fox in the traditional "ride to the hounds" by the "Nantucket Harriers." These people rode on horseback dressed in full English fox hunting

BLACK-TAILED
JACKRABBIT

Nantucket Island
—15 miles—

regalia and chased the jackrabbits with their hounds. A second batch of hares was introduced on the island in the early 1940s.

In New England the species occurs only on Nantucket Island, where it is a game animal.

**Ecology.** On Nantucket Island, the jackrabbit has adapted well to the beach grass, open fields, and cultivated areas along with a natural succession of stabilized beach dunes. The species scratches out a simple depression or form in the ground as a place to rest or nest. The form is located at the base of some large plants that provide overhead cover. A form may be used regularly by the same animal or by other jackrabbits.

Man is the chief predator of black-tailed jackrabbits. Large birds of prey may capture the leverets, and fires, floods, and storms also take their toll. This species is host to fleas, ticks, lice, cestodes, nematodes, and botfly larvae and may be susceptible to tularemia.

**Behavior.** Very little is known on the biology of black-tailed jackrabbits on Nantucket Island, but the biology of the species has been studied elsewhere in the United States.

Black-tailed jackrabbits are alert and skillful in evading capture. If danger is imminent the animals may freeze or flee. They can leap in long, graceful bounds, taking strides at 35 miles per hour or more for short distances. These long leaps are interrupted now and then by high observation leaps. These hares are active mainly at dusk and at night but may be seen during daylight (Lechleitner 1958).

The length of the breeding season varies according to latitude. In Arizona the season is from December through September (Vorhies and Taylor 1933); in California, from late January into August with sporadic matings in any month of the year (Lechleitner 1959); and in Kansas, from late January into August (Bronson and Tiemeier 1959).

Females produce from three to four litters annually and from one to six leverets per litter in Arizona (Vorhies and Taylor 1933). The gestation period is from 41 to 47 days, averaging 43 days in California (Haskell and Reynolds 1947).

**Specimens Examined.** MASSACHUSETTS, total 3: Nantucket County, Nantucket Island 3 (collection of Mr. James J. McDonough).

# European Hare

*Lepus europaeus* Pallas

*Lepus europaeus* Pallas, 1778. *Novae species quadrupedum e glirum ordine,* . . . p. 30.
Type Locality: Burgundy, France
Dental Formula:
 I 2/1, C 0/0, P 3/2, M 3/3 × 2 = 28
Names: Brown hare, German hare, Belgian hare, and jackrabbit. The meaning of *Lepus* = a hare, and *europaeus* = of Europe.

**Description.** The European hare is larger than the snowshoe hare and has relatively larger and wider ears and longer and more powerful hind legs and feet. The head is relatively small, tail moderately long, and eyes large, yellowish, and set high in the skull. Females have six mammae.

The skull is elongated, with a long, slender rostrum. The mandibular condyle is weak and strongly inclined backward, and the auditory bullae are small and inflated. The total length of the skull is usually more than 100 mm (3.90 in).

The fur on the back is wavy and slightly curly. The sexes are colored alike and show seasonal variation. In summer the coloration above is rich tawny buff mixed with blackish hairs, and on the sides is paler and yellower. The neck and shoulders are yellowish buff. The individual hairs on the back are white at the base, followed by a black band and tipped with dull brownish yellow. The underparts are white, and there is a white or buff stripe across the eyes. The ears are grizzled, being mixed yellow and brownish

black, white behind and black at the tip. The legs and feet are like the sides of the body above and white or yellowish below. The tail is black above, white below. The winter pelage is paler and grayer, with whitish tones, but the pelage does not turn white as in the snowshoe hare. The leverets are paler and more russet than adults. Color variants have not been recorded.

The sexes are equal in size. Measurements range from 640 to 700 mm (25.0–27.3 in); tail, 70 to 100 mm (2.7–3.9 in); and hind foot, 160 to 170 mm (6.2–6.6 in). Weights vary from 3.3 to 4.5 kg (7.3–10 lbs).

**Introduction into North America.** European hares were introduced into North America solely as game animals for coursing. There appear to have been five or more importations. Probably the most successful in terms of subsequent spread and increase occurred about 1893. In that year a well-to-do resident of Millbrook, Dutchess County, New York, im-

EUROPEAN HARE

ported hares from Hungary in shipments of up to 500. These importations and releases continued at about five-year intervals until 1910 or 1911. There was a gradual buildup in hare abundance and an extension of range into neighboring counties and eastward, extending into southern Vermont, eastward 20 or 30 miles into Massachusetts and Connecticut, south to central New Jersey, and to a limited extent westward into extreme eastern Pennsylvania. During the peak populations at the turn of the century it was estimated that there were 20 to 40 hares per square mile in Sheffield, Berkshire County, Massachusetts (Silver 1924).

**Distribution.** The European hare occurs from southern Ontario, Michigan, east in the Hudson River valley of eastern New York and in western Connecticut. In Vermont, a European hare was taken at Fair Haven, Rutland County, in 1915 (Silver 1924). Several hares have been taken in southwestern Vermont (Osgood 1938). European hares were introduced into Connecticut shortly before 1920 and into Massachusetts in the late

1920s (Presnall 1958). At present the hares do not seem to occur in Vermont or Massachusetts. The species is a game animal in Connecticut and is taken in Fairfield and Litchfield counties.

**Ecology.** This species prefers open dairy country and low rolling hills with sparse vegetation which provide a good view. It may be encountered in open hardwood forests with a sparse understory.

Man, foxes, and bobcats are the main predators of the European hare in New England. Large hawks and owls can catch and hold it, and smaller raptors and carnivores may prey on leverets. This hare may harbor the same parasites as other hares and rabbits. It is often killed by the cutting bars of high-speed farm machines, since it has a tendency to freeze when they approach.

**Behavior.** European hares are mainly crepuscular and more or less nocturnal but are sometimes active during the daytime. During the day they rest partially concealed in a form that they scratch out in a small clump of grass, weeds, or bushes, near boulders, or in a leaf drift. The form is situated in a spot where the view and sound are not obstructed. Although they do not dig burrows, they will dig into the snow to get food or to gain shelter from strong winds.

European hares rely greatly on their senses of smell and hearing and on their great speed and endurance to escape their enemies. They make best speed uphill with their long hind legs, and always start up an incline if possible, to gain advantage over pursuers. Occasionally they fall over themselves when running downhill.

When chased by dogs European hares often travel over ice just thick enough to hold themselves but too thin to hold the hounds, and they will frequently backtrack. During a chase, they often waltz on their hind toes or make "spy hops" to study the dogs' progress. MacLulich (1936) reported the top speed of a European hare as 30 miles per hour for a quarter of a mile along a road. Millias (1906) reported that the hare's normal stride is about 4 feet, and when surprised its leap may be extended to 10 to 12 feet; it can jump over a wall 5 feet high.

European hares are good swimmers and do not hesitate to cross rivers. Eabry (1970) reported that the animal has been seen swimming the Hudson River above Troy, New York, with no apparent reason except to reach the other side. They are usually silent, but the female may scream in fear and sometimes will emit a faint cry in calling her young for nursing. Eabry

(1970) estimated that the home range is a radius of 1.9 miles or 11 square miles and probably is the same for both sexes in New York.

This species eats grass, clover, corn, winter wheat, apples, blackberries, raspberries, twigs and buds of thorn apple, and bark of elm, maple, sumac, and butternut trees. Like other lagomorphs it practices refection. Saunders (1929) reported instances of European hares killing cottontail rabbits by biting the neck and head, but the hares did not eat the cottontails.

Little is known about the reproduction of European hares. Adult bucks usually fight vigorously among themselves during the breeding season. In New York, the breeding season begins in early January, and as early as mid-March some females are in their second pregnancy. Breeding usually occurs again within 36 hours after conception. Females may produce one to three young per litter (Dell 1957); there seems a high embryonic loss in all litters and as much as 80 percent in the first litter (Raczynski 1964). The gestation period is about 42 days (Bourlière 1954). The doe scatters her young, placing one (sometimes two) in different forms, nursing each at different times of night. The female usually drives them away and forces them to shift for themselves after they are weaned (Millais 1906). Longevity in the wild is probably 12 years (Burt 1957).

**Specimens Examined.** CONNECTICUT, total 1: New Haven County, Southbury 1 (UCT).

## REFERENCES

Adams, Lowell. 1959. An analysis of a population of snowshoe hares in northwestern Montana. *Ecol. Monogr.* 29(2):141–70.

Bailey, James A. 1969. Trap responses of wild cottontails. *J. Wildl. Manage.* 33(1):48–58.

Bangs, Outram. 1895. The geographical distribution of the eastern races of the cotton-tail (*Lepus sylvaticus* Bach.) with a description of a new subspecies, and with notes on the distribution of the northern hare (*Lepus americanus* Erxl.) in the east. *Proc. Boston Soc. Nat. Hist.* 26:404–14.

Beule, John D. 1940. Cottontail nesting study in Pennsylvania. *Pennsylvania Game News* 11(2):10–11, 28.

Beule, John D., and Studholme, Allan T. 1942. Cottontail rabbit nests and nestlings. *J. Wildl. Manage.* 6(2):133–40.

Bider, J. Roger. 1961. An ecological study of the hare, *Lepus americanus. Canadian J. Zool.* 39(1):81–103.

Bookhout, Theodore A. 1964. Prenatal development of snowshoe hares. *J. Wildl. Manage.* 28(2):338–45.

Bothma, J. du P.; Teer, James G.; and Gates, Charles E. 1972. Growth and age determination of the cottontail in south Texas. *J. Wildl. Manage.* 36(4):1209–21.

Bourlière, François. 1954. The natural history of mammals. New York: Alfred A. Knopf.

Bronson, Franklin H., and Tiemeier, Otto W. 1958. Reproduction and age distribution of black-tailed jack rabbits in Kansas. *J. Wildl. Manage.* 22(4):409–14.

Burt, William Henry. 1957. *Mammals of the Great Lakes region.* Ann Arbor: University of Michigan Press.

Carson, J. D., and Cantner, D. E. 1963. West Virginia cottontails. *West Virginia Dept. Nat. Res. Bull.* 5:1–24.

Casteel, David A. 1966. Nest building, parturition, and copulation in the cottontail rabbit. *Amer. Midland Nat.* 75(1):160–67.

Chandler, Edwin H. 1952. The geographical and ecological distribution of Massachusetts cottontails. M.S. thesis, University of Massachusetts.

Chapman, Joseph A. 1975. *Sylvilagus transitionalis. Mammalian Species* 55:1–4. Amer. Soc. Mammalogists.

Chapman, Joseph A., and Morgan, Raymond P. II. 1973. Systematic status of the cottontail complex in western Maryland and nearby West Virginia. *Wildl. Monogr.* 33:1–54.

Chapman, Joseph A., and Trethewey, Donald E. C. 1972. Movements within a population of introduced eastern cottontail rabbits. *J. Wildl. Manage.* 36(1):155–58.

Cox, W. T. 1936. Snowshoe rabbit migration, tick infestation, and weather cycles. *J. Mamm.* 17(3):216–21.

Dalke, Paul D. 1937. A preliminary report of the New England cottontail studies. *Trans. N. Amer. Wildl. Conf.* 2:542–48.

———. 1942. The cottontail rabbits in Connecticut. *Connecticut Geol. Nat. Hist. Surv. Bull.* 65:1–97.

Dalke, Paul D., and Sime, Palmer R. 1938. Home and seasonal ranges of the eastern cottontail in Connecticut. *Trans. N. Amer. Wildl. Conf.* 3:659–69.

———. 1941. Food habits of the eastern and New England cottontails. *J. Wildl. Manage.* 5(2):216–28.

Dell, Joseph. 1957. The European hare in New York. *New York State Conservationist* 11(4):28–29.

Dice, L. R. 1927. The transfer of game and fur-bearing mammals from state to state, with special reference to the cottontail rabbit. *J. Mamm.* 8(2):90–96.

Dodds, Donald G. 1965. Reproduction and productivity of snowshoe hares in Newfoundland. *J. Wildl. Manage.* 29(2):303–15.

Eabry, H. Stephen. 1968. An ecological study of *Sylvilagus transitionalis* and *S. floridanus* of northeastern Connecticut. M.S. thesis, University of Connecticut.

———. 1970. A feasibility study to investigate and evaluate the possible future directions of European hare management in New York. Federal Aid P-R Proj. W-84-R-17 (multilith).

Edwards, W. R. 1963. Fifteen cottontails in a nest. *J. Mamm.* 44(3):416–17.

Fay, Francis H. 1951. Studies of the management, ecology and distribution of Massachusetts cottontails. M.S. thesis, University of Massachusetts.

Fay, Francis H., and Chandler, Edwin H. 1955. The geographical and ecological distribution of cottontail rabbits in Massachusetts. *J. Mamm.* 36(3):415–24.

Green, R. G., and Evans, C. A. 1940a. Studies on a population cycle of snowshoe hares on the Lake Alexander area. I. Gross annual censuses, 1932–1939. *J. Wildl. Manage.* 4(2):220–38.

———. 1940b. Studies on a population cycle of snowshoe hares on the Lake Alexander area. II. Mortality according to age groups and seasons. *J. Wildl. Manage.* 4(3):267–78.

———. 1940c. Studies on a population cycle of snowshoe hares on the Lake Alexander area. III. Effect of reproduction and mortality of young hares on the cycle. *J. Wildl. Manage.* 4(4):347–58.

Green, R. G., and Larson, C. L. 1938. A description of shock disease in the snowshoe hare. *Amer. J. Hygiene* 28(2):190–212.

Green, R. G.; Larson, C. L.; and Bell, J. F. 1939. Shock disease as the cause of the periodic decimation of the snowshoe hare. *Amer. J. Hygiene* 30(3):83–102.

Hale, James B. 1949. Aging cottontail rabbits by bone growth. *J. Wildl. Manage.* 13(2):216–25.

Hall, E. Raymond, and Kelson, Keith R. 1959. *The mammals of North America.* 2 vols. New York: Ronald Press.

Haskell, H. S., and Reynolds, H. G. 1947. Growth, development, food requirements and breeding activity of California jack rabbit. *J. Mamm.* 28(1):129–36.

Hewitt, C. Gordon. 1921. *The conservation of the wildlife in Canada.* New York: Charles Scribner's Sons.

Hill, Edward P. 1967. Homing by a cottontail rabbit. *J. Mamm.* 48(4):648.

Holden, Henry E., and Eabry, H. Stephan. 1970. Chromosomes of *Sylvilagus floridanus* and *S. transitionalis. J. Mamm.* 51(1):166–68.

Huber, James J. 1962. Trap response of confined cottontail populations. *J. Wildl. Manage.* 26(2):177–85.

Jackson, Hartley H. T. 1961. *Mammals of Wisconsin.* Madison: University of Wisconsin Press.

Jackson, Steven Neil. 1973. Distribution of cottontail rabbits (*Sylvilagus* spp.) in northern New England. M.S. thesis, University of Connecticut.

Johnston, Joseph E. 1972. Identification and distribution of cottontail rabbits in southern New England. M.S. thesis, University of Connecticut.

Keith, Lloyd B., and Meslow, E. Charles. 1967. Juvenile breeding in the snowshoe hare. *J. Mamm.* 48(2):327.

Kirkpatrick, Charles M. 1960. Unusual cottontail litter. *J. Mamm.* 41(1):119–20.

Lechleitner, R. R. 1958. Certain aspects of behavior of the black-tailed jack rabbit. *Amer. Midland Nat.* 60(1):145–55.

———. 1959. Sex ratio, age classes and reproduction of the black-tailed jack rabbit. *J. Mamm.* 40(1):63–81.

Linkkila, Timothy E. 1971. Influence of habitat upon

changes with interspecific Connecticut cottontail populations. M.S. thesis, University of Connecticut.

Llewellyn, Leonard M., and Handley, C. O. 1945. The cottontail rabbits of Virginia. *J. Mamm.* 26(4):379–90.

Lord, Rexford D., Jr. 1959. The lens as an indicator of age in cottontail rabbits. *J. Wildl. Manage.* 23(3):358–60.

———. 1961a. Mortality rates of cottontail rabbits. *J. Wildl. Manage.* 25(1):33–40.

———. 1961b. Potential life span of cottontails. *J. Mamm.* 42(1):99.

McDonough, James J. 1952. Census of rabbit populations on management area. Massachusetts Div. Fisheries and Game. Job Compl. Rept. P-R Proj. 22-R-1 (mimeographed).

———. 1960. *The cottontail in Massachusetts.* Boston: Massachusetts Division of Fisheries and Game.

MacFarlane, R. 1905. Notes on the mammals collected and observed in the northern Mackenzie River district, Northwest Territories of Canada, with remarks on explorers and explorations of the far north. *Proc. U.S. Nat. Mus.* 28:673–764.

MacLulich, D. A. 1936. Running speeds of skunk and European hare. *Canadian Field-Nat.* 50(5):92.

Madson, John. 1959. *The cottontail rabbit.* East Alton, Illinois: Olin Mathieson Chemical Corp.

Marsden, Halsey M., and Conaway, C. H. 1963. Behavior and reproductive cycle in the cottontail. *J. Wildl. Manage.* 27(2):161–70.

Meslow, E. Charles, and Keith, Lloyd B. 1968. Demographic parameters of a snowshoe hare population. *J. Wildl. Manage.* 32(4):812–34.

———. 1971. A correlation analysis of weather versus snowshoe hare population parameters. *J. Wildl. Manage.* 35(1):1–15.

Millais, John Guille. 1906. *The mammals of Great Britain and Ireland.* Vol. 3. London: Longmans, Green.

Nelson, E. W. 1909. The rabbits of North America. *N. Amer. Fauna* 29:1–314.

Nottage, Edward J. 1972. Comparative feeding trials of *Sylvilagus floridanus* and *Sylvilagus transitionalis.* M.S. thesis, University of Connecticut.

Nugent, Richard F. 1968. Utilization of fall and winter habitat by the cottontail rabbits of northwestern Connecticut. M.S. thesis, University of Connecticut.

Olmstead, Donald L. 1970. Behavioral comparisons of two species of cottontails *Sylvilagus floridanus* and *Sylvilagus transitionalis. Trans. N.E. Sect.*

Osgood, Frederick L. 1938. The mammals of Vermont. *J. Mamm.* 28(1):13–16.

Palmer, C., and Armstrong, R. 1967. Chromosome number and karyotype of *Sylvilagus floridanus,* the eastern cottontail. *Mammalian Chromos. Newsltr.* 8(4):282–83.

Pelton, Michael R. 1969. The relationship between epiphyseal groove closure and age of the cottontail rabbit, *Sylvilagus floridanus. J. Mamm.* 50(3):624–25.

Petrides, George A. 1951. The determination of sex and age ratios in the cottontail rabbit. *Amer. Midland Nat.* 46(2):312–36.

Presnall, Clifford C. 1958. The present status of exotic mammals in the United States. *J. Wildl. Manage.* 22(1):45–50.

Pringle, Laurence P. 1960. A study of the biology and ecology of the New England cottontail, *Sylvilagus transitionalis,* in Massachusetts. M.S. thesis, University of Massachusetts.

Raczynski, Jan. 1964. Studies of the European hare. V. Reproduction. *Acta Theriologica* 9(19):305–52.

Rongstad, Orrin J. 1966. A cottontail rabbit lens-growth curve from southern Wisconsin. *J. Wildl. Manage.* 30(1):114–21.

Rongstad, Orrin J., and Tester, John R. 1971. Behavior and maternal relations of young snowshoe hares. *J. Wildl. Manage.* 35(2):338–46.

Rowan, W. M., and Keith, L. B. 1956. The reproductive potential and sex ratios of snowshoe hares in Alberta. *Canadian J. Zool.* 34:273–81.

Saunders, W. E. 1929. Carnivorous habits of the European hare. *J. Mamm.* 10(2):170.

Schwartz, Charles W., and Schwartz, Elizabeth R. 1959. *The wild mammals of Missouri.* Columbia: University of Missouri Press.

Severaid, Joe Harold. 1942. The snowshoe hare: Its life history and artificial propagation. Augusta: Maine Department of Inland Fisheries and Game.

———. 1945. Breeding potential and artificial propagation of the snowshoe hare. *J. Wildl. Manage.* 9(4):290–95.

Shadle, A. R.; Austin, T. S.; and Meyer, F. Z. 1940. Five cottontail rabbits up a tree. *J. Mamm.* 21(4):462–63.

Siegler, Hilbert R. 1954. Late breeding snowshoe hare. *J. Mamm.* 35(1):122.

Silver, James. 1924. The European hare, *Lepus europaeus* Pallas, in North America. *J. Agric. Res.* 28(11):1133–37.

Terres, J. Kenneth. 1941. Speed of the varying hare. *J. Mamm.* 22(4):453–54.

Thomsen, Hans P., and Mortensen, Otto A. 1946. Bone growth as an age criterion in the cottontail rabbit. *J. Wildl. Manage.* 10(2):171–74.

Tiemeier, Otto W. 1965. The black-tailed jack rabbit in Kansas. I. Bionomics. *Kansas Agr. Exper. Sta. Tech. Bull.* 140:1–37.

Vorhies, Charles T., and Taylor, Walter P. 1933. The life histories and ecology of jack rabbits, *Lepus alleni* and *Lepus californicus* spp., in relation to grazing in Arizona. *Univ. Arizona Tech. Bull.* 49:471–587.

Wing, L. W. 1935. Wildlife cycles in relation to the sun. *Trans. N. Amer. Wildl. Conf.* 21:345–63.

# Order Rodentia

THE ORDER RODENTIA contains the rodents, or gnawing mammals, the largest order of mammals in both number of species and number of individuals. Rodents are characterized by prominent, chisellike incisors which grow continuously from persistent pulps. The canines are absent and in their place is a wide gap, or diastema, separating the incisors from the grinding cheek teeth. The cheek teeth may be either low- or high-crowned, with many patterns of enamel and dentine.

Rodents are a complex group with great morphological diversity, and the order has undergone considerable adaptive radiation. Rodents may be terrestrial, such as field mice; arboreal, as tree squirrels; volant, or gliding, as flying squirrels; fossorial in varying degrees, as woodchucks and pocket gophers; or semiaquatic, as beavers and muskrats. Most rodents are nocturnal or crepuscular except for tree squirrels, which are diurnal. Rodents are generally plantigrade (walking on the soles of their feet), but some have adapted to leaping, such as the jumping mice. Some species are primarily plant eaters while others, such as the grasshopper mice, are mainly flesh eaters.

Rodents are economically important, as valuable fur-bearers and as pests. They have a high reproductive rate and constitute the main item in the diets of many predators.

The order Rodentia contains three well-defined functional suborders of rodents, classed mainly by the cranial characters of the infraorbital foramen and the zygomatic arch. In the *sciuromorphs* or squirrellike rodents, the foramen is not slitlike and is not enlarged for passage of the masseter muscle. The zygomatic arch is slender, mainly formed by the jugal, which is not supported by the maxillary zygomatic process. In the *myomorphs*, or ratlike rodents, the foramen is slitlike and enlarged for passage of a portion of the masseter muscle. The zygomatic arch is broadened, and the jugal seldom extends far forward, usually being supported by the maxillary zygomatic process. In the *hystricomorphs*, or porcupinelike rodents, the foramen is greatly enlarged for passage of the masseter muscle and blood vessels. The zygomatic arch is stout, and the jugal is not supported below by a continuation of the maxillary zygomatic process.

Rodents are distributed nearly worldwide except for some arctic and oceanic islands, including New Zealand, and Antarctica (Anderson 1967). The Rodentia is represented in New England by six families.

## SUBORDER SCIUROMORPHA

### FAMILY SCIURIDAE (SQUIRRELS)

The family Sciuridae, or squirrels, includes the marmots, ground squirrels, prairie dogs, chipmunks, and tree squirrels. Squirrels typically have bushy tails, and some have internal cheek pouches. Some species are terrestrial and semifossorial, others are arboreal, and still others are volant, or gliders.

Terrestrial sciurids make their nests in burrows. They are typically more chunky and have shorter legs and less bushy tails than other sciurids. Most of them become dormant for varying periods of time each year—some up to half a year. Their breeding season begins shortly after hibernation. The tree and gliding squirrels are very agile, remarkably active, and well adapted to arboreal life. They seek refuge in nests in hollow trees or on high limbs. Their long, bushy tails help them balance while they are jumping through treetops. Flying squirrels are extremely well adapted to gliding; a furred flap of skin stretching along the sides of the body from the front legs to the hind legs, when flattened out, gives a broad gliding surface.

Representatives of this family occur worldwide except for the Australian region, Madagascar, southern South America, the polar regions, and certain desert regions such as Arabia and Egypt (McLaughlin 1967). There are six species of squirrel in New England.

# Eastern Chipmunk

*Tamias striatus*

Two subspecies of eastern chipmunk are recognized in New England:
*Tamias s. lysteri* (Richardson). *Sciurus (Tamias) lysteri* Richardson, 1829. *Fauna Boreali-Americana* 1:181, pl. 15 (June) *Tamias s. lysteri* Merriam, 1886. *Amer. Nat.* 20:242.
Type Locality: Penetanguishene, Ontario
*Tamias s. fisheri* A. H. Howell, 1925. *J. Mamm.* 6(1):51.
Type Locality: Merritts Corner, 4 miles west of Ossining (Sing Sing), Westchester County, New York (Hall and Kelson 1959)
Dental Formula:
I 1/1, C 0/0, P 1/1, M 3/3 × 2 = 20
Names: Common chipmunk, striped ground squirrel, chipping squirrel, chippie, hackie, and rock squirrel. The meaning of *Tamias* = a collector and storer of provisions, and *striatus* refers to the prominent body stripes.

**Description.** The eastern chipmunk is a small and graceful, beautifully colored squirrel about one-third the size of the gray squirrel. The head is short and somewhat rounded, with short rounded and erect ears; the tail is somewhat flattened, well-haired but not bushy. The limbs are short and the forefeet have four clawed toes plus a small thumb covered by a soft rounded nail; each hindfoot has five clawed toes. There are prominent internal cheek pouches for carrying food and excavated earth. Females have eight mammae—two pectoral, four abdominal, and two inguinal. Musk glands are situated on either side of the anus in both sexes.

The narrow skull has a tapering rostrum and a flattened, oval-shaped cranium. The postorbital process is small and weak, and the infraorbital foramen is relatively larger than most sciurids and does not form a canal. The upper incisors are short, and the upper third premolar is absent.

The fur is short, dense, and moderately soft. The coloration of both sexes is similar and shows some seasonal variation. In summer the coloration above is reddish brown with a mixture of black and white hairs, blending to a bright orange brown or rusty on the rump and flanks. The cheeks and sides of the body are grayish brown to tawny brown. The sides of the face have two buffy or whitish stripes above and below the eyes and a black stripe across the eyes. There are five distinct dark brown to blackish stripes from the shoulders to the rump, separated by moderately broad white stripes. The belly and lower sides of the body are whitish, and the feet are buff. The tail is blackish above and rusty below, fringed with white or pale gray. The winter pelage is paler. Albino and melanistic chipmunks are occasionally seen.

In adults molting occurs in late spring and early summer and again in late fall or early winter, extending into late winter in some individuals. Immatures begin their first molt at about 2 months of age. Those born in April and May molt in late June through late August, and young born in August molt in mid-October and are probably finished by early December. Molting in adults is diffuse, and the dates apparently vary in different localities (Yerger 1955).

The sexes are equal in size. Measurements range from 215 to 247 mm (8.4–9.6 in); tail, 77 to 113 mm (3.0–4.4 in); and hind foot, 32 to 38 mm (1.2–1.5 in). Weights vary from 70 to 110 g (2.5–3.9 oz).

**Distribution.** The eastern chipmunk occurs from Quebec south through the eastern half of the United States into Georgia, western Florida, and northeastern Louisiana, and west to eastern Oklahoma, Kansas, North and South Dakota, and Saskatchewan.

**Ecology.** The eastern chipmunk is common throughout its range. It is found in the borders of open deciduous woods where thick understory and brier grow among old logs, on rocky ledges covered with brush and vines, on loose stone walls in rocky hillsides, in old forests without undergrowth, in old farm woodlots, on brushland and cutover land along brushy fencerows, in rubbish heaps, old buildings, and camps, and sometimes in parks and around gardens.

The species is largely terrestrial and rarely climbs trees. It makes its den in the ground. The den is situated in a burrow system that may be simple or extensive, and the entrance is usually well hidden under a rotten log, old tree stump, rock, or old stone wall. The entrance of the den is plugged during cold weather.

Man, hawks, mink, weasels, martens, foxes, bobcats, house cats, raccoons, Norway rats, red squirrels, and large snakes kill chipmunks, and forest fires and floods take their toll. Fleas, lice, mites, cestodes, nematodes, hornyhead worms, and botfly larvae are parasites of eastern chipmunks.

**Behavior.** Eastern chipmunks are diurnal, and are less active during hot weather in late summer. Although they do not disappear underground, they are seen and heard much less often in summer than in spring and autumn. Females are more active than males during the summer lull, foraging quietly (Schooley 1934; Yerger 1955). The morning activity is slightly less than the afternoon activity; both decrease from June to July to August, to resume gradually in the fall and taper off until the chipmunks enter their winter inactivity period. High ambient temperatures do not seem to have a direct effect on the summer lull activity of chipmunks,

EASTERN CHIPMUNK

but unusually high winds and heavy rains depress their activities (Dunford 1972).

Eastern chipmunks are not typical deep hibernators, though some of them may be completely dormant for varying lengths of time in severe winter weather. The date of onset of winter inactivity apparently depends on latitude. As a rule eastern chipmunks retire to their burrows in late October to early November and emerge to breed in late February or early March, but they may become inactive again before spring. Severe winter weather may cause them to go underground.

The degree of torpidity in eastern chipmunks also appears to vary among different populations. Panuska (1959) reported that eastern chipmunks subjected to ambient temperatures of 1.5°C to 3.0°C either became torpid or semitorpid or remained active throughout the 10-month experimental period. Scott and Fisher (1972) found that approximately 63% of the chipmunks they observed experienced some degree of topor. Brenner (1975) found that approximately 30% of chipmunks became torpid and about 60% were classified as nonhibernators. Panuska (1959) reported that torpidity ranged from 1 to 6 days at a time. Similarly, Brenner (1975) observed that torpidity in eastern chipmunks ranged from 1 to 8 days. Eastern chipmunks do not accumulate fat before hibernating but survive the winter months partially or entirely on stored food (Cade 1964; Brenner 1975).

Eastern chipmunks have three notes, a loud "chip" similar to a robin's note, a soft "cuck-cuck" note that may be repeated for several minutes, and a combination of the loud chip with a trill on the end—like "chip-r-r-r-r." Dunford (1970) remarked that chipmunk vocalizations seem to function as alarm calls, one of which, "chipping," may be an agonistic signal, helping to maintain the spatial organization of the population by advertising the presence of a chipmunk in its dominance area, thereby dissuading neighboring animals from intruding.

As a rule eastern chipmunks are solitary animals and are agonistic toward one another, except for females with young. These animals exhibit a hierarchy correlated with size rather than with sex, though females in estrus or with young in the nest are aggressive. A dominant chipmunk just beginning to feed may tolerate a closer approach by a lower-ranking individual than one which had been feeding for some time (Wolfe 1966). Eastern chipmunks exhibit high and low states of alertness. In low alertness the animal is generally relaxed as it explores or feeds alone. In high alertness the chipmunk exhibits jerky movements of the body, bristled tail, flattened ears, and an "alert" or "frozen" posture, which generally precedes attack or escape (Wolfe 1969).

The home range of adult males is slightly larger than that of adult females, and the range is smaller for immatures of both sexes. In New York, Yerger (1953) calculated the average home range to be 0.37 square acres for adult males; 0.26 for adult females; and 0.18 for immature males and females. The author believed that any movement of 100 yards or more was a significant change in home range, provided recaptures were in the new area.

Eastern chipmunks returned to their original home site after being released 775 feet away by Layne (1957). Juveniles do not home as accurately as adults, probably because they have a greater instinctive dispersal tendency than adults. Females home more often than males, particularly over longer distances (Seidel 1961).

There seems to be a broad overlap of home ranges, with each animal defending a territory centered on its burrow entrance (Dunford 1970). Adult females display territoriality around their den sites (Blair 1942; Manville 1949; Yerger 1953).

Eastern chipmunks feed on nuts, including acorns, and on a long list of seeds of woody and herbaceous plants such as box elder, striped maple, sugar maple, shadbush, dogwood, viburnum, American yew, ragweed, wintergreen, wild geranium, Canada mayflower, and wild buckwheat. They are fond of certain mushrooms, cultivated berries and cherries, corn fungi, sunflower seeds, and the flesh and seeds of watermelon, apples, pears, peaches, cantaloupe, and squash.

The amount of food a chipmunk can carry in its cheek pouches is remarkable. Klugh (1923) counted a total of 31 kernels of corn crammed in a chipmunk's cheek pouches, and Fraleigh (1929) recorded 13 prune pits at one time and approximately 70 sunflower seeds at another time, while Allen (1938) collected a chipmunk that held 32 beechnuts in its pouches. Sometimes the animal eats on the spot. Such places are recognized by the accumulation of shelled seeds or nut fragments.

Eastern chipmunks also eat insects and insect larvae, earthworms, snails, slugs, cutworms, wireworms, millipedes, butterflies, dragonflies, and salamanders, and occasionally star-nosed moles, young mice, sparrows, juncos, bank swallows, starlings, bird's eggs, frogs, and small snakes.

Eastern chipmunks breed in spring and summer. In New York, the spring estrous period is from mid-March to early April and from early June to mid-July. Some females possibly may be in estrus as late as mid-August, but these are believed to have been born in early spring (Yerger 1955). Males may be sexually active as early as late February (Condrin 1936; Allen 1938).

The gestation period is 31 days. The young are born from about 16 April to 10 May, and summer litters are born between 20 July and 10 August. One to eight young may be produced per litter, but four or five is the usual number. A female may produce two litters in a year. The spring-breeding females consist of old females, yearlings, and some 8-month-old females born the preceding summer, whereas the summer females consist of old females which may or may not have bred in the spring, yearlings, and a few 3-month-old females from spring litters (Yerger 1955).

Although in New York a few females from spring litters become sexually mature at the age of 3 months, and a few from summer litters may mature at 7 to 8 months, most females probably attain sexual maturity at the age of 1 year. Males from spring litters become sexually mature at the age of 1 year, while those born in summer mature at 7 or 8 months of age (Yerger 1955).

At birth the young are naked, reddish, and blind. They are approximately 64 mm (2.5 in) long and weigh 3 g (0.11 oz). White vibrissae begin to show on the fifth day, and by the eighth day the lateral stripes are noticeable. The eyes open at 30 to 31 days. They shed their milk teeth when 3 months old. Longevity is at least 2 to 3 years in the wild and 5 to 8 years in captivity (Allen 1938).

**Age Determination.** Allen (1938) has described the development of the dentition of known-age eastern chipmunks from birth to age 3 months, and Yerger (1955) determined the age of chipmunks from 1 month to 1 year based on characteristics of the skull and epiphyses of the humerus, femur, and tibia.

**Specimens Examined.** MAINE, total 113: Androscoggin County, Lincoln 1 (UCT); Aroostook County, Madawaska 26 (HMCZ); Cumberland County, North Bridgeton 2 (USNM), Portland 1 (UME); Franklin County, Dryden 4 (AMNH), Farmington 22 (UME); Hancock County, Mt. Desert Island 15 (HMCZ); Oxford County, Gilead 1 (UCT), Norway 1 (HMCZ), South Waterford 1 (AMNH), Umbago Lake 2–6 (AMNH–HMCZ), Upton 6 (HMCZ); Penobscot County, Holden 1 (UME),

Millinocket 1 (UCT), Penobscot River (East Branch) 1 (USNM), Orono 3 (UME), South Twin Lake 6 (AMNH); Piscataquis County, Greenville 3 (HMCZ), Mt. Katahdin 1 (DC), Sebec Lake 1 (USNM); Sagadahoc County, East Harpwell 1; Somerset County, Anson 2 (UCT), Crocker Pond 1, Enchanted Pond 2 (AMNH), Lakewood 2 (UCT).

NEW HAMPSHIRE, total 114: Carroll County, Bartlett 1, Jackson 8 (UCT), Kearsage 1 (HMCZ), Ossipee 4–12 (UCT–USNM), Tamworth 1, Wolfeboro 1 (UCT); Cheshire County, Dublin 1, Stoddard 1 (HMCZ), Walpole 1; Coos County, Carroll 1 (UCT), Mt. Washington 2 (HMCZ); Grafton County, Canaan 4, Enfield 1, Franconia 1, Hanover 16 (DC), Livermore 7 (UCT), Lyme 1 (DC), Warren 7; Hillsborough County, Antrim 6 (HMCZ), Hancock 1 (UCT), Manchester 1; Merrimack County, Webster 13 (HMCZ); Rockingham County, Exeter 1–1 (UCT–UNH), Hampton 2 (UCT), Kingston 2–1 (UCT–UNH), Rye 1 (UCT); Strafford County, Durham 3 (UNH); Sullivan County, Charlestown 9 (USNM), Claremont 1, Newport 1 (UCT).

VERMONT, total 182: Addison County, Cornwall 1, Lake Dunmore 1 (UVT), New Haven 1 (UCT), Orwell 1 (HMCZ); Bennington County, Landgrove 2 (UVT), Readsboro 19 (AMNH), West Shaftsbury 1; Chittenden County, Burlington 1, Colchester 6, Essex Junction 6 (UVT), Hinesburg 1 (UCT), Jericho 7, Shelburne 3, Underhill 1, West Milton 1, Weybridge 1, Williston 3; Franklin County, Bakersfield 1, Georgia 2 (UVT), Highgate 3; Grand Island County, Isle La Motte 1 (UCT); Lamoille County, Eden 1 (DC), Mt. Mansfield 11 (USNM); Orange County, Bradford 1 (URI), Braintree 1 (UVT), Wakefield 1 (URI); Rutland County (no location) 28 (UCT), Brandon 2 (USNM), Castleton 1, Chittenden 7, Claredon 2 (DC), Killington Peak 2 (HMCZ), Mendon 10 (DC), Pico Peak 1 (HMCZ), Rutland 11–1 (AMNH–USNM), Sherburne Center 1 (DC), Wallingford 2–1 (AMNH–UCT); Washington County (no location) 3 (UCT), Warren 3 (UVT); Windham County, Newfane 7 (AMNH), Vernon 1 (UCT), Whitingham 4 (AMNH); Windsor County, Norwich 3 (DC), Pomfret 1–9 (AMNH–DC), Royalton 1, Springfield 1 (UCT), Woodstock 2 (HMCZ).

MASSACHUSETTS, total 163: Barnstable County, Sagamore 1 (USNM), Sandwich 1 (UMA), Woods Hole 1 (USNM); Berkshire County, Monterey 7 (UMA), Mt. Greylock 1 (HMCZ); Essex County, Lynn 1; Bristol County, Fall River 2 (USNM); Franklin County, Charlemont 6, Montague 1, Mt. Toby 1, Orange 2, Shutesbury 2, Sunderland 3 (UMA); Hampden County, Brimfield 1 (UCT), Feltonville 3 (HMCZ); Hampshire County, Amherst 12 (UMA), Easthampton 2 (USNM), Leeds 1 (UMA), Montgomery 1 (UCT), Pelham 1 (UMA), Southwick 1 (UCT), Ware 2, Williamsburg 1 (UMA); Middlesex County, Burlington 3 (USNM), Cambridge 7 (HMCZ), Concord 2 (UCT), Newton 26 (HMCZ), Pepperell 1 (UCT), Waltham 1 (HMCZ), Wilmington 11 (USNM); Norfolk County, Quincy 1 (HMCZ), Stoughton 2 (USNM), Walpole 1 (HMCZ), Wellesley 1 (USNM), Westwood 1 (UCT), Weymouth 1 (UMA); Plymouth County, Wareham 17 (HMCZ); Worcester County, Brookfield 1 (UCT), Harvard 19–2 (HMCZ–USNM), Lunenburg 6, Mt. Wachusett 2 (USNM), Shrewsbury 1, Spencer 1 (UCT), Worcester 2 (UMA).

CONNECTICUT, total 165: Fairfield County, Easton 1 (UCT), Noroton 6 (HMCZ), Stamford 1 (UCT), Westport 1–10 (AMNH–UCT); Hartford County, Avon 2, Hartford 1; Litchfield County, Barkhamsted 2, Kent 6, Litchfield 2 (UCT), Macedonia Park 1 (AMNH), Northfield 1 (UCT), Sharon Mt. 2 (AMNH); Middlesex County, East Hampston 1, Malborough 1, Middletown 1 (UCT), Portland 3–2 (UCT–USNM), Westbrook 1 (AMNH); New Haven County, Bethany 1 (UCT), Cheshire 3 (USNM), Guilford 1, Hamden 1 (AMNH); New London County, Franklin 1 (UCT), Liberty Hill 3 (HMCZ), Lyme 4 (AMNH), Stonington 4–1 (AMNH–UCT); Tolland County, Coventry 2, Hebron 1, Mansfield 6, Storrs 5, Tolland 4, Union 65, Willington 2; Windham County, Ashford 2, Chaplin 1, Eastford 1, Natchaug 1 (UCT), Plainfield 4 (USNM), Pomfret 1, Scotland 1, Warrensville 1, Willimantic 1 (UCT), Woodstock 3 (AMNH).

RHODE ISLAND, total 5: Kent County, Coventry 1 (UCT); Providence County (no location) 2 (USNM), Providence 2 (AMNH).

# Woodchuck

*Marmota monax*

**Description.** The woodchuck is the largest member of the squirrel family in New England. It is large, robust, and generally fat, with short, powerful legs and a short, bushy, and almost flattened tail. The ears are low and round and can close over the ear openings to keep out dirt. Each front foot has four well-developed toes with strong claws well adapted for digging and a rudimentary thumb with a small flat nail. Each hind foot has five clawed toes. The cheek pouches are rudimentary and lack retractor muscles. Females have eight mammae—four pectoral, two abdominal, and two inguinal.

The woodchuck has three white, nipplelike glands just within the anus which secrete a musky odor but without the repelling quality of the skunk.

The skull is nearly straight in profile, with the interorbital region much wider than the postorbital region; the postorbital processes project at

There are three subspecies of woodchucks recognized in New England:

*Marmota m. canadensis* (Erxleben). [*Glis.*] *canadensis* Erxleben, 1777. *Systema regni animalis. . .*, 1:363.

Type Locality: Based primarily on the Quebec marmot of Pennant, from "Canada et ad fretum Hudsonis," but type locality fixed as Quebec, Quebec, Canada, by A. H. Howell, 1915. *N. Amer. Fauna* 37:31.

*Marmota m. canadensis* Trouessart, 1904. *Catalogus mammalium . . .* suppl., p. 344: *Marmota m. preblorum* A. H. Howell, 1914. *Proc. Biol. Soc. Washington* 27:14.

Type Locality: Wilmington, Middlesex County, Massachusetts

*Marmota m. rufescens* H. H. Howell, 1914. *Proc. Biol. Soc. Washington* 27:13.

Type Locality: Elk River, Sherbune County, Minnesota (Hall and Kelson 1959)

Dental Formula:

I 1/1, C 0/0, P 2/1, M 3/3 × 2 = 22

Names: Groundhog, chuck, and whistle pig. The meaning of *Marmota* = a marmot or mountain mouse, *monax* = a digger.

right angles to the long axis of the skull or slightly forward of the right angle. The palate is abruptly truncated at the posterior border, the large rostrum gradually depresses anteriorly, and the broad braincase is somewhat flattened. The posterior extent of the nasals is noticeably wider than the premaxillae, and the incisive foramen is narrowest anteriorly. The cheek teeth are high crowned and the fourth premolar is as large as or larger than the first molar. Malocculusions or malformed incisors, resulting from deflected upper or lower incisors, occasionally occur.

The sexes are colored alike and show slight seasonal variation. There is much color variation among individuals, but the general coloration above is grizzled brownish, with a pinkish, cinnamon cast. The tops of the head, face, legs, and tail are dark brown, and the sides of the face, nose, chin, and lips are buffy white. The underparts are light buff. Immatures are paler than adults. Pure white or nearly all black individuals occasionally occur.

The annual molt of adults and subadults begins in May or June and ends in August or September, and in immatures the molt runs from early July through early October. Molting is usually completed in three and a half weeks. It begins with the tail, then appears about the facial region, followed by the hindquarters, from which it progresses both anteriorly and posteriorly and laterally.

Males are slightly larger than females. Measurements range from 450 to 630 mm (17.6–24.6 in); tail, 100 to 157 mm (3.9–6.1 in); hind foot, 69 to 88 mm (2.7–3.4 in). Weights vary considerably with the individual, the amount of fat, and the season. When just out of hibernation in the spring the weight is about 2.8 kg (6.2 lb), but it rises to nearly 4.5 kg (10 lb) just before hibernation in the fall. Unusually large woodchucks may weigh up to 6.4 to 6.8 kg (14–15 lb).

**Distribution.** The woodchuck occurs from Labrador and Nova Scotia south to the northern parts of Georgia, Alabama, and northwestern Louisiana, west to eastern Kansas, and northwestward into Alaska.

**Ecology.** Woodchucks are seldom found in dense woods. They prefer the edges of brushy

woodlands and open high, rolling fields, such as cow pastures and meadows studded with stumps, and they frequent fencerows and gullies with a good growth of vegetation. They may be encountered along the banks of streams or lakes, near roads and railroads, and occasionally near barns, other buildings, and lumber piles.

Woodchucks are chiefly fossorial rodents and are excellent diggers. They dig both simple and complex burrow systems, the depth and length depending on the type of soil. A burrow system may be at least 6 feet underground in loose, sandy soil or 2 to 4 feet in gravelly soil. An extensive burrow may extend more or less parallel to the surface, varying in length from 10 to 40 or more feet. It frequently has several blind pockets, which often contain fecal remains partially covered by dirt. Most burrows have at least one enlarged hibernation chamber and a nest chamber some 15 inches in diameter and about half as high, lined with dead leaves and dried grass.

Woodchucks may have two dens, one for winter and another for summer. The winter den is situated in a wooded area, often on sloping, well-drained ground, and serves for hibernation, but it may be used throughout the year. It usually has only one entrance, rarely more than two. The summer den is found in open flat or gently rolling areas and is occupied from late spring to early fall. It is usually characterized by a main entrance with a mound of fresh earth and a small plunge, or escape, hole. Often the main entrance is beneath a tree stump, stone wall, or among rocks, though it may be in open areas. It may be recognized by the mound of fresh earth that is regularly brought up from within the burrow. The plunge hole is well concealed in vegetation; it is dug from within the main tunnel system, leaving little, if any, soil on the surface.

Man, dogs, foxes, black bears, large hawks and owls, bobcats, mink, weasels, and occasionally large rattlesnakes kill woodchucks. Fleas, mites, ticks, and nematodes parasitize them, and they are susceptible to tularemia, rabies, hepatoma, arteriosclerosis, and nose and ear infestations by *Cuterebra* larvae. Fibroma may occur in woodchucks.

Flooding and heavy rains sometimes drown young and hibernating woodchucks in the burrows. Incisor malocclusions often cause the animal to starve, since the deflected incisors cannot cut plants adequately, and a continuously growing incisor not worn away by the opposing tooth may penetrate the skull and cause death.

**Behavior.** Woodchucks are usually solitary, but they are somewhat social, and at times two or three animals may be seen feeding or sunning themselves together. Bailey (1965) demonstrated that woodchucks are relatively sociable after arousal from hibernation and relatively unsociable just before hibernation. Bronson (1964) described postreproductive agonistic behavior in woodchucks as nonterritorial but noted that dominant animals are avoided by subordinate individuals. Merriam (1966) found that the frequency of interburrow movements by woodchucks was low after they emerged in the morning, declined less than total activity at midday, and continued later in the evening than all activity. Merriam (1971) also noted that movements within a cluster of burrows were freer and more rapid than movements between clusters.

The home-range size seems to be directly dependent upon the location of the den in relation to the food supply. The home range size is often restricted to within 20 yards of the den in clover and alfalfa fields. There seem to be three kinds of seasonal movements in woodchucks: in spring, shortly after emergence from hibernation, when the animals move into the fields during the mating season; in summer, when the young disperse from the home dens; and in fall, when the animals move from the summer dens to winter dens. Males tend to travel farther than females, particularly in search of mates (Grizzell 1955).

Woodchucks are not very intelligent: they do not learn to avoid traps even when repeatedly trapped. But they have great strength and courage and are fierce fighters if cornered. They are curious and often peek from their holes, cautiously raising their heads until they can look around. Then they stand on the top of the burrow mound and, if satisfied that it is safe, drop on all fours and forage. They normally walk slowly to examine a tender shoot, or sit upright to scan the field for foes. When alarmed they can lope or gallop as fast as 10 miles per hour for short distances.

WOODCHUCK

Woodchucks emit a shrill whistle or alarm call, preceded by a low, abrupt "phew" and often followed by a low, rapid warble that sounds like "tchuck, tchuck" and fades slowly away. The voice is usually heard at the entrance of the burrow, and sometimes several woodchucks may whistle in succession. Another call, made by the grinding of the teeth, denotes anger or fear. Contented captive woodchucks usually grunt or make a soft purring noise.

Although woodchucks are primarily terrestrial, they easily climb trees up to 15 feet or more to escape an enemy and infrequently climb to obtain fruits and succulent leaves or to enjoy the sun. It is not uncommon to see a woodchuck stretched out on a log, a stump, the limb of a slanting tree, or a stone wall, basking in the sun. Woodchucks take to the water and swim well, with the top of the head and the tail above water.

Woodchucks have clean habits. They deposit their scats in a dry pocket off from the main tunnel shaft of the burrow or, more often, bury them in the mound by the entrance. Female woodchucks cover their young's waste with new bedding placed directly over the old. When the nest becomes too bulky or unsanitary, the animals remove the matted material and add fresh bedding.

Woodchucks spend almost all their time resting or sleeping in the den. They are primarily diurnal but may be active at night in early March. With the advent of warmer spring weather, woodchucks are more likely to be seen sunning themselves during the warmest part of the day. As spring progresses, the level of daily activity gradually changes from the warmest part of the day to early morning hours and to evening in summer. During late summer and early autumn the animals tend to be most active in late afternoon or early evening. On cool, rainy, and windy days they stay in their burrows.

Woodchucks do not cache food. As summer progresses they voraciously eat succulent plants and accumulate fat, an important source of energy during hibernation and for a time after emergence when food is scarce. Woodchucks gain and lose weight periodically, gaining steadily for about six months, then losing for the next six. Snyder, Davis, and Christian (1961) found that the rate of weight loss was much higher immediately after hibernation than during hibernation; that woodchucks attain progressively higher weights each year at least for the first three years of life; and that males vary considerably more in weight than females. Bailey (1965) stated that the degree of social interaction and visual contact or isolation in woodchucks influences their annual bodyweight cycle. Davis (1967a) has shown that woodchucks born in captivity and kept under a constant 16 hours of light and 8 hours of darkness at 20°C lost weight in autumn and gained in spring. Davis and Finnie (1975) demonstrated that captive woodchucks shipped to Australia from Pennsylvania showed a shift of endogenous annual rhythm, and that hyperphagia responded in two years to some seasonal change in the southern hemisphere, presumably length of day. Davis (1967b) found that slim wood-

chucks become torpid as readily as fat ones, and that woodchucks are torpid for only 2 months.

Woodchucks do not awake easily from their deep sleep, and the time required is usually dependent on ambient temperatures. Some woodchucks may emerge from hibernation during spells of warm weather in midwinter, but with a drop in temperature they resume their hibernation. The time of entering and emerging from hibernation depends on latitude, and is earlier in the south than in the north. Males usually emerge from hibernation earlier than females and subadult males.

Woodchucks feed chiefly on alfalfa, clover, and a variety of grasses, on leaves and buds of agrimony, dandelion, common chickweed, wild onion, plantain, goldenrod, aster, daisy, fleabane, wild mustard, bitter buttercup, wild lettuce, buckwheat, sheep sorrel, blackberries, cherries, and raspberries, and staghorn sumac, on bark of hickory and maples, and on other wild plants. They are fond of beans, peas, corn, grains, and apples. Occasionally they eat grasshoppers, June bugs, and other insects.

Food consumption of captive adults varies from 3.7 ounces per day in March to a maximum of 8.9 ounces in May and 5.1 ounces through the summer to October (Fall 1971).

The breeding season extends from early March to middle or late April, and the young are born from early April to mid-May, after a gestation period of about 32 days. Copulation often occurs near the burrow entrance but may take place in the grass. Woodchucks produce one litter annually. The litter may contain from two to eight or nine young, the usual number being four or five.

The newborn are blind and helpless, naked, and colored dark pink. The forelegs are more developed than the hind legs. The total length is approximately 105 mm (4.1 in); tail, 16 mm (0.62 in); and hind foot, 13 mm (0.51 in); with an average weight of 26.5 g (0.93 oz). When they are 2 weeks old, soft, grizzled gray fur appears on the body and head. The eyes open at 4 weeks and the young begin to appear at the entrance of the burrow for brief periods. By this time the female brings them green food (Hamilton 1934). Lactation lasts approximately 44 days. At 5 weeks of age the young emerge from the burrow to romp and feed, and they begin to disperse by early July. Longevity is probably 3 to 4 years in the wild and 4 to 5 years in captivity. Woodchucks are able to breed at 1 year old but do not commonly breed until they are 2 years old (Snyder and Christian 1960).

**Age Determination.** Immatures (young of the year), subadults (yearlings), and adults (2 years and older) can be recognized by using a combination of external and internal characters such as teeth, pelage, and condition of testes until about 15 April, or later in some cases (Davis 1964). Immatures gain 19 g (0.7 oz) per day from June through September. Young woodchucks begin their molt early in July, in contrast to late May in subadults and adults, so that their fine, short pelage is striking enough to be distinguished from that of older animals even in September (Snyder, Davis, and Christian 1961).

**Specimens Examined.** MAINE, total 10: Androscoggin County, Auburn 1; Franklin County, Oquossoc 1 (USNM); Knox County, West Rockport 1; Penobscot County, Milford 1, Orono 2 (UME); Washington County (no location) 1 (UCT), Columbia Falls 1 (USNM); York County, Cornish 1 (UME), Eliot 1 (USNM).

NEW HAMPSHIRE, total 18: Carroll County, Ossipee 2 (USNM); Coos County, Jefferson 1 (UCT), Mt. Washington 2 (AMNH); Grafton County, Canaan 3, Hanover 3, Lebanon 1, Lyme 1 (DC); Hillsborough County, Mt. Monadnock 1 (USNM); Strafford County, Durham 2 (UNH); Sullivan County, Charlestown 2 (USNM).

VERMONT, total 29: Addison County, New Haven 1 (UVT); Bennington County, Bennington 1 (DC); Chittenden County, Hinesburg 1, Shelburne 1 (UVT), Williston 1 (URI); Lamoille County, Mt. Mansfield 1 (USNM); Orange County, Strafford 1 (DC), Thetford Center 1 (AMNH), Union Village 1 (DC); Rutland County, Clarendon 1 (UCT), Hydeville 1 (AMNH), Rutland 2–1–1 (AMNH–DC–UCT), Saxton's River 1 (UCT), Sherburne 1; Windham County, Putney 3; Windsor County, Hartland 1 (AMNH), Norwich 5, Pomfret 2, White River Junction 1 (DC).

MASSACHUSETTS, total 48: Berkshire County, Mt. Washington 1 (AMNH); Essex County, Burlington 10, Haverhill 1 (USNM); Hampden County, Ludlow 2; Hampshire County, Amherst 10 (UMA), Easthampton 2 (USNM), Goshen 1 (UMA); Middlesex County, Lincoln 1, Sherborn 1 (USNM), Sudbury 1 (UCT), Wilmington 8, Woburn 1; Norfolk County, Stoughton 6 (USNM); Worcester County, Clinton 3 (UMA), Petersham 1 (UCT).

CONNECTICUT, total 65: Fairfield County, Fairfield 2 (UCT), Stamford 1 (AMNH), Weston 1 (UCT), Wilton 1 (AMNH); Hartford County, Berlin 1, Bloomfield 1, East Windsor 1, Enfield 1, Hartford 1, Manchester 1, Simsbury 1; Litchfield County, Cornwall 1, Kent 1, Litchfield 1 (UCT), Macedonia Park 1 (AMNH), New Milford 2 (UCT), Salisbury 1–1 (AMNH–UCT), Sharon 1; Middlesex County, Westbrook 1; New Haven County, East Wallingford 3 (AMNH), Meriden 1, New Haven 1, Orange 1 (UCT); New London County, New London 1 (AMNH), Stonington 1–2 (UCT–USNM); Tolland County, Bolton 1, Coventry 2, Mansfield 5, Stafford 1, Storrs 13, Vernon 1, Willington 1; Windham County, Ashford 1, Brooklyn 1, Chaplin 1, Hampton 1 (UCT), Plainfield 1 (USNM), Scotland 1 (UCT), South Woodstock 1 (AMNH), Willimantic 1, Windham 1 (UCT).

RHODE ISLAND, total 5: Providence County, Scituate 1 (AMNH); Washington County, Charlestown 1 (URI), Kingston 1–1 (UME–URI), Sanderstown 1 (USNM).

# Gray Squirrel

*Sciurus carolinensis pennsylvanicus* Ord

**Description.** The gray squirrel is a medium-sized, slender tree squirrel with a broad, flattened, bushy tail about half the total length of the animal. The hind limbs are much larger than the front limbs, an adaptation for leaping. Females have eight mammae—four pectoral, two abdominal, and two inguinal.

The skull is broad and rather long, with a large braincase which is deeply depressed posteriorly. The infraorbital foramen forms a canal; the broad palate ends immediately behind the tooth rows; the jugal has an angular process on its upper surface, and the rostrum and nasals are somewhat short. The incisors are large and long, and their anterior surfaces are orange. The third upper premolar is vestigial, and the upper fourth premolar is well developed and triangular while the upper first, second, and third molars have well-marked cusps and are nearly quadrate.

The sexes are colored alike and show some seasonal variation. In summer the upperparts are usually yellowish brown, slightly grayish on the sides of the neck, shoulders, and thighs. The face is cinnamon buff; the forelegs are gray above, the hind legs reddish; and the tail is brown at base, banded with black and tan, and broadly tipped with white. The chin, throat, and

*Sciurus pennsylvanicus* Ord, 1815. In [Guthrie], *A new geographical, historical, and commercial grammar;* . . . Philadelphia, 2(2):292.

*Sciurus carolinensis pennsylvanicus,* Rhoads, 1894. *A Reprint of the North American Zoology, by George Ord, 1815, Appendix, p. 19.*

Type Locality: Pennsylvania, west of the Allegheny Ridge

Dental Formula:

I 1/1, C 0/0, P 2/1 or 1/1, M 3/3 × 2 = 20 or 22

Names: Timber squirrel, black squirrel, migrating squirrel, cat squirrel, and silvertail. The meaning of *Sciurus* = shadow tail or shade of tail, alluding to the squirrel's habit of sitting in the shade of its tail; *carolinensis* = of Carolina; and *pennsylvanicus* = of Pennsylvania.

underparts are whitish. There is a white buffy ring around the eye and white on the backs of the ears. The winter fur is more silver gray, with heavily frosted white-tipped hairs.

Color phases include blacks, reds, albinos, and blendings of reds and blacks. Black phases tend to be more common in northern than southern latitudes. Normal grays and blacks, or various gradations between grays and blacks, occur in the same litter.

Molting occurs in the spring and autumn. The spring molt starts in March or April and is completed by June, beginning at the top of the head and progressing toward the tail. The autumn molt begins by mid-September and is completed in October, starting at the rump and progressing toward the head. The hairs of the tail are molted in the late stages of this molt.

The sexes are equal in size. Measurements range from 460 to 530 mm (17.9–20.7 in); tail, 165 to 250 mm (6.4–9.8 in); and hind foot, 60 to 75 mm (2.3–2.9 in). Weights vary from 340 to 680 g (11.9–23.8 oz).

**Distribution.** Gray squirrels typically occur in forested regions from southern Quebec, New Brunswick, and Ontario to Florida, west into eastern Texas, and north into southern Manitoba. They have been introduced into Washington, British Columbia, and Vancouver Island.

**Ecology.** Gray squirrels occur in hardwood and mixed coniferous-hardwood forests, particularly where there are nut-producing oaks, beeches, hickories, butternuts, and black walnuts, in brushy undergrowth along slopes and bluffs, and in river bottoms near water. They also occur in small woodlots and farm woodlands, along wooded fencerows, and in city parks and wooded residential areas. They den in tree cavities if available or build leaf nests among the branches. Tree cavities are used as permanent residences, whereas leaf nests are used for temporary shelter.

Gray squirrels construct their bulky, globular leaf nests on a supporting platform of twigs. The outside is firmly woven with small twigs and leaves, and the inner chamber is lined with shredded bark, moss, grass, ferns, sedges, paper, cloth, and other soft material. The nest usually has a single entrance facing the main tree trunk

or the nearest limb. The outside of the leaf nest is from 12 to 14 inches in diameter, and the chamber is from 4 to 6 inches; the nest may weigh between 6 and 7 pounds. The nest may be situated 70 or more feet above ground.

Man, foxes, raccoons, bobcats, house cats, hawks and owls, and tree-climbing pilot black snakes prey on gray squirrels. Black bass, pickerel, and pike sometimes catch swimming squirrels. Fires, automobiles, and electrical wires kill many.

Fleas, lice, mites, cestodes, nematodes, and trematodes, mange, scabies, rabies, tetanus, and multiple skin fibromas plague gray squirrels. Occasionally they die from shock when they become comatose after being caught and confined for long periods in live traps. Pack, Mosby, and Siegal (1967) observed that shock mortality was somewhat higher among low-ranked squirrels than among dominant individuals.

**Behavior.** Gray squirrels are alert, nervous, wily, and wary in the wild but are rather playful and friendly in parks. Their senses of smell, sight, touch, and hearing are well developed. Although they are diurnal, they are especially active at dawn and in late afternoon and spend most of the day in a nest or resting on a limb. They do not hibernate, but they will remain in a nest during extreme cold and stormy winter weather and may not be seen for several days.

These tree squirrels are mainly arboreal, but they venture to the ground to feed, frequently stopping to sniff the air or investigate an object. When startled they can run as fast as 10 miles per hour for short distances and jump at least 10 feet (Jackson 1961). Gray squirrels can descend from a tree almost as fast as they can ascend.

They are strong swimmers and can swim for at least 2 miles in calm waters, with the head and rump out of water and the tail floating high behind in the air. In rough water they swim slowly with the tail dropped into the water (Jackson 1961).

Gray squirrels have several calls: apprehension, fright, after-feeding, fussing, a variety of clucking calls, and a death squeal (Sharp 1960).

Gray squirrels are somewhat gregarious and appear to live together amicably with little strife, though they may be antagonistic to each other when the food supply is concentrated.

These animals have social hierarchies and probably exhibit territorial behavior (Flyger 1955). A female will vigorously defend against other females the den tree where her young are nesting (Robinson and Cowan 1953). Pack, Mosby, and Siegal (1967) found that the social rank of gray squirrels increased with age, that adult males were dominant over females and younger males, that the size of the minimum home range increased slightly with social rank, and that most social interactions were settled by bluffing, though the few actual combats observed were intense.

Gray squirrels appear to be a sedentary species and may live their entire lives within 1.0 square acre (Seton 1909). Often they customarily use what appear to be identical routes through trees. Flyger (1955) found that gray squirrels released in a strange environment did not know where to run, whereas resident squirrels disappeared in short order when released. Flyger (1960) found no significant difference between the sizes of the home areas of adult males, adult females, and immatures. Pack, Mosby, and Siegal (1967) estimated the average minimum home range to be 1.24 square acres. Doebel and McGinnes (1974) determined that the minimum home range was 1.20 square acres and that males had a larger home range than females. Hungerford and Wilder (1941) reported that gray squirrels returned home from a distance of 2.8 miles during a period of 4 weeks, based on the return of 6 of 15 squirrels they released.

In the autumn gray squirrels readily shift from one local food area to another. These shifts of home ranges do not involve traveling more than a few hundred yards and are not extensive enough to be characteristic of any considerable migration (Sharp 1960). However, gray squirrels sporadically emigrate from their home area in late summer or early autumn either in small groups or in masses, usually moving in the same general direction and covering considerable distances. Just what causes this phenomenon is not known, though overpopulation and failure of the nut crop are strongly suspected as the main reasons. Goodwin (1935) reported an extensive emigration in Connecticut in 1933 when a thousand or more gray squirrels swam across the Connecticut River between Hartford and Essex. Osgood (1938) documented an east-to-west migration in Vermont in 1936 that extended over most of the state. Larson (1962) found a general north-to-south migration in Massachusetts and Connecticut during September 1960.

Birds have been known to use ants for some undetermined type of stimulation: a bird may either hold ants in its bill and rub them into its plumage in sweeping or jabbing thrusts or may

GRAY SQUIRREL

disturb an anthill and then settle down over it, permitting the ants to crawl through its feathers. In these instances the bird often appears to be highly excited. Bagg (1952) reported an instance of similar anting behavior by a gray squirrel in Massachusetts, and Hauser (1964) observed anting in juvenile gray squirrels. The squirrels would dig into an anthill, then settle down passively on top of it. Suddenly they would roll, leap wildly, and turn somersaults over and over, then they would repeat the whole process many times. Nichols (1958) saw a gray squirrel clear a patch of loose earth and then roll on it for a minute or so with seeming enjoyment.

Gray squirrels seem to be able to detect wormy nuts and fruits. Their feeding sites are often marked with gnawed shells, clippings of twigs, and dislodged fruits. Gray squirrels commonly cache acorns and other nuts or foods, burying each item shallowly in random places. Habeck (1960) observed gray squirrels caching butternuts in cavities of pine saplings 6 to 13 feet above ground. When foraging in winter, gray squirrels wander about and sniff until they find a buried nut.

Gray squirrels feed on acorns and other nuts, fruit, the buds of oaks, hickories, beeches, walnuts, butternuts, and mulberries, the inner bark and sap of maples, elms, and basswoods, and on various wild and cultivated cherries, grapes, and apples. They are fond of mushrooms and other fungi and of grains, especially field corn. Sometimes they eat insects and their larvae, insect galls, and young birds and bird's eggs. They gnaw on bones, shed antlers, turtle shells, and salt blocks. Often they will return to the same bone over a period of several weeks and sometimes will hide one at the base of a tree or in the hollow of a limb.

Gray squirrels are promiscuous. They breed during midwinter and sometimes in May and June, depending upon age, the physical condition of the females, and the availability of food. Uhlig (1956) reported three tagged females as having bred before they were 7.5 to 9.0 months old. Smith and Barkalow (1967) recorded females that had bred when only 124 days old and would have given birth at about 168 days of age. Allen (1954) reported that females from spring litters can produce a litter the following spring but seldom follow up with another litter

that summer, while females from summer litters may produce a litter the following summer, and adult females in good condition may have litters in both spring and summer. Uhlig (1955a) reported that females can bear young during the same mating season in which they were born. Barkalow and Soots (1975) found that the reproductive life of the female is at least 8 years, and possibly 12½ years in the wild.

The gestation period lasts about 44 days. The number of young per litter varies from one to nine, the usual being two or three. At birth the young are naked and helpless, with eyes and ears closed, but have well-developed claws. They weigh about 15 g (0.53 oz) and grow slowly. At 3 weeks of age they are covered with short hair, their ears begin to open, and the lower incisors begin to erupt through the gums. The eyes open between 4 and 5 weeks of age, and the young begin to venture from the nest when 6 to 7 weeks old. At 8 weeks they are half-grown and fully furred, with bushy tails, and they are weaned at about 8 or 10 weeks old. At this time the young forage with the adults in family groups.

The spring litter usually remains with the female until she has a second litter, and the summer litter often stays with her during the winter. The female cares for and defends her young. If the nest is disturbed she will move them to another nest, carrying them by the legs and tail.

Longevity in the wild is at least 12½ years in females and 9 years in males (Barkalow and Soots 1975), and in captivity gray squirrels live up to 15 years (Uhlig 1955a).

**Age Determination.** Gray squirrels can be separated into three age classes—juveniles, subadults, and adults—on the basis of constant characters of the ventral side of the tail pelage. There are two types of hairs in the tail: uniformly long primary hairs that make up the basic framework and shorter secondary hairs that add to the tail's bushy appearance. The primary hairs grow laterally from the sides of the tailbone and impart the characteristic flattish appearance to the tail, while the secondary hairs develop along the dorsal side but do not extend to the tail border. Juvenile and subadult age classes have two, or sometimes three, dark lines

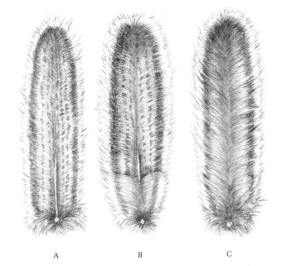

Ventral view of tails of gray squirrels taken in November. *A*, juvenile; the shorter secondary hairs are absent on the lower side of the tail bone. *B*, subadult; short appressed hairs are present on the lower third of the tail bone. *C*, adult; the appressed hairs obscure the outline of the tail bone (after Sharp 1958).

or bands running through the reddish brown primary hairs of the tail. But in juveniles the short appressed hairs on the tail are absent, and in subadults the lower or proximal third of the tail is covered with short, appressed hairs. In adults the tail bone is obscured by appressed secondary hairs that radiate over and partly obscure the long primary hairs of the tail. The prominent lines or bars in the young of the year weaken in color intensity with age and are obscure in the tails of adults (Sharp 1958).

In the fall and winter gray squirrels can be separated into nestlings, juveniles, subadults, and adults, based on the winter pelage characters in the rump region, the characters of the tail pelage, and the molt of juveniles. In the nestling there is no evidence of molting on the head, neck, or back, and the pelage of the body is very fine, soft, dull, and often shaggy in appearance. When the pelage of the rump is separated and laid flat by the thumbs, one sees no yellow prebasal band in the black underfur, and all the banded guard hairs will be black-tipped. In the juvenile there is new hair at least on the nose and crown, with a coarser and more glossy appearance. The body fur often retains the same soft texture as in the nestling. Using the same

thumb technique, one can see that the yellow prebasal band in the black underfur is indistinct and uneven, and the banded guard hairs are black-tipped. In the subadult the yellow prebasal band is diffuse and indistinct, and the tips of the guard hairs are typically white, as in the adult. In the adult the yellow prebasal band is distinct and the tips of all the guard hairs are white (Barrier and Barkalow 1967).

Juveniles can also be separated from adults by X-rays of the epiphyseal cartilage of the radius and ulna between 1 February and 1 December (Petrides 1951; Carson 1961). The cartilage remains obviously open through the 18th week of life and can be detected until the 12th month, but thereafter it is closed and replaced by bone (Carson 1961).

The lens-weight method, when used alone, is a sufficient criterion for separating juveniles from adults and for estimating year-classes of adults (Fisher and Perry 1970).

**Specimens Examined.** MAINE, total 28: Franklin County, Farmington 18 (UME); Lincoln County, Boothbay Harbor 1 (UCT); Oxford County, Norway 2 (HMCZ); Penobscot County, Brewer 1, Holden 1, Orono 4 (UME); Washington County, Calais 1 (HMCZ).

NEW HAMPSHIRE, total 42: Carroll County, Ossipee 1 (USNM); Grafton County, Hanover 23 (DC), Wentworth 1 (AMNH); Hillsborough County, Peterborough 1; Merrimack County, Webster 9 (HMCZ); Rockingham County, Kingston 1 (UCT); Strafford County, Durham 6 (UNH).

VERMONT, total 72: Addison County, Addison 1, Cornwall 2, Hancock 7, New Haven 1; Bennington County, Arlington 2–6 (HMCZ–UVT), Readsboro 1 (AMNH), Shaftsbury 1; Caledonia County, Lyndon 1; Chittenden County, Burlington 8, Charlotte 1, Colchester 4, Essex Junction 2, Huntington 1, Richmond 3, Shelburne 2, Williston 1; Franklin County, Enosburg Falls 1 (HMCZ), Georgia 2, St. Albans 1; Grand Isle County, Grande Isle 1 (UVT), North Hero 1 (UCT); Rutland County, Middletown 3 (AMNH), Rutland 2–1–1 (AMNH–DC–UVT), Shrewsbury 1 (DC); Washington County, Barre 1; Windham County, Putney 1 (UVT), Saxton's River 1 (UCT); Windsor County, Bethel 1 (UCT), Norwich 2 (DC), Pomfret 1–5 (AMNH–DC), Royalton 1, Woodstock 1 (HMCZ).

MASSACHUSETTS, total 101: Barnstable County, Barnstable 1 (HMCZ), Falmouth 1 (USNM); Bristol County, Taunton 1; Dukes County, Martha's Vineyard 2; Essex County, Danvers 1 (HMCZ), Newburyport 3 (USNM), Saugus 1 (HMCZ); Franklin County, Leverett 1, Montague 3, North Sunderland 2, Shutesbury 7 (UMA);

Hampden County, Feltonville 3 (HMCZ), Westfield 5 (4 melanistic) (UMA); Hampshire County, Amherst 42–1–1–1 (UMA–UCT–UNH–USNM), Easthampton 2–1 (UMA–USNM), Hadley 4; Middlesex County, Belmont 1 (HMCZ), Cambridge 3–2 (HMCZ–UMA), Newton 9, Watertown 1 (HMCZ), Wilmington 6 (USNM); Norfolk County, Brookline 13 (HMCZ), Canton 1 (USNM), South Braintree 1 (HMCZ), Stoughton 1 (USNM), Westwood 2 (UCT); Plymouth County, Middleboro 2, South Hanson 2 (USNM), Wareham 4; Suffolk County, Boston 4 (HMCZ); Worcester County, Petersham 1 (UMA), West Brookfield 1, Worcester 1 (UCT).

CONNECTICUT, total 144: Fairfield County, Bridgewater 2 (UCT), Cos Cob 1 (USNM), Fairfield 3 (UCT), Greenwich 2–1 (AMNH–UCT), New Canaan 1, Newton 1 (UCT), Noroton 19 (HMCZ), Norwalk 3, Stamford 1, Trumbull 1, Weston 1, Wilton 1; Hartford County, Bristol 1 (UCT), East Hartford 3 (HMCZ), Elmwood 1, Hartford 1, West Hartford 1, Windsor 1; Litchfield County, Litchfield 3; Middlesex County, East Haddam 1, East Hampton 1 (albino) (UCT), Middletown 1–1 (HMCZ–UCT), Westbrook 3 (AMNH); New Haven County, Guilford 1, Meriden 1, Milford 4 (UCT), New Haven 1–2 (AMNH–USNM), West Haven 3 (UCT); New London County, East Lyme 1, Franklin 2, Lebanon 2 (UCT), Liberty Hill 9 (HMCZ), Mystic 2, Norwich 1, Pachaug 1, Preston 1 (UCT); Tolland County, Andover 1 (USNM), Bolton 1, Columbia 1, Coventry 2, Mansfield 15, Stafford 1, Spring Hill 1, Storrs 21, Union 3, Vernon 1, Willington 3; Windham County, Ashford 3, Sterling 1, Warrensville 4 (UCT).

RHODE ISLAND, total 19: Providence County (no location) 1; Washington County, Chepachet 15 (USNM), Kingston 1–1 (UME–URI), Narragansett 1 (UCT).

# Red Squirrel

*Tamiasciurus hudsonicus*

There are two subspecies of red squirrel recognized in New England:
*Tamiasciurus h. gymnicus* (Bangs). *Sciurus h. gymnicus* Bangs, 1899. *Proc. New England Zool. Club* 1:28; *Tamiasciurus h. gymnicus* F. L. Osgood, 1938. *J. Mamm.* 19(4):438.
Type Locality: Greenville, near Moosehead Lake, Piscataquis County, Maine
*Tamiasciurus h. loquax* (Bangs). *Sciurus h. loquax* Bangs, 1896. *Proc. Biol. Soc. Washington* 10:161; *Tamiasciuris h. loquax* A. H. Howell, 1936. *Occa. Pap. Mus. Zool. Univ. Michigan* 338:1 (Hall and Kelson 1959).
Type Locality: Liberty Hill, New London County, Connecticut
Dental Formula:
I 1/1, C 0/0, P 2/1 or 1/1, M 3/3 × 2 = 20 or 22
Names: Bang's red squirrel, chickaree, chatterbox, boomer, pine squirrel, rusty squirrel, and southern red squirrel. The meaning of *Tamias* = a storer; *sciurus* = shadow of a tail; *hudsonicus* = of Hudson Bay; and *loquax* = talkative.

**Description.** The red squirrel is about half the size of the gray squirrel. The bushy tail is nearly as long as the head and body, but rather narrow for a tree squirrel. There are four toes on each front foot and five on each hind foot. Anal glands are present. Females have eight mammae—four pectoral, two abdominal, and two inguinal.

The skull is moderately broad, and its dorsal surface is comparatively flat. The rostrum is rather short, and the braincase is shallow, with the postorbital processes rather short and the zygomatic arches slightly arched and nearly parallel to the axis of the skull. The anterior upper premolar is rudimentary or absent and covered by the crown of the last premolar. Males have an os baculum.

The sexes are colored alike and show marked seasonal variation. In summer the coloration above is rusty reddish gray vermiculated with black; the sides are olive gray and the legs, feet, tail, and ears are reddish ochraceous. The lips, chin, throat, belly, inner parts of the limbs, and a ring around the eye are whitish or faintly yellowish, and there is a narrow or wide black lateral stripe extending along the sides of the body, separating the reddish upper color from the whitish belly. The tail is reddish above and yellowish gray to dull rusty below, and the hairs above are bordered with black and tipped with yellowish.

In winter the coloration above is chestnut blending to olive gray on the sides, which are slightly flecked with black. The ears have prominent tufts of hair. The black lateral stripes on the sides of the body are either absent or only faintly indicated, and the belly and inner parts of the limbs are grayish white. Immatures are darker and paler than adults. White-spotted, all-white, and albino individuals occasionally occur, but black specimens are rare.

Red squirrels molt twice annually, in spring and in winter. The spring molt begins in late March or early April and starts either around the nose and under the chin, on the front feet, or in both areas at the same time. It progresses slowly to the back and rump and is completed by August. The summer pelage is brightest during September. The winter molt is the reverse of the spring molt, appearing first on the rump just above the tail and back. It begins in October and is completed by late December. The tail hairs are apparently replaced once annually.

The sexes are equal in size. Measurements range from 280 to 350 mm (10.9–13.7 in); tail, 114 to 152 mm (4.4–5.9 in), and hind foot, 46 to 52 mm (1.8–2.0 in). Weights vary from 165 to 240 g (5.8–8.4 oz).

**Distribution.** Red squirrels are transcontinental from Alaska to northern Quebec, occurring south to South Carolina and Tennessee in the

Appalachian Mountains and in the Rockies into New Mexico, Arizona, and across California into British Columbia.

**Ecology.** This species occurs mainly in coniferous forests of deep spruce, hemlock, pine, and balsam, to a lesser extent in deciduous woods, and occasionally in barns and deserted buildings in wooded rural areas. The animal frequently builds its nest among branches against a tree trunk and in cavities of trees, in deserted holes of the yellow-shafted flicker, in abandoned crow's or hawk's nests, and occasionally in crevices in stone walls or in burrows. In trees, the bulky outside nest generally straddles the base of one or several branches and is supported by a platform of twigs. The nest exterior is made of cones, leaves, and twigs, and the inner lining is of shredded bark, moss, dried leaves, and grass to provide warmth. The nest generally contains no food waste and occasionally is free from scats, but it harbors many parasites. The outside is 10 to 20 inches in diameter and the inner chamber 3 to 6 inches. The nest usually has a single entrance on one side near the trunk of the tree. It is often about 5 feet from the top of the tree and as high as 65 feet from the ground.

Man, martens, fishers, mink, bobcats, house cats, foxes, weasels, large hawks, owls, and tree-climbing snakes prey on red squirrels, and snapping turtles, pickerel and pike sometimes pull them down while they are swimming. Fires and automobiles kill some squirrels, and fleas, mites, ticks, cestodes, and nematodes parasitize them. Diseases appear to be unknown for this species, though it may be susceptible to rabies and tularemia.

**Behavior.** Red squirrels are active primarily between dawn and dusk, but may also be heard at night. They are active throughout the year except during stormy weather. In the subarctic, red squirrels become subterranean and subnivian in winter, digging extensive tunnel systems under the snow and rarely appearing on the surface when the ambient temperature is below −25°F (Pruitt 1958).

Although primarily arboreal, red squirrels spend much time on the ground and occasionally burrow in loose soil, making short tunnel systems that usually contain a nest and several storerooms situated under rotten logs, stumps, or boulders.

Red squirrels are aggressive and unsociable, defending their territories and food caches ferociously against other squirrels and against birds. The home range is about 1 square acre (Hamilton 1939). Red squirrels appear to have two distinct types of territories: defended winter food caches which are abandoned during the summer, and "prime" territories, in which a specific area is defended year round (Kemp and Keith 1970). The much publicized aggressiveness of red squirrels toward gray squirrels may be a function of territorial behavior rather than an inherent characteristic of the species (Ackerman and Weigl 1970).

Red squirrels are vociferous and frequently become so enraged at an intruder that they stamp their feet, jerk their tails frantically, and angrily chatter, uttering a long, vibrant staccato "chir-r-r-r, chir-r-r-r" in rapid sequence. More often, they chatter cheerfully, and they may remain silent for hours. They may stretch out along a limb or just sit in a fork with the head resting on a limb. At times two or three red squirrels are seen chasing each other in play, but these may be young from the same litter. Red squirrels are curious and prone to investigate a strange animal or object.

Red squirrels are agile among trees and have been known to leap 5 feet with a rise of 3 feet, 8

RED SQUIRREL

feet with a drop of 2 feet (Klugh 1927), and 6 feet with a drop of 3 feet, from a moving branch (Hatt 1929). Sometimes red squirrels miss their intended landing place. Shufeldt (1920) described how a red squirrel fell 140 feet to the ground, spreading its legs far apart to distribute the shock of impact. They can run at a maximum speed of about 14 miles per hour (Jackson 1961).

Red squirrels take to water readily and are strong swimmers. They have been seen swimming at least one-half mile from a lakeshore (by Cole 1922), in the Penobscot River in Maine (by Pope 1924) and in the middle of Lake Champlain (by Merriam 1884). They can dive a foot or two below the surface of the water (DeKay 1842). They drink water frequently and even eat snow.

Unlike gray squirrels, red squirrels do not spread their food about at random but bury it in large underground caches. Sometimes a cache may contain a bushel or more of food. Red squirrels habitually feed at a particular spot, accumulating large midden heaps.

Red squirrels eat green seeds of white pine, red pine, pitch pine, jack pine, scotch pine, American larch, European larch, red spruce, black spruce, white spruce, Norway spruce, balsam fir, eastern hemlock, and northern white cedar. In hardwood forest, red squirrels feed on almost any kind of nuts, and on seeds, buds, tender leaves, inner bark, and flowering parts of the willow, butternut, black walnut, yellow birch, European white birch, American beech, white oak, red oak, bur oak, American elm, pear, apple, black cherry, pin cherry, plum, sugar maple, red maple, silver maple, basswood, white ash, and American hazelnut. At times they eat the berries of the European barberry, pasture gooseberries, red currants, red chokecherries, wintergreen berries, blueberries, partridge berries, and common elderberries. Abbott and Belig (1961) noted that in western Massachusetts red squirrels feed heavily on seeds of the common juniper from January through March.

Red squirrels relish the sap of maples during spring and often cut out or strip away the tender bark of limbs until the sap flows freely, sometimes hanging on the undersides of limbs to lick the sap. They also feed on grasses and on mushrooms and other fungi, and commonly eat the poisonous *Amanita muscaria* mushroom (fly agaric) without ill effect. Red squirrels also eat insects, snails, and occasionally birds and their eggs, and they may prey on young gray squirrels and on cottontail rabbits.

The breeding season is from mid-January to late September, peaking in mid-February and March and again in June and July, with most of the litters born in early spring and summer. A second litter may be born in August or September. Females born in spring mate the following spring, and females born in late summer and fall probably mate the following summer.

Red squirrels are promiscuous. The male seizes and mounts the female without preliminary courtship, and shortly after coitus the pair become antagonistic to one another. The gestation period is 36 to 40 days. From one to seven young are born in a litter, but four or five is the usual number.

At birth the young are naked and pink, blind and helpless, and weigh about 7.5 g (0.26 oz); they develop slowly. At 9 or 10 days old they have fine fur; the eyes open in about 27 days; and they become fully furred at approximately 1 month old. At this time they begin to venture outside the nest for short periods, and they are weaned shortly thereafter. The young generally remain with the female as a family group until they disperse in late summer or early fall. Longevity may be 2 years or more.

**Specimens Examined.** MAINE, total 137: Aroostook County, Madawaska 12 (HMCZ); Franklin County, Farmington 29 (UME); Hancock County, Acadia National Park 1, Brooklin 1 (USNM), Mt. Desert Island 9–1 (HMCZ–USNM), Prospect Harbor 1 (UVT); Kennebec County, Manchester 1–2 (HMCZ–UME), Oakland 1 (USNM), Winthrop 1 (UME); Lincoln County, Boothbay Harbor 1 (UCT), Bristol 1 (USNM); Oxford County, Norway 3, Upton 21 (HMCZ); Penobscot County, Brewer 1 (UME), Burlington 2 (UCT), Milford 1, Orono 16, Patten 2 (UME); Piscataquis County, Greenville 6 (HMCZ), Kokadjo 3 (USNM), Milo 2 (HMCZ), Moosehead Lake 1 (HMCZ), Mt. Katahdin 4 (USNM); Sagadahoc County, Small Point 3 (USNM); Somerset County, Brassua Lake 2 (HMCZ); Washington County, Columbia Falls 3 (USNM), East Machias 2, West Pembroke 2 (HMCZ); York County, Eliot 2 (USNM).

NEW HAMPSHIRE, total 93: Belknap County, Gilford 1 (UCT); Carroll County, Bartlett 2 (HMCZ), Intervale 1, Ossipee 15 (USNM); Cheshire County, Keene 1 (UCT); Coos County, College Grant (Errol) 4, Mt. Washington 6 (DC); Grafton County, Franconia 2 (HMCZ), Hanover

21, Lyme 4 (DC), Pierce Bridge 4 (USNM); Hillsborough County, Amherst 6, Antrim 1 (HMCZ), Hancock 1, Hudson 1; Merrimack County, Pittsfield 1 (UCT), Webster 14 (HMCZ); Rockingham County, Kingston 1 (UCT), Londonderry 1 (HMCZ), Portsmouth 2 (USNM); Sullivan County, Charleston 3 (USNM), Goshen 1 (UCT).

VERMONT, total 186: Addison County, Addison 3, Cornwall 2, Lincoln 1, Middlebury 1 (UVT), Salisbury 1 (DC), Whiting 1; Bennington County, Arlington 1 (UVT), Readsboro 8 (AMNH); Chittenden County, Burlington 1–2 (UCT–UVT), Huntington 1, Jericho 11, Richmond 4, Shelburne 8 (DC), Underhill 4 (UCT), Westford 2, Williston 4 (UVT); Essex County, Island Pond 1 (DC); Franklin County, Enosburg Falls 1 (HMCZ), Swanton 4 (UCT); Lamoille County, Cambridge 1 (UVT), Mt. Mansfield 4 (USNM); Orange County, Bradford 3, Braintree 2 (UVT), Randolph 2–1 (UCT–UVT), Thetford Center 1 (DC); Orleans County, Barton 1 (UVT); Rutland County, Clarendon 1 (UCT), East Wallingford 8 (DC), Killington Park 2 (HMCZ), Mendon 16–1 (AMNH–UCT), North Clarendon 2 (AMNH), Rutland 2–3–11 (AMNH–DC–UCT), Sherburne 2 (HMCZ), Shrewsbury 1 (DC), Sudbury 1, Wallingford 1 (UCT); Washington County, Montpelier 6 (UVT), Plainfield 4 (UCT), Waterbury 2 (UVT); Windham County, Newfane 5, Putney 1 (AMNH), Rockingham 1 (UCT), Saxton's River 4–1–8 (AMNH–DC–UCT), Whitingham 1 (AMNH), Wilmington 1 (UVT); Windsor County, Barnard 2 (HMCZ), Cavendish 1, Ludlow 1, Norwich 6 (UVT), Pomfret 4–8 (AMNH–UVT), Quechee 1 (DC), South Woodstock 1 (UVT), Woodstock 2 (HMCZ).

MASSACHUSETTS, total 182: Barnstable County, Barnstable 1 (HMCZ), Falmouth 4, Provincetown 1 (USNM); Bristol County, Easton 1, Taunton 2; Berkshire County, Windsor 2 (UCT); Essex County, Lynn 1, Manchester 1 (HMCZ), Newburyport 5 (USNM), North An-dover 1; Franklin County, Charlemont 1, Greenfield 1, Shutesbury 1, Sunderland 4; Hampden County, Chester 1 (UCT), Chicopee Falls 1 (UMA), Feltonville 3 (HMCZ), Southampton 1 (UMA); Hampshire County, Amherst 1–7–1 (AMNH–UMA–USNM), Easthampton 1–3 (UMA–USNM), Leverett 1, Pelham 1, Ware 1 (UMA); Middlesex County, Arlington 1 (HMCZ), Cambridge 19–1 (HMCZ–USNM), East Lexington 1 (HMCZ), Framingham 1 (UMA), Littleton 1, Malden 6, Newton 26, Newtonville 1, Sherborn 2, Tyngsboro 2; Norfolk County, Bellingham 3, Brookline 1 (HMCZ), Canton 1, Milton 1 (USNM), Ponkapoag 1 (HMCZ), Stoughton 3, Wellesley 1; Plymouth County, Plymouth 1, South Hanson 1 (USNM), Wareham 22; Suffolk County, Boston 1 (HMCZ); Worcester County, Athol 1, Clinton 2 (UMA), Harvard 29, Mt. Wachusett 1 (HMCZ), Northboro 2 (UMA), Petersham 2 (HMCZ).

CONNECTICUT, total 112: Fairfield County, Cos Cob 1 (AMNH), Noroton 14 (HMCZ), Norwalk 1 (USNM), Wilton 1 (UCT); Hartford County, East Hartford 10–8 (HMCZ–USNM); Litchfield County, Barkhamsted 1 (UCT), Bear Mt. 2 (USNM), Litchfield 1, Macedonia Park 2 (UCT), Pleasant Valley (AMNH), Warren 1 (UCT), Winsted 1 (AMNH); Middlesex County, Portland 2; New Haven County, Guilford 1 (UCT), Milford 1 (HMCZ), New Haven 1 (AMNH); New London County, Colchester 1, Lebanon 2 (UCT), Liberty Hill 16 (HMCZ), Stonington 1 (USNM); Tolland County, Coventry 2, Gurleyville 1, Mansfield 15, Storrs 1, Union 2, Vernon 4, Willington 3; Windham County, Ashford 4, Canterbury 1, Chaplin 1, Hampton 1 (UCT), Plainfield 3 (USNM), Pomfret 2, Scotland 1 (UCT), South Woodstock 1 (AMNH), Warrensville 2 (UCT).

RHODE ISLAND, total 6: Providence County, Johnston 3 (UCT); Washington County, Charlestown 1 (URI), Kingston 2 (UCT).

# Southern Flying Squirrel

*Glaucomys volans* (Linnaeus)

[*Mus*] *volans* Linnaeus, 1758.
  *Systema naturae*, 10th ed., 1:63.
*Glaucomys volans*, A. H. Howell,
  1915. *Proc. Biol. Soc. Washington*
  28:109.
Type Locality: Virginia
Dental Formula:
  I 1/1, C 0/0, P 2/1, M 3/3 × 2 = 22
Names: Eastern flying squirrel,
  white-furred flying squirrel, glider
  squirrel, and fairy diddle. The
  meaning of *Glaucomys* = gray
  mouse, and *volans* = volant, or
  flying.

**Description.** The southern flying squirrel is recognized by its gliding membrane—a loose fold of skin fully furred on both sides, which extends from the outside of the wrist on the front leg to the ankle of the hind leg—and by a broad, flattened tail, almost parallel-sided, with a rounded tip. The gliding membrane is supported at the wrist by a cartilaginous spur that permits it to extend beyond the outstretched leg. The membrane lies in loose folds when the squirrel is running about or at rest, but when the animal springs into the air with its limbs outstretched the membrane forms a wide, flattened gliding surface that acts as a parachute, while the tail acts as auxiliary parachute, rudder, and stabilizer during the glide. The head is blunt and rounded, with a short, upturned nose. The eyes are large and dark and reflect a lustrous red. The ears are rather large and the whiskers are long. The legs are moderately long, but the feet are small. The forefoot has four toes and the hind foot five. The claws are curved and sharp. Females have eight mammae—two pectoral, two abdominal, and four inguinal.

The skull is highly arched, with a short rostrum. The nasals are abruptly depressed at the tip, the frontals are narrow and long, and the zygomatic arches are nearly vertical. The interorbital and postorbital regions are narrow and the postorbital processes slender. The braincase

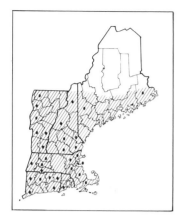

is depressed posteriorly. The incisors are slender and not recurved as in most tree squirrels.

The pelage is soft, dense, and long. The sexes are colored alike and there is slight seasonal color variation. In winter the coloration above is drab to pinkish cinnamon or fawn brown. The sides are darker and almost black along the edge of the membrane. The hairs above are slate gray at the base. The underparts are white or tinged creamy white with the hairs entirely white. The head is grayish or buff, and the ears are light brown. The tail is brownish gray above and light pinkish cinnamon below. The feet are gray above and whitish below, and the toes are white. In summer the pelage is darker and redder. There is apparently one molt annually, from early September to late November. Color phases are rare.

Adults are equal in size. Measurements range from 210 to 255 mm (8.2–10.0 in); tail, 80 to 110 mm (3.1–4.3 in); and hind foot, 25 to 35 mm (1.0–1.4 in). Weights vary from 50 to 70 g (1.8–2.5 oz).

**Distribution.** This species occurs from Nova Scotia, southern Ontario, central Minnesota, Wisconsin, Michigan, and northern New York south to southern Florida, the northern Gulf Coast, eastern Texas, northern Arkansas and Oklahoma, and west to eastern Nebraska and Kansas.

**Ecology.** Southern flying squirrels are found in mature deciduous woods, usually near water (Muul 1968; Madden 1974). They prefer hardwood forests of red maple, oak, aspen, beech, walnut, and birch, where there is usually an abundance of nuts, but they may also be encountered in mixed coniferous-hardwood areas, including old orchards. They occasionally are found in attics and lofts, or under eaves, particularly where oak and hickory trees predominate. They prefer abandoned woodpecker cavities for their nests but may build an outside nest in a hole in snags, stubs, or hollow limbs in dead and decaying trees. They infrequently utilize birdhouses. Goertz, Dawson, and Mowbray (1975) reported that these squirrels burrow under the bark of trees and completely cover themselves.

Man, bobcats, house cats, foxes, weasels, great horned owls, and tree-climbing snakes prey on southern flying squirrels. Forest fires and accidents undoubtedly kill some. They are hosts to lice, mites, chiggers, and nematodes. Sheldon (1971) reported that a group of southern flying squirrels developed partial alopecia, or hair loss, when kept in captivity for an unknown period of time and fed on sunflower seeds and peanuts, but the animals completely regrew their hair within the following 11 months on a diet of mouse chow.

**Behavior.** In spite of its name the flying squirrel cannot fly, because the membrane acts only as a plane, imparting no forward motion. The squirrel travels from tree to tree by volplaning through the air like a soaring glider. It climbs high in a tree then, gauging the glide distance and the intended landing site, lowers its head and gathers its legs for as strong a leap as possible. It launches itself by jumping with legs and membranes outstretched and sails through the air in a descending curve toward another tree or the ground.

Depending on the air current, the squirrel may glide downward 150 feet or more from a height of 60 feet. The angle of the glide is usually between 30 and 50 degrees. The squirrel can turn easily at right angles from the glide line, and in a short glide of 5 to 6 feet it is able to maintain an almost horizontal position (Sollberger 1940). Females may carry their young while gliding (Stack 1925). Sometimes the squirrel accidentally lands in water, and at best it swims poorly. In high grass the squirrel is almost helpless and usually attempts to hide rather than to escape by running.

Southern flying squirrels are active all night and are rarely seen in daylight, though at times they may venture out in search of food after sunset or in late afternoon on cloudy days. They are sociable and active all year, though during cold weather they may become temporarily inactive. At such times several squirrels will cluster together in snug balls in a nest for protection and warmth.

These squirrels are delicate, gentle, and shy. A slight wound or rough handling may kill them. They rarely attempt to bite even if disturbed in the nest or captured. They utter little notes, usually when feeding, that resemble the weak chirp of a small bird, but when irritated they

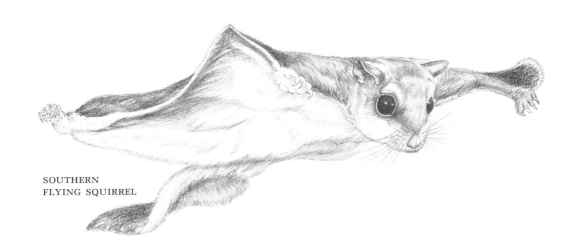

SOUTHERN
FLYING SQUIRREL

intensify the call and may stamp their feet. They are clean animals and seldom deposit feces in the nest.

McCabe (1947) reported that a marked female southern flying squirrel returned in 6 days to her point of release. The female traveled 1 mile in a straight line, volplaning through the trees for a total distance of 1¾ miles, and at one point crossed 250 yards of treeless terrain. Madden (1974) found that in Long Island, New York, the home range averaged 0.99 square acres for females and 1.31 square acres for males.

Territoriality is restricted to breeding females (Sollberger 1943; Muul 1968). Madden (1974) noted that adult females were markedly aggressive and defended an entire home range, whereas males were not territorial, and that females both in the wild and in captivity frequently moved their young.

Southern flying squirrels eat a host of plant material, including hickory nuts, beechnuts, black walnuts, acorns, seeds, berries, tree buds and blossoms, maple sap, sunflower seeds, corn, and mushrooms, as well as peanut butter, sugar, and suet. They prey on moths, beetles, insect larvae, and birds and their eggs. They seem to be the most carnivorous of tree squirrels. Stoddard (1920) reported that a captive flying squirrel killed and ate a yellow-bellied sapsucker that was placed in the same roomy cage with the squirrel. The squirrel already had ample food. Southern flying squirrels are sometimes caught in traps baited with meat.

Nuts opened by southern flying squirrels have a roughly circular or elliptical opening showing fine tooth marks, and the edges have the appearance of being ground off instead of broken into pieces like the shells of nuts eaten by other tree squirrels. Opened nuts beneath an old woodpecker hole usually attest to the presence of one or more flying squirrels. These animals often store surplus nuts in a den but may also bury them in the ground. Muul (1968) found that flying squirrels store food in caches during late summer, autumn, and winter. Flying squirrels drink water frequently except when succulent foods are available.

This species has two mating seasons; one in February through March and the other from June through July. The gestation period is 40 days. Infrequently, young are born as late as early October (Goertz 1965), but in Massachusetts the peaks in the number of litters occur in April and August. The number of young per litter varies from two to six, but three or four is the usual number (Muul 1969b).

At birth the young are naked and dark pink. The eyes and ears are closed, and the folds of the gliding membrane are evident. They weigh approximately 4 g (0.14 oz), and are nearly 62 mm (2.4 in) long. The young double their weight by the end of their first week, and by the end of 6 weeks they are almost adult size. At 3 weeks of age the ears open and the young are almost completely furred. At 28 days of age their eyes open and the body is well furred. They are weaned shortly after 5 weeks of age and at 6 weeks are able to forage for themselves, though they may remain with the female until the next litter arrives. Although the female is solicitious of her young, she may sometimes kill and eat them (Sollberger 1943). Squirrels born during summer are capable of breeding the first winter after birth. Longevity is probably 5 years in the wild and 8 to 10 years in captivity (Jordan 1956).

**Specimens Examined.** MAINE, total 2: Oxford County, Norway 1; Washington County, Eastport 1 (HMCZ).

NEW HAMPSHIRE, total 14: Coos County (no location) 1 (UCT); Grafton County, Hanover 5 (DC); Hillsborough County, Hancock 3; Merrimack County, Webster 2 (HMCZ); Rockingham County, Kingston 1 (UNH), Nottingham 1 (HMCZ); Strafford County, Durham 1 (UNH).

VERMONT, total 25: Addison County, Middlebury 3; Chittenden County, Shelburne 1 (UVT); Franklin County, Fairfax 2; Rutland County (no location) 5 (UCT), Mendon 1 (AMNH), Rutland 4 (DC); Windham County (no location) 2 (UCT), Saxton's River 1–2

(AMNH–DC), Vernon 1 (UCT); Windsor County, Norwich 1 (DC), Pomfret 1 (AMNH), Woodstock 1 (HMCZ).

MASSACHUSETTS, total 27: Barnstable County, Hatchville 1; Berkshire County, Mt. Greylock 1 (HMCZ); Essex County, Ipswich 1 (USNM); Franklin County, North Charlemont 1; Hampden County, Brimfield 2, Wales 1 (UCT); Hampshire County, Amherst 5, Leverett 1, Pelham 2 (UMA); Middlesex County, Concord 1 (HMCZ), Midford 2 (UMA), Newton 1, Waltham 1, Waverley 1 (HMCZ), Wilmington 2 (USNM); Norfolk County, Wellesley 1; Plymouth County, Wareham 1 (HMCZ); Suffolk County, Boston 1 (USNM); Jamaica Plains 1 (HMCZ).

CONNECTICUT, total 65: Fairfield County (no location) 1 (UCT), Noroton 18, Stamford 1 (HMCZ); Hartford County, East Hartford 2–2 (HMCZ–USNM), Granby 1 (UCT); Litchfield County, Kent 1–1 (AMNH–UCT), Salisbury 2 (UCT), Sharon 4 (AMNH); New London County, Liberty Hill 8 (HMCZ); Tolland County, Columbia 1, Coventry 1, Mansfield 5, Tolland 2, Union 7, Willington 1; Windham County, Ashford 6, Damelson 1 (UCT), South Woodstock 1 (AMNH).

RHODE ISLAND, total 2: Washington County, Peace Dale 1, Wakefield 1 (URI).

# Northern Flying Squirrel

*Glaucomys sabrinus macrotis* (Mearns)

*Sciuropterus sabrinus macrotis* Mearns, 1898. *Proc. U.S. Natl. Mus.* 21:353. G[laucomys] s[abrinus] macrotis, A. H. Howell, 1915. *Proc. Biol. Soc. Washington* 28:111.
Type Locality: Hunter Mountain, Catskill Mountains, Greene County, New York
Dental Formula:
I 1/1, C 0/0, P 2/1, M 3/3 × 2 = 22
Names: Mearns's flying squirrel, big flying squirrel, and Canadian flying squirrel. The meaning of *Glaucomys* = gray mouse, *sabrinus* = of the River Severn (the type locality of the species), northwestern Ontario, Canada, and *macrotis* = big ear.

**Description.** The northern flying squirrel resembles the southern flying squirrel but is somewhat larger, darker, and redder above, and the base of the white hairs on the belly is slate colored instead of pure white. Females have eight mammae—two pectoral, two abdominal, and four inguinal. The skull is similar to that of the southern flying squirrel but somewhat larger, and the frontal region is more elevated, while the braincase is slightly more rounded.

The sexes are equal in size. Measurements range from 250 to 295 mm (9.8–11.5 in); tail, 115 to 135 mm (4.5–5.3 in); and hind foot, 35 to 40 mm (1.4–1.6 in). Weights vary from 57 to 125 g (2.0–4.4 oz).

**Distribution.** This species occurs from southeastern Alaska across northern Canada, south through most of New England to western Pennsylvania, south into the Appalachian Mountains to North Carolina and eastern Tennessee; west to eastern North Dakota; in the Rocky Mountains south into Utah; along the Pacific coast, south to northern California; and from western California south into western Nevada and the San Bernardino Mountains of California.

**Ecology.** The northern flying squirrel occurs at altitudes of 1,000 feet or more in cool, dense, wooded areas of mixed mature conifers and deciduous trees, preferring pure stands of hemlock-birch or hemlock-maple forests. It is found less often in open hardwood forests of beech, oaks, maples, and pines.

The nest is usually in a natural cavity or in an abandoned woodpecker hole. Occasionally a squirrel builds a nest in the attic of a deserted building, and they also build outside nests. At times the animals may use abandoned crow's nests or squirrel's nests. The hollow-tree nest is used during winter and the outside nest during summer. The outside nest is built close to the tree trunk, 5 to 30 or more feet above ground, usually in coniferous trees. The outside of the nest is composed of coarse strips of bark and twigs, and the interior is lined with shredded bark, grass, and other plant material. The entrance is near the trunk of the tree, and in some instances there is a second entrance at the top of the nest.

Man, bobcats, house cats, martens, fishers, weasels, great horned owls, and tree-climbing snakes prey on flying squirrels. Fires, lightning,

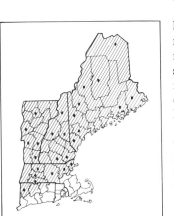

NORTHERN FLYING SQUIRREL

and traps set for other mammals take their toll, and a flying squirrel may die by becoming entangled in a barbed-wire fence. Fleas, lice, chiggers, cestodes, and nematodes are parasites of flying squirrels.

**Behavior.** Little is known about behavior of the northern flying squirrel, but its general behavior may be similar to that of the southern flying squirrel. Northern flying squirrels are chiefly nocturnal and are gentle, sociable, and gregarious in winter. The voice is somewhat softer and lower than that of the southern flying squirrel.

This species is active on the snow at temperatures at least as low as 10°F. On windy winter nights the animal usually stays in its den. This squirrel often burrows vertically downward through the snow at least 10 inches in search for food, and in some instances it makes horizontal snow tunnels somewhat resembling tunnels made by weasels (Connor 1960).

Northern flying squirrels eat acorns and other nuts, seeds, winter buds of hemlock, maples, and beech, catkins of speckled alder, fruits, berries, insects, slugs, small birds and their eggs, small mammals, lichens, and mushrooms and other fungi.

They seem to produce two litters a year, one in late March to early July and the other in late August or early September. In New York, Connor (1960) found females lactating in late May and in mid-September. Muul (1969a) reported that a captive female gave birth to a litter of four young on July 6 and that the gestation period was about 37 days. The number of young per litter varies from two to six, but four or five is usual.

**Specimens Examined.** MAINE, total 44: Aroostook County (no location) 2 (UCT), Madawaska 1, Quimby 2 (HMCZ); Cumberland County (no location) 1, Cape Elizabeth 1 (UCT); Franklin County, Farmington 4 (UME); Hancock County, Bucksport 5, Mt. Desert Island 2; Oxford County, Norway 1, Upton 1; Piscataquis County, Greenville 8–1 (HMCZ–USNM), Milo 2, Moosehead Lake 1 (HMCZ), Mt. Katahdin 1–1 (DC–UME); Somerset County, Enchanted Pond 3 (AMNH); Washington County, Baring 4, East Machias 1, Steuben 1; York County, Wells 1 (HMCZ).

NEW HAMPSHIRE, total 31: Carroll County, Ossipee 4 (USNM), Sandwich 1 (HMCZ), Tamworth 1 (UNH); Coos County, College Grant (Errol) 1–9 (AMNH–UNH), Mt. Washington 5 (DC); Grafton County, Alexandria 1 (HMCZ), Etna 1, Hanover 2 Hillsborough County, Hancock 1, Peterboro 2; Rockingham County, Epping 1, Nottingham 2 (HMCZ).

VERMONT, total 31: Addison County, Hancock 1 (HMCZ); Lamoille County, Cambridge 1 (UVT); Rutland County, East Wallingford 2 (DC), Mendon 7–2 (UCT–UVT), Rutland 1–3 (DC–UCT), Tinmouth 1 (UCT); Windham County (no location) 5 (UCT), Londonderry 1 (DC), Saxton's River 1–1 (DC–UCT), Westminister Station 2 (URI); Windsor County, Norwich 1, Pomfret 1, White River Junction 1 (DC).

MASSACHUSETTS, total 18: Essex County, Andover 1 (HMCZ); Franklin County, Birch Hill 1 (UMA);

Hampden County, Feltonville 1 (HMCZ); Hampshire County, Amherst 1, Leeds 2 (UMA); Middlesex County, Carlisle 1 (UCT), Hudson 1, Weston 1 (HMCZ), Wilmington 3 (USNM); Worcester County, Harvard 2 (HMCZ), Lunenburg 1 (USNM). Royalston 1 (UMA), Winchendon 1 (USNM), Worcester 1 (UMA).

CONNECTICUT, total 2: Litchfield County, Riverton 1; Tolland County, Stafford 1 (UCT).

## FAMILY CASTORIDAE (BEAVERS)

Beavers are easily recognized by their large, horizontally flattened and scaly tails and webbed hind feet. The beaver is the largest rodent in North America and is exceeded in size only by the largest rodent in the world, the capybara of Panama and South America.

The beaver occurs throughout North America excepting parts of northern Alaska and northern Canada, central Nevada, parts of California, western Utah, Florida, and most of Mexico and southward (McLaughlin 1967). Only one species occurs in New England.

# Beaver

*Castor canadensis* Kuhl

*Castor canadensis* Kuhl, 1820. *Beitr. Zool. Vergleich. Anat.* 1:64.
Type Locality: Hudson Bay
Dental Formula:
I 1/1, C 0/0, P 1/1, M 3/3 × 2 = 20
Names: Canadian beaver, American beaver, bank beaver, flat-tail, swamp engineer, and chisel-tooth. The meaning of *Castor* = a beaver, and *canadensis* = of Canada.

**Description.** The beaver is a compact, heavy-bodied animal with powerful muscles and short legs, each foot having five clawed toes. The tail is broad and muscular, greatly flattened dorsoventrally, nearly hairless, and covered with large hard, leathery scales. The head is relatively small, broad, and rounded. The nose, ears, and mouth are valvular, closing automatically when the animal dives and opening again as it surfaces. The loose, hairy lips close behind the chisellike incisors, permitting gnawing underwater, while the cheek teeth behind the closed lips can be used for chewing.

The small, handlike forefeet are unwebbed, with strong claws which serve to carry mud, dig burrows, handle material, hold food, and comb fur. The forefeet are not used in swimming but are held close to the sides of the body so as to offer the least resistance to the water and to ward off underwater obstacles. The large hind feet are fully webbed for swimming and to support the beaver on muddy ground. The beaver swims at slow speeds by paddling alternately with the hind feet, and at faster speeds it uses an undulating motion of the body and the tail. The two inner toes on the hind feet are cleft, or double, specialized for combing the fur and removing ectoparasites, as well as for spreading oil over the fur. The innermost toe has a double-edged claw, or "slit nail," which clamps over a soft horny lobe, and extends from the base of the toe on its ventral side to form a "coarse comb." The second toe has a similar claw, but possessing a sharp, finely serrated edge, which forms a "fine comb." Water-repellent oil is obtained from two

abdominal oil glands situated at the sides of the urogenital opening. The exact function of these glands is unknown; they may not be used totally for hair care. These oil glands should not be confused with the much larger castor, or anal scent, glands (larger in males) which produce castoreum, a pungent yellow-brown oil which the beaver deposits on scent mounds made of mud, ranging in size from a mere "mud pie" to a pile at least 2 feet high.

The tail serves for temperature regulation (Steen and Steen 1965; Coles 1967) and fat storage (Aleksuik 1970). It also is used as a rudder and sculling oar, helps the beaver submerge quickly, and acts as a strong rear balance when the animal walks upright with its front paws full of material and as a support when it cuts a tree. The beaver gives warning by slapping its tail against the water just before it dives, apparently to signal other beavers. It does not use its tail as a trowel or hod for carrying mud, as is sometimes supposed.

The sexes are difficult to distinguish except that pregnant and lactating females can be recognized by the four conspicuous pectoral mammae, but at other times the mammae are inconspicuous and difficult to find. Beavers have a single cloaca-like opening within which are situated the anal and urogenital orifices. The paired castor glands and the smaller oil glands lie under the skin at the base of the tail and also open into the cloacal structure.

Other remarkable anatomical features are the rudimentary, highly variable bicornuate uterus masculinus in males, a nonfunctional organ resembling the female uterus. The baculum is present.

The skull is stout, broad, and flattened dorsoventrally, with a deep rostrum, broad nasals, narrow braincase, strong zygomatic arches, and no postorbital processes. The infraorbital canals are inconspicuous, opening on the side of the rostrum anterior to the zygomatic plates. The basioccipital region has pitlike depressions, and the angle of the mandible is rounded. The cheek teeth are hypsodont, rootless, and high crowned, with hard long-lasting enamel folds. The incisors are strongly developed and are deep orange on their anterior faces.

The pelage consists of extremely dense, short, durable, soft, waterproof underfur overlain by sparse, coarse guard hairs that are long and shiny. The sexes are colored alike and show no seasonal variation. They are rich brownish black or yellowish brown above and paler below, and the ears, feet, and tail are blackish. Immatures are colored like adults. White, black, and silvering beavers have been reported. There is only one molt annually.

The sexes are nearly equal in size. Measurements range from 81.3 to 122.0 cm (32.0–48.1 in); tail, 30.5 to 52.1 cm (12.0–20.5 in); and hind foot, 14.0 to 19.1 cm (5.5–7.5 in). Weights vary from 12.3 to 30.4 kg (27.0–67.0 lb).

**Distribution.** The beaver formerly ranged most of the forested regions of North America, from Alaska across most of Canada, and south to central California, northern Nevada, northern Mexico, the Gulf coastal plain, and extreme northern Florida. It was exterminated in many areas and later reintroduced into some sections of its range.

**Ecology.** The beaver occurs along gently sloping streams and rivers and quiet small lakes and marshes bordered by stands of small timber such as quaking aspens, poplars, birches, maples, and willows. This rodent is one of the most industrious of our native animals, constructing dams, lodges, burrows, and canals. All members of a colony except the young, called kits, help to keep the dams in repair.

The dam is probably the most conspicuous and impressive example of the beaver's work. It provides the depth of water needed to float food and building material to the lodge and furnishes protection. In building a dam the beavers work from the upstream side. They lay sticks, leaves, grass, sod, and mud across a stream until the flow is checked and the water level begins to rise. The beavers then push sticks over the top and allow them to lie crisscross on the lower slope until the dam is high and strong enough to hold the impounded water at the desired level, and as the water rises the ends of the dam are extended. The final shape and position of the dam are often a result of long tests of strength and endurance, experiments, failures, and

BEAVER

changes. Some of the larger dams are the work of many generations of beavers; they continually add something.

The average height of beaver dams in Maine is about 2½ or 3 feet, and the largest reported was 10 feet high (Hodgdon and Hunt 1953). As a rule, longer dams are found in more level terrain and shorter ones in mountainous country. Dams vary from 100 feet long to 200 or more. Beavers usually construct secondary dams on nearby tributaries of a main stream to make feeding and transporting material easier.

The beaver lodge is usually surrounded by water, but it may be built against a bank or over the entrance to a burrow in a bank. It is usually started on some elevation on the bottom of the pond where the water is shallow but may also begin with a burrow covered with mud and sod on which sticks are laid, much as in building a dam. As the water level rises more brush and mud are added, and the interior is kept hollowed out as the work progresses. The lodge is roughly conical. Above water it may be up to 8 feet high and nearly 40 feet wide, but as a rule it is 5 to 6 feet high and 20 to 30 feet across. The single chamber is situated inside and just above the water level and is connected with the outside by one or more tunnels opening underwater. The chamber is usually 6 to 8 feet wide and about 2 feet high, and may be further enlarged from within. It is kept clean of droppings. The floor is semidry and bedded with dry grasses, sedges, moss, or shredded wood. The compact wall of the chamber is usually 2 to 3 feet thick at the bottom of the chamber, and much thinner and looser at the top, where an opening provides ventilation. In occupied lodges the snow melts around this opening, which trappers call the "smoke hole," as body heat from beavers inside the lodge filters up through the hole. Stephenson (1969) found that changes in lodge temperatures were associated with a combination of extreme cold periods and lack of snow for insulation, and that despite cold weather beavers could maintain a relatively stable environmental temperature within their lodge. Beavers may build bank burrows along large, swift streams or where conditions for a dam or a lodge are not favorable. These structures are small and temporary and probably serve as refuges.

Beavers build canals to float logs to a pond and also build dams in these canals to hold the water at the desired level. A canal is from a foot to a yard or more wide and from a few inches to 3 feet or more deep, and its length may vary from a few feet to several hundred feet depending on the terrain. There are usually no canals in rocky or hilly areas. Beavers dig submerged canals in the bottom of a shallow pond, deepening the pond and raising the dam at the same time. These canals often run from the dam to the lodge and an underwater food cache. Canals may be seen exposed in beaver ponds that have been drained.

At the end of a canal there may be one or more beaver trails or paths through the long, rank vegetation. These trails serve as avenues between ponds not connected by water and are used for transporting timber cut at a distance from the end of a canal. Beavers do not seem to make definite attempts to build trails, except in some instances where logs or roots interfere with their travels. Trails are formed as they repeatedly use a route, often dragging branches and twigs. Some trails may become so worn from continuous usage that where the land slopes steeply they may end in mud slides at the water's edge. Tevis (1950) found that trails never paralleled the shore but often led at right angles to it, and that a long trail sometimes ended at an anthill. Seton (1909) speculated that beavers might use ants to pick off parasites.

Scent mounds are situated near the water's edge and vary from two to seven or more per colony. Castoreum and mud are deposited on the scent mounds at frequent intervals, sometimes daily. Beavers appear to maintain a system of territorial rights, and there seems to be no evidence of overlapping colonies. The scent mounds may function as a means of communication among beavers from adjacent colonies or between established colonies and transient beavers by delineating occupied territory and preventing further colonization (Aleksiuk 1968). Castoreum was formerly used by Indians as a medicine, and trappers use it for bait, since the odor is a strong attractant to other mammals.

Man, black bears, coyotes, foxes, dogs, Canada lynxes, bobcats, and probably fishers prey on beavers, and large hawks and owls may take unwary kits on land. Occasionally a beaver is killed or trapped by a falling tree it has cut.

Beavers are parasitized by tiny bloodsucking beetles, screwworms, nematodes, and trematodes. They are sometimes infected with tularemia, rabies, lung fungus, and lumpy jaw.

**Behavior.** Beavers live in family units consisting of an adult male and female and their yearlings and kits. A unit may contain up to twelve members, five or six being the usual number. Within a family beavers are generally tolerant of one another, though they have a social hierarchy where the adult female is dominant, but the family may attack and kill a transient beaver who fails to recognize the family territory.

Beavers are active throughout the year. They are mainly nocturnal but may be seen during the day in summer. In autumn beavers become active earlier in the afternoon, repairing dams and lodges and gathering food for the winter. In winter they obtain their food from their underwater food caches and clean the lodge of debris. Novakowski (1967) found that food caches were not sufficient to supply the energy requirements of beavers during their long icebound period in subarctic Canada and that the animals survived this period by their heavy fur and well-insulated lodge, plus reduced activity, longer periods of dormancy, and huddling. Patric and Webb (1960) noted a decrease in the size of the beaver's tail during the winter, indicating depletion of this fat depot.

During the spring, before the next litter is born, the sexually mature 2-year-olds are either driven from the lodge by the parents or leave of their own accord, traveling several miles in search of a suitable habitat for homesites and mates. This emigration provides for a systematic dispersal of the vigorous 2-year-olds into new areas. Hodgdon and Hunt (1953) reported that 12 Maine beavers moved an average of 7 stream miles, with a maximum of 29 miles.

Beavers emerge from the lodge independently of one another, and the adult female usually emerges before the adult male (Hodgdon and Larson 1973). In contrast to their emerging behavior, beavers often congregate to loaf and play together around the lodge before retiring. They sometimes groom each other, and they seem drawn together by natural affection.

These animals spend more time in feeding than in any other activity, consuming most of their food in the first half of the night. Tevis (1950) and Hodgdon and Larson (1973) observed that when a beaver that was feeding squatted partly out of the water was approached by another, it would usually surrender the food and leave if it had already been eating for a few minutes. Beavers dislike hard rain and may retreat to the lodge to escape it.

Beavers swim gracefully but slowly, about 2 miles per hour. The large lungs and liver let the animal hold enough air and oxygenated blood to stay underwater for up to 15 minutes (Irving and Orr 1935). Under the ice beavers obtain oxygen from air bubbles and air holes. On land they walk slowly, with a heavy, shuffling gait, but they can run as a fast as a person walks.

Only one beaver works at cutting down a tree at any given time, though beavers may sometimes work on the same tree at different times. A beaver can cut an average of one tree in two days. The time required depends on the kind of tree—an aspen or cottonwood takes much less time than an oak or birch. A small tree is generally gnawed through from one side, but a larger one may be gnawed on two sides or all around the base of the tree. Chips are cut out from above and below, and the tree is cut into an hour-glass shape. The animal cannot choose a desired direction in felling a tree, and trees fall randomly according to the wind, the shape and weight distribution of the tree, or the shape and location of the cuts the beaver happens to make. A good number of trees fall against other trees and become lodged so they cannot be used, and sometimes a tree may fall on a beaver.

Once a small tree has been felled, several beavers may feed on the bark. They gnaw off the branches and section off the trunk in lengths of 3 to 6 feet or more which they drag to the feeding grounds or food cache or add to the dam or lodge. Hodgdon and Hunt (1953) reported that beavers usually travel no more than 300 feet from water to cut trees. Hiner (1938) found that tree hauls were up to 450 feet, and Longley and Moyle (1963) reported that one beaver cut a tree 467 feet from the water. On land the beavers drag the wood by grasping the end with their strong incisors, the head turned to one side. In water they often grasp the wood with their front paws while they swim at one side and steer with their broad tails.

During early autumn beavers begin to build their underwater food cache by piling up green trimmed branches and small logs. The cache is usually in some deep part of the pond near the lodge and often reaches to the surface of the water. A food cache may be from 3 to 10 feet high and 20 to 40 feet in diameter. Beavers carry heavier pieces of timber to the bottom and partly bury them in mud, where they soon become waterlogged and remain sunk even though uncovered. As food is needed the animals swim to the cache, gnaw off a piece of wood, and transport it to the lodge, where the bark is eaten. They often use the remains of logs to repair the dam or lodge the following spring.

Beavers generally prefer the bark of deciduous trees, but will eat herbaceous plants and grasses. They are not known to eat flesh. In spring and summer beavers feed on bulrushes, sedges, coontails, pond lily roots, the bark of poplar, alder, willow, paper birch, gray birch, red oak, beaked hazelnut, red maple, white cedar, hemlock, black spruce, red spruce, white pine, pitch pine, balsam fir, American larch, cherry, and viburnum.

Aldous (1938) found that captive beavers ate between 22 and 33 ounces of food per day. Stegeman (1954) found the average daily food consumption of aspen was 4.5 pounds per day, and Brenner (1962) determined that the daily consumption of woody vegetation was 23.4 ounces and that waste amounted to 1.03 ounces a day.

Beavers are monogamous and apparently pair for life, though the male may breed with other females. The breeding season is from mid-January to mid-March, peaking in mid-February. The gestation period is about 106 days (Hediger 1970). The single annual litter may contain from one to eight or nine kits, but the usual is three to five. Huey (1956) found a relationship between food and litter size: 2.06 young per female were produced when willow was the principal food; 2.75 were produced in cottonwood areas and 4.2 in aspen areas. Gunson (1967) found a similar relationship: 2.8 embryos were produced in softwood communities, 3.9 in mixed-wood areas, and 4.6 in predominately hardwood areas. Very young and very old females produce smaller litters than middle-aged females (Longley and Moyle 1963).

Most beavers are born from mid-May to early June. At birth the precocious kits are fully furred, with eyes slightly open and incisors well-formed but still covered with a slight layer of tissue. They weigh from 227 to 624 g (8–22 oz) and are approximately 30.5 cm (12 in) long. Bailey (1927) reported that a captive kit went into the water when only 4 days old. Tevis (1950) noted that kits 2 to 3 weeks old begin eating leafy vegetation which the female brings them. Seton (1909) reported that the kits were weaned at 6 weeks of age, and Hodgdon and Hunt (1953) saw them outside the lodge as early as 30 May and feeding on bark during August.

Kits often ride on the back of a swimming mother, though she sometimes dives to force them to swim. They rarely follow the female on land, preferring to wait in the water for her return. All beavers vocalize, but the kits are much more vociferous and make soft whines and cries. The normal sound is a whine, which tends to decrease with age. Hodgdon and Larson (1973) reported that the young beavers directed most of the vocalization toward females of the colony rather than males. Kits grow rapidly in summer. Hodgdon and Hunt (1953) found that in Maine kits average nearly 7.3 kg (16 lb) by November; but during winter they grow much more slowly, so that by early spring they average only 7.7 kg (17 lb).

Beavers attain sexual maturity at 1½ to 2 years of age and are capable of breeding in the second breeding season after birth (Larson 1967; Henry and Bookhout 1969). When the young remain in the colony, breeding may be delayed until the third season after birth. Longevity is from 15 to 20 years in captivity. Larson (1967) reported a female beaver that was 20½ to 21 years old.

**Remarks.** Since beavers alter the areas they inhabit, they influence other animals and may be either beneficial or detrimental to man. The adverse effect of beavers upon trout waters has been argued for years, mainly on the more or less unwarranted assumptions that dams act as barriers to fish migration; beavers cause detrimental high water temperatures in summer; and beavers contribute to water stagnation, toxicity, and siltation of the water. However, much depends upon the condition of the individual stream or pond, or upon topography. In some cases dams in sluggish streams do spread the water out over vegetation, forming shallow, warm ponds with decaying plants which are unsuited to trout. But such streams are not usually important trout streams even before damming by beavers. In the rapid, cold streams that are more suitable to trout, the ponds seldom become too warm or stagnant and may provide trout with more spawning grounds and deeper resting and hiding pools. During spring floods and summer freshets the water pours over the dam, often in an unbroken stream, so that trout can pass freely up and down. Trout also feed on the insects that live in shallow quiet waters in beavers ponds, and more insects are dropped into the water by beavers' cutting trees.

Many species of wildlife are attracted to beaver ponds. In New England such areas provide excellent nesting and resting areas for waterfowl. Muskrats are particularly attracted to beaver flowages, even living in the same lodges with beavers, and beaver flowages provide food and water for many other mammals and birds. Beaver dams stabilize stream flow, conserve water, prevent rapid runoff, and control soil erosion, and they form many rich new meadows.

**Sex Determination.** It is difficult to determine the sex of live beavers, except for lactating females, but one field technique is palpating or feeling for the testes and the small penis. Hold the beaver in an upright position with its head covered, then gently insert the index finger about an inch into the cloaca and move it from side to side. If the beaver is a male, the baculum of the penis can be felt.

Davidson (1966) discovered from beaver blood smears that the nuclei of the polymorphonuclear neutrophil leucocytes showed deep-staining, chromatin-rich drumsticklike structures in the female beaver only, and Larson and Knapp (1971) found drumsticks in neutrophils in females only.

**Age Determination.** Various criteria have been used as estimators of beaver age classes based on a combination of morphological characteristics: os baculum length and size (Friley 1949); total body weight or pelt size (Buckley and Libby 1955; Brenner 1964); early tooth development (Cook and Maunton 1954); correlation between skull characteristics and reproductive condition (Bond 1956); changes in size of the tail and body weight (Patric and Webb 1960); and relationship between whole and skinned body weights (Longley and Moyle 1963). These criteria permit separation of 1- and 2-year-old beavers from adults, but cannot be used effectively on older beavers.

The most reliable technique is based on the development of the mandibular cheek teeth in dead beavers. Age can be estimated by noting the replacement of the temporary premolar, the progressive closure of the basal tooth openings, and the annual rings laid down in the cementum at the base of the first molar. The opening at the root base of the tooth decreases progressively as dentin begins to be deposited during the beaver's third summer. It is added in annual layers throughout the life of the animal, each layer consisting of a relatively broad summer layer and an usually narrower, lighter colored winter layer (Van Nostrand and Stephenson 1964; Larson and Van Nostrand (1968).

*½–1 year:* At 6 months the three molars are fully erupted and the double-rooted deciduous premolar, if present, forms a cap over the emerging single-rooted permanent premolar, which becomes fully erupted at about 1 year.

*2½–3 years:* The basal cavity of the premolar typically has two openings, while the second and third molars each have only one small, restricted basal opening. Though the basal opening of the first molar is markedly restricted in size, some teeth from 2½–3-year-old beavers are as wide open as those from 1½–2-year-olds. The 2½–3-year-old teeth can be distinguished only by the cementized lips of the openings. Longitudinal sectioning reveals a single layer of cementum on the base of all molars.

*3½–4 years:* The opening of the basal cavity of the first molar is usually closed, although small openings may remain in the bases of the other cheek teeth until the beaver is 4 years old. A second annual layer of cementum has been added on all cheek teeth; it is easily seen on the first and second molars and on the premolar in many beavers.

*4½–8 years:* By 4½–5 years all the basal openings are closed and a third cementum layer has been added to all teeth. In a few cases the layering of cementum occurs more often than the usual twice a year. In such cases the true annual layers are more distinct and have more consistent depths.

*Over 9 years:* There is some error in estimating the age of these older animals, owing to the apparent diffusion of the older layers. However, most are quite distinct and an error of 1 or 2 years in a few of these older classes is less critical in evaluating age ratios than in the more numerous and significant 4½–8 group.

**Specimens Examined.** MAINE, total 34: Aroostook County, Patten 1; Franklin County, Farmington 6; Hancock County, Acadia National Park 1 (UME), Mt. Desert Island 1; Oxford County, Upton 1 (HMCZ); Penobscot County, Bradley 1, Hudson 1, Lincoln 1, Milford 1 (UME); Piscataquis County, Brownsville 15, Moosehead Lake 1 (HMCZ); Sagadahoc County, Swan Island Refuge 2; Somerset County, West Forks 2 (UME).
NEW HAMPSHIRE, total 2: Grafton County, Wentworth 1 (AMNH); Strafford County, Barrington 1 (UNH).
VERMONT, total 10: Franklin County, Fairfax 10 (UVT).

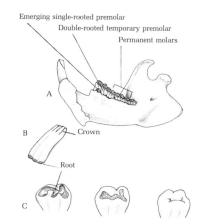

Age criteria for beaver. *A*, lingual view of lower jaw, age ½ to 1 year; *B*, permanent molar; *C*, progressive closure of basal tooth opening as dentine is deposited around the root as the beaver ages.

MASSACHUSETTS, total 13: Franklin County, New Salem 9; Worcester County, Petersham 4 (UMA).
CONNECTICUT, total 34: Fairfield County, Sherman 1; Litchfield County, Winsted 1; Tolland County, Mansfield 4, Storrs 2, Willington 1; Windham County, Chaplin 22, Eastford 2, Pomfret 1 (UCT).

## SUBORDER MYOMORPHA

### FAMILY CRICETIDAE (CRICETID MICE AND RATS)

The Cricetidae or New World mice, rats, voles, hamsters, lemmings, and so forth, is the largest family in number of species. The animals in this group differ widely in morphology, ecology, and behavior. Cricetids live in a wide variety of habitats from the desert and rain forests to the Arctic tundra.

Because of many differences in body structure and behavior, the family is divided into two subfamilies, the Cricetinae and the Microtinae. The cricetine rodents are slender to robust, with pointed muzzles and long tails. The limbs are not quite hidden in the body integument. The cheek teeth are tuberculate, the tubercles arranged in two longitudinal rows. The microtines are stocky and robust, with blunted muzzles and short tails. The limbs are mostly hidden in the body integument so that the animals appear short-legged. The cheek teeth are flat-crowned, markedly prismatic, and usually rootless in most genera.

Cricetids are almost cosmopolitan in distribution. In New England the subfamily Cricetinae comprises the deer mouse, white-footed mouse, and eastern woodrat. The subfamily Microtinae consists of the red-backed mouse, meadow vole, beach vole, pine vole, rock vole, southern bog lemming, northern bog lemming, and muskrat.

## Deer Mouse

*Peromyscus maniculatus*

**Description.** The deer mouse and the white-footed mouse may be difficult to distinguish where they occur sympatrically in New England, but the two species can be identified by the following external characters. In the deer mouse, the pelage is soft and luxuriant, grayish, and usually with only a slight middorsal stripe. The tail is usually as long as or longer than the

Two subspecies of deer mice are recognized in New England:

Peromyscus m. abietorum Bangs.
Peromyscus canadensis abietorum Bangs, 1896. Proc. Biol. Soc. Washington 10:49. Peromyscus m. abietorum Osgood, 1909. N. Amer. Fauna 28:45.
Type Locality: James River, Nova Scotia, Canada
Peromyscus m. gracilis (Le Conte). H[esperomys] gracilis Le Conte, 1855. Proc. Acad. Nat. Sci. Philadelphia 7:442. Peromyscus m. gracilis Osgood, 1909. N. Amer. Fauna 28:42 (Hall and Kelson 1959).
Type Locality: Michigan
Dental Formula:
I 1/1, C 0/0, P 0/0, M 3/3 × 2 = 16
Names: Wood mouse, Canadian deer mouse, white-bellied deer mouse, and long-tailed deer mouse. The meaning of Peromyscus = pouched little mouse, for the internal cheek pouches, and maniculatus = small-handed, referring to the small paws.

head and body, usually more distinctly white on the ventral surface, and frequently has a well-defined pencil marking. In the white-footed mouse, the pelage is seldom soft or luxuriant, but is reddish and usually has a well-defined middorsal stripe. The tail is seldom as long as head and body, is usually darker than white on the ventral surface, and seldom has a well-defined pencil marking.

The length of the tail is not a diagnostic character, because tails of large individual white-footed mice are as long as or longer than those of small individual deer mice. However, the ratio of length of the tail to length of head and body in adult deer mice approaches or exceeds 1:1, whereas in adults of white-footed mice a ratio of considerably less than 1:1 is the general rule. The only taxonomically useful cranial character separating the two species in New England is the breadth of the rostrum, which is greater in the white-footed mouse than in the deer mouse (Choate 1973).

The ears of the deer mouse are medium-size and thinly haired; the eyes are prominent and blackish. There are four clawed toes and a minute nailed thumb on the front foot, and five clawed toes on the hind foot. There are two small internal cheek pouches. The tail is well haired, with annulated scales, and tipped with a small tuft of long hairs. The tail serves as a tactile organ, as a prop when the mouse climbs vertical objects, and as an aid to balance when the animal runs along a horizontal branch, and is prehensile to a limited degree. Females have six mammae—two pectoral, four inguinal.

The skull is delicate and smooth, with a slender, short, and tapered rostrum and a somewhat arched and well-inflated braincase. The tympanic bullae are small to moderate. The zygomatic arch is slender and the anterior palatine foramen long. The molars are tuberculate, with tubercles in two longitudinal series. The upper molars are three-rooted, the lower molars two-rooted.

The sexes are colored alike and show slight seasonal variation. The coloration is rich brownish gray and slightly darker, mixed with darker hairs along the middorsal region from the shoulders to the base of the tail. The ears are dusky, edged with grayish, and there is a small tuft of white hairs on the anterior base of the

ears. The underparts and feet are white. The tail is bicolored, brownish black above, white below. Subadults are grayish above and almost blackish on the middorsal region. Immatures are gray and become buffier with succeeding molts. White-sided and piebald mutations occur occasionally.

The single annual molt begins as early as June and is completed in late summer or early fall, depending on when it began. The hairs are replaced gradually from the head toward the tail.

The sexes are equal in size. Measurements range from 175 to 215 mm (6.8–8.4 in); tail, 79 to 106 mm (3.1–4.1 in); and hind foot, 19 to 22 mm (0.74–0.86 in). Weights vary from 16 to 28 g (0.56–0.98 oz).

**Distribution.** Deer mice occur from Alaska across Canada, south in the Appalachians to northern Georgia, west through Tennessee, Arkansas, southward to central Mexico and Baja California, and northward along the Pacific coast.

In New England the species seems to have a limited distribution. It ranges southward from the boreal forests of northern Maine through mesic, sub-boreal habitats on the White Mountains of New Hampshire, the Green Mountains of Vermont, and the Berkshires in western Massachusetts and adjacent northwestern Connecticut. It may occur east of the Connecticut River in either Massachusetts or Connecticut, but it does not occur naturally in the coastal plains of southernmost Maine, southeastern New Hampshire, eastern Massachusetts, Connecticut, or Rhode Island (Choate 1973).

**Ecology.** Deer mice prefer habitats associated with coniferous forests consisting mainly of red, white, and black spruce and balsam fir, or northern hardwood forests comprising mainly beech, sugar maple, and yellow birch. In a few localities in southwestern Massachusetts and northwestern Connecticut the species occurs in isolated habitats interspersed among restored hemlock, white pine, and hardwood, and mainly northern red oak forest (Choate 1973).

This species is encountered along fencerows, on field borders, and in places where trees are small and ground cover is dense. The nest is

DEER MOUSE

usually built in stone walls, in old cabins or farm buildings, under tree roots and stumps, in hollow logs, granaries, corn shocks, under a rock or board, in abandoned burrows of small mammals, or even in abandoned bird nests. The nest is made of grass, leaves, or other plant materials and is lined with finely shredded plants, hair, feathers, cotton, rags, and other soft material.

Man, foxes, house cats, weasels, mink, short-tailed shrews, hawks, owls, and snakes prey on deer mice, and fires and floods kill many. Fleas, ticks, chiggers, mites, botfly larvae, cestodes, nematodes, and trematodes parasitize them.

**Behavior.** Deer mice are active throughout the year, except during severe winter weather, and are primarily nocturnal. Several biologists have studied the daily activity of this species. Gentry and Odum (1957) trapped more mice on warm, cloudy nights than on cold, clear nights. Hammer (1969) reported that wind, or an interaction of wind and temperature, did not significantly affect their activities at night when they were normally active. Marten (1973) reported that activity of deer mice diminished and was compressed into the early part of the night as summer progressed, and that neither wind nor moonlight had a strong effect upon activity within a night.

Deer mice have well-developed senses of hearing, smell, and taste, and good vision, and can discriminate differences in light intensity. Howard and Cole (1967) indicated that they detect seeds by smell and suggested that odor may be an important factor in the palatability of various seeds.

Deer mice are semiarboreal and are capable climbers. They do not dig burrows, although they will dig small excavations for refuge under rocks, stumps, and logs. They may use the burrows and trails of other small mammals.

Deer mice are strong swimmers but will not readily enter water, as white-footed mice will. Young deer mice 10 to 14 days old are able to swim (King 1961). Adults can run about 5.7 miles per hour for short distances (Jackson 1961).

These animals are rather placid, but when disturbed they may stamp their front feet rapidly. They produce high-pitched squeaks and make a shrill buzzing sound for short periods.

They are sociable but not colonial, exhibiting a loose social structure.

Adult males generally have larger home ranges than adult females (Blair 1940a; Stickel and Warbach 1960), and immatures usually have ranges smaller or nearly the same size as those of adults (Storer, Evans, and Palmer 1944). Manville (1949) reported that home-range size varied from 0.10 to 0.31 square acre for adult males, from 0.12 to 0.25 square acre for adult females, and from 0.10 to 0.27 square acre for juveniles of both sexes in Michigan. Stickel and Warbach

(1960) found that deer mice traveled more when food was scarce than when it was plentiful. Bovet (1968) demonstrated that some deer mice displaced far from their homesites and released on snow followed a homeward-oriented route, though they traveled a zigzag path. Murie (1963) reported that a male homed 550 yards in daylight within 8 hours, another male homed 800 yards at night, 8 hours after release, one mouse homed 1,600 yards in 9 days, and none of the 14 mice released 2,500 yards from home returned. Broadbooks (1961) recaptured a male deer mouse that was successful in finding its way home from two different directions, traveling a quarter mile in each case. Rawson and Hartline (1964) reported that an adult male homed from a release point 656 feet southwest of its nest and from a second release point 984 feet north of its nest. The longest homing journey recorded by deer mice was slightly over 2 miles in 2 days by a juvenile (Murie and Murie 1931). Furrer (1973) found that although males were better homers on the average, females traveled the longest distances.

In autumn, deer mice cache ripening seeds, mast, and other nuts in hollow trees, logs, stumps, abandoned bird's nests, and burrows near the nest chamber. They either carry food in the mouth or in the small internal cheek pouches or drag it by holding on with the incisors and bracing their legs.

Deer mice eat a wide variety of seeds, berries, buds, nuts, and grains, including corn, acorns, pin cherries, blueberries, Juneberries, raspberries, partridgeberries, beaked hazelnuts, and wintergreen, as well as other wild small fruits and fungi, including mushrooms. They also eat grubs, larvae, worms, snails, caterpillars, centipedes, millipedes, and sometimes the carrion of dead birds and other mice.

The breeding season normally begins in March and extends through October, though some females may mate in winter under favorable conditions. The gestation period is about 23 days, and a female may produce three or four litters annually. The litter size is from three to seven young, though four to six are common. The female ovulates immediately after parturition (Layne 1968).

Newborn deer mice are blind, naked, and pinkish, weighing from 1.1 to 2.4 g (0.04–0.08 oz), and their total length is about 48.2 mm (1.9 in). The young are very vocal. They grow rapidly and are weaned at between 3 and 4 weeks. The gray juvenile pelage is complete by 4 weeks of age. Each litter remains with the female for approximately 21 days (Layne 1968).

Nursing young hold tenaciously onto the nipples of the female. Under normal conditions the female can disengage the suckling young by stretching her body and slowly moving off. When a female is startled and makes a quick exit from the nest, the young may be roughly dragged and bounced around while still clinging to the nipples.

Both parents often occupy the natal nest together with the young, and the male may even share in their care (Horner 1947). Females sometimes share a nest and nurse their litters collectively (Hansen 1957).

In general, females become sexually mature at between 40 and 50 days of age and males about 10 days later (Layne 1968). Longevity in the wild is at least 1 year (Snyder 1956). Some captive deer mice have lived up to 8 years and were fertile up to 4 years of age (Dice 1933).

**Specimens Examined.** MAINE, total 169: Aroostook County, T11–R10 11, T11–R11 2, T12–R16 4, T16–R12 1, Garfield 4 (UCT), Madawaska 13 (HMCZ); Cumberland County, Portland 1; Franklin County, Carthage 1 (UCT), Farmington 46 (UME), Oquossoc 3 (USNM), Weld 6; Hancock County, Dixfield 3, Mt. Desert Island 7 (UCT); Oxford County, North Woodstock 3 (UMA); Penobscot County, 5 (UCT), Orono 7 (UME); Piscataquis County, T3–R9 1, T3–R11 4, T7–R9 5 (UCT), Baxter State Park 3 (UMA), Big Squaw 1, Greenville 1, Lake Sebec 1 (UCT), Mt. Katahdin 17–8 (DC–UCT); Somerset County, Lakewood 1, Madison 1, Taunton-Raynhen 2; Washington County, Baring 3, Edmunds 2, Vanceboro 2 (UCT).

NEW HAMPSHIRE, total 360: Carroll County, Albany 1 (HMCZ), Bartlett 32, Hart's Location 8, Jackson 29 (UCT), Ossipee 13 (USNM); Cheshire County, Dublin 1 (HMCZ); Coos County, Carroll 20, Colebrook 4 (UCT), College Grant 7 (DC), Dixville 1 (HMCZ), Gorham 8 (DC), Jefferson 2 (UCT), Milan 3 (DC), Mt. Washington 24–4–13 (DC–UCT–USNM), Randolph 14; Grafton County, Benton 1, Canaan 4, Dorchester 5, Etna 3 (DC), Franconia 8–1 (DC–HMCZ), Hanover 59 (DC), Livermore 44 (UCT), Lyme 3, Mt. Moosilauke 27, Piermont 2 (DC), Profile Mt. 1, Warren 1 (HMCZ), Woodstock 1 (DC); Hillsborough County, Hancock 1 (UCT); Merrimack County, New London 2 (DC); Sullivan County, Springfield 13 (UCT).

VERMONT, total 302: Addison County, Middlebury 1 (UVT), Ripton 3 (HMCZ), Starksboro 1 (UVT); Bennington County, Peru 3–1 (DC–UCT), Caledonia County, Brighton 4, Groton 1, Hardwick 3, Lyndon Center 2 (UCT); Chittenden County, Burlington 2, Huntington 18, Jericho 1, Shelburne 2, Underhill (Nebraska Notch) 1–3 (URI–UVT); Essex County, Island Pond 4, Maidstone 19; Lamoille County, Cambridge 19, Eden 9, Mt. Mansfield 2–3–16 (DC–HMCZ–USNM); Orange County, Chelsea 2 (UVT), Williamstown 7 (UCT); Orleans County, Starksboro 1 (UVT); Rutland County, Danby 1–3 (UMA–UVT), East Wallingford 2 (DC), Killington 1, Mendon 34 (UCT), Rutland 1, Sherburne 5–13–2 (DC–UCT–USNM), Shrewsbury 5 (DC); Washington County, Barre 7, Duxbury 39 (UVT), Montpelier 1, Plainfield 10, Woodbury 2; Windham County, Jamaica 1, South Londonderry 1; Windsor County, Ludlow 1 (UCT); Norwich 12, Plymouth Union 2, Pomfret 27 (DC), Royalton 3 (UCT), Woodstock 1 (DC).

MASSACHUSETTS, total 76: Berkshire County, Hinsdale 6 (UCT), Mt. Greylock 2, Mt. Washington 2 (HMCZ), North Savoy 4 (UMA), Peru 8 (UCT), West Charlemont 7, Windsor 2 (UMA); Franklin County, Ashfield 12 (UCT), Charlemont 3 (UMA), Deerfield 2, Whately 1; Hampden County, Chesterfield 17, Montgomery 1 (UCT); Hampshire County, Amherst 5; Norfolk County, Braintree 3; Worcester County, Holden 1 (UMA).

# White-footed Mouse

*Peromyscus leucopus*

There are three subspecies of white-footed mice recognized in New England:

*Peromyscus l. noveboracensis* (Fischer). *Mus sylvaticus noveboracensis* Fischer, 1829. *Synopsis mammalium*, p. 318. *Peromyscus l. noveboracensis* Miller, 1897. *Proc. Boston Soc. Nat. Hist.* 28:22 (April).
Type Locality: New York
*Peromyscus l. ammodytes* Bangs, 1905. *Proc. New England Zool. Club* 4:14 (28 February).
Type Locality: Monomoy Island, Barnstable County, Massachusetts. Known only from Monomoy Island.
*Peromyscus l. fusus* Bangs, 1905. *Proc. New England Zool. Club* 4:13 (28 February).
Type Locality: West Tisbury, Martha's Vineyard, Dukes County, Massachusetts. Known only from island of Martha's Vineyard (Hall and Kelson 1959).
Dental Formula:
I 1/1, C 0/0, P 0/0, M 3/3 × 2 = 16
Names: Deer mouse, wood mouse, and vesper mouse. The meaning of *Peromyscus* = pouched little mouse, for the internal cheek pouches, and *leucopus* = whitefooted.

**Description.** The white-footed mouse resembles the deer mouse but is distinguished by the external characters discussed under that species. The sexes are colored alike and show slight seasonal variation. The coloration of the upperparts is rich reddish brown, darker on the middle of the back. The underparts and the feet are white. Immatures are a uniform deep gray, and subadults are brownish gray to brown. There seems to be one annual molt. White spotting and yellow mutations occasionally occur.

The sexes are equal in size. Measurements range from 157 to 191 mm (6.1–7.4 in); tail, 62 to 86 mm (2.4–3.4 in); and hind foot, 18 to 22 mm (0.70–0.86 in). Weights vary from 16 to 29 g (0.56–1.02 oz).

**Distribution.** White-footed mice occur in forests and brush habitats of the eastern United States north of a line drawn from southeastern Mississippi to northeastern North Carolina, west to western Montana and eastern Arizona, and south through eastern Mexico to Yucatan.

In New England the species does not seem to have as limited an ecological distribution as does the deer mouse. The range of the white-footed mouse frequently overlaps that of the deer mouse. The species seems to occur naturally east of Penobscot Bay and the Penobscot River or north of the transition of hardwood to coniferous forest in Maine (Choate 1973).

**Ecology.** White-footed mice are primarily shrubland and woodland inhabitants. They prefer slightly more arid habitats than the deer mouse, primarily in areas of hemlocks, white pines, and oaks, with occasional populations extending into adjacent more mesic, northern hardwood areas or the slopes of the White Mountains, Green Mountains, and Berkshires (Choate 1973). Specimens have been trapped at between 5,000 and 6,200 feet, at least 1,000 feet above the treeline, among rocks and grassy vegetation in grass-sphagnum stunted spruce habitat on Mt. Washington (Starrett and Starrett 1952). In summer the species is often encountered within woodland borders, in ravines, on the margins of streams, in pastures with bushes and hedgerows, in farm buildings, and in houses.

This species seems to prefer nest sites above ground in trees near the edge of woods at all seasons, though some individuals may nest on the ground when there is snow cover (Nicholson 1941). M'Closkey and Fieldwick (1975) found that arboreal activity of white-footed mice was independent of the presence or absence of meadow voles, but Getz (1961a) indicated that white-footed mice have no innate habitat orientation but are excluded from certain areas by meadow voles. Bowker and Pearson (1975) found a population shift of white-footed mice, resulting in an increasing population of meadow voles.

The nest may be built in stone walls or cavities in ledges, under old stumps and logs, in hollow trees, under a board or slab, or in corn shocks, beehives, or abandoned squirrel's and bird's nests. The slightly globular nest is about 8 to 12 inches in diameter and 4 to 8 inches deep. It is made of dried grasses, leaves, moss, and

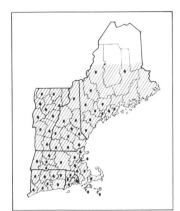

fibrous plant material and lined with lichen, milkweed, shredded grasses, cedar bark, and sometimes hair and feathers. The single entrance is usually near the bottom of the nest and is frequently closed. A pair of mice may build two or more nests. These mice frequently eat food and deposit their scats inside the nest. These animals change nests often, particularly females with litters. Like deer mice, white-footed mice do not make burrows but use runways of other small mammals. This species seems to have the same predators and parasites as the deer mouse.

**Behavior.** White-footed mice are active year in and year out, except in the cold winter weather. They are somewhat sociable but not gregarious, though two or more mice may live together. They are active from dusk to dawn and have a marked daily periodicity of activity even in continuous darkness. Getz (1959) found that from a total of 60 captures, 59 mice were active during the night and only 1 during the day. Orr (1959) found that white-footed mice were most active when ambient temperatures were between 40° and 50°F and that activity decreased at higher and lower temperatures.

White-footed mice have a larger home range in upland pine-oak woods, where trees are small and ground cover dense, than in bottomlands where trees are large and ground cover sparse (Stickel 1948). Both sexes have approximately the same home range size (Blair 1940a; Hirth 1959; Ruffer 1961). Burt (1940) determined that the home range size was 0.16 to 0.54 square acre for adult males and 0.06 to 0.37 square acre for adult females and found that males wandered farther than females. Wolfe (1968) determined the minimum home range varied from 0.11 to 0.86 square acre. Hamilton (1937a) reported that white-footed mice returned repeatedly to their home sites from a quarter mile away.

White-footed mice are good swimmers and can easily swim for half an hour, traveling up to 765 feet in calm water, although swimming is interrupted by floating rest periods (Sheppe 1956).

This species feeds on seeds of hemlock, acorns, small fruits, clover, insects, grasshopper eggs, grubs, larvae, and caterpillars and on small dead birds and mice. White-footed mice cache food for future use. Hamilton (1941b) estimated that the animals can eat about 30% of their body weight daily, or 6 g (0.21 oz) of food. Sealander (1952) found that white-footed mice generally consume 30% more food at 47°F than at 77°F.

The breeding season is from late February or early March into November and possibly lasts throughout the year in the southern latitudes. The gestation period is 22 to 25 days. The number of young per litter varies from three to seven, with five the most common. At birth the young are blind, helpless, and pinkish; they average approximately 2 g (0.07 oz) and are about 47 mm (1.8 in) in total length. The postnatal growth of this species is generally similar to that of the woodland deer mouse. The young grow rapidly and become sexually mature at 6 to 7 weeks of age.

Dewsbury (1975) observed that captive male white-footed mice displayed a basic copulatory pattern with no lock, no intravaginal thrusting, multiple intromissions before ejaculation, and multiple ejaculations.

**Specimens Examined.** MAINE, total 134: Androscoggin County, Auburn 2; Cumberland County, Gray 1, New Gloucester 4 (UCT), Portland 1; Franklin County, Farmington 5 (UME); Kennebec County, Oakland 4–1 (UCT–USNM), Waterville 12; Knox County, Rockland 9; Lincoln County, Barter Island 8, Boothbay Harbor 25 (UCT); Oxford County, Fryeburg 1 (AMNH); Penobscot County, Brewer 2, Orono 14 (UME); Sagadahoc County, East Harpswell 26 (AMNH), Freeport 1 (UME), Small

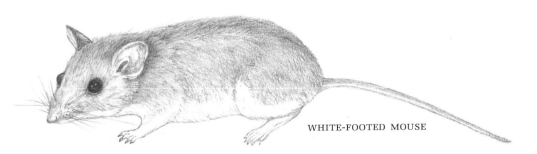

WHITE-FOOTED MOUSE

Point 1 (USNM); Somerset County, Detroit 1, Lakewood 5 (UCT), Seboonook 6 (UME), Skobegan 1 (UCT); York County, Eliot 4 (USNM).

NEW HAMPSHIRE, total 202: Belknap County, Alton 4; Carroll County, Bartlett 3, Hart's Location 1, Jackson 5 (UCT), Ossipee 4–35 (UCT–USNM), Wolfboro 1; Cheshire County, Alstead 1 (UCT), Dublin 1 (HMCZ); Coos County, Brenton Woods 1 (USNM), Carroll 19, Colebrook 3 (UCT), Mt. Washington 1–1–1 (HMCZ–UCT–URI); Grafton County, Benton 2, Bridgewater 1, Canaan 7, Dorchester 1, Hanover 29, Littleton 1, Lyme 5, Mt. Moosilauke 14 (DC); Hillsborough County, Antrim 5 (HMCZ), Brookline 1, Francistown 19; Merrimack County, Loundon 1 (UCT), Webster 11 (HMCZ); Rockingham County, Stratham 1 (UNH), Rye 2 (UCT); Strafford County, Durham 5–10 (UCT–UNH); Sullivan County, Springfield 4 (UCT), Sunapee Harbor 2 (USNM).

VERMONT, total 363: Addison County, Cornwall 3, Ferrisburg 1, Starksboro 2 (UVT); Caledonia County, Lyndon Center 5 (UCT); Chittenden County, Burlington 3–38 (USNM–UVT), Cedar Beach 3 (DC), Colchester 8, Essex Junction 4 (UVT), Hinesburg 1, Jericho 2, Richmond 15, Shelburne 5 (UCT), Underhill 1 (URI), Westford 1, West Milton 2, Williston 1 (UVT); Franklin County, St. Albans 1 (UCT); Orange County, Bradford 2 (URI), North Thetford 4 (UCT), Randolph 3 (UVT), Wells River 2 (HMCZ), Williamstown 1 (UCT); Orleans County, Jay 2, Westfield 1, Westmore 5 (UVT); Rutland County, East Wallingford 2 (DC), Mendon 19–2–15 (AMNH–DC–UCT), Mt. Tabor 3, Pittsford 3 (UCT), Rutland 4–7–33–2 (AMNH–DC–UCT–URI), Sudbury 2, Wallingford 4 (UCT); Washington County, Barre 1 (UVT), Montpelier 1, Plainfield 7 (UCT), Waterbury 2 (UVT); Windham County, Brattleboro 5 (AMNH), Londonderry 2 (UCT), Saxton's River 3–17 (AMNH–UCT), Vernon 5 (UCT), Whitingham 2; Windsor County, Bridgewater 3, Hartland 15 (AMNH), Norwich 21, Pomfret 39 (DC), Royalton 6 (UCT), Taftsville 3 (AMNH), Woodstock 4–3–8 (AMNH–DC–HMCZ).

MASSACHUSETTS, total 1,361: Barnstable County, Barnstable 29–2 (HMCZ–UCT), Cape Cod 3 (UCT), Chatham 3, Dennis 39 (HMCZ), Falmouth 19, Hatchville 1 (USNM), Mashpee 4, Monomoy Island 6 (UCT), North Truro 2 (USNM), Pantuit 2 (UCT), Provincetown 22 (USNM), Sandwich 3 (UCT), South Wellfleet 23–1 (HMCZ–UCT); Berkshire County, Adams 2, Ashley Falls 2, Barrington 1, Hinsdale 11 (UCT), Mt. Washington 1–22 (AMNH–UCT), North Monterey 4, North Savoy 1 (UMA), Otis 1–2 (UCT–UMA), Peru 5 (UCT), Pittsfield 2, (UMA), Sheffield 4 (AMNH), West Charlemont 2 (UMA); Bristol County, Lakeville 1 (UCT), Somerset 6 (UMA), Taunton 1 (HMCZ); Dukes County, Chilmark 12–4 (UCT–UMA), Edgartown 25–3 (UCT–UMA), Martha's Vineyard 19–15 (UCT–UVT), Naushan Island 1 (USNM), Nonomesset Island 6 (UCT), Oak Bluffs 1 (UMA), Tis-

bury 2–2 (UCT–UMA); Essex County, Andover 1 (UMA), Beverley 1 (HMCZ), Gloucester 1–2 (AMNH–HMCZ), Lawrence 2 (UMA), Lynnfield 2 (UCT), Newburyport 1 (UMA), Parker River National Wildlife Refuge 4 (USNM); Franklin County, Ashfield 12 (UCT), Charlemont 9, Greenfield 1 (UMA), Leverett 15–4 (UMA–UCT), Northfield 8, Orange 18, Shutesbury 1 (UMA), Sunderland 1 (UNH), Warwick 18, Whately 29; Hampden County, Brimfield 78, Chester 4 (UCT), Feltonville 1 (HMCZ), Holland 5, Montgomery 55, Southwick 3 (UCT), Springfield 3 (HMCZ), Wales 13 (UCT), Westfield 2 (UMA); Hampshire County, Amherst 1–50–2 (UCT–UMA–UNH), Chesterfield 6 (UCT), Easthampton 26–1 (UCT–UMA), Hadley 3, Hatfield 1, Huntington 6, Leeds 2, Leverett 1 (UMA), Mt. Tom 2–6 (UCT–UMA), Ware 2 (UMA); Middlesex County, Arlington Heights 1 (HMCZ), Ashby 1 (UCT), Ayer 1 (UMA), Bedford 11, Belmont 1 (HMCZ), Cambridge 98–31 (HMCZ–USNM), Concord 1 (AMNH), East Lexington 12 (HMCZ), Hopkinton 1 (UMA), Hudson 2, Lexington 6, Malden 2 (HMCZ), Marlboro 1 (UCT), Newton 15, Tynsboro 4 (HMCZ), Wakefield 5 (USNM), Weston 1, Wilmington 44 (HMCZ); Nantucket County, Monomoy Island 7–80–5 (AMNH–HMCZ–USNM), Muskeget Island 11 (HMCZ), Nantucket Island 17–71 (HMCZ–UCT), Siasconset 9 (UCT); Norfolk County, Braintree 1 (UMA), Milton 3, Quincy 5, Randolph 1 (HMCZ), South Weymouth 3 (UMA), Wrentham 2; Plymouth County, Hanson 30 (UCT), Hingham 1 (HMCZ), Middleboro 1–5–3 (AMNH–HMCZ–USNM), South Hanson 4 (USNM), Wareham 47 (HMCZ), West Bridgewater 5 (UMA); Suffolk County, Jamaica Plain 2 (AMNH); Worcester County, Fitchburg 1 (UMA), Harvard 25–2–4 (HMCZ–UCT–USNM), Holden 2 (UMA), Mt. Wachusetts 7 (HMCZ), Petersham 27 (UCT), Southboro 2, Southbridge 4 (UMA), Sterling 3 (UCT), Upton 2 (UNH), Uxbridge 1, West Brookfield 2 (UMA), Westboro 1 (UCT), Whitinsville 1 (UMA), Worcester 1–5 (UCT–UMA).

CONNECTICUT, total 1,867: Fairfield County, Bridgeport 1 (UCT), Cos Cob 2 (USNM), Easton 4, Fairfield 2 (UCT), Greenwich 3 (AMNH), Monroe 3 (UCT), Noroton 19 (HMCZ), Norwalk 5, Shelton 4, Stamford 1, Trumbull 1, Weston 5 (UCT), Westport 2–55 (AMNH–UCT), Wilton 5–23 (AMNH–UCT); Hartford County (no location) 7, Avon 3, Bristol 2, Canton 2 (UCT), East Hartford 62–22 (HMCZ–UCT), Farmington 4, Glastonbury 21 (UCT), Hartford 1–10 (UCT–USNM), Manchester 2 (UCT), Windsor 38 (AMNH), Windsor Locks 4; Litchfield County (no location) 23, Barkhamsted 38 (UCT), Bear Mt. 11 (AMNH), Cromwell 2–1 (AMNH–UCT), Goshen 5 (UCT), Kent 14–36 (AMNH–UCT), Lime Rock 4 (UCT), Macedonia Park 11 (AMNH), New Milford 4, Norfolk 2, Plymouth 5, Salisbury 16 (UCT), Sharon 19–3 (AMNH–UCT), Thomaston 3 (UCT); Middlesex County, Clinton 11 (AMNH), East Haddam 4, East Hampton 1, Marlborough 5, Middlesex 1, Portland

10; New Haven County, Bethany 7, Hamden 1, Meriden 5, Middlebury 8, Milford 4, Naugatuck 11, New Haven 6, North Haven 2, Oxford 13, Seymour 4, Southbury 10, Wallingford 7, Waterbury 57, Wolcott 11; New London County, Barn Island 2, Black Point 5, Bozrah 1, Colchester 1, Griswold 109, Groton 3, Hopeville 4, Lebanon 3, Ledyard 4 (UCT), Lyme 1–7 (DC–UCT), Niantic 6–3 (AMNH–UCT), Norwich 5, Patchaug State Forest 66, Preston 4 (UCT), Stonington 169–7 (UCT–USNM), Voluntown 23 (UCT), Waterford 30–4 (AMNH–UCT); Tolland County (no location) 110, Bolton 2, Columbia 6, Coventry 54, Eagleville 1, Hebron 2, Mansfield 190, Stafford 8, Storrs 30, Tolland 12, Union 113, Vernon 24, Willington 1; Windham County, Ashford 30, Brooklyn

10 (UCT), Chaplin 5–5 (AMNH–UCT), Eastford 3, Hampton 4 (UCT), Mt. Riga 4 (AMNH), Nachaug Forest 48, Scotland 5, Sterling 6, Thompson 8, Warrensville 15, Windham 6, Willimantic 3 (UCT), Woodstock 17 (AMNH).

RHODE ISLAND, total 195: Kent County, West Warwick 1 (URI); Newport County, Block Island 12–23 (HMCZ–UCT), Conanicut Island 27 (USNM), Fort Adams 5–9 (HMCZ–USNM), Newport 3–6 (HMCZ–USNM); Providence County, Chepachet 5; Washington County (no location) 75 (USNM), Charlestown 3 (UCT), Exeter 1 (URI), Great Cedar Swamp 3 (UCT), Kingston 16, Narragansett 1, Snug Harbor 1 (URI), South Kingston 1, Westerly 3 (UCT).

# Eastern Woodrat

*Neotoma floridana magister* Baird

N[*eotoma*] *magister* Baird, 1858. Mammals. In *Repts. Expl. Surv.* . . ., 8 (1):498.
*Neotoma floridana magister* Schwartz and Odum, 1957. *J. Mamm.* 38(2):197–206.
Type Locality: Cave near Carlisle, Cumberland County, or near Harrisburg, Dauphin County, Pennsylvania
Dental Formula:
   I 1/1, C 0/0, P 0/0, M 3/3 × 2 = 16
Names: Cave rat, pack rat, trade rat, Allegheny woodrat, mountain rat, and brush rat. The meaning of *Neotoma* = a new kind of species with cutting incisors, distinct from the genus *mus*, in which it was originally placed; *floridana* = of Florida; and *magister* = pertaining to a harvester of objects.

**Description.** The eastern woodrat closely resembles the Norway rat but may be distinguished by its longer, distinctly bicolored, well-haired tail which is sometimes bushy and not scaly; by its larger, conspicuous naked ears, larger, bright eyes, and longer whiskers that reach to the shoulders; and by the softer, longer, denser pelage. Also, the molar teeth of the two species differ—not tuberculate in the woodrat, tuberculate in the Norway rat.

The front foot of the woodrat has four clawed toes and a small thumb, and the hind foot has five clawed toes. Females have four inguinal mammae, males have a baculum, and both sexes have a ventral skin gland which, particularly during the breeding season, secretes a strong musky odor, especially in males.

The skull is elongated, with marked interorbital constriction and relatively small auditory bullae. The palate, lacking a posterior median spine, ends about even with the front of the first molar, which is high, flat-crowned, and prismatic. The coronoid process of the mandible is well developed.

The sexes are colored alike and show slight seasonal variation. In winter the coloration above is buffy gray to pale cinnamon, heavily overlaid with black. The head and the sides of the body are buffy gray, while the axillae, or armpits, are creamy buff. The throat, belly, and feet are white, the fur being white to the roots except along the sides of the belly, where the basal color is pale leaden gray. The tail is sharply bicolored, blackish brown above, white below. The summer pelage is slightly paler and

shorter. Immatures are grayer than adults, particularly on the belly. Color phases are not known. The single annual molt occurs from November to January and appears to progress uniformly, without showing a contrasting pattern between the worn, faded hair and fresh areas.

Males average slightly larger than females. Measurements range from 355 to 447 mm (13.8–17.4 in); tail, 160 to 210 mm (6.2–8.2 in); and hind foot, 33 to 50 mm (1.3–2.0 in). Weights vary from 394 to 500 g (13.8–17.5 oz).

**Distribution.** The eastern woodrat occurs from extreme western Connecticut through southeastern New York, most of pennsylvania, in southern Ohio and Indiana, northern and western Virginia, the southeastern United States, except parts of North and South Carolina and Georgia, south to central Florida, west to eastern Texas, through eastern Oklahoma to southern Missouri, west through Kansas and eastern Colorado, and from Nebraska into southern South Dakota.

Goodwin (1932) reported woodrats on the high ridges of Litchfield County, Connecticut, and caught one specimen from Scattergoat Mountain, near Kent in Litchfield County. The species is probably an animal of the past in New England.

**Ecology.** This species occurs among rugged rock ledges or rock formations, in clefts and slides of ridges and fissures of cliffs and caves, and on bare patches on slopes. It is rarely found in lowland swamps or open areas. Occasionally it is found in abandoned buildings in mountainous regions.

The animal constructs a bulky nest in deep crevices of large rocks and boulders, on ledges in caves, and in abandoned buildings. The nest is open above, though it may have a roof in winter. It is normally used all year and often for the entire life of the animal. The woodrat continually adds material to the structure, so that it may become large enough to nearly fill a bushel basket. The outside dimension is from 5 to 12 inches in diameter, and the cavity is 3 to 4 inches in diameter. The coarser outside material of the nest is composed of leafy twigs, long strips of bark, rootlets, leaves, grasses, moss, oakum, and other available material. The inner lining is made of bits of shredded bark and soft fiber. The woodrat approaches its nest from several directions, and there is often a large pile of debris and scats nearby. The nest is seldom occupied by more than one adult.

Man, owls, foxes, weasels, skunks, racoons, wildcats, and large bull snakes prey on woodrats. Mites, fleas, chiggers, cuterebra larvae, cestodes, and nematodes are known to infest them.

**Behavior.** Perhaps the most conspicuous work of the woodrat is the midden, which is composed of a great variety of plant material and a wide assortment of objects such as shiny bits of tin, bottle caps, nails, coins, belt buckles, china, keys, watches, spectacles, shotgun shells, cartridges, paper, rags, rubber bands, rubber from old tires, feathers, scraps of leather, bones and small skulls, and even fox, cow, and horse dung. This propensity for packing off such items has given the animal the name pack rat. In camping areas this inquisitive animal will often trade a stick or two for a more appealing shiny object, which has earned it the name trade rat.

Woodrats are usually solitary and unsociable. When annoyed or excited they produce a chattering sound with their teeth, and they often thump their hind feet and rapidly vibrate the tail up and down on the ground.

Woodrats are chiefly nocturnal, though they may be seen during daylight, particularly in caves and in deep crevices. They are active throughout the year except during storms and extremely cold weather. They are very agile and can climb trees. They do not ordinarily travel far; Poole (1940) indicated that the home range is from 100 to 200 square yards.

Woodrats eat mushrooms, fungus puffballs, rhododendron twigs, twigs and seeds of hemlock, mountain ash, flowering dogwood, sweet birch twigs and bark, black cherry, pin cherry, American elder, bear oak, staghorn sumac, and a host of other plants. They also eat snails, insects, and carrion. Most of the food is eaten where it is found, though some berries, nuts, and seeds are often cached.

The reproduction of this species is imperfectly known. The breeding season occurs from midwinter or early spring until autumn, the young being born from mid-March to early September (Poole 1940). The gestation period is

EASTERN WOODRAT

between 30 and 39 days (Pearson 1952). There may be two or three litters annually, and one to four young usually are born per litter.

At birth the young are naked and pink, with eyes and ears closed, and are about 89 to 102 mm (3.5–4.0 in) long and weigh nearly 15 g (0.53 oz). From birth through most of the first 21 days of life the young cling tenaciously to the mother's teats (Hamilton 1953), and if she is disturbed she may drag them about still holding on. Almost from birth the young have strongly decurved incisors with an oval hole between them into which the elongated teat fits. To release their hold the mother simply pinches the young on the back, jaw, or neck with her teeth and twists them off with her feet (Schwartz and Schwartz 1959).

The young grow very fine, silky gray hair. The ears unfold at 9 days old, and the eyes open by the 15th day. Weaning occurs after the 20th day (Rainey 1956).

**Specimens Reported.** Connecticut; Litchfield County, Scattergoat Mt., near Kent (Goodwin 1932; Goodwin 1935).

# Gapper's Red-backed Mouse

*Clethrionomys gapperi*

Two subspecies of red-backed mouse are recognized in New England:

*Clethrionomys g. ochraceus* (Miller). *Evotomys gapperi ochraceus* Miller, 1894. *Proc. Boston Soc. Nat. Hist.* 26:193. *Clethrionomys gapperi ochraceus* Sheldon, 1930. *J. Mamm.* 11(3):318.

Type Locality: Alpine Garden, near the head of Tuckerman's Ravine, Mount Washington, Coos County, New Hampshire

*Clethrionomys g. gapperi* (Vigors). *Arvicola gapperi* Vigors, 1830. *Zool. J.* 5:204 *Clethrionomys g. gapperi* Green, 1928. *J. Mamm.* 9(3):255.

Type Locality: between York, Toronto, and Lake Simcoe, Ontario, Canada (Hall and Kelson 1959).

Dental Formula:

I 1/1, C 0/0, P 0/0, M 3/3 × 2 = 16

Names: Boreal red-backed mouse and red-backed mouse. The meaning of *Clethrionomys* = a kind of dormouse, and *gapperi* = for Gapper, a zoologist.

**Description.** The red-backed mouse may be distinguished from most other New England mice by its short tail and reddish color. It resembles the pine vole in both these characters but differs in that the reddish coloration is confined to the back. Also, the ears of the red-backed mouse are larger and the tail much longer than in the pine vole, and the fur is longer and coarser.

The front feet have four toes and the hind feet five toes. Females have eight mammae—four pectoral and four inguinal. The skull is smooth, with the auditory bullae inflated and the zygomatic arches slender. The upper incisors are without grooves and are colored orange. The posterior border of the palate is straight.

The sexes have the same striking coloration and show seasonal color variation. In winter there is a broad stripe of bright chestnut from the forehead to the base of the tail, sprinkled with black hairs, contrasting with the ochraceous buff of the nose and the sides of the head and body. The underparts are pale buff to silvery. The feet are gray, and the tail is bicolored, brownish above, whitish below, and tufted at the tip. The summer coloration is slightly darker and duller. Immatures are more subdued in coloration and the dorsal stripe is less distinct. Albinos and melanistics occur rarely. A dark phase occasionally occurs in which the dorsal stripe is replaced by brown or grayish black and the general coloration is darker.

The sexes are equal in size. Measurements range from 125 to 155 mm (4.9–6.0 in); tail, 30 to 50 mm (1.2–2.0 in); and hind foot, 16 to 20 mm (0.62–0.78 in). Weights vary from 16 to 32 g (0.56–1.12 oz).

**Distribution.** This species occurs across Canada, south into the Appalachians to western North Carolina, in northern Michigan, most of Wisconsin, Minnesota, and North Dakota, and through the Rocky Mountains south to southwestern New Mexico.

**Ecology.** This woodland mouse is found in cool forests of deep hemlock and red spruce, preferring damp areas strewn with mossy rocks, stumps, rotten logs, and root systems in loose forest litter, sphagnum, or fern-filled depressions or around old white cedar swamps with sparse to medium cover. It is also found in shaded mixed hardwood areas along old rock walls, rocky outcrops, and talus slopes, on old farms, and occasionally in grassy clearings.

The burrows of red-backed mice are not as elaborate as those of meadow voles. The burrows may be under deep cover or out in open spots. The runways often come to the surface and disappear into seepage areas. Most often red-backed mice occupy burrows dug by moles, voles, and other small mammals. The nest, which may be situated among rootlets of small trees or on top of the ground underneath a root of a stump, is about 4 inches in diameter and about 2 inches deep, composed of fragments of dry grasses, leaves, bark, twigs, and hemlock cones.

Man, birds of prey, coyotes, foxes, skunks, weasels, mink, opossums, cats, and snakes prey on red-backed mice. Floods, droughts, extreme cold temperatures, and fires also take their toll. They are hosts to ticks, fleas, cestodes, and nematodes.

This species may be afflicted with waltzing, a condition in which the animal makes rapid circling movements to left and right and jerks its head when handled or disturbed (Benton 1966). Waltzing has been reported for several other species of rodents (Warnock 1964).

**Behavior.** Red-backed mice are active throughout the year, at all hours, but are more active at night than during daytime. Getz (1968) demonstrated that temperature and humidity conditions modified but did not alter the basic daytime activity of this species. McManus (1974) showed that the animals ate more food when the ambient temperature was at 10°C and less at 30°C.

Red-backed mice are irritable and nervous and will fight when placed together. Most of the time they are silent, but when disturbed they utter a shrill chattering "chur-r-r-r."

These animals can jump 6 to 8 inches (Jackson 1961). They are agile climbers and have been trapped 7 and 10 feet above ground (Muul and Carlson 1963); and they can walk along stems only 5 mm (0.2 in) in diameter (Getz and Ginsberg 1968). They are fairly good swimmers.

The home ranges of adult males and females and of immatures are not significantly different (Manville 1949). Blair (1941) estimated the home range size at 3.56 square acres. Getz (1968) reported that the local distribution of the species is correlated with available free water rather than with suitable food or microclimate.

Red-backed mice feed mainly on the leaves and tender stems of many weeds and grasses and on seeds, various berries, nuts, roots, green buds, ferns, and fungi. They also eat insects,

spiders, and snails. During lean times they gnaw on the bark of shrubs and trees, and during severe winter they feed on seeds and tubers cached in burrows.

The breeding season begins as early as mid-January and continues into late November. The gestation period is 17 to 19 days, and at least two or three litters are produced annually. From two to eight young may be born per litter, though four to six are usual.

At birth the young are blind and hairless, weighing about 2 g (0.07 oz). They are capable of vocalization within 1 day after birth, hair appears in 4 days, at 5 days they begin to crawl, and at 13 days the eyes open. The young are weaned in 17 days if another litter follows immediately, otherwise nursing may continue for at least 3 weeks. When nursing the young hold tightly to the female's nipples (Benton 1955b).

**Specimens Examined.** MAINE, total 284: Aroostook County, T10–R7 1, T10–R8 2, T11–R9 5, T11–R10 23, T11–R11 6, T12–R15 4, T12–R16 32, T16–R12 1 (UCT); Cumberland County, Harrison 1 (AMNH); Franklin County, Dixfield 1 (UCT), Dryden 4 (AMNH), Farmington 47, Kingfield 2 (UME), Weld 4 (UCT); Hancock County, Bar Harbor 1 (USNM), Castine 2 (AMNH), Mt. Desert Island 1; Kennebec County, Oakland 2; Lincoln County, Boothbay Harbor 16 (UCT), Medomak 4 (UVT); Oxford County, Fryeburg 1 (AMNH), Upton 2; Penobscot County (no location) 3, Holden 3, Orono 11 (UME), South Twin Lake 41 (AMNH); Piscataquis County (no location) 2, T7–R9 5, Frost Pond 3, Greenville 2 (UCT), Mt. Katahdin 9–6 (DC–USNM), Sebec Lake 1 (USNM); Somerset County, Enchanted Pond 9, Grace Pond 9 (AMNH), Lakewood 2, Tauton 1; Washington County, Baring 2 (UCT), Calais 2–5 (DC–UCT), Columbia Falls 3 (USNM), Vanceboro 3 (UCT).

GAPPER'S RED-BACKED MOUSE

NEW HAMPSHIRE, total 651: Belknap County, Alton 11 (UCT); Carroll County, Albany 1–5 (UCT–UNH), Bartlett 28, Harts Location 61, Jackson 178 (UCT), Ossipee 6–4 (UCT–USNM); Coos County (no location) 7, Carroll 11 (UCT), Fabyans 20 (USNM), College Grant (Errol) 1–12 (AMNH–DC), Mt. Washington 18–31–15–1–6 (AMNH–DC–UCT–UME–USNM), Randolph 17 (DC), Twin Mt. 17 (UCT); Grafton County, Canaan 5, Dorchester 1, Etna 2, Franconia Notch 1 (DC), Hanover 2–20 (AMNH–DC), Lake 12 (DC), Lincoln 4 (UCT), Littleton 1 (DC), Livermore 53 (UCT), Mt. Moosilauke 87, Piermont 4 (DC); Strafford County, Durham 1–4 (UCT–UNH); Sullivan County, Springfield 4 (UCT).

VERMONT, total 231: Addison County, Lincoln 1; Bennington County, Peru 2 (UCT), Readsboro 8 (AMNH), Styles Peak 1; Caledonia County, Lyndon Center 5, Hardwick 11 (UCT); Chittenden County, Essex Center 2, Milton 2, Richmond 3, Shelburne 3, Underhill (Nebraska Notch) 8 (UVT); Essex County, Brighton 2, Ferdinand 1 (UCT), Island Pond 1, Maidstone 21 (DC), Victory 2 (UCT); Lamoille County, Cambridge 19, Eden 10 (DC), Mansfield 3–14–1 (DC–USNM–UVT), Moscow 2 (UCT); Orleans County, Westmore 6; Rutland County, Danby 1 (UVT), East Wallingford 2 (DC), Mendon 26–2–32 (AMNH–DC–UCT), Mt. Killington 4–1 (DC–UCT), Sherburne 5–2 (UCT–USNM), Wallingford 1; Washington County, Duxbury 5 (UCT), Plainfield 9; Windham County, Londonderry 1, Saxton's River 4, South Windham 1, Vernon 1; Windsor County, Ludlow 5 (UCT), Pomfert 1 (AMNH).

MASSACHUSETTS, total 377: Barnstable County, Barnstable 5 (UCT), Falmouth 5 (USNM), Hardwick 2 (AMNH); Berkshire County, Adams 1, Hinsdale 15 (UCT), Lenox 1 (UMA), Mt. Greylock 2–4 (UCT–UMA), Mt. Washington 1–4 (AMNH–UCT), North Monterey 2, North Savoy 14 (UMA), Otis 2, Peru 16, Sheffield 1 (UCT), West Charlemont 2; Bristol County, Somerset 1 (UMA), Taunton 1 (UCT); Dukes County, Martha's Vineyard 2 (UVT); Franklin County, Ashfield 19 (UCT), Charlemont 36, Leverett 7 (UMA), Northfield 24 (UCT), Shelburne 1, Shutesbury 4, Sunderland 1 (UMA), Warwick 14, Whately 8; Hampden County, Brimfield 11, Chester 13 (UCT), Chicopee Falls 1 (UMA), Holland 5, Montgomery 19, Southwick 2 (UCT); Hampshire County, Amherst 34 (UMA), Chesterfield 21, Easthampton 7, Huntington 1 (UCT), Shutesbury 4; Middlesex County, Framingham 1, Hopkinton 1 (UMA), Wilmington 8 (USNM); Plymouth County, Middleboro 1–1 (AMNH–USNM); Worcester County, Fitchburg 1 (UMA), Harvard 3 (UCT), Lunenburg 1 (USNM), Petersham 43–1 (UCT–UMA), Southbridge 1 (UMA), Sterling 2 (UCT).

CONNECTICUT, total 338: Fairfield County, Monroe 1, Weston 2; Hartford County, East Granby 1, Glastonbury 1; Litchfield County, Barkhamsted 11, Goshen 4, Kent 31 (UCT), Mt. Sharon 2 (AMNH), Norfolk 2, Salisbury 7, Winchester 1; Middlesex County, Portland 1; New Haven County, Southbury 1; New London County, Barn Island 5, Black Point 3, Griswold 20, Hopewell 4, Lebanon 1, Stonington 6, Voluntown 35 (UCT), Waterford 2 (AMNH); Tolland County (no location) 31, Charter Marsh 2, Coventry 6, Mansfield 42, Stafford 1, Storrs 19, Tolland 1, Union 34, Vernon 1, Willington 1; Windham County, Ashford 21, Brooklyn 8, Chaplin 1, Eastford 3, Hampton 1, Scotland 3, Sterling 2, Thompson 1, Warrenville 1, Westford 10, Willimantic 8 (UCT).

RHODE ISLAND, total 26: Providence County, Chepachet 10; Washington County (no location) 16 (USNM).

# Meadow Vole

*Microtus pennsylvanicus*

**Description.** The meadow vole is a thick-bodied microtine with a short, scaly, scantily haired tail about twice the length of the hind foot. The eyes are beady and black and the ears are short, rounded, furred along their borders, and practically hidden in fur. Females have ten mammae —four pectoral and six inguinal.

The meadow vole resembles the pine vole, but has larger ears, a longer tail, and coarser pelage. It differs from Gapper's red-backed mouse in its larger size, lack of the red dorsal stripe, and more brownish color. It is similar in appearance to the southern bog lemming but has shorter, coarser pelage, a much longer tail, and ungrooved upper incisors.

The skull is slightly angular, with a short rostrum and broad braincase. The auditory bullae are moderately large and rounded. The squamosals have more or less developed crests, and the palate usually ends anterior to the posterior end of the molar row. The anterior faces of the yellow upper incisors are simple and ungrooved, and the roots of the lower incisors extend far behind and on the outer side of the molars. The molars are rootless and the crown of the middle upper molar has four triangles and a posterior loop.

The sexes are colored alike, and seasonal variation is slight. In summer the coloration of the upperparts is dull chestnut brown or yel-

There are two subspecies recognized in New England:

*Microtus p. pennsylvanicus* (Ord).
   *Mus pennsylvanica* Ord, 1815. In Guthrie, *A new geographical, historical, and commercial grammar . . .,* Philadelphia, 2d Amer. ed., 2:292. *Microtus pennsylvanicus* Rhoads, 1895. *Amer. Nat.* 29:940.
Type Locality: Meadows below Philadelphia, Pennsylvania
*Microtus p. provectus* Bangs.
   *Microtus provectus* Bangs, 1908. *Proc. New England Zool. Club* 4:20 (March). *Microtus p. provectus* Chamberlain, 1954. *J. Mamm.* 34(4):587–89.
Type Locality: Block Island, Newport County, Rhode Island. Known only from Block Island (Hall and Kelson 1959).
Dental Formula:
   I 1/1, C 0/0, P 0/0, M 3/3 × 2 = 16
Names: Eastern meadow mouse, Block Island meadow vole, field mouse, and ground vole. The meaning of *Microtus* = small ear, and *pennsylvanicus* = of Pennsylvania.

lowish brown, darkest along the middle of the back. The underparts are grayish white or buffy white, the feet are grayish brown, and the tail is dusky above and paler below. The winter pelage is grayer. Immatures are darker than adults, and in some the feet and tail are almost black. All-white, white-spotted, albino, melanistic, yellow, and cinnamon specimens occasionally occur.

Males average slightly larger than females. Measurements range from 150 to 195 mm (5.9–7.6 in); tail, 33 to 65 mm (1.3–2.5 in); and hind foot, 18 to 24 mm (0.70–0.94 in). Weights vary from 20 to 64 g (0.70–2.28 oz).

**Distribution.** This species occurs from Quebec and New Brunswick south into Georgia, and west into Nebraska, South Dakota, and North Dakota. The subspecies Block Island meadow vole, *Microtus p. provectus*, is restricted to Block Island, Rhode Island.

**Ecology.** The meadow vole is one of the most abundant mammals in New England. It occurs in grass-sedge marshes, salt marshes, wooded swamps and sphagnum bogs, along streams and lakes, and in orchards, open woodland, corn shocks, haystacks, or other grain cover.

Meadow voles build extensive, interlaced shallow subterranean runways and aboveground pathways through dense grass and other vegetation. There is no definite design for the runway systems. The animal cuts its way through succulent grass and moist ditches as it threads along. The runways are used for escape and often lead to underground nests and occasionally to open water. The runways are about 1½ inches in diameter. In sparse or more open cover runways are not well defined because the voles travel freely between the scattered plant stems.

The nest is globular and bulky, about 7 to 8 inches in diameter, and is made of dried vegetation. It may be situated in a dense tussock of grass, in a runway, or underground. It is usually kept clean. The winter underground nest is smaller and composed of similar material. Some meadow voles may use bird's nests.

Man, hawks, owls, crows, jays, gulls, herons, shrikes, house cats, bobcats, lynxes, raccoons, mink, weasels, dogs, foxes, skunks, opossums,

short-tailed shrews, bears, bass, pickerel, and snakes prey on meadow voles. Floods and fires occasionally kill many young voles. Meadow voles are host to fleas, mites, lice, cestodes, nematodes, and trematodes. Some inherit the recessive trait of waltzing, exhibiting rapid circling to left and right, with jerky head movements.

**Behavior.** Meadow voles are gregarious, fierce, and aggressive. They are active day and night and throughout the year. Getz (1961*b*) reported that meadow voles are less active during the day when the ambient temperature is above 70°F, and become less active at all times when the temperature is below 0°F.

These rodents can run as fast as 5 miles per hour for short distances. They are good swimmers and divers; Blair (1939) saw a meadow vole swim underwater for about 10 feet and on the surface for 80 to 90 feet.

Meadow voles defend a territory and only occasionally travel outside their home range (Blair 1940*b*; Getz 1961*b*; Van Vleck 1968). These authors reported that the size of the home range may be density-dependent; the lower the density, the larger the home range. Grant (1971) believed that intraspecific interaction, associated with high density, induced movement from meadows to woodland areas. Van Vleck (1968) reported that on making a sortie outside the home range, an alien vole is forced to move on if it is confronted with another vole, but when not confronted tends to remain there and might establish a new home range. Stickel and Warbach (1960) demonstrated that meadow voles traveled up to 470 feet and suggested that possibly the greater distances represented range shifts. Linduska (1950) found that the animals migrated from wetter areas to drier upland fields just before the first killing frost and that fall invasion of upland fields was a regular occurrence. Hamilton (1937*a*) found that the home range was seldom more than 0.7 square acres in New York. In Connecticut, Getz (1961*c*) found that the mean monthly range size of males was larger than that of females in both moist and dry grasslands. Ambrose (1973) found that a meadow vole placed in a new environment established the rough outer limits of its new home range within 3 to 5 hours.

MEADOW VOLE

Meadow voles are primarily vegetarians but will eat meat when available. They feed mainly on grasses, sedges, legumes, seeds, grains, tubers, and roots, and on inner bark and tender cambium from shrubs, trees, and vines, often completely girdling the trunk. Occasionally they store bulbs, tubers, roots, seeds, and grains during times of surplus food.

Meadow voles are promiscuous and among the most prolific of mammals. A female can produce seventeen litters per year (Bailey 1924; Hamilton 1941*a*). As a rule a female produces from eight to ten litters in high population years, but rarely more than five or six litters in years when food is scarce.

This species has a protracted breeding season and will breed throughout the year when food and cover are available, though breeding activity drops off from January to March (Beer and MacLeod 1961). The gestation period is about 21 days, and a litter comprises from one to nine young, the usual being four or five. Newborns are pink, naked, and blind, but very active, and weigh from 1.6 to 3.0 g (0.06–0.11 oz). They grow rapidly and gain 0.2 to 0.5 g (0.007–0.018 oz) per day during their first 3 to 4 weeks of age (Barbehenn 1955). By the 4th or 5th day their backs are covered with soft gray hair; incisors usually erupt on the 6th or 7th day, and the eyes and ears open on the 8th day. The young are weaned before they are 2 weeks old and by the 3rd week become independent. Females may become sexually mature when only 4 weeks old and males when 5 weeks old (Hamilton 1941*a*). Longevity may be less than one year in the wild and somewhat longer in captivity.

Under optimum food and cover conditions meadow vole populations tend to increase significantly within a single year, reaching their maximum about every four years. They then die off or "crash" within a few months' time to a low level, only to build up and renew the population cycle. The die-offs are phenomenal and may be somewhat local in character, but the reasons are poorly known, though the effect of epizootics, food shortages, or physiological changes due to the tensions of overcrowding are suspected.

**Specimens Examined.** MAINE, total 157: Aroostook County, T16–R12 4 (UCT), Allegash 1 (USNM), Madawaska 22 (HMCZ); Cumberland County, Gray 5 (UCT), Westbrook 2 (HMCZ); Franklin County, Dryden 6 (AMNH), Farmington 14 (UME); Hancock County, Bar Harbor 1 (USNM), Casture 1 (AMNH), Mt. Desert Island 17 (HMCZ), Smith Cove 12 (AMNH); Knox County, Oakland 1 (USNM); Lincoln County, Barter Island 2, Boothbay Harbor 2 (UCT); Oxford County, Fryeburg 1 (AMNH), Upton 9 (HMCZ); Penobscot County, Bangor 5 (UME), Benedicta 1 (UCT), Brewer 1 (UME), Burlington 2 (UCT), Holden 2, Lincoln 1, Orono 4 (UME), South Twin Lake 11 (AMNH); Piscataquis County, T7–R9 7 (UCT), Dover–Foxcraft 1 (UME), Frost Pond 1 (UCT), Mt. Katahdin 2, Sebec Lake 4 (USNM); Sagadahoc County, East Harpswell 3; Somerset County, Enchanted Pond 1 (AMNH); Washington County, Addison 1, Calais 1 (USNM), Carthage 1 (UCT), Columbia Falls 1 (USNM), Columbus Island 2, Edmunds Unit 1, Moosehorn 4 (UCT).

NEW HAMPSHIRE, total 195: Carroll County, Albany 1 (DC), Bartlett 3, Jackson 13, Silver Lake 1 (UCT), Ossipee 14–1 (AMNH–UCT); Cheshire County, Keene 1 (DC); Coos County, Carroll 62 (UCT), College Grant (Errol) 5, Gorham 15 (DC), Mt. Washington 2–1 (DC–UCT); Grafton County, Canaan 2, Dorchester 1 (DC), Hanover 17–5

(DC–UNH), Mt. Moosilauke 27–6 (DC–HMCZ); Hillsborough County, Hollis 2 (HMCZ), Pembroke 3 (UCT); Rockingham County, Nottingham 2; Strafford County, Durham 11 (UNH).

VERMONT, total 267: Addison County, Bridgeport 7 (HMCZ), Cornwall 1, Middlebury 1 (UVT); Bennington County, Bennington 1 (UMA), Peru 2; Caledonia County, Groton 3, Hardwick 18, Lyndon Center 5 (UCT); Chittenden County, Burlington, 28, Essex Junction 2, Richmond 5, Salisbury 1, Shelburne Falls 1, Westford 1, West Milton 4, Winooski 2 (UVT); Essex County, Brighton 1 (UCT), Island Pond 4, Maidstone 1 (DC), Victory 1; Franklin County, Swanton 2; Grand Isle County, Grand Isle 1 (UCT); Lamoille County, Cambridge 6 (DC); Orange County, Ainsworth 1 (UCT), Fairlee 1 (UVT), Mt. Mansfield 1–11 (HMCZ–USNM), Williamstown 5 (UCT); Rutland County, East Wallingford 2 (DC), Mendon 5–8 (AMNH–UCT), Rutland 15–5–34 (AMNH–DC–UCT), Sherburne 3 (UCT), Waban 1 (AMNH); Washington County, Plainfield 12; Windham County, Brighton 1, Brooklyn 1, Jamaica 1 (UCT), Putney 1 (AMNH), Saxton's River 3–10 (AMNH–UCT), Victory 1 (UCT); Windsor County, Dummerston 1 (DC), Hartland 8 (AMNH), Norwich 6 (DC), Pomfret 2–3 (AMNH–DC), Royalton 12 (UCT), Weathersfield 1, Woodstock 2 (HMCZ).

MASSACHUSETTS, total 629: Barnstable County, Barnstable 8–1 (HMCZ–UCT), Chatham (Tern Island) 1 (AMNH), Monomoy Island 14–4–6 (HMCZ–UCT–UMA), South Dennis 1 (HMCZ), South Wellfleet 7–7 (HMCZ–UCT), Wellfleet 4; Berkshire County, Adams 1, Ashley Falls 3, Hinsdale 5, Peru 3; Bristol County, Raynham 1 (UCT), Taunton 1–5 (HMCZ–UCT); Dukes County, Elizabeth Island 12 (UCT), Martha's Vineyard 32–4–4–4 (HMCZ–UCT–UMA–USNM–UVT), Naushon Island 1 (AMNH); Essex County, Bedford 3 (HMCZ), Gloucester 4–2 (AMNH–HMCZ), Ipswich 1 (HMCZ), Marblehead 1 (AMNH), Methuen 1 (HMCZ), Parker River National Wildlife Refuge 4 (USNM), Rockport 1; Franklin County, Ashfield 3 (UCT), Charlemont 4 (UMA), Chesterfield 1 (UCT), Deerfield 2, Mt. Toby 2, North Colrain 1, Northfield 2, Orange 2 (UMA), Shelburne 1–3 (UCT–UMA), Shutesbury 1, Sunderland 3 (UMA), Warwick 1, Whately 10; Hampden County, Chester 1, Holland 1 (UCT), Springfield 18 (HMCZ), West Springfield 1 (AMNH); Hampshire County, Amherst 45, Northampton 1 (UMA), Worthington 1; Middlesex County, Ashby 3 (UCT), Belmont 4, Cambridge 48 (HMCZ), Groton 13 (UCT), Hudson 4–14 (HMCZ–UCT), Lexington 3, Malden 5, Manchester 1 (HMCZ), Marlboro 1 (UCT), Newton 10, Stoneham 1 (HMCZ), Wilmington 13 (USNM); Nantucket County, Nantucket Island 26–59–1 (HMCZ–UCT–USNM); Norfolk County, Milton 1, Randolph 2 (HMCZ), Wrentham 2 (UCT); Plymouth County, Brocton 1 (UMA), Clark's Island 2 (AMNH), Duxbury 2 (HMCZ), Halifax 1, Hanson 1 (UCT), Marshfield 2 (USNM), Middleboro 2 (HMCZ), Plymouth 2–4 (AMNH–HMCZ), Wareham 60 (HMCZ); Worcester County, Auburn 2 (UMA), Bolton 1–4 (AMNH–UCT), Brookfield 2, Hardwick 2 (UMA), Harvard 17–3 (HMCZ–USNM), Leominster 3, Marlboro 17, Northboro 32 (UCT), Petersham 1, Princeton 1 (AMNH), Mt. Wachusett 1 (HMCZ), Southbridge 1 (UMA), Sterling 4 (UCT), Ware 1 (UMA).

CONNECTICUT, total 583: Fairfield County, Cos Cob 5 (USNM), Easton 1 (UCT); Fairfield County, Fairfield 1–1 (AMNH–UCT), Greenwich 2–1 (AMNH–UCT), Noroton 3 (HMCZ), Norwalk 2, Redding 1, Stratford 9, Weston 5, Westport 3, Wilton 2 (UCT); Hartford County, East Hartford 14–2 (HMCZ–USNM), Glastonbury 17, Granby 1, Hartford 2, Newington 1, Simsbury 6, South Windsor 2, Wethersfield 5, Windsor Locks 3; Litchfield County, Barkhamsted 4, Canaan 1, Kent 2, Lakeville 1, Lime Rock 2, Litchfield 1 (UCT), Macedonia Park 1–1 (AMNH–UCT), Norfolk 5, Salisbury 8 (UCT), Sharon 6–12 (AMNH–UCT); Middlesex County, Clinton 8 (AMNH), Essex 1, Hadden 1, Marlborough 1 (UCT), Westbrook 2 (AMNH); New Haven County, Cheshire 5 (USNM), Guilford 1, Middlebury 3, Milford 2, Oxford 1, Southbury 1, Waterbury 2, Wolcott 1; New London County, Barn Island 62, Bozrak 1, East Lyme 1, Griswold 13, Hopeville 2, Lebanon 5, Ledyard 1 (UCT), Liberty Hill 1 (HMCZ), Manacoke Island 2, Mystic 6 (UCT), Niantic 1 (AMNH), Noank 1, Norwich 2, Preston 1, Stonington 41 (UCT), Waterford 7–9 (AMNH–UCT), Westchester 1; Tolland County, Bolton 4, Columbia 1, Coventry 11, Mansfield 58, Storrs 44, Vernon 27, Willington 8; Windham County, Ashford 5, Brooklyn 71, Chaplin 3, Scotland 4, Thompson 12, Warrenville 8, Willimantic 3, Windham 2 (UCT), Woodstock 3 (AMNH).

RHODE ISLAND, total 49: Newport County, Block Island 17–21 (AMNH–UCT), Newport 1–5 (AMNH–HMCZ), Providence County, Burrillville 1; Washington County, East Beach 1, Grass 2, South Kingston 1 (UCT).

# Beach Vole

*Microtus breweri* (Baird)

**Description.** The beach vole is distinguished from the meadow vole, the most closely related insular and mainland species, by its larger size proportionately shorter tail, and paler and coarser fur. Also, when the skull is viewed from above, the nasals appear wider anteriorly, the interparietal is longer, and the braincase is longer, narrower, and less highly arched, while

*Arvicola breweri* Baird, 1857. Mammals In *Rept. Expl. Surv. Railr. to Pacific* 8(1):525.
*Microtus breweri*, Miller, 1896. *Proc. Boston Soc. Nat. Hist.* 27:83
Type Locality: Muskeget Island, west of Nantucket Island, Massachusetts
Dental Formula:
I 1/1, C 0/0, P 0/0, M 3/3 × 2 = 16
Names: Brewer's vole, beach mouse, and Muskeget Island beach vole. The meaning of *Microtus* = small ear, and *breweri* = for T. M. Brewer, its discoverer.

Muskeget Island
—2 miles—

the occiput is depressed and widened. The posterior upper molar has fewer closed triangles than that of the meadow vole.

The sexes are colored alike, with no apparent seasonal differences; in winter the fur is strikingly pale or light gray throughout, sprinkled with longer blackish hairs on the back and sides. The basal four-fifths of the hairs everywhere are grayish. The belly is whitish and the feet gray or grayish white, and the tail is indistinctly bicolored, brownish above, whitish below. There is usually a white blaze on the forehead and less often on the chin and throat. Immatures are colored much like adults.

Three molts are recognized: juvenile, subadult, and adult. Adult molt occurs seasonally; whereas the other two are age-dependent. The diffuse molt pattern in adults seems to depend on reproductive activity, peaking when reproductive activity is reduced. The juvenile molt is of a dorsad type, involving a uniform sequence of replacement; subadult and adult molts are dorsad, but are also diffuse, characterized by an irregular and blotchy pattern of replacement (Rowsemitt, Kunz, and Tamarin 1975).

Measurements range from 175 to 215 mm (6.8–8.4 in); tail, 41 to 58 mm (1.6–2.3 in); and hind foot 22 to 25 mm (0.86–0.98 in). Weights were not available.

**Distribution.** The beach vole occurs only on Muskeget Island, an island situated off the west coast of Nantucket Island, Nantucket County, Massachusetts. The former distribution of the species included Adam's Island and South Point Island.

Muskeget Island is 8 miles southeast of the extreme eastern point of Martha's Vineyard and only 6 miles northwest of the western end of Nantucket Island and lies just within the terminal moraine of the last glaciation.

**Ecology.** Muskeget Island is a desolate low, sandy island 1.2 by 0.5 miles. Its highest elevation is approximately 15 feet above high-water mark. The surface is irregularly ridged and furrowed, with some deeper hollows containing freshwater ponds or marshes. There are also saltwater marshes and areas of bare sand (Miller 1896). In spite of its barrenness, the island has varied flora: common and abundant plants are eel grass, narrow-leaved cattail, beach grass, marsh grass, bulrush, sedges, bayberry, wild peppergrass, sea rocket, beach plum, beach pea, poison oak, common evening primrose, and beach goldenrod. The island serves as a major breeding ground for herring gulls and great black-backed gulls.

Beach voles construct frail nests or forms anywhere on loose soil or under some shelter. The nest, large enough to hold only one animal, is usually open on the top and is made with fine shreds of beach grass. The natal nest is constructed under the protecting stalks of luxuriant beach goldenrod or under cover of a fragment of wood. When no such convenient shelter is found, the animals build a short nesting burrow from 1 to 2 feet long into the sand at a steep slope of about 45 degrees. In this way the burrow penetrates quickly through the dry, crumbling surface sand to the moister, more compact sand beneath. The end of the burrow is filled with a

BEACH VOLE

bulky nest. Apparently beach voles prefer the lower, wetter, grassy areas.

Beach meadow voles have been preyed upon heavily by house cats (Miller 1896) and by short-eared owls and marsh hawks (Starrett 1958). Much of the beach vole habitat has been destroyed by erosion after hurricanes and storms, and by construction. Muskeget Island has been designated a refuge for nesting terns, and this measure naturally may benefit the beach vole.

**Behavior.** Little is known about the behavior of this species. Beach voles seek cover among the sparse beach vegetation and fragments of driftwood and wreckage. They are adapted for digging in sand, in which they make extensive runways through the beach grass. The runways are less distinct in open areas and seem to be more common in winter.

Beach voles feed mainly on the tender stalks of beach grass, which they obtain either by burrowing short tunnels or by scratching away the sand until the soft parts of the grass are exposed. In autumn they store much grass for winter food by burying it in the sand, where it stays damp and fresh. The entire grass stalk is cached, and the tender parts are selected for food later. Miller (1896) found several caches that contained nearly a peck each of green vegetation.

Very little is known about the reproduction of this species. Miller (1896) reported that beach voles probably breed throughout the warm seasons and produce litters of four or five young.

**Remarks.** The specific status of the beach vole has been questioned by some authorities since its description in 1857. Allen (1869) reduced the beach vole to a race of the meadow vole, because some beach voles were of the ordinary color and some meadow voles inhabiting the sand dunes at Ipswich, Massachusetts, had a light color similar to that of the beach vole. Merriam (1888) again elevated the beach vole to species status. Starrett (1958) supported a specific status for the beach vole on the basis of pelage color and cranial characteristics. Wetherbee, Coppinger, and Walsh (1972) report that they have interbred beach and meadow voles in the laboratory, but they give no further details. It is doubtful whether this is ample evidence for changing the species status of this insular taxon. However, Fivush, Parker, and Tamarin (1975), on the basis of the examination of the karyotype of eighty-two specimens of beach voles taken at Muskeget Island, concluded that "the relatively recent isolation of Muskeget Island has not provided sufficient time for karyotype differentiation of the beach vole, and perhaps the period of isolation has been too short for true species differentiation to have occurred."

**Specimens Examined.** MASSACHUSETTS, total 92: Dukes County, Muskeget Island 34–52 (HMCZ–UCT), South Point Island 6 (HMCZ).

# Rock Vole

*Microtus chrotorrhinus* (Miller)

*Arvicola chrotorrhinus* Miller, 1894.
  *Proc. Boston Soc. Nat. Hist.* 26:190.
*Microtus chrotorrhinus,* Bangs, 1896.
  *Proc. Biol. Soc. Washington* 10:49
Type Locality: Head of Tuckerman's
  Ravine, Mount Washington, Coos
  County New Hampshire
Dental Formula:
  I 1/1, C 0/0, P 0/0, M 3/3 × 2 = 16
Names: Yellow-nosed vole and fern
  vole. The meaning of *Microtus* =
  small ear, and *chrotorrhinus* =
  yellow-orange color of the nose, or
  yellow nose.

**Description.** The rock vole looks much like the meadow vole, except for the yellow-orange nose and to a lesser extent a yellowish color below the ears and rump. The pattern of the molar teeth is different—the last upper molar (M3) has five closed triangles. The skull is thin and smooth, with the incisive foramen short and wide and the auditory bulla large.

In molting, adults pass through a sacral subtype of a cephalosacral (head to rump) sequence before progressing through the diffuse type of molt sequence considered characteristic of the genus *Microtus* (Martin 1973).

The sexes are equal in size. Measurements range from 140 to 178 mm (5.5–6.9 in); tail, 45 to 50 mm (1.8–2.0 in); and hind foot, 19 to 22 mm (0.74–0.86 in). Weights vary from 30 to 40 g (1.05–1.40 oz).

**Distribution.** This species occurs from Labrador, eastern Quebec, northern New Brunswick, and central Ontario west to the east side of Lake Superior and northeastern Minnesota, and from Maine, New Hampshire, and Vermont into the Adirondacks and Catskills in New York and Pennsylvania and southward along the crest of the Appalachians into North Carolina, Tennessee, and probably Kentucky.

The apparent rarity of this species in scientific collections is probably due to the limited

amount of trapping within its restricted ecological niche rather than to actual scarcity of numbers.

**Ecology.** Rock voles are found in deep, cool, very damp crevices along streams among moss-covered rocks in areas with moderate growth of spruce and fir, occasional birches, and much understory. They are generally encountered at elevations above 3,000 feet—at 5,300 feet on Mount Washington, the type locality—and as low as 1,300 feet in "ice cave" areas where ice and snow persist under the rocks into summer.

Man, short-tailed shrews, and bobcats are known to prey on rock voles. Fleas, cestodes, and nematodes parasitize them.

**Behavior.** Little is known regarding the life history of rock voles. They do not appear to hibernate even in extreme weather conditions as on Mount Washington. They seem to be wholly diurnal, with feeding activity greatest in the morning. Rock voles tend to establish sanitary stations on the surface, rather than dropping their scats at random like most other native voles. They seem to require high humidity and much surface water for survival.

Rock voles cut the leaves of plants and carry them between rocks or under logs to eat. They eat alpine goldenrod, mountain avens (Peck's gum), bunchberry, blackberry, false miterwort, mayflower, violet, moss, and grasses (Martin 1965).

The breeding season is from late March to mid-October. Two or three litters may be produced during the summer. The number of young per litter is one to seven, usually four, and litter size tends to be largest in June and larger in more northerly latitudes. Females are generally sexually mature at above 140 mm (5.5 in) total length and 30 g (1.05 oz) body weight, and males are generally sexually mature above 150 mm (5.9 in) total length and 30 g (1.05 oz) body weight (Martin 1971).

**Specimens Examined.** MAINE, total 1: Somerset County, Enchanted Pond 1 (AMNH).

NEW HAMPSHIRE, total 59: Cheshire County, Keene 1 (DC); Coos County, Mt. Washington 3–8–4–1–22–12 (AMNH–DC–HMCZ–UCT–UNH–USNM); Grafton County, Franconia 1–1 (DC–HMCZ), Mt. Moosilauke 2–3 (AMNH–DC), Profile Mt. 1 (HMCZ).

VERMONT, total 7: Chittenden County, Underhill (Nebraska Notch) 1 (UVT); Essex County, Island Pond 5; Lamoille County, Cambridge 1 (DC).

**Records and Reports.** MAINE, total 10: Somerset County, Mt. Coburn 3 (Wyman 1923), Enchanted Pond 1 (Martin 1975); Piscataquis County, Mt. Katahdin 2 (Heinrich 1953); Franklin County, Mt. Sugarloaf 4 (Martin 1975).

ROCK VOLE

# Pine Vole

*Microtus pinetorum scalopsoides*
(Audubon and Bachman)

**Description.** The pine vole resembles the meadow vole but may be distinguished by its shorter tail (less than 25 mm [1 in] long), shorter ears, and smooth, velvety russet fur which gives the animal its molelike appearance. The front feet are enlarged for digging. The front foot has four toes and a small thumb, and the hind foot has five toes. Anal glands are present and well-developed in breeding males. Females have four mammae, all inguinal.

The skull is flat, wide, and weak, with a short rostrum, small squamosal crests, a braincase that is quadrate posteriorly, and small auditory bullae; palatal slits end forward the cheek-tooth row. Although the upper incisors are not grooved on the anterior surfaces, Fish and Whitaker (1971) found a female pine vole with well-developed grooves on the interior surfaces of the upper incisors. The narrow, unrooted molars are persistently growing. The upper third molars have two closed triangles, the lower first molars have three closed and two

*Arvicola scalopsoides* Audubon and Bachman, 1841. *Proc. Acad. Nat. Sci. Philadelphia* 1:97.

*Microtus pinetorum scalopsoides,* Batchelder, 1896. *Proc. Boston Soc. Nat. Hist.* 27:187.

Type Locality: Long Island, New York

Dental Formula:

I 1/1, C 0/0, P 0/0, M 3/3 × 2 = 16

Names: Woodland vole, pine mouse, and mole mouse. The meaning of *microtus* = small ear; *pinetorum* = belonging to the pines (a misnomer, because this vole is seldom found in pine woods, although the type specimen was taken in pine woods of Georgia); and *scalopsoides* = to dig, or digging animals.

open triangles, the lower second molars have the anterior pair of triangles confluent, and there are three transverse loops in the lower third molars.

The sexes are colored alike and show some seasonal variation. Some individuals are much darker or paler than others. The coloration of the upperparts is bright chestnut and the underparts are slate gray or silvery gray. The back and rump occasionally have black-tipped hairs. The tail is indistinctly bicolored, brownish above and paler below, the feet are brownish gray. Immatures are much paler or more grayish than adults. Buffy and white-spotted, cream-colored, orange-yellow, and albino individuals occasionally occur.

The biannual molt is more marked than in the meadow vole. The darker winter fur is usually molted in May and June, although molting may occur throughout the summer. Molting starts on the head and proceeds backward to the shoulders, beginning simultaneously on the belly. The rump and haunches are usually the last to molt. The autumn molt begins in November and is completed in early December. The juvenile fur is replaced during the third week of life, when molting begins on the sides near the forelimbs, proceeds forward and upward over the shoulders and head, backward along the sides, belly, and back, and ends on the rump. The adult pelage may be attained at 7 to 10 weeks.

The sexes are nearly equal in size. Measurements range from 110 to 135 mm (4.3–5.3 in); tail, 15 to 26 mm (0.59–1.01 in); and hind foot, 14 to 20 mm (0.55–0.78 in). Weights vary from 20 to 37 g (0.70–1.30 oz).

**Distribution.** Pine voles occur from north central New England west to southern Ontario, west

central Wisconsin, and southern Minnesota, south through eastern Kansas, eastern Oklahoma, and central Texas, east to central Georgia, and north along the Atlantic coast. In New England the northernmost range is Essex County in northeastern Vermont. The species is probably extending its range northward into southern Maine and southern Quebec.

**Ecology.** Pine voles are found in a wide range of habitats, from sea level to over 2,500 feet. They prefer woodland and grassland habitats but have also been taken in rocky areas, marshes, and swamps. These voles are the most fossorial of our microtine rodents, digging their burrows in light, loose, well-drained humus soils. In digging a vole uses its head, neck, and forefeet to loosen the dirt, which is thrown backward by the hind feet. When the vole has proceeded a foot or more it turns around and pushes the loose dirt out of the burrow with its head. Voles seldom leave the burrow, and they feed there almost exclusively.

The burrows are extensive and may be dug just beneath the thick leaf mold, but more often they are dug 3 to 4 inches below the surface, and rarely to a depth of a foot or more. The burrows open frequently to the surface and are visible as short ridges similar to the burrow ridges of star-nosed moles. Caches of fruits, tubers, and rootstocks, some containing a gallon or more, may be found in burrows. The burrows often run along roots of trees, which may be stripped of their bark, and sometimes become complex systems around the trunk of a tree completely girdling it to a depth of a foot or more. The diameter of the burrows varies from ¾ of an inch to 2 inches, depending on the type of soil. Surface and subsurface runways are usually

PINE VOLE

found within the drip line of a tree. Pine voles often use burrows of other voles, mice, shrews, and moles.

The globular nests are made of shredded plant stems, leaves, or rootlets of available plants or of cloth material. They are from 6 to 7 inches in diameter. They are usually found under rocks, logs, stumps, or debris near the surface of the ground or slightly below, while in orchards they are often found at depths of several inches to a foot near the trunk of a tree. There may be three or four exits leading away from a nest. The nests harbor great numbers of insects, such as springtails, staphylinid beetles, and parasitic mites.

Man, short-tailed shrews, house cats, skunks, weasels, dogs, foxes, coyotes, raccoons, mink, owls, hawks, and snakes prey on pine voles. They are parasitized by mites, lice, fleas, nematodes, and acanthocephalan worms.

**Behavior.** Pine voles spend most of their time underground but occasionally venture above ground, particularly during the summer. They are active throughout the year; apparently they are more active by day underground and more likely to appear above ground at night.

Pine voles are gregarious and aggressive and often fight among themselves. They often eat their own kind caught in traps but rarely kill one another in combat. They chatter harshly when fighting and as an alarm emit a single or double note like that of a wood thrush.

These voles can run up to 3.8 miles per hour (Layne and Benton 1954). They are poor climbers but good swimmers. Pine voles may often be found with meadow voles, field mice, lemmings, jumping mice, shrews, and moles. Although Hamilton (1938) and Linduska (1942) reported that pine vole populations are cyclic, Benton (1955a) found no evidence of cyclic tendencies, reporting that the animals do not remain in the same area from year to year but emigrate slowly into new areas.

Pine voles do not range far. Home range diameters have been reported to be: 38 yards (Burt 1940; Stickel 1954); 37.1 yards (Miller and Getz 1969); 21 yards (Benton 1955a); 20.8 to 34.2 yards (Paul 1966). Benton (1955a) recorded that a pine vole released 150 feet from its point of capture returned to its home tree within 24

hours. Fitch (1958) found that 70% of pine voles recaptured moved less than 10 yards between captures. Stickel and Warbach (1960) recorded movements of less than 40 yards for 14 of 16 voles trapped four or more times.

Pine voles feed on a wide variety of succulent bulbs, tubers, seeds, berries, acorns, roots, bark, leaves, and fruits, and occasionally on insects and larvae and other animal matter. In orchards they burrow just beneath the thick carpet of grass to reach fallen apples, then burrow back up to the surface to eat the fruit.

Pine voles are less prolific than meadow voles. The breeding season extends from mid-February to mid-November in southern New England (Miller and Getz 1969). The gestation period is approximately 24 days. Pine voles have a low reproductive potential and each year produce one small litter of two to four young, rarely more.

During courtship the female is the aggressor; she initiates copulation by seizing the male by the flank or hip with her teeth, and at times she may drag him backward. A vaginal plug forms after copulation (Benton 1955a).

The newborns are similar in appearance to meadow voles. At birth they weigh 2.2 g (0.08 oz). They grow rapidly, and at 9 days of age they crawl about vigorously and are mostly furred, and their eyes begin to open. By the 17th day they are weaned and begin eating solid food. At 24 days of age the young are completely furred and very active and weigh 15 g (0.53 oz). After this period they grow about 2.0 g (0.07 oz) per day until they attain their adult weight (Benton 1955a). They become sexually mature at 2 months of age (Hamilton 1938). Longevity is not known.

**Specimens Examined.** NEW HAMPSHIRE, total 5: Carroll County, Ossipee 1 (UCT); Rockingham County, Brentwood 1; Strafford County, Durham 3 (UNH).

VERMONT, total 19: Caledonia County, Hardwick 3, Lyndon 9; Essex County, Brighton 1; Windham County, Saxton's River 6 (UCT).

MASSACHUSETTS, total 108: Barnstable County, Yarmouth 1; Berkshire County, Peru 1; Franklin County, Whately 1; Hampden County, Brimfield 3 (UCT), East Longmeadow 2 (UMA), Montgomery 2; Hampshire County, Amherst 2, Belchertown 3, Leverett 1, Pelham 1 (UMA); Middlesex County, Ashby 21, Groton 16, Hudson 11, Marlboro 8; Worcester County, Harvard 16,

Leominster 1, Northboro 13 (UCT), Oxford 4 (UMA), Upton 1 (UNH).

CONNECTICUT, total 80: Fairfield County, Cos Cob 3 (USNM), Fairfield 4 (AMNH); Hartford County, Windsor Locks 1; Litchfield County, Barkhamsted 2 (UCT), Litchfield 4 (UMA); Middlesex County, Clinton 8 (AMNH); New Haven County, Cheshire 3 (USNM); New London County, Griswold 8, Leyard 2; Tolland County, Mansfield 5, Storrs 26, Tolland 1, Union 1, Vernon 1; Windham County, Ashford 1, Brooklyn 10 (UCT).

# Muskrat

*Ondatra zibethicus* (Linnaeus)

[*Castor*]*zibethicus* Linnaeus, 1766.
 *Systema naturae*, 12th ed., 1:79.
*Ondatra* z[*ibethicus*] *zibethicus*, Davis
 and Lowery, 1940. *J. Mamm.*
 21(2):212.
Type Locality: Eastern Canada
Dental Formula:
 I 1/1, C 0/0, P 0/0, M 3/3 × 2 = 16
Names: Musk beaver and mudcat.
 The meaning of *Ondatra* = the
 Iroquois name for the muskrat, and
 *zibethicus* = musky odor of the
 animal.

**Description.** The muskrat is about the size of a house cat. It has a long, thick, laterally compressed tail, sparsely haired and covered with small scales, which serves as an efficient rudder and scull for swimming. The head is broad and blunt, with small beady black eyes and short ears, covered with hair, that barely extend beyond the fur. The mouth is valvular, like a beaver's, permitting the muskrat to gnaw underwater. The small forefeet have four sharp claws and a thumbnail for digging. The hind feet are large and broad, with five clawed toes which are partly webbed and fringed with short stiff hairs. A pair of well-developed perineal musk glands, situated under the skin near the anus, give off a musky odor that is particularly strong during the breeding season, accounting for the name muskrat. Females have six mammae—two pectoral and four inguinal.

The skull resembles that of a meadow vole but is much larger and more massive. The incisors are not grooved, and the molars are rooted. The first lower molar has six triangles (the first not closed) between the anterior loop and posterior loop. The third lower molar has three outer salient angles.

The pelage consists of dense, soft, practically waterproof gray underfur, overlain by long brown, glossy guard hairs. The sexes are colored alike and show slight seasonal variation. The coloration of the upperparts is a rich brown, darkest on the head, nose, and back. The sides are grayish brown to russet and the underparts are considerably lighter, varying from pale gray to bright cinnamon. The tail is blackish brown. Immatures are uniformly dusky on the back and paler on the sides and belly. Color phases occur as fawn and yellow and as silver; albinos are less frequent. Dark brown and nearly all black animals occur more commonly. The muskrat apparently molts once each year, during the summer months.

The sexes are about equal in size. Measurements range from 55 to 64 cm (21.7–25.3 in); tail, 25 to 32 cm (9.9–12.6 in); and hind foot, 8 to 9 cm (3.2–3.6 in). Weights vary from 0.68 to 1.82 kg (1.5–4.0 lb).

**Distribution.** This species occurs over most of North America north of Mexico, except for Florida, a coastal strip in Georgia and South Carolina, and most of Texas and California.

MUSKRAT

**Ecology.** Muskrats are semiaquatic rodents. They are found in heavy growths of herbaceous vegetation near slow-running water, swamps, marshes, and bogs, and along shores of creeks, canals, ponds, and lakes that have no large or sudden fluctuations in depth. Muskrats may dig a den in a bank or build a house with aquatic plants. Muskrats living along the edge of woods, stream banks, and drainage ditches often do not build houses, but burrow into the stream banks. A bank den may have one or more dry chambers, well above water and ventilated by small holes hidden under a pile of roots, shrubs, trees, or thick vegetation. One or more underwater entrances open into tunnels, of various lengths and diameters, which slope upward to the chamber. The chamber is several feet within the bank and contains a bulky nest composed of dried vegetation. Sometimes channels or leads extend from burrow entrances out into deeper water, preventing ice from closing the passages. Such channels vary in length and are prominent when the water level drops in dry periods.

When the banks are too low and shallow for a den, a house is built where the water is about 2 feet deep. It may be constructed on the foundation of a submerged old stump, brush pile, or log, or built directly on the bottom of the marsh or swamp. Sometimes a house is not surrounded by open water but is connected to the water by short underground water tunnels. The cone-shaped house is constructed of piled-up cattail stalks, roots, remains of plants, as well as mud brought to the site from an area of 10 feet around the house foundation. In this way the foundation is gradually built up and the surrounding water deepened and cleared of vegetation. As the foundation rises above the water level, less mud and more vegetation is used in the construction. The muskrats gnaw a tunnel from the bottom upward through the foundation, in most cases clear to the top of the house. More vegetation and mud are then heaped on top of the surface chamber. As the roof settles, the animals gnaw away the encroaching ceiling until the mass becomes stabilized so that the irregularly shaped chamber is enlarged to a foot or more in diameter and is high enough so the animals can move about freely.

Occasionally muskrats may make two separate chambers, each inhabited by a different family. Litter accumulates on the floor, which gradually becomes elevated. The chamber is usually a few inches above high-water level and is lined with fine grasses which are at all times kept clean. The walls of the chamber are at least a foot thick. In summer the flimsy house can easily be broken open, but after a few days of freezing weather it becomes a fortress against predators and gives protection from severe winter weather.

Each house usually has two or more underwater entrances, of which one or two are in the form of a plunge pool used for escape. There is considerable variation in the size of the houses, depending on the kind and amount of vegetation available and the number of seasons the house is occupied. The average house may be from 1 to 4 feet high and 8 to 10 feet in diameter at the water level. Houses that have been occupied for several seasons are much larger than a newly built house.

The muskrat continues building throughout the year. New houses seen in late summer and early fall are often the work of young muskrats.

Muskrats construct one or several feeding huts or shelter huts, called "feeders," surrounding the house at a distance. A feeder is simply a roofed feeding platform made of aquatic plants where the animal can bring food and eat without interference from predators or inclement weather. A feeder is roughly circular, much smaller than the house, and several runs lead to it from underneath the water. The feeders are used throughout the winter in the south and during less severe winter weather in the north.

Often, as soon as the ice forms muskrats cut a 4- to 5-inch hole through the ice and push up through it 12- to 18-inch lumpy, globular piles of fine fibrous roots, water weeds, and other submergent plants. These piles are different from the bulkier stems and leaves of emergent plants used in constructing feeding stations. Within the pile the animal excavates a cavity at ice level, which serves as shelter and breathing space and which may be used as a feeding station during inclement weather. These small "pushups" or "breather houses" are not prominent in deep snow but become distinct after the snow disappears. They collapse when the ice melts. Breather houses appear to be largely confined to the deeper channels, the edges of rivers, and lakes.

Man, mink, dogs, coyotes, red foxes, raccoons,

skunks, river otters, weasels, bobcats, house cats, great horned owls, marsh hawks, snapping turtles, large snakes, pickerel, large-mouth bass, and northern pike prey on muskrats. Some die from fires, floods, drought, freezing weather, automobiles, and intraspecific strife. Fleas, cestodes, nematodes, and trematodes parasitize muskrats, and many die from infectious hemorrhagic disease. They are susceptible to tularemia, yellow-fat disease, and porocephaliasis.

**Behavior.** Muskrats are active throughout the year; although they are mainly nocturnal, they may be seen during the daytime building houses, swimming, or sunning themselves on logs or on muskrat houses. They are wary and aggressive and may attack men or other muskrats that come near them.

Muskrats tend to be territorial—particularly females with young. Some females may even kill a few newly weaned young while persuading them to leave home before a new litter is born (Errington 1961). Intraspecific strife between muskrats leads to cannibalism (Errington 1940). Although muskrats live singly or with mates for most of the year, several muskrats may share a house during the winter.

As a rule muskrats spend their entire lives near their houses, but they may be forced to move when living conditions become unfavorable: for example, when the water level drops too low, when a marsh dries up, or when there is overcrowding. Their normal home range is within 200 yards of their house or den (Errington and Errington 1937) and tends to be more or less circular toward the center of marshes and "strip-shaped" along banks, extending out from bank burrows several hundred feet into deeper water (Errington 1937). Muskrats living in small ponds are more likely to wander than those in larger ponds, and muskrats inhabiting streams tend to move out along the stream, whereas muskrats in ponds tend to move from pond to pond (Shanks and Arthur 1952).

Muskrats move mainly in the spring and fall; they may travel overland more than 20 miles from their homesites in search of more favorable habitats (Errington 1939). They can be expected to emigrate before winter in marshes subject to late summer drought, and many of these wanderers are killed by automobiles in autumn.

Muskrats may also be forced to emigrate by abnormally high water in late winter or early spring, at the advent of the breeding season. During droughts muskrats may occupy woodchuck burrows and other holes, but eventually they return to water (Errington 1961).

For survival, muskrats spend much of their lives in their tunnels, burrows, houses, and feeding shelters, or in the water. They cannot endure severe cold or freezing wind and are vulnerable to sudden changes of air temperature about the time of winter thaw. A drop to subzero temperatures kills many exposed muskrats. Adults can maintain themselves for weeks in hot weather with no water except what they get from moist foods, but immatures and subadults may die of thirst under similar conditions (Errington 1939).

These animals are rather quiet except during copulation. Salinger (1950) heard repeated loud, birdlike squeals from a female muskrat during an entire copulation which lasted 5 minutes.

Muskrats swim and dive much like a beaver. The powerful hind feet stroke alternately to propel the animal along, and the sinewy tail acts as a scull. Muskrats swim 1 to 3 miles per hour and can even swim backward for some distance (Peterson 1950). They can swim underwater for some 150 feet and can stay submerged for 15 minutes or more (Jackson 1961). They make a loud warning splash by slapping the water with the tail, somewhat as beavers do.

Muskrats feed mainly on cattails, arrowhead or duck potato, bur reed, bulrushes, pondweed, swamp loosestrife, duckweed, pickerelweed, and water lilies; occasionally, they also feed on smartweed, dandelion roots, and a host of other tender and succulent plants. They also eat corn, clover, alfalfa, soybeans, carrots, apples and other fruits, insects, crayfish, freshwater clams, snails, mussels, frogs, reptiles, young turtles, minnows, sluggish fish such as bullheads, and young birds and carrion. O'Neil (1949) reported that muskrats eat about one-third of their weight per day.

In Louisiana the breeding season is year-round (O'Neil 1949), but in New York muskrats probably breed in March, the first litter being born in April or May, a second in June or July, and possibly a third in August (Alexander 1956). In Maryland ovulation begins about the first of

February, and three litters may be produced by late May (Enders 1939). In Manitoba, females become sexually active earlier in the year than males and may pass through three or four estrous cycles before the males display sexual interest (McLeod and Bondar 1952).

The gestation period appears to be 28 or 30 days. Muskrats may produce one to five litters annually, but three or four litters is most common. The number of young per litter varies from one to fourteen, and six or seven is usual. Younger females tend to produce smaller litters, though older females may produce small litters when unfavorable conditions exist. In spring the young are usually born in a burrow or house; in summer they are born in rather bulky nests made of dry aquatic plants in a brush pile above high-water level.

Newborns are blind, helpless, and almost naked, approximately 102 mm (4.0 in) long, and weigh nearly 22 g (0.77 oz). They cling tightly to the mother's nipples. If disturbed the female may plunge into the water, and the clinging young are thus often pulled underwater and drowned. If some of them are torn loose the female usually attempts to retrieve them. The young are noisy and squeal when disturbed; the female will carry them away from danger. They grow rapidly, and their eyes open between 14 and 16 days after birth. At this time they are well-furred, begin to nibble on succulent plants, and are able to swim, dive, and climb. They are weaned at about 2 months old. They become almost adult size at 6 or 7 months of age. Females born in early spring may mate in autumn of the same year. Longevity is about 3 to 4 years.

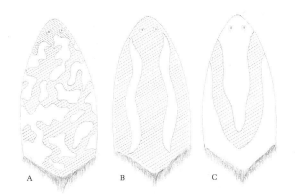

Dorsal view of stretched raw pelt primeness in muskrats taken during autumn and early winter. *A*, adult, with typical blotchy pattern; *B* and *C*, juvenile, showing classes of pelt patterns. Shaded areas denote unprimed sections of pelt; white areas are primed.

**Sex Determination.** The sex of muskrats can be determined by the presence or absence of the penis. Grasp the urinary papilla between forefinger and thumb and strip posteriorly. If the penis is present it will be either felt or exposed. The presence or absence of visible nipples will indicate sex in young muskrats not yet fully furred. Females show teat marks on the flesh side of pelts, making it relatively easy to determine the sex of the pelt.

**Age Determination.** Several reliable methods have been used to determine age in muskrats: the fluting of the first upper molar distinguishes three age classes—juveniles, subadults, and adults (Sather 1954; Olsen 1959), and the zygomatic breadth separates subadults and adults (Alexander 1951; Alexander 1960). Ossification of the baculum has been used by Elder and Shanks (1962), and primeness pattern on the dorsal side of cased pelts by Shanks (1948).

**Specimens Examined.** MAINE, total 22: Aroostook County, Ashland 1 (USNM), Madawaska 2; Hancock County, Bucksport 4 (HMCZ); Kennebec County, Randolph 1; Penobscot County, Holden 1, Lincoln 2 (UME), Norcross 1 (HMCZ), Old Town 1, Orono 1 (UME); Piscataquis County, Chamberlain Lake (head of) 1 (HMCZ), Kokadjo 4 (USNM); Sagadahoc County, Richmond 1 (UME); Washington County, East Machias 2 (HMCZ).

NEW HAMPSHIRE, total 15: Coos County, Errol 1; Strafford County, Durham 1 (UNH); Grafton County, Hanover 10 (DC); Merrimack County, Webster 1 (HMCZ); Sullivan County, Charlestown 2 (USNM).

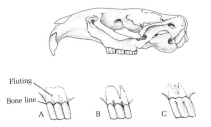

Dental pattern of cheek teeth in the muskrat. *A*, immature, with very little root development, fluting extending deep into alveolar socket; *B*, subadult, with moderate root development, end of fluting near point of emergence from bone line; *C*, adult, with roots well developed, end of fluting extending well below bone line.

VERMONT, total 21: Addison County, Ferrisburg 2, Ripton 1; Chittenden County, Bolton 1, Milton 1 (UVT); Franklin County, Enosburg Falls 1 (HMCZ), Fairfax 2, Missisquois National Wildlife Refuge 1 (UVT); Grand Isle County, Grand Isle 7 (HMCZ); Rutland County, Rutland 1 (DC); Washington County, Barre 1, Marshfield 1 (UVT); Windham County, Saxton's River 1 (UCT); Windsor County, Bridgewater 1 (USNM).

MASSACHUSETTS, total 97: Barnstable County, Barnstable 2, North Truro 12, Provincetown 1; Bristol County, Swansea 1 (UCT); Dukes County, Martha's Vineyard 1; Essex County, Hamilton 1 (HMCZ), Ipswich 1, Newberryport 2 (USNM), Peabody 1 (HMCZ); Franklin County, Greenfield 1; Hampden County, Westfield 4; Hampshire County, Amherst 8, Hadley 1, Northampton 2 (UMA); Middlesex County, Belmont 2–16 (HMCZ–USNM), Cambridge 6 (HMCZ), Concord 2 (UMA), Wilmington 8, Woburn 2 (USNM), Wyoming 2 (HMCZ); Norfolk County, Stoughton 1 (USNM), West Quincy 2;

Plymouth County, Eastham 1 (HMCZ), Middleboro 2 (USNM), Wareham 12 (HMCZ); Worcester County, Berlin 1 (UMA), Dudley 1 (UCT), Harvard 1 (HMCZ).

CONNECTICUT, total 338: Fairfield County, Fairfield 1, Monroe 1 (UCT), Ridgefield 1 (AMNH), Westport 3; Hartford County, Bloomfield 1 (UCT), East Hartford 4–1–2 (HMCZ–UCT–USNM), Hartford 2 (UCT), Windsor 1 (AMNH); Litchfield County, Litchfield 1 (UCT), Plymouth 1, Watertown 1 (AMNH); Middlesex County, Portland 3; New Haven County, Derby 1 (UCT), New Haven 2 (USNM), Waterbury 1; New London County, East Lyme 204 (UCT), Liberty Hill 1 (HMCZ), Stonington 1, Waterford 1; Tolland County, Mansfield 64, Storrs 6, Stafford 1, Tolland 3, Willington 2; Windham County, Brooklyn 2, Canterbury 3, Chaplin 8 (UCT), South Woodstock 1 (AMNH), Thompson 14 (UCT).

RHODE ISLAND, total 20: Newport County, Newport 16; Washington County (no location) 3 (USNM), Kingston 1 (URI).

# Southern Bog Lemming

*Synaptomys cooperi*

Two subspecies of southern bog lemming are recognized in New England:
*Synaptomys c. cooperi* Baird. *Synaptomys cooperi* Baird, 1858. *Mammals In Repts. Expl. Surv...*, 8(1):558.
Type Locality: fixed by Bole and Moulthrop, 1943, *Sci. Publ. Cleveland Mus. Nat. Hist.* 5:146, at Jackson, Carroll County, New Hampshire
*Synaptomys c. stonei* Rhoads. *Synaptomys cooperi stonei* Rhoads, 1893. *Amer. Nat.* 27:53. *Synaptomys c. stonei* Rhoads, 1897. *Proc. Acad. Nat. Sci. Philadelphia* 49:305.
Type Locality: May's Landing (on Egg River), Atlantic County, New Jersey (Hall and Kelson 1959)
Dental Formula:
I 1/1, C 0/0, P 0/0, M 3/3 × 2 = 16
Names: Cooper's bog lemming, lemming vole, lemming mouse, and bog mouse. The meaning of *Synaptomys* = a link of physical characters joining mice and true lemmings and *cooperi* = for William Cooper, a zoologist.

**Description.** The southern bog lemming resembles the eastern meadow vole but is smaller and has a much shorter tail. Also, it differs in cranial and dental characters. The skull has well-developed supraorbital ridges forming median interorbital crests in adults; the squamosal crests are well developed, the rostrum is very short (less than one-fourth the total skull length); the auditory bullae are large; and the zygomatic arches are heavy. The length of the cheek tooth row is greater than the length of the anterior palatine foramina. The upper incisors are grooved, and there are closed triangles on the outer surface of the mandibular molars.

The eyes are small and the ears are nearly concealed in the long, loose, shaggy fur. The lips close behind the orange incisors as the animal gnaws. The tail is scarcely longer than the hind foot. The forefoot has four toes and the hind foot five. Females have six mammae—two pectoral, four inguinal.

The sexes are colored alike, with no apparent seasonal variation. The upperparts are brown to chestnut, with a grizzled appearance. The sides and underparts are silvery, with no sharp line of demarcation on the sides. The tail is indistinctly bicolored, brownish above and whitish below, and the feet are brownish black. Old males may have white hairs growing from the center of the hip glands. Immatures are darker than adults. Albino and melanistic individuals are rare.

There appears to be a single annual molt from spring to autumn.

The sexes are equal in size. Measurements range from 115 to 135 mm (4.5–5.3 in); tail, 18 to 24 mm (0.7–0.9 in); and hind foot, 18 to 20 mm (0.70–0.78 in). Weights vary from 20 to 40 g (0.7–1.4 oz).

**Distribution.** This species occurs from Godbout, Quebec, west to southeastern Manitoba, south to southwestern Kansas, east through northern Arkansas and Kentucky, south to western North Carolina and Virginia and western Maryland to the Atlantic coastal plain of Maryland, Delaware, and New Jersey, and northward through New England to Cape Breton Island.

**Ecology.** Southern bog lemmings mainly inhabit sphagnum bogs, but they are sometimes found in woodland habitats, including beech-maple, oak-hickory, pines, and hemlocks, in dense stands of bluegrass or other herbaceous vegetation, in orchards and marshes and in shocked corn. They prefer habitats of deep, thick leaf mold. The burrows form complex series of short tunnels with side chambers which are used for feeding, resting, and storing food. The burrows are from 1 to 2 inches in diameter, and are 6 to 12 inches below the surface; they sometimes run through burrows of hairy-tailed moles.

The surface runways are well-defined and

SOUTHERN BOG LEMMING

maintained and often cross one another. They may contain scattered little piles of cut grass stems and bright green oval scats.

The globular nest is usually an enlarged section of the burrow, lined with dried leaves of grasses and sedges and sometimes with bits of fur and feathers. Its outside diameter is from 6 to 8 inches, and its inside diameter is about 2 or 3 inches. It may have two to four entrances. In winter the nest may be found 4 to 6 inches underground; in summer it is often concealed in tussocks of grass or amid other surface cover.

Man, numerous predatory mammals, hawks and owls, and snakes prey on southern bog lemmings. They are hosts to fleas, lice, cestodes, and nematodes.

**Behavior.** Little is known about the life history of this species. Southern bog lemmings live in small local colonies. They are often found with red-backed mice, meadow voles, moles, shrews,

white-footed mice, and deer mice. They are active throughout the year, and though mainly nocturnal they may be seen during daylight. They are gentle animals who generally travel slowly, but when frightened they can move fast. They swim well. Buckner (1957) reported that the home range is between 0.8 and 1.0 square acre or an equivalent cruising radius of from 105 to 118 feet.

Southern bog lemmings feed mainly on succulent leaves, stems, and seeds of grasses and sedges, and occasionally on fungi, moss, stems of woodland sedges, ground pine, bark, and insects.

The breeding season lasts throughout the year, and the gestation period is 21 to 23 days. Several litters are produced annually which may contain from one to eight young, but two to five is usual. Connor (1959) indicated that the average female produces a litter every 67 days during the spring and summer in New Jersey. Newborns are blind, naked, and helpless and weigh about 2.5 g (0.09 oz). Hair appears by the 6th day and the eyes open on the 12th day after birth.

**Specimens Examined.** MAINE, total 4: Aroostook County, T11–R11 1, T12–R15 2, T12–R15 1 (UCT).

NEW HAMPSHIRE, total 2: Coos County, Mt. Washington 1 (AMNH); Grafton County, Lyme 1 (DC).

VERMONT, total 14: Rutland County, East Wallingford 1 (DC), Mendon 1–1–7 (AMNH–DC–UCT), Rutland 2 (UCT); Windham County, Saxton's River 2 (UCT).

MASSACHUSETTS, total 1: Suffolk County, Boston 1 (AMNH).

CONNECTICUT, total 6: New London County, Griswold 1; Tolland County, Mansfield 1, Storrs 2; Windham County, Ashford 1, South Windham 1 (UCT).

# Northern Bog Lemming

*Synaptomys borealis sphagnicola*
Preble

**Description.** This species differs from the southern bog lemming in lacking closed triangles on the outer surface of the mandibular molars, and having a palate with sharp median projection pointed backward. Also, females have eight mammae instead of six as in the southern bog lemming. The sexes are colored alike, and there appears to be no seasonal variation. The coloration of the upperparts is dull brown, brighter on the rump and more grizzled anteriorly. The underparts are grayish. The tail is bicolored—

brownish above, paler below. The feet are dark grayish to almost black.

The sexes are equal in size. Measurements range from 118 to 135 mm (4.6–5.3 in); tail, 19 to 27 mm (0.7–1.1 in); and hind foot, 16 to 22 mm (0.6–0.9 in). Weights vary from 32 to 34 g (1.1–1.2 oz).

**Distribution.** This northern species occurs from Alaska and British Columbia south to northern Washington and Idaho, across Canada to north-

NORTHERN BOG LEMMING

*Synaptomys (Mictomys) sphagnicola* Preble, 1899. *Proc. Biol. Soc. Washington* 13:43.

*Synaptomys borealis sphagnicola*, A. B. Howell, 1927. *N. Amer. Fauna* 50:9.

Type Locality: Fabyans, near base of Mount Washington, Coos County, New Hampshire

Dental Formula:
I 1/1, C 0/0, P 0/0, M 3/3 × 2 = 16

Names: Preble's bog lemming. The meaning of *Synaptomys* = a link of physical characters joining mice and true lemmings; *borealis* = northern; and *sphagnicola* = of sphagnum bogs.

ern Wisconsin, and northward to Labrador. The disjunct subspecies *sphagnicola* is found in northern New Hampshire, north to and from Mount Katahdin, from Maine to New Brunswick and in the portion of Quebec east and south of the Saint Lawrence River.

**Ecology.** Northern bog lemmings occur in cold sphagnum bogs, in bluegrass fields matted with weeds, and in dense hemlock and beech woods. The animals may construct crisscrossing runways above ground or may burrow just beneath the leaf mold. The nest is lined with dried leaves and grasses and sometimes with fur. It may be found several inches below the ground or may be on the surface, concealed in vegetation. The

enemies of this species are probably similar to those that prey on and parasitize the southern bog lemming.

**Behavior.** Very little is known about the behavior of northern bog lemmings. They are sociable animals and may be found in colonies and in the burrows of other small mammals. They are known to eat raspberry seeds and the fungus *Endogone*. Their reproductive behavior is probably similar to that of the southern bog lemming.

**Specimens Reported.** MAINE, Piscataquis County, Mt. Katahdin (Hall and Cockrum 1953).

NEW HAMPSHIRE, Coos County, Mt. Washington (type Locality).

### FAMILY MURIDAE (OLD WORLD RATS AND MICE)

These exotic rodents have been introduced into North America from Europe. Members of this family have three cheek teeth both above and below on each side of the jaw. The cheek teeth are not prismatic as in microtines but are either laminated or cuspidated. When laminated, the molars are not separated by wide

folds or valleys as in Cricetidae but are pressed tightly together. When cuspidated, the cusps are arranged in three longitudinal rows; the inner row may be vestigial.

These mammals are nearly cosmopolitan through introduction by man. They are represented in New England by the black rat, Norway rat, and house mouse.

## Black Rat

*Rattus rattus* (Linnaeus)

**Description.** The black rat is more slender than the Norway rat, with a more pointed snout, larger, longer ears, and a much longer tail, which is longer than the head and body combined. Females usually have ten mammae, but some may have twelve.

The skull of the black rat can be distinguished

from that of the Norway rat by the diastema, which is considerably less than twice the length of the cheek teeth row; by the well-developed temporal ridges, which are bowed decidely outward for a considerable distance on each side of the cranium; and by the shorter, rounded braincase.

[Mus]rattus Linnaeus. 1785. *Systema naturae*, 10th ed, 1:61.

*Rattus rattus*, Hollister, 1916. *Proc. Biol. Soc. Washington* 29:126.

Type Locality: Uppsala, Sweden

Dental Formula:
    I 1/1, C 0/0, P 0/0, M 3/3 × 2 = 16

Names: Roof rat and Alexandrine rat. The meaning of *Rattus* = a rat.

The sexes are colored alike and show slight seasonal variation. The coloration of the upperparts is grayish black, darker and glossier along the middle of the back. The sides and underparts are slate gray, and the tail is colored somewhat like the body.

The sexes are equal in size. Measurements range from 320 to 480 mm (12.5–18.7 in); tail, 150 to 255 mm (5.9–10.0 in); and hind foot, 30 to 35 mm (1.2–1.4 in). Weights vary from 150 to 540 g (5.3–19.0 oz).

**Distribution.** The black rat was introduced into the United States as early as 1609 with the early colonists. The species appears to have prospered until the 1700s in most parts of its new range, except in the northern regions where the larger, more aggressive Norway rat drove it out. The black rat seems to prefer warmer climates, and is common in the Gulf Coast and south Atlantic ports. Where both species of rats occur, the Norway rat has forced the black rat to occupy the upper portions of buildings.

The black rat was recorded in Boston in the late 1800s by Mearns (1900), and has been reported in Vermont from Wallingford (Green 1936) and from Rockingham and Woodstock (Osgood 1938). A specimen was taken in West Greenwich, Rhode Island, in 1903 (Cronan and Brooks 1968). Apparently black rats do not occur in New England today, though newcomers may gain a temporary foothold in the vicinity of a seaport.

**Ecology.** Black rats are essentially arboreal animals and seldom inhabit burrows. They live chiefly in colonies in attics, rafters, walls, and enclosed spaces in buildings, but seldom inhabit basements and sewers.

Man and Norway rats are their main predators; they are probably infested with parasites similar to those of Norway rats.

**Behavior.** In general the life history of the black rat is similar to that of the Norway rat, except that the black rat is a more agile and adept climber.

**Specimens Examined.** NEW HAMPSHIRE, total 10: Hillsborough County, Antrim 1 (September 1893), Peterboro 8 (no date); Sullivan County, Meriden 1 (no date) (HMCZ).

MASSACHUSETTS, total 17: Middlesex County, Hudson 4 (February 1870); Suffolk County, Boston 1 (summer 1885); Worcester County, Princeton 12 (November 1893) (HMCZ).

CONNECTICUT, total 6: New London County, Liberty Hill 6 (March 1895) (HMCZ).

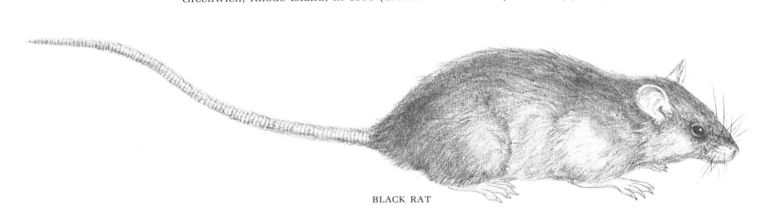

BLACK RAT

# Norway Rat

*Rattus norvegicus* (Berkenhout)

**Description.** The Norway rat is similar to the black rat but is larger and stockier, with a shorter tail and smaller ears that are partly hidden in fur. The tail is scaly, nearly hairless, and shorter than the length of the head and body combined. The forefoot has four clawed toes and a small clawless thumb. The hind foot has five clawed toes. Females have twelve mammae—six pectoral, two abdominal, and four inguinal.

The skull is flattened, with supraorbital and

Mus norvegicus Berkenhout, 1769.
   *Outlines of the natural history of
   Great Britain and Ireland*, 1:5.
Rattus norvegicus, Cabera, 1932. *Trab.
   Mus. Nac. Cien. Natl., Madrid, ser.
   Zool.* 57:264.
Type Locality: England
Dental Formula:
   I 1/1, C 0/0, P 0/0, M 3/3 × 2 = 16
Names: Brown rat, gray rat, house rat,
   barn rat, dump rat, sewer rat, dock
   or wharf rat, and common rat. The
   meaning of *Rattus* = a rat, and
   *norvegicus* = of Norway.

temporal ridges. The palate extends well beyond the last molars. The molars are laminate or with cusps arranged in three longitudinal rows. The skull of the Norway rat can be distinguished from that of the black rat by the length of the parietal bone, which, measured along the temporal ridge, is not less than the distance between the temporal ridges; by the length of the diastema, almost twice the length of the cheek tooth row; and by the first molar, which is without distinct outer notches on the first row of cusps.

The fur is coarse. The sexes are colored alike, and there appears to be no seasonal variation. The coloration of the upperparts and sides varies from reddish brown to grayish brown, mixed with scattered black hairs. The underparts are silvery or yellowish white. The tail is bicolored, dark gray above, lighter below, and the feet are grayish or dull white above. Immatures are paler than adults. Color phases include whites, melanistics, pieds or blotched gray-blacks, and the familiar commercial albino laboratory rats.

Males are somewhat larger than females. Measurements range from 300 to 460 mm (11.7–18.0 in); tail, 150 to 225 mm (5.9–8.8 in); and hind foot, 40 to 45 mm (1.6–1.8 in). Weights vary from 285 to 490 g (10.0–17.2 oz). An unusually heavy rat may weigh as much as 680 g (24 oz).

**Distribution.** The Norway rat apparently originated in Asia. It arrived in North America on the north Atlantic seaboard about 1775. The species occurs throughout North America, and its abundance varies according to climate, habitat, land use, and sanitation. In urban areas Norway rat infestation is heavier in commercial than in residential sections, especially in restaurants and grocery stores.

**Ecology.** Norway rats are extremely adaptable and are found wherever suitable food, water, and shelter are available. They are encountered in, under, or adjacent to buildings and dumps, along streams and rivers, in marshy areas and sewers, on waterfronts, and occasionally in open fields.

These animals are excellent burrowers and dig extensive burrow systems. The burrows are from 2 to 3 inches in diameter, normally 12 inches deep and about 36 inches long. Each burrow system has one or more entrances and emergency exits called "bolt holes" which are well hidden under boards or grass. They are lightly plugged with weeds or loose earth at the outer end. The den is from 4 to 6 inches in diameter and is used as a refuge, a breeding nest, or a place to eat. The nest is lined with shredded paper, cloth, vegetation, or almost any material available.

Man, large hawks and owls, foxes, dogs, house cats, mink, skunks, weasels, and snakes prey on Norway rats. Perhaps the chief enemies of rats are the diseases salmonellosis, tularemia, leptospiral jaundice, Haverhill fever, and murine typhus fever, which kill large numbers when population is dense. Norway rats are hosts to fleas, mites, ticks, cestodes, nematodes, and trematodes.

**Behavior.** Norway rats are active throughout the year; they are mainly nocturnal but may be seen during the daytime. They have high light sen-

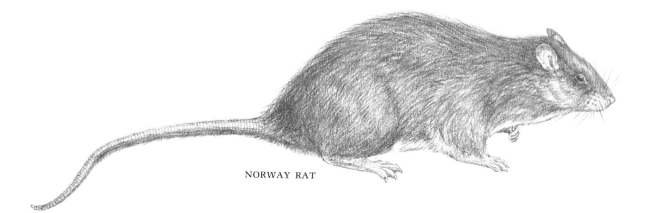

NORWAY RAT

sitivity but poor visual acuity, although they can discriminate between simple patterns and objects of different size and can recognize motion at distances up to 30 feet. They have good depth perception up to about 3 feet and are able to gauge accurately the effort needed for varying jumps. The senses of smell, taste, hearing, and touch are highly developed. Norway rats continually sniff. They leave odor trails which are followed by other rats.

Brooks (1969) reported that Norway rats can discriminate between plain bait and the same bait containing as little as 2 parts per million (ppm) of an estrogen. Similarly, Richter and Clisby (1941) found that rats refused drinking water that contained 3 ppm of phenylthiocarbamide, a bitter toxic substance. This unusual ability of rats to detect minute quantities of unpleasant or toxic substances is what leads to "bait shyness." This phenomenon was first clearly demonstrated by Rzoska (1953), who found that bait shyness induced by poisons was associated with the bait base, not the poison itself. This avoidance is due mainly to the rats' association of ill effects with the particular food bait which has made them ill rather than their recognition of the poison used in the bait.

Gould and Morgan (1941) reported that rats can hear sounds in the ultrasonic range, 20 to 40 thousand cycles per second. Anderson (1954) has shown that laboratory rats emit ultrasonic sounds, and Riley and Rosenzweig (1957) demonstrated that blinded rats use echolocation in learning to run a maze.

Norway rats are inquisitive and constantly explore and reexplore their surroundings. They quickly detect any new objects and avoid them. These animals prefer to travel over established, protected routes. They are fairly good climbers, especially when young, but they are not inclined to climb wires, pipes, trees, or other objects as black rats do, preferring to travel over flat surfaces. Norway rats are excellent swimmers and divers and are somewhat semiaquatic by habit. Jarvis (1927) reported a Norway rat that swam at least 80 feet underwater, and Richter (1958) found that one rat swam 60 to 72 hours in a tank of water at 95°F before it drowned.

Norway rats are gregarious and sociable. They form colonies composed of several families and share nesting and feeding areas. Members of each colony live amicably together but vigorously fight members of adjacent colonies that invade their home grounds.

Norway rats are territorial and hierarchical. Dominant males occupy the most favorable places close to the food supply, establish territories there, and defend a system of burrows containing several females. As a population of rats increases and social competition forces lower-ranking rats, especially males, into fringe areas, all changes in social rank are downward. Animals from the lower-ranking groups are smaller, grow more slowly, and pass on their low rank to their offspring. Reproductive success in low social-rank groups is poor because the sexually receptive females are harrassed by packs of males (Calhoun 1962).

Norway rats often move from fields and ditches to buildings with the advent of cold weather and back into the open in spring. Normally, however, they remain in deep burrows in dumps the year round, though occasionally they may enter nearby buildings in cold weather. Some of them are extensive wanderers and frequently cover large areas. If food and water are far apart, the rat moves a great deal, and if they are removed the rat migrates. Davis (1948) determined that the range of a colony of rats is about 100 feet in diameter and that when rats are liberated in strange places they may wander far away. Creel (1915) captured marked rats as far as 4 miles from the point of release, and Stewart (1946) captured rats 9 days later, 11 miles from where they were released.

Norway rats eat practically anything they can get, eating at least one-third of their body weight in 24 hours and wasting much more. They seem to thrive on meats and fish, vegetables, grains, fruits, nuts, garden crops, eggs, and roots. Garbage appears to offer them a fairly balanced diet and also satisfies their water requirements. They are known to kill poultry, wild birds, young pigs and lambs, black rats, rattlesnakes, fingerling fishes in hatcheries, and even their own young unprotected in nests. Norway rats often gnaw through lead pipes to get to water, and they chew on leather, cloth, soap, paint, books, and all sorts of packaged goods. They may cache their food but most often eat it directly at the source. They require about one and one-half ounces of water daily.

Norway rats are polygamous and breed throughout the year, with peaks of breeding activity in the spring and autumn. The female comes into estrus about every five days and can mate with several males within a day after parturition (Davis and Hall 1948).

The gestation period is 21 to 22 days. The prolific female may produce from three to twelve litters annually, though six is usual. The number of young per litter varies from six to twenty-two, with seven or eight the usual. A female may wean twenty to thirty young per year.

The newborns are naked and pink, helpless and blind, with ears closed, and are approximately 51 mm (2.0 in) long and weigh about 6.0 g (0.21 oz). They grow rapidly, acquiring a soft gray coat before they are weaned at about 3 weeks of age. Norway rats attain sexual maturity by 80–85 days of age. Longevity is about 3 years (Calhoun 1962).

**Remarks.** Norway and black rats are ubiquitous pests throughout the world. Beyond merely being repulsive, they have created a menace to health and a huge economic burden. More human deaths can be attributed to them than to all other vertebrate enemies of man combined. Rats and their parasites have caused worldwide epidemics of fatal diseases as well as minor illnesses. The most terrible disease spread by rats is the bubonic plague, which, as the "Black Death" of the fourteenth century, killed 25 million Europeans, and from 1898 to 1923 killed over 10 million people in India. The plague bacterium, *Pasteurella pestis*, is transmitted from rats to other rodents and from rats to man by the bite of the oriental rat flea, *Xenopsylla cheopis*. Plague outbreaks are rare in the United States, mainly owing to efficient rat suppression.

**Specimens Examined.** MAINE, total 28: Hancock County, Mt. Desert Island 1 (HMCZ); Lincoln County, Barter's Island 1 (UCT); Penobscot County, Orono 26 (UME).

NEW HAMPSHIRE, total 22: Carroll County, Jackson 5 (UCT); Grafton County, Hanover 5, Lyme 1 (DC); Rockingham County, Exeter 1; Strafford County, Durham 10 (UNH).

VERMONT, total 17: Addison County, Middlebury 1 (UVT), Caledonia County, Hardwick 1 (UCT); Chittenden County, Burlington 4, Shelburne 1; Orange County, Randolph 1 (UVT); Rutland County, Mendon 1; Windham County, Rockingham 3, Saxton's River 2 (UCT); Windsor County, Bethel 1 (UVT), Norwich 1 (DC), Pomfret 1 (AMNH).

MASSACHUSETTS, total 33: Barnstable County, Barnstable 2 (HMCZ); Franklin County, Charlemont 1; Hampshire County, Amherst 7, Granby 2, Northampton 1, Westfield 1 (UMA); Middlesex County, Belmont 5, Newton 1 (HMCZ); Nantucket County, Nantucket Island 1 (UCT); Norfolk County, Stoughton 1 (USNM); Plymouth County, Wareham 2; Suffolk County, Boston 5, Dorchester 1, Forest Hills 1; Worcester County, Harvard 2 (HMCZ).

CONNECTICUT, total 68: Fairfield County, Bridgeport 2, Norwalk 35, Weston 1 (UCT); Hartford County, East Hartford 2 (HMCZ), Elmwood 1, Newington 2; Litchfield County, Cromwell 1, New Milford 1; Middlesex County, East Hampton 1 (UCT); New Haven County, Cheshire 1, New Haven 1 (USNM); New London County, Liberty Hill 9 (HMCZ); Tolland County, Bolton 1, Coventry 6, Mansfield 1, Storrs 2; Windham County, Brooklyn 1 (UCT).

# House Mouse

*Mus musculus Linnaeus*

*Mus musculus* Linnaeus, 1758.
 *Systema naturae*, 10th ed., 1:62.
Type Locality: Uppsala, Sweden
Dental Formula:
 I 1/1, C 0/0, P 0/0, M 3/3 × 2 = 16
Names: Domestic mouse, feral mouse, gray mouse, and common mouse. *Mus* comes from the ancient Sanskrit word *mush* = thief, or to steal, and *musculus* = muscular.

**Description.** The small gray house mouse, with its long pointed nose, is well known and needs no detailed description, except that it may be confused with the native deer mouse and white-tailed mouse of the genus *Peromyscus*. In *Peromyscus* the fully furred tail is distinctly bicolored, dark above and white below, and the abdomen is pure white. The tail of the house mouse has only scanty, fine hairs and scaly annulations, and like the abdomen is tinged yellowish gray. Also, the molars of the house mouse have three longitudinal rows of tubercles like those of the Norway rat and black rat, whereas in *Peromyscus* the molars have only two longitudinal rows of tubercles.

The skull of the house mouse is generally flat in profile. The border of the zygomatic plate is cut back above, and there is usually a small knob on its lower border. The occlusal surface of the upper incisors is notched. The first upper molar is larger than the second and third combined.

The coloration of the sexes is grayish and shows no noticeable seasonal color variation. Immatures are somewhat grayer than adults.

Some individuals are piebald, agouti, leaden or other shades of gray, all white, or melanistic. The commercial white mouse is the albino mutation. Some mutations are bald "rhinos," or only partially haired, because of a genetic anomaly. These mutations are hairless except for whiskers. They have an excessive growth of wrinkled or corrugated skin and lack pigment except for the eyes and tips of the ear and tail, which are dark gray. Other color phases are "misty," in which individuals are a dilute brown with a white tail tip and a unilateral rectangular-shaped white belly spot.

The sexes are equal in size. Measurements range from 160 to 200 mm (6.2–7.8 in); tail, 70 to 95 mm (2.7–3.7 in); and hind foot, 15 to 20 mm (0.59–0.78 in). Weights vary from 15 to 30 g (0.53–1.05 oz).

**Distribution.** This Old World species undoubtedly preceded the Norway rat into North America, probably about the time the first settlers arrived or a little later. The species occurs along the coast of Alaska, from southern Canada throughout the United States, and through Mexico to Panama.

**Ecology.** Like the Norway rat the house mouse is most often encountered in or near human habitations, but it is also found in fields, shocks of hay, wheat ricks, and corncribs. The nest may be hidden in buildings and is made of shredded paper, cloth, grass, or anything available. House mice may construct communal nests.

Man, birds of prey, foxes, coyotes, house cats, shrews, weasels, meadow voles, skunks, rats, and snakes prey on house mice. They harbor many parasites and serve as transmitters of human bacterial and viral diseases. House mice are known as carriers of murine typhus fever, rickettsial pox, leptospirosis, and salmonellosis, or food poisoning. Mange and certain epizootics sometimes plague house mice.

House mice are known to waltz by twisting, prancing, and pivoting around on one foot. These abnormally behaving mice apparently are the result of inbreeding. Aside from being deaf and having the dance habit, the animals appear normal and make attractive pets (Stebbins 1932). Another behavioristic mutation in house mice is the shaker; these animals make nervous head movements, circle, and are deaf. This anomaly is probably due to a different factor or gene from the one that causes waltzing (Lord and Gates 1929).

**Behavior.** House mice are active throughout the year and are mainly nocturnal. Some of them move indoors in autumn and return outdoors in spring. In a house, they scurry about and gnaw at the woodwork, and they drop their little black scats on pantry shelves or wherever they forage.

Outside of buildings the average home range of both male and female feral house mice is 1,316 to 1,560 square feet where high populations of meadow voles exist (Lidicker 1966; Quadagno 1968); but where the meadow vole population is negligible, the mean home range size of house mice is 3,925 square feet (Quadagno 1968). Young, Strecker, and Emlen (1950) found that indoor house mice traveled an average distance of only 12 feet under optimal conditions of food and cover.

This species is not as colonial as the Norway rat, though in feral forms and in captivity loose organizations are formed. In these colonies, males are very aggressive and fighting is common. Aggressive males called "patrollers" make rounds, going in and out of nests and often attacking and chasing subordinates (Brown 1953). Southwick (1955) observed that aggression increased directly with population density, but Lloyd and Christian (1967) reported that aggression may be dependent upon social structure rather than on density alone. Crowcroft (1955) reported that social hierarchies develop in small colonies under certain conditions and that fighting tends to disperse the population into spatially distinct mating areas initially occupied by one male and one or two females, suggesting territoriality.

House mice are nervous and curious. When foraging they usually walk slowly, stopping often to sniff and to poke here and there. They usually run from place to place and may travel at speeds of up to 8 miles per hour. They are excellent climbers and jumpers and can swim well. They generally have good dispositions.

House mice are normally quiet, though they squeak when trapped or injured. Occasionally, however, there are house mice that sing. The reason for their singing is unknown. The re-

HOUSE MOUSE

markable song approximates the chirp of a cricket but is much more continuous. Carrighar (1942) reported that the song is audible 15 feet away, and that the mouse he observed usually sang when the radio was on. The Chinese are known to have kept singing house mice in cages as Westerners keep canaries.

The senses of smell, taste, vision, touch, and hearing are well developed. The long whiskers and guard hairs are sensitive to touch, and house mice apparently can readily detect motions and changes of light intensity. Hopkins (1953) found that house mice can perceive some protective cover at least 45 feet away in almost all cases.

House mice frequently groom themselves by licking and pawing, but they commonly defecate and foul their environment. Welch (1953) reported that captive house mice formed urinating posts made of an accumulation of droppings, dust, and debris.

House mice eat almost any kind of food, particularly sweet and high-protein foods. They eat grains, seeds, fleshy roots, cereal, many kinds of vegetables, pea weevils, roaches, and meat. Mills (1947) found that house mice eat lightly, intermittently, and erratically, consuming about 10% of their body weight each day; they feed 15 to 20 times a day, taking only about 0.1 to 0.2 g of food at each feeding. Spencer (1946) noted that they are able to exist for more than 4 months without free water, though they eat moist foods. Strecker (1955) reported that house mice tend to eat less when the ambient temperature decreases.

House mice are known to eat glue, paste, soap, and other household articles, and often damage much more than they can eat. They also gnaw on paper, books, leather, wood, cardboard, lead, and most plastics, and often damage clothing. They may cache food at times.

House mice are prolific, producing from five to eight litters annually, though six are usual. The number of young per litter varies from three to twelve, but the usual is four or five. The breeding season occurs throughout the year, peaking in early spring and again in late summer. The gestation period is from 19 to 21 days. Newborns are naked and pink, blind, and helpless. They grow rapidly and in 10 days are fully furred. They are weaned at 3 weeks of age, and then they disperse. Females attain sexual maturity at 8 weeks of age. Occasionally two or more females may produce young in the same nest at the same time, and as many as fifty young may be found in a communal nest. Longevity is about 1 or 2 years in the wild and at least 6 years in captivity.

**Specimens Examined.** MAINE, total 34: Aroostook County, Madawaska 8 (HMCZ); Franklin County, Farmington 22 (UME); Hancock County, Mt. Desert Island 1; Oxford County, Upton 2 (HMCZ); Piscataquis County, Sebec Lake 1 (USNM).

NEW HAMPSHIRE, total 55: Carroll County, Ossipee 24–2 (UCT–USNM); Grafton County, Canaan 11, Hanover 11; Hillsborough County, Greenfield 1, Peterborough 1 (DC); Merrimack County, Webster 3 (HMCZ); Strafford County, Durham 2 (UNH).

VERMONT, total 44: Caledonia County, Hardwick 2 (UCT); Chittenden County, Burlington 3–1 (DC–USNM), Shelburne 3, West Milton 2; Franklin County, Bakersfield 1; Orange County, Randolph 3 (UVT); Rutland County, Mendon 1 (UCT), Pawlet 1 (UVT), Rutland 1–11 (DC–UCT); Washington County, Cabot 1, Waterbury 2 (UVT); Windham County, Dummerston Center 2 (AMNH), Saxton's River 1–6 (AMNH–UCT), Whitingham 1 (AMNH); Windsor County, Pomfret 1 (DC), Woodstock 1 (UVT).

MASSACHUSETTS, total 73: Barnstable County, Barnstable 1 (HMCZ), Falmouth 1 (USNM); Franklin County, Whately 1 (UCT); Hampshire County, Amherst 19, Sunderland 1, Westport 1 (UMA); Middlesex County, Cambridge 4–5 (HMCZ–USNM), Hudson 6,

Newton 1 (HMCZ); Plymouth County, Middleboro 1–5 (UCT–USNM), Wareham 8; Suffolk County, Boston 5, Forest Hills 12, Jamaica Plains 1 (HMCZ); Worcester County, North Brookfield 1 (UMA).

CONNECTICUT, total 27: Fairfield County, Greenwich 1 (AMNH), Stratford 12 (UCT); Hartford County, East Hartford 1–1 (UCT–USNM), Windsor 1 (AMNH); New Haven County, New Haven 2 (USNM); New London County, East Lyme 3 (DC), Stonington 1 (UCT), Waterford 3 (AMNH); Tolland County, Storrs 2 (UCT).

RHODE ISLAND, total 11: Kent County, West Warwick 2 (URI); Newport County, Block Island 1 (HMCZ), Newport 2 (AMNH); Washington County, Lake Worden 6 (USNM).

## FAMILY ZAPODIDAE (JUMPING MICE)

This family contains small to medium-sized mouselike rodents that have tails much longer than the head and body and large hind feet modified for jumping in the subfamily Zapodinae. Each upper incisor has a lengthwise groove only in the subfamily Zapodinae. Members of the Zapodidae occur in Eurasia and in North America from the southern arctic area of Canada south to approximately 35 degrees north latitude. The family is represented by two species in New England.

# Meadow Jumping Mouse

*Zapus hudsonius*

There are two subspecies of meadow jumping mice recognized in New England:

*Zapus hudsonius acadicus* (Dawson). Meriones acadicus Dawson, 1856. Edinburgh *New Philos. J.*, n.s., 3:2. Zapus h. acadius Anderson, 1942. *Ann. Rept. Provancher Soc. Nat. Hist.*, Quebec 52(1941):38.
Type Locality: Nova Scotia, Canada

*Zapus h. americanus* (Barton), *Dipus americanus* Barton, 1799. *Trans. Amer. Philos. Soc.* 4:115. Zapus h. americanus Batchelder, 1899. *Proc. New England Zool. Club* 1:6.
Type Locality: Near Schuykill River, a few miles from Philadelphia, Pennsylvania (Hall and Kelson 1959)

Dental Formula:
I 1/1, C 0/0, P 1/1, M 3/3 × 2 = 18

Names: Canadian jumping mouse, Hudson Bay jumping mouse, and long-legged mouse. The meaning of *Zapus* = foot, referring to the large hind feet, and *hudsonius* = for Hudson Bay, where the first specimen was named.

**Description.** The meadow jumping mouse is characterized by its greatly elongated hind legs, which are much longer than the forelegs and modified for jumping, and by a long, slender, tapering, scaly tail, which is longer than the head and body. The small eyes are situated midway between the nose and ears. The ears are large and scantily haired and are partially hidden in long, coarse fur. The whiskers are conspicuous. The upper lip has a median groove. Although Sheldon (1934), Krutzsch (1954), and Walker et al. (1964) reported the presence of external cheek pouches in *Zapus* and *Napaeozapus*, Klingener (1971) reported that he never found cheek pouches in either of these species. Females have eight mammae—two pectoral, four abdominal, and two inguinal.

The skull is delicate, with a narrow braincase, large, rounded infraorbital foramen, depressed zygomatic arches, and nasals that project noticeably beyond the incisors. The enamel of the molars is much folded. The upper and lower incisors are deep orange or yellow, and the upper incisors are deeply grooved. There are four upper molariform teeth, with the first reduced in size.

The sexes are colored alike and show slight seasonal variation. The guard hairs are gray basally with darker black or brown at the tips. The hairs of the underfur are grayish or whitish basally and yellowish brown distally. In summer the coloration of the upperparts is yellowish olive or yellowish brown mixed with prominent, black-tipped guard hairs, forming a dark, well-defined dorsal band on the back from the head to tail. The sides are paler, with fewer black hairs. The underparts are white, sometimes washed with yellow.

The tail is sharply bicolored, grayish brown above and yellowish white below, with a terminal tuft of black hairs, though some individuals have white-tipped tails (Schorger 1951; Whitaker 1963a). The ears are edged with white or buff. The winter pelage is duller and more suffused. Immatures are duller than adults, with a less well-defined dorsal stripe. Color variants are rare, though white spotting has been reported.

The molt occurs throughout the summer. It begins in mid-June and lasts through mid-October before hibernation in New York. Molting starts on the head and anterior parts of the body and proceeds to the posterior parts without well-defined molt lines.

Females are slightly larger than males. Measurements range from 187 to 255 mm (7.3–9.9 in); tail, 108 to 155 mm (4.2–6.0 in); and hind foot, 28 to 35 mm (1.1–1.4 in). Weights vary from 13.5 to 23.0 g (0.5–0.8 oz).

**Distribution.** This species occurs from Alaska and northern Canada south into Colorado, eastern Alabama, northern Georgia and South Carolina.

**Ecology.** The meadow jumping mouse occurs primarily in moist, abandoned grassy and brushy fields with stands of touch-me-nots, and in thick vegetation along streams, ponds, and marshes. It is sometimes taken in rank herbaceous vegetation of wooded areas.

Getz (1961a) found that the meadow jumping mouse avoided sparse vegetation and concluded that the animal had a definite affinity for moist areas. Whitaker (1963a) concluded that populations of meadow jumping mice were often greater in grassy vegetation and that the distance to large bodies of water and the amount of soil moisture seemed to have little or no direct effect on the mice except that they indirectly influenced vegetation. Bider (1968) found that the activity of the mice increased with rainfall.

The nest may be found on a hummock of grass, under rotten logs or planks, in hollow logs, or at least 6 inches underground. It is made of grass, leaves, and other soft vegetation. The outside of the nest is about 6 inches in diameter, and the inside cavity is about 3 inches across. There is a small entrance near the top of the nest.

The snug hibernating nest may be 3 or more inches underground where the temperature seldom is below freezing. It may be found under a log, or in a gravel pit, a sand bank, layers of compacted wood ash, a woodchuck den, a coal-ash pile, dikes of cranberry bogs, a potato hill, a small mammal burrow, or clumps of bayberry bushes.

Man, coyotes, foxes, mink, house cats, red-tailed hawks, barn owls, and long-eared owls, snakes, northern pike, and black bass prey on meadow jumping mice. Fires, floods, and farm machines occasionally kill some. Fleas, ticks, mites, cestodes, nematodes, trematodes, and botfly larvae, *Cuterebra* spp., are known parasites of meadow jumping mice. Nakamura (1950) reported tularemia in a jumping mouse from Wyoming.

**Behavior.** Meadow jumping mice are essentially solitary and generally not antagonistic toward their own kind. They are mainly active at night but may sometimes be encountered during the daytime. They are among the most profound of hibernators and remain in hibernation as long as or longer than most other mammals. In central New York Whitaker (1963a) found that the ani-

mals began to accumulate fat around the first of September and continued until they presumably entered hibernation by about 20 October. In Minnesota Quimby (1951) found that eight of nineteen mice entered hibernation between 17 September and 1 October, whereas the remainder had shown no sign of hibernation by this time. Later records of when meadow jumping mice enter hibernation are 25 October and 2 November, Washington, D.C. (Bailey 1923); 13 and 18 November, New York (Hamilton 1935); 25 October and 15 November, Pennsylvania (Richmond and Roslund 1949); and 22 October and 3 November, New York (Whitaker 1963a). In New York the mice presumably emerge from hibernation in late April or early May (Hamilton 1935; Whitaker 1963a). Males emerge first and females 2 or 3 weeks later (Sheldon 1938a; Quimby 1951). Manville (1956) trapped an adult male on 12 February in Michigan.

Meadow jumping mice hibernate singly or in pairs huddled close together. They roll themselves tightly into a ball, with the nose buried in the belly, ears flattened against the head, eyes closed tightly, and the long tail coiled around or partially under the body.

Townsend (1935) and Sheldon (1938a) suggested that meadow jumping mice tend to wander. According to Townsend (1935) the movement may be associated with the animals' seeking moist areas during the dry part of summer. Quimby (1951) suspected that the movement is due to a relatively unstable home range and that the shape of the home range is determined by terrain. He found that the home range varied from 0.19 to 0.87 square acres with a mean of 0.38 in females, and from 0.14 to 1.10 square acres with a mean of 0.43 in males. Quimby (1951) also found meadow jumping mice to have essentially no homing tendencies, based on releases of thirteen individuals 0.2 to 0.5 miles from their original home ranges.

The common name "jumping mice" is a misnomer. The animals do not normally travel by jumping; more often they progress slowly through the grass or else take little hops of 1 to 6 inches. They frequently crawl through the grass or under it, sometimes flattening the body to the ground and moving on all fours. When startled they usually take a few jumps or may stop abruptly and remain motionless. Quimby (1951)

MEADOW
JUMPING MOUSE

and Sheldon (1934) felt that this motionless state was their chief means of protection. Whitaker (1963a) thought that when motionless the mice produced less odor than other small mammals and that the dark dorsal stripe might help to conceal them. Quimby (1951) and Whitaker (1963a) suspected that records of long leaps over 3 feet are probably errors and thought these records probably referred to woodland jumping mice.

Meadow jumping mice are good swimmers both on the surface and underwater. They swim only with the hind legs, head held high and tail arched high as a counterbalance. They can dive to a depth of 18 inches (Sutton 1956) and remain submerged for some 30 seconds (Priddy 1949).

These mice are able to climb with ease over brush and grass stems. They do not cache food. It is not known whether they normally drink water or get it from green food in the wild, but in captivity they drink regularly. They often wash their faces and feet, and their long tails, grasping them in their forepaws, and passing them completely through their mouths.

Meadow jumping mice are usually silent, except the young. Adults on occasion make a clucking sound, and a series of short chirps like a sparrow. They also produce a drumming noise by vibrating their tails rapidly against a surface. These mice can dig in loose soil, and they occasionally use burrows of other small mammals.

Meadow jumping mice feed on the seeds of grasses, fleshy fruits of shrubs, berries, nuts, tomatoes, melons, sunflower seeds, tender roots, subterranean fungi, *Endogone* spp., earthworms, insects, spiders, and slugs.

The breeding season begins shortly after the females emerge from hibernation, extending from nearly May to early October, with peak numbers of births in the first three weeks of June, the first three weeks of August, and occasionally in the first two weeks of September in New York. Most females have two litters annually, but some have three (Whitaker 1963a). The gestation period is approximately 18 days, or slightly longer if the female is lactating. The number of young per litter varies from two to eight, and five or six are usual (Quimby 1951).

Newborns are naked, blind, and helpless, weigh about 0.8 g (0.03 oz), and are from 30 to 39 mm (1.2–1.5 in) long. During the first week the whiskers become noticeable, the claws appear, the ears unfold, and the tail becomes bicolored. Tawny yellow hairs appear about the 9th day, and white incisors erupt about the 13th day. The young are fully furred by the 17th day, and the dorsal stripe becomes evident between the 15th and 19th days. The auditory meatus opens about the 19th day, and the young can by then support themselves on their legs. Between the 22d and 25th day the eyes open, and gradual weaning begins about this time. The juvenile pelage is replaced during the 4th week, and the young become independent. Mice born in early litters may mate the same year, but those born late usually mate in the spring following their birth. Longevity is nearly 2 years (Whitaker 1963a).

**Specimens Examined.** MAINE, total 76: Aroostook County, T12–R15 2, T12–R16 2, T12–R17 1 (UCT), Madawaska 42 (HMCZ), Cumberland County, Gray 1

(UCT); Franklin County, Farmington 2; Hancock County, Bucksport 1 (UME), Mt. Desert Island 11 (HMCZ); Penobscot County, Holden 1 (UME), Orono 1–3 (HMCZ–UME); Piscataquis County, Greenville 1 (UCT), Mt. Katahdin 3 (DC); Somerset County (no location) 1; Washington County, Baring 1, Calais 1, Edmunds Unit 1, Porcupine Lake 1 (UCT).

NEW HAMPSHIRE, total 45: Carroll County, East Sandwich 1, Jackson 7, Ossipee 1, Tuftonboro 2, Wolfeboro 1 (UCT); Cheshire County, Keene 1 (DC); Coos County, Carroll 4, Colebrook 1 (UCT), Gorham 4 (DC), Mt. Washington 1 (AMHN), Twin Mt. 1 (UCT); Grafton County, Hanover 3 (DC), Mt. Moosilauke 2–3 (AMNH–DC); Hillsborough County, Peterborough 1–1 (DC–HMCZ); Rockingham County, Epping 1–1 (UCT–UNH), Durham 1 (UNH); Sullivan County, Claremouth 1, Springfield 5 (UCT).

VERMONT, total 65: Bennington County, Landgrove 1 (UVT), Peru 1 (UMA); Caledonia County, Lyndon Center 6, Hardwick 3 (UCT); Chittenden County, Burlington 1, Westford 1, Williston 2; Lamoille County, Stowe 1 (UVT); Orange County, Bradford 1 (URI), Chelsea 1 (UVT), Williamstown 1; Rutland County, East Wallingford 1 (DC), Killington Peak 1 (UCT), Mendon 6–5 (AMNH–UCT), Rutland 5–14 (DC–UCT), Sherburne 1–1 (DC–UCT), Wallingford 1; Washington County, Plainfield 1; Windham County, Saxton's River 1, South Londonderry 7 (UCT), Whitingham 2 (AMNH).

MASSACHUSETTS, total 62: Barnstable County, Barnstable 3 (UCT), Falmouth 1, Provincetown 1 (USNM), South Wellfleet 2; Berkshire County, Ashley Falls 3 (UCT), Monterey 1 (UMA); Dukes County, Martha's Vineyard 1 (USNM); Essex County, Plum Island 1–3 (UMA–USNM), Wenham 1 (HMCZ); Franklin County, Whately 2 (UCT); Hampden County, Springfield 2 (HMCZ); Hampshire County, Amherst 3 (UMA), Northampton 1 (HMCZ), Sunderland 1 (UMA); Middlesex County, Belmont 2, Cambridge 10, Concord 3, Hudson 1, Lexington 1, Newton 3 (HMCZ); Nantucket County, Nantucket Island 2; Plymouth County, Hanson (UCT), Marshfield 6 (USNM); Worcester County, Harvard 4, Lancaster 1 (HMCZ), Petersham 1 (UCT), Worcester 1 (UMA).

CONNECTICUT, total 53: Fairfield County, Redding 1, Stratford 3; Hartford County, Farmington 1, Glastonbury 4 (UCT), Windsor 1 (USNM); Litchfield County, East Norfolk 1 (UCT), Macedonia Park 2, Sharon Mt. 5; Middlesex County, Clinton 1 (AMNH), Portland 1 (UCT); New Haven County, Cheshire 1 (USNM), Guilford 1, Southbury 2; New London County, Stonington 9; Tolland County, Charter Marsh 3, Eagleville 2, Mansfield 6, Storrs 3, Willington 1; Windham County, Scotland 1 (UCT), South Woodstock 3 (AMNH), Warrenville 1 (UCT).

RHODE ISLAND, total 2: Washington County, Naragansett 1 (URI), West Kingston 1 (UCT).

# Woodland Jumping Mouse

*Napaeozapus insignis* Miller

*Napaeozapus insignis* Miller, 1899.
  *Bull. New York State Mus. Nat.*
  *Hist., Albany* 6:330.
Type Locality: Restigouche River,
  New Brunswick, Canada
Dental Formula:
  I 1/1, C 0/0, P 0/0, M 3/3 × 2 = 16
Names: The meaning of
  *Napaeozapus* = a woodland
  nymph, and *insignis* = insignia,
  referring to the distinguishing dark
  dorsal stripe on the back.

**Description.** The woodland jumping mouse is distinguished from the meadow jumping mouse by its white-tipped tail, larger ears, more fulvous coloration, and the absence of the two small premolars in the upper jaw. It is somewhat larger than the meadow jumping mouse. The skull characters are similar to those of the meadow jumping mouse, and the upper incisors are grooved. Females have eight mammae—two pectoral, four abdominal, and two inguinal.

The sexes are colored alike and show little seasonal variation. The coloration is a bright yellow to orange on the flanks and about the face. A broad, brownish black dorsal stripe runs the length of the back. The underparts are white; the tail is distinctly bicolored, grayish brown above, white below, and virtually always white at the tip. Apparently color variants have not been reported.

Adults molt once a year, from mid-June to early September (mostly in August). The new hair generally appears first on the cheeks and sides of the neck, then spreads to the rostrum, eyes, shoulders, and forelimbs. The molt progresses posteriorly along the sides, spreading later to the back and belly and finally to the rump and hind limbs. Subadults appear not to molt. Young from spring litters exhibit new hair in late August and September (at 2 months of age), whereas those from summer litters molt in late September and October or the following spring (Wrigley 1972).

Females average slightly larger than males. Measurements range from 204 to 259 mm (8.0–10.1 in); tail, 115 to 160 mm (4.5–6.2 in); and hind foot, 28 to 34 mm (1.1–1.3 in). Weights vary from 15 to 28 g (0.5–1.0 oz).

**Distribution.** This species occurs from Cape Breton Island, Prince Edward Island, Nova Scotia,

Quebec, south to northern New Jersey, eastern West Virginia and western Maryland, and north to northeastern Ohio.

**Ecology.** Woodland jumping mice inhabit moist, cool woodlands with herbaceous vegetation along banks of mountain streams. They are seldom found in open fields or in marshes where there are no woods nearby.

These mice make a few shallow runways but mostly use burrows and trails made by moles, shrews, and other small mammals. Nests may be in brush piles, under stumps and rotting logs, or other shelter, or several inches underground. Sometimes the entrance of the nest is concealed when the mouse is in the burrow. The nest is made of dry grass and leaves and is about 5 to 6 inches in outside diameter.

Man, house cats, foxes, weasels, skunks, hawks, and owls prey on woodland jumping mice. Mites, fleas, nematodes, and cuterebrid larvae are known to parasitize them.

**Behavior.** Woodland jumping mice are mainly nocturnal but may also be active in late morning and early evening and on cloudy days. They are profound hibernators and accumulate fat before entering hibernation. Preble (1956) reported that in New Hampshire weather affects when they begin and end hibernation. Some years the mice enter hibernation in early October and other years in early November, and they emerge from late April in some years to mid-May in others. Wrigley (1969) reported that immature males enter hibernation after the adults and emerge before them. He found that in southern Quebec most males emerged from hibernation in early May, and most females emerged in late May. Klein (1957) found that withholding food and water from captive mice induced them to go into a state of torpidity. These mice do not cache food, though occasionally one may carry seeds to its nest to eat.

Woodland jumping mice travel on all four feet (quadrupedal walk) when moving slowly and take quadrupedal hops for greater speed. They take several moderate leaps, stop abruptly under the nearest cover, and remain motionless unless pursued. They can jump a maximum of 6 feet (Wrigley 1972). They are partly arboreal in habits (Saunders 1921; Sheldon 1934).

Woodland jumping mice are elusive, quiet, gentle, tolerant of their kind, and easily tamed. Captives utter a soft clucking sound or squeal if disturbed while sleeping. They commonly vibrate their tails (Sheldon 1934).

Although woodland jumping mice occur mainly in woods and meadow jumping mice occur primarily in meadows, the two species commonly are found together, particularly in forest edges or shrubs. Preble (1956) found meadow jumping mice invading the habitats of woodland jumping mice, but not the reverse. The two species have little or no direct effect on each other's distribution (Whitaker 1963b).

WOODLAND JUMPING MOUSE

The home range is from 1.0 to 6.5 square acres for females and 1.0 to 9.0 square acres for males (Blair 1941). One male traveled 117 yards in 24 hours (Sheldon 1938).

Woodland jumping mice feed on blueberries, huckleberries, raspberries, strawberries, buttercups, mayapple, mitrewort, wintergreen, small seeds and nuts, basal parts of plants, green leaves, fronds of spleenwort, rootstocks, fungi *Endogene* spp., centipedes, butterflies, grasshoppers, crane flies, dragonflies, hawk moths, spiders, caterpillars, beetles, larvae of lepidoptera and diptera, grubs, small soil worms, and millipedes.

The breeding season seems to be protracted. Pregnancies occur from late May to late August in New Hampshire (Preble 1956). Usually one litter is produced each year, though a second litter may be born in September (Hamilton 1935). The gestation period is between 21 and 25 days. Litter size varies from one to eight, but the usual is five; Stupka (1934) reported a litter of nine young in Maine. Newborns are naked, blind, and helpless; they are from 35 to 44 mm (1.4–1.7 in) long and weigh about 1 g (0.04 oz). Weaning occurs between 22 and 25 days after birth. Sexual activity is evident at 38 days (Layne and Hamilton 1954).

**Specimens Examined.** MAINE, total 59: Aroostook County, T11–R10 1 (UCT), Macwahoc 1 (UNH); Cumberland County, Gray 1 (UCT); Franklin County, Farmington 6 (UME); Hancock County, Mt. Desert Island 4 (HMCZ); Oxford County, Dixfield 2 (UCT), Fryeburg 1 (AMNH); Penobscot County, Enfield 3 (UNH), Holden 1 (HMCZ), Lincoln 3, Orono 3 (UME), Penobscot River (East Branch) 1 (USNM); Piscataquis County, Big Squaw 2, Greenville 1 (UCT), Mt. Katahdin 18–4–1–3 (DC–UCT–UME–USNM); Somerset County, Raynhen 1 (UCT), Seeboonock 1 (UME); Washington County, Vanceboro 1 (UCT).

NEW HAMPSHIRE, total 230: Carroll County, Bartlett 19, Harts Location 26 (UCT), Intervale 1 (USNM), Jackson 73, Ossipee 1, Tuftonboro 2 (UCT); Cheshire County, Dublin 3 (HMCZ), Keene 2 (DC); Coos County, Carroll 10 (UCT), Fabyan 3 (USNM), Gorham 4 (DC), Lower Branch (Grant) 1 (UCT), Pickham Notch 6 (AMNH), Mt. Washington 8–12–2 (AMNH–DC–USNM), Randolph 9 (DC); Grafton County, Bethlehem 1 (HMCZ), Franconia Notch 1–4–10 (AMNH–HMCZ–USNM), Hanover 3 (DC), Livermore 9 (UCT), Mt. Moosilauke 8 (DC), Profile Lake 4, Warren 1; Hillsborough County, Antrim 1 (HMCZ); Rockingham County, Epping 1; Strafford County, Barrington 1 (UNH); Sullivan County, Springfield 4 (UCT).

VERMONT, total 244: Bennington County, Peru 1, Styles Peak 11 (UVT); Caledonia County, Lyndon Center 28 (UCT); Chittenden County, Burlington 1, Richmond 1, Westford 2 (UVT); Essex County, Island Pond 8, Maidstone 11 (DC), Victory 4 (UCT); Lamoille County, Cambridge 8, Eden 4 (DC), Mt. Mansfield 8–15 (DC–USNM); Orange County, Ainsworth 1 (UCT), Chelsea 2 (UVT), Williamstown 1; Rutland County, Brighton 5 (UCT), Danby 3 (UMA), Mendon 1–23 (AMNH–UCT), Rutland 8 (UCT), Sherburne 9–1 (UCT–USNM); Washington County, Barre 1, Duxbury 3 (UCT), Hardwick 27 (UCT), North Duxbury 8 (UVT), Plainfield 2, Woodbury 24; Windsor County, Londonderry 3, Ludlow 1 (UCT), Norwich 4 (USNM), Pomfret 2 (AMNH), South Londonderry 11 (UCT), Woodstock 1–1 (DC–USNM).

MASSACHUSETTS, total 76: Berkshire County, Adams 2, Ashley Falls 1 (UCT), Barrington 1, Monterey 10 (UMA), Mt. Greylock 1 (HMCZ), Mt. Washington 1 (AMNH), Pittsfield 1, Savoy Center 8; Franklin County, Montague 5, Orange 1, Quabbin Reservoir 2, Sunderland 1 (UMA); Hampden County, Holland 1, Montgomery 1 (UCT); Hampshire County, Amherst 22, Pelham 2, Shutesbury 3 (UMA); Worcester County, Harvard 2 (UCT), New Salem 1 (UMA), Petersham 10 (UCT).

CONNECTICUT, total 91: Fairfield County, New Fairfield 1; Hartford County, Simsbury 1 (UCT); Litchfield County, Bear Mt. 3 (AMNH), Colebrook 2 (HMCZ), Lime Rock 1, Litchfield 2 (UCT), Macedonia Park 4 (AMNH), Sharon Mt. 6 (AMNH); Middlesex County, Marlborough 1, Portland 3; New Haven County, Southbury 1; Tolland County, Charter Marsh 2, Eagleville 1, Mansfield 11, Storrs 8, Union 15, Vernon 1; Windham County, Ashford 3, Brooklyn 2 (UCT), South Woodstock 3 (AMNH), Woodstock 1 (UCT).

## SUBORDER HYSTRICOMORPHA

### FAMILY ERETHIZONTIDAE
### (AMERICAN PORCUPINES)

This family contains the arboreal or New World porcupines and consists of four genera: *Erethizon*, *Coendou*, *Chaetomys*, and *Echinoprocta*.

Porcupines are large, robust rodents with long, sharp quills that can be raised and lowered by muscular contraction. The quills vary in distribution on the body; they are primarily on the head, back, and tail in *Erethizon* and cover most of the body in *Coendou*. Locomotion is plantigrade—slow, clumsy, and deliberate both on land and in trees. The limbs are short. The soles of the hind limbs are heavy and calloused, with the hallux replaced by a broad movable pad adapted for climbing. Porcupines have an elaborate courtship before mating.

The genus *Erethizon* occurs in North America from the Arctic Ocean south to Sonora, the upper Mississippi Valley, and the Appalachian Mountains, and the three genera occupy tropical forests. There is only one species of porcupine in New England.

# Porcupine

*Erethizon dorsatum dorsatum*
(Linnaeus)

[*Hystrix*] *dorsata* Linnaeus, 1758.
   *Systema naturae*, 10th ed., 1:57.
*Erethizon dorsatum dorsatum*, Miller,
   1912. *U.S. Natl. Mus. Bull.* 79:289.
Type Locality: Eastern Canada, Province of Quebec
Dental Formula:
   I 1/1, C 0/0, P 1/1, M 3/3 × 2 = 20
Names: Eastern porcupine,
   hedgehog, porky hog, prickle pig, quill pig, quiller, and silver cat. The meaning of *Erethizon* = irritable, or to excite or irritate, probably referring to the animal's bristling up its quills, and *dorsatum* = back, alluding to the armor or quills on the back.

**Description.** The porcupine has a robust, high-arched body with long, sharp quills in the pelage. The head is small, with a rounded, blunt muzzle, hairy lips, small ears almost hidden in fur, and dull black, beady eyes that do not reflect light at night. The limbs are short and bowed. The forefoot has four toes and the hind foot five, all with strong curved claws. The tail is short, stout, and muscular, with quills above and stiff bristles below. Females have four mammae—two pectoral, and two abdominal.

The pelage is composed of three kinds of hairs: long, soft, woolly underfur overlaid with long, stiff glistening guard hairs mixed with still longer quills—though in winter the guard hairs are longest. The guard hairs are very sensitive, and the slightest touch on their tips causes prompt erection of the quills. The slender quills are from 76 to 102 mm (3–4 in) long, thickest on the hollow main shaft. The basal portion is constricted, while the tip is needlelike, covered with many hundreds of minute overlapping diamond-shaped scales that slant backward and act as barbs. The number of quills varies with the individual; some porcupines may have as many as 30,000, while others may have many fewer. The quills are loosely attached to the skin and dislodge easily. They are erected and lowered by special muscles, but they cannot be thrown. Old quills are so loosely attached to the skin that they eventually are shed or dislodged at the slightest flip of the tail. The quills occur all over the upperparts and sides of the body from the crown of the head to the end of the tail. The feet and underparts of the body and tail are quilless.

A cornered porcupine erects its quills, tucks its head between its front legs or under a protective object, turns its rear to the adversary, and rapidly swings its tail or makes a rolling lunge of the body. Once a quill is imbedded it absorbs moisture and expands slightly. Contractions of the victim's muscles engage the tips of the scales and draw the quill in deeper.

A quill or quill fragment may travel at a rate of an inch per day, causing intense suffering, and may reach the heart, arteries, or lungs and cause death. Some quills travel back out through the skin, pass out in the droppings, come to rest against bone, or become absorbed or encapsulated.

The porcupine's skull is heavy and compact, with a short, broad rostrum and broad frontals that are heavily ridged and converge posteriorly to form the sagittal crest. The zygomatic arches are simple, and the jugal is deeper anteriorly than posteriorly. The large auditory bullae are rotund, and the postorbital processes are absent. The mandible has low coronoid and angular processes, and the lower border of the angular process is inflated. The palate is short and narrow, extending posteriorly to the anterior border of the last molar. The cheek teeth are rooted, flat-crowned, and subhypsodont, and the molar enamel pattern is intricately folded. The incisors are heavy and colored deep orange.

The sexes are colored alike and show some seasonal variation. In summer the coloration is glossy black or brownish black all over, with or without white-tipped guard hairs. The bases of the quills are yellowish white and the tips are brown or black. The tail is colored like the back. The winter pelage is longer, blacker, and duller, particularly on the shoulders. The underfur is blackish to gray like the summer pelage, but much more woolly. Immatures are darker than adults.

The guard hairs are molted and replaced en masse, as contrasted with the gradual loss and replacement of the underfur. Lost quills soon grow back, but damaged quills are not replaced till the next molt. Molting of the winter pelage begins during mid-spring and is completed by late July or early August (Po-Chedley and Shadle 1955).

Males are heavier than females. Measurements range from 63.5 to 102 cm (25–40 in); tail, 14.5 to 29.7 cm (5.7–11.7 in); and hind foot, 7.4 to 8.9 cm (2.9–3.5 in). Weights generally vary from 2.3 to 7.6 kg (5.0–16.8 lb), but extremely large porcupines may weigh 11.4 kg (25.0 lb) or more.

**Distribution.** The porcupine occurs from Nova Scotia, Quebec, and Alaska across boreal Canada, south through the northeastern states, south along the Appalachian Mountains of western Maryland, in West Virginia (and probably Virginia and northeastern Tennessee), west through Indiana and Iowa, south through northwestern Texas and northern Mexico, and northward through the Pacific states.

The porcupine is found throughout Maine, New Hampshire, Vermont, most of east central and western Massachusetts, and Connecticut. It is uncommon in Rhode Island.

Changes in land use, such as cutting and burning of many sections of woodlands over many years, apparently have led to reduction in the range of the porcupine in New England. In some areas, however, the animal probably has returned to its former range and gradually increased its population in farm woodlots and orchards as many marginal farms have been bought by urban people seeking seclusion.

**Ecology.** Porcupines are found in mixed hardwood-conifer woodlands where suitable den sites are available. They prefer hemlock and sugar maple associations in New England and are found in valleys as well as on mountaintops. The den may be in a cave, a deep crevice among rocks, a hollow log, a deserted fox den, a dry, caved-in beaver burrow, a hole under a stump, or in an abandoned building.

Man and fishers are the chief predators of porcupines. Other predators are lynx, bobcats, coyotes, gray foxes, dogs, and owls. Fires, automobiles, and accidents occasionally kill some porcupines. Lice, ticks, cestodes, and nematodes infest them, and they suffer from mange and occasionally from the "snuffles"—a green mucous exudation from the nose—which is usually followed by pneumonia or death.

**Behavior.** Porcupines are active throughout the year; they are mainly nocturnal and spend most of the day perched in a high tree or hidden in a ground den. They are amiable and make interesting pets. They rely greatly on hearing and smell, and much of their so-called stupid behavior stems from the fact that they are myopic. Porcupines react quickly to touch, and jarring the earth often makes them look around and bristle their quills.

They are sure-footed and persistent, though slow-moving, and walk with a waddle, swinging the hips and tail from side to side. They gallop clumsily at about 2 miles per hour for short distances. They climb upward skillfully and deliberately, arm over arm, but descend clumsily backward. Although they do not like water, they will swim short distances. Dean (1950) saw a porcupine swim about 15 feet directly to some water lilies, which it cut and brought back to shore to eat.

Porcupines have a wide range of vocalizations, producing moans, shrieks, owllike "hoo, hoos," shrill screeches, barks, whines, grunts, coughs, sniffs, chatters, snorts, low-pitched mews, and sobs like a child crying.

These rodents are solitary for most of the year, but in winter several occasionally den together in caves and deep crevices. Dodge (1967) reported as many as 100 porcupines in a 4- to 5-square acre area in western Massachusetts where there were many ledges, and he found six animals in various parts of an abandoned house in New Hampshire. These situations are probably unusual.

The den serves primarily for protection from snow, wind, and predators. It is not defended, and many different porcupines may use it on a rotating basis. A den is often recognized by large piles of porcupine droppings. Porcupines make no bedding site. They may spend the winter in a "station tree," usually of hemlock or white spruce. Station trees can be recognized by their gnawed bark and by a strong urine odor, especially on mild, humid days. Porcupines may defend their winter feeding trees, and occasionally paired animals are found in the same tree. In summer, porcupines often climb trees to avoid mosquitoes.

Porcupines do not travel far. They spend most of their time in trees, especially during fall and winter. Curtis and Kozicky (1944) noted that the

PORCUPINE

distance between den and feeding grounds was approximately 330 feet. Shapiro (1949) determined that the cruising radius of feeding averaged 430 feet and the winter range size 13.3 square acres. Faulkner and Dodge (1962) reported that the winter range size averaged about 6 acres, with a mean cruising radius of approximately 300 feet depending on food availability. Dodge (1967) reported that the average distance between a winter den site and the point of capture in summer was 4,923 feet. Some individuals wander throughout the year. Schoonmaker (1938) tracked a porcupine in winter for about three-fourths of a mile. Marshall, Gullion, and Schwab (1962) reported that the average spring and summer movements of two adult females between dawn and dusk were 264 to 267 feet; maximum diurnal movements were 620 to 685 feet; and home range size 32 to 36 square acres. The authors found that porcupines varied their daytime activity from resting in a fork or branch of a tree during the entire day to foraging on the ground for most of the daylight hours. Costello (1966) reported some indications of local wandering and migrations of porcupines. Dodge (1967) found a major shift in the food habits of porcupines between winter and summer months.

Porcupines feed on a variety of grasses and sedges, leaves, yellow pond lilies, apples, sweet corn and other farm crops, tender twigs, forbs, roots, buds, flowers, catkins, seeds of sugar maple, bark of hemlock, white cedar, larch, balsam, white pine, basswood, beech, American elm, American chestnut, gray birch, paper birch, yellow birch, large tooth aspen, quaking aspen, pin cherry, poplar, red maple, red oak, white ash, and other hardwoods.

Porcupines are fond of salt or salt solutions such as perspiration and will do almost anything to get salt. They also like to gnaw on shed antlers.

The breeding season usually begins in early November through early December, and some females may breed in late March or in April (Shadle and Ploss 1943). During late August and September the testes of males descend into the scrotum, and spermatogenesis reaches its highest level during October. The female seems to be seasonally polyestrous and will recycle in 25 to 35 days if fertilization does not occur at the time of the first ovulation. This recycling may continue into late January in New England. Estrus lasts for 8 to 12 hours (Dodge 1967). The gestation period is 205 to 217 days, averaging 210 days (Shadle 1948).

The courtship of porcupines is elaborate, with much vocalization. The excited male does a "three-legged walk" usually with the left forepaw holding onto his genitals. The male may hold a long stick in his forepaws, straddling and riding it as the female does, or rub his genitals on objects and "sing" in a low whine. The male usually urinates on the female (Shadle, Smelzer, and Metz 1946).

Porcupines do not copulate in a standing position, belly to belly. Copulation occurs by the usual rear mount. Within a few minutes after copulation the female discharges a bluish white vaginal plug, about half an inch long, from the vulva (Shadle 1946; Burge 1966). There is usually only a single young born. The porcupette is generally born in a ground den or shelter in April, May, or June. At birth it is precocial, weighs from 340 to 567 g (12–20 oz), and is about 305 mm (12 in) long (Shadle 1946). It is well covered with long black hairs and has quills about 25.4 mm (1 in) long, which soon become functional. It feeds on green plants by about the 2d week but continues to nurse for 52 days and may attempt to nurse until 3½ months or longer (Shadle and Ploss 1943). The female usually takes little or no care of the porcupette beyond a week or two after birth. By about 6 months of age the porcupette and mother drift apart.

Porcupines become sexually mature at 15–16 months. Longevity may be at least 10 years in captivity (Shadle 1951; Burge 1966) and at least 10.1 years in the wild (Brander 1971).

**Specimens Examined.** MAINE, total 49: Hancock County, Bucksport 7, West Pembroke 4; Oxford County, Norway 12, Umbagog Lake 3, Upton 9; Penobscot County, Bangor 1; Piscataquis County, Moosehead Lake 1 (HMCZ); Somerset County, Lakewood 1 (UCT); Washington County, Baring 1, East Machias 3, Jonesboro 1 (HMCZ), Moosehorn National Wildlife Refuge 1 (UCT), Steuben 4 (HMCZ); York County, Kittery 1 (UCT).

NEW HAMPSHIRE, total 19: Carroll County, Ossipee 5 (USNM), Tamworth 2; Cheshire County, Dublin 1 (HMCZ); Coos County, Pickham Notch 1 (AMNH), Pittsburg 1 (UCT); Grafton County, Piermont 1 (DC); Hillsborough County, Bennington 1; Merrimack County, Newbury 1, Webster 4 (HMCZ); Strafford County, Barrington 1, Durham 1 (UNH).

VERMONT, total 22: Addison County, Granville 1–3 (HMCZ–USNM), Ripton 1 (UVT); Caledonia County, Passumpsic 1 (DC); Chittenden County, Underhill 1, Williston 1 (UVT); Rutland County, East Wallingford 1 (DC), Killington Peak 7 (HMCZ); Washington County, Plainfield 1–1 (UCT–UVT); Windham County, Newfane 3 (AMNH); Windsor County, Bridgewater 1 (DC).

MASSACHUSETTS, total 8: Franklin County, New Salem 1–1 (UMA–UNH), Montague 1 (UMA); Hampden County, Holyoke 1; Hampshire County, Northampton 1 (UCT); Middlesex County, Pepperell 3 (HMCZ).

CONNECTICUT, total 4: Hartford County, Manchester 1; Tolland County, Mansfield 1; Windham County, Ashford 1, Chaplin 1 (UCT).

## REFERENCES

Abbott, Herschell, and Belig, William H. 1961. Juniper seed: A winter food of red squirrels in Massachusetts. *J. Mamm.* 42(2):240–44.

Ackerman, Ralph, and Weigl, Peter D. 1970. Dominance relations of red and grey squirrels. *Ecology* 51(2):332–34.

Aldous, Shaler E. 1938. Beaver food utilization studies. *J. Wildl. Manage.* 2(4):215–22.

Aleksiuk, Michael, 1968. Scent-mound communication, territoriality, and population in beaver *Castor canadensis* Kuhl. *J. Mamm.* 49(4):759–62.

——. 1970. The function of the tail as a fat storage depot in the beaver *Castor canadensis*. *J. Mamm.* 51(1):145–48.

Alexander, Maurice M. 1951. The aging of muskrats on the Montezuma National Wildlife Refuge. *J. Wildl. Manage.* 15(2):175–86.

——. 1956. *The muskrat in New York State.* Syracuse: State University of New York, College of Forestry.

——. 1960. Shrinkage of muskrat skulls in relation to aging. *J. Wildl. Manage.* 24(3):326–29.

Allen, Elsa G. 1938. The habits and life history of the eastern chipmunk (*Tamias striatus lysteri*). *New York State Mus. Bull.* 314:1–122.

Allen, Joel A. 1869. Mammalia of Massachusetts. *Bull. Mus. Comp. Zool. Harvard Coll.* 1:143–252.

Allen, John M. 1954. Gray and fox squirrel management in Indiana. *Indiana Dept. Conserv. Pittman-Robertson Bull.* 1:1–112.

Ambrose, Harrison W., III. 1973. An experimental study of some factors affecting the spatial and temporal activity of *Microtus pennsylvanicus*. *J. Mamm.* 54(1):79–110.

Anderson, J. W. 1954. The production of ultrasonic

sounds by laboratory rats and other mammals. *Science* 199:808–9.

Anderson, Sydney. 1967. Introduction to the rodents. In *Recent mammals of the world*, ed. S. Anderson and J. K. Jones, Jr., pp. 206–9. New York: Ronald Press.

Bagg, Aaron M. 1952. Anting not exclusively an avian trait. *J. Mamm.* 33(2):243.

Bailey, E. D. 1965. The influence of social interaction and season on weight change in woodchucks. *J. Mamm.* 46(3):438–45.

Bailey, Vernon. 1900. Revision of American voles of the genus *Microtus*. *N. Amer. Fauna* 17:1–19.

——. 1923. Mammals of the District of Columbia. *Proc. Biol. Soc. Washington* 36:103–38.

——. 1924. Breeding, feeding, and other life habits of meadow mice *Microtus*. *J. Agric. Res.* 27(8):523–36.

——. 1927. Beaver habits and experiments in beaver culture. *U.S. Dept. Agric. Tech. Bull.* 21:1–39.

Barbehenn, Kile R. 1955. A field study of growth in *Microtus pennsylvanicus*. *J. Mamm.* 36(4):533–43.

Barkalow, F. S., Jr. 1967. A record gray squirrel litter. *J. Mamm.* 48(1):141.

Barkalow, F. S., and Soots, R. F., Jr. 1975. Life span and reproductive longevity of the gray squirrel, *Sciurus c. carolinensis* Gmelin. *J. Mamm.* 56(2):522–24.

Barrier, M. J., and Barkalow, F. S., Jr. 1967. A rapid technique for aging gray squirrels in winter pelage. *J. Wildl. Manage.* 31(4):715–19.

Beer, James R., and MacLeod, Charles F. 1961. Seasonal reproduction in the meadow vole. *J. Mamm.* 42(4):483–89.

Benton, Allen H. 1955a. Observations on the life history of the northern pine mouse. *J. Mamm.* 36(1):52–62.

——. 1955b. Notes on the behavioral development of captive red-backed mice. *J. Mamm.* 36(4):566–67.

——. 1966. Waltzing in the red-backed mouse. *J. Mamm.* 47(2):357.

Bider, J. R. 1968. Animal activity in uncontrolled terrestrial communities as determined by a sand transect technique. *Ecol. Monogr.* 38:269–308.

Blair, W. Frank. 1939. A swimming and diving meadow vole. *J. Mamm.* 20(3):375.

——. 1940a. A study of prairie deer-mouse populations in southern Michigan. *Amer. Midland Nat.* 24(2):273–305.

——. 1940b. Home ranges and populations of the medaow vole in southern Michigan. *J. Wildl. Manage.* 4(2):149–61.

——. 1941. Some data on the home ranges and general life history of the short-tailed shrew, red-backed mouse, and woodland jumping mouse in northern Michigan. *Amer. Midland Nat.* 25(3):681–85.

——. 1942. Size of home range and notes of the life history of the woodland deer-mouse and eastern chipmunk in northern Michigan. *J. Mamm.* 23(1):27–36.

Bond, Charles F. 1956. Correlations between reproductive condition and skull characteristics of beaver. *J. Mamm.* 37(4):506–12.

Bovet, Jacques. 1968. Trails of deer mice (*Peromyscus maniculatus*) traveling on the snow while homing. *J. Mamm.* 49(4):713–25.

Bowker, Leslie S., and Pearson, Paul G. 1975. Habitat orientation and interspecific interaction of *Microtus pennsylvanicus* and *Peromyscus leucopus*. *Amer. Midland Nat.* 94(2):491–96.

Brander, Robert B. 1971. Longevity in wild porcupines. *J. Mamm.* 52(4):835.

Brenner, Fred J. 1962. Foods consumed by beavers in Crawford County, Pennsylvania. *J. Wildl. Manage.* 26(1):104–7.

——. 1964. Reproduction of the beaver in Crawford County, Pennsylvania. *J. Wildl. Manage.* 28(4):743–47.

——. 1975. Effect of previous photoperiodic conditions and visual stimulation on food storage and hibernation in the eastern chipmunk. *Amer. Midland Nat.* 93(1):227–34.

Broadbooks, Harold E. 1961. Homing behavior of deer mice and pocket mice. *J. Mamm.* 42(3):416–17.

Bronson, F. H. 1964. Agonistic behavior in woodchucks. *Animal Behavior* 12(4):470–78.

Brooks, Joe E. 1969. *Behavior of the Norway rat and its significance in control programs.* Natl. Pest Control Assoc. Tech. Release.

Brown, Robert Z. 1953. Social behavior, reproduction, and population changes in the house mouse, *Mus musculus* L. *Ecol. Monogr.* 23(1):217–40.

Buckley, J. L., and Libby, W. L. 1955. Growth rates and age determination Alaskan beaver. *Trans. N. Amer. Wildl. Conf.* 20:495–505.

Buckner, C. H. 1957. Home range of *Synaptomys cooperi*. *J. Mamm.* 38(1):132.

Burge, Betty Lou. 1966. Vaginal casts passed by captive porcupines. *J. Mamm.* 47(4):713–14.

Burt, William Henry. 1940. Territorial behavior and population of some small mammals in southern Michigan. *Misc. Publ. Mus. Zool. Univ. Michigan* 45:1–58.

Cade, T. J. 1964. The evolution of torpidity in rodents. *Ann. Acad. Sci. Fenn. Ser. A. IV Biol.* 71:77–112.

Calhoun, John B. 1962. The ecology and sociology of the Norway rat. *U.S. Dept. of Health, Education, and Welfare Public Health Serv. Publ.* 1008:1–288.

Carrighar, Sally. 1942. A singing mouse. *J. Mamm.* 23(4):445–46.

Carson, James D. 1961. Epiphyseal cartilage as an age indicator in fox and gray squirrels. *J. Wildl. Manage.* 25(1):90–93.

Choate, Jerry R. 1973. Identification and recent distribution of white-footed mice (*Peromyscus*) in New England. *J. Mamm.* 54(1):41–49.

Christian, John J. 1971. Fighting, maturity, and population density in *Microtus pennsylvanicus*. *J. Mamm.* 52(3):556–67.

Cole, Leon J. 1922. Red squirrels swimming a lake. *J. Mamm.* 3(1):53–54.

Coles, Richard W. 1967. Thermoregulation of the beaver. Ph.D. diss., Harvard University.

Condrin, John M. 1936. Observations on the seasonal and reproductive activities of the eastern chipmunk. *J. Mamm.* 17(3):231–34.

Connor, Paul F. 1959. The bog lemming (*Synaptomys cooperi*) in southern New Jersey. *Publ. Mus. Michigan State Univ. Biol. Ser.* 1(5):161–248.

———. 1960. The small mammals of Otsego and Schoharie counties, New York. *New York State Mus. and Sci. Serv. Bull.* 382:1–84.

Cook, A. H., and Maunton, E. R. 1954. A study of criteria for estimating the age of beavers. *New York Fish and Game J.* 1(1):27–46.

Costello, D. F. 1966. *The world of the porcupine*. Philadelphia: J. B. Lippincott Co.

Creel, R. H. 1915. The migratory habits of rats with special reference to the spread of plague. *U.S. Public Health Rept.* 30(23):1679–85.

Cronan, John M., and Brooks, Albert. 1968. *The mammals of Rhode Island*. Wildlife Pamphlet no. 6. Providence: Rhode Island Department of Agriculture and Conservation. Division of Fish and Game.

Crowcroft, Peter. 1955. Territoriality in wild house mice (*Mus musculus*) L. *J. Mamm.* 36(2):299–301.

Curtis, James D., and Kozicky, Edward L. 1944. Observations on the eastern porcupine. *J. Mamm.* 25(2):137–46.

Davidson, W. M. 1966. Sexual dimorphism in nuclei of polymorphonuclear leucocytes in various animals. In *The sex chromatin*, ed. K. L. Moore, pp. 59–75. Philadelphia: Saunders.

Davis, David E. 1948. The survival of wild brown rats on a Maryland farm. *Ecology* 29(4):437–48.

———. 1964. Evaluation of characters for determining age in woodchucks. *J. Wildl. Manage.* 28(1):9–15.

———. 1967a. The annual rhythm of fat deposition in woodchucks (*Marmota monax*) *Physiol. Zool.* 40:391–402.

———. 1967b. The role of environmental factors in hibernation of woodchucks, *Marmota monax*. *Ecology* 48(4):683–89.

Davis, David E., and Finnie, E. P. 1975. Entrainment of circannual rhythm in weight of woodchucks. *J. Mann.* 56(1):199–203.

Davis, David E., and Hall, Octavia. 1948. The seasonal reproductive condition of male brown rats in Baltimore, Maryland. *Physiol. Zool.* 21(3):272–82.

Dean, Howard J. 1950. Porcupine swims for food. *J. Mamm.* 31(1):94.

Dekay, James E. 1842. *Zoology of New York; or, The New York fauna.* Part 1. *Mammalia.* Albany; W. and A. White and J. Visscher.

Dewsbury, Donald A. 1975. Copulatory behavior of white-footed mice (*Peromyscus leucopus*). *J. Mamm.* 56(2):420–28.

Dice, Lee R. 1933. Longevity in *Peromyscus maniculatus gracilis*. *J. Mamm.* 14(2):147–48.

Dodge, Wendell E. 1967. The biology and life history of the porcupine, *Erethizon dorsatum*, in western Massachusetts. Ph.D. diss., University of Massachusetts.

Doebel, John H., and McGinnes, Burd S. 1974. Home range and activity of a gray squirrel population. *J. Wildl. Manage.* 38(4):860–67.

Dunford, Christopher. 1970. Behavioral aspects of spatial organization in the chipmunk, *Tamias striatus*. *Animal Behavior* 36(3):215–31.

———. 1972. Summer activity of eastern chipmunks. *J. Mamm.* 53(1):176–80.

Elder, William H., and Shanks, C. E. 1962. Age changes in tooth wear and morphology of the baculum in muskrats. *J. Mamm.* 43(2):144–50.

Enders, Robert K. 1939. The corpus luteum as an indicator of the breeding of muskrats. *Trans. N. Amer. Wildl. Conf.* 4:631–34.

Errington, Paul L. 1937. Food habits of Iowa red foxes during a drought summer. *Ecology* 18(1):53–61.

———. 1939. Reactions of muskrat populations to drought. *Ecology* 20(2):168–86.

———. 1940. Natural restocking of muskrat-vacant habitats. *J. Wildl. Manage.* 4(2):173–85.

———. 1961. *Muskrats and marsh management.* Harrisburg: Stackpole Co.

Errington, Paul L., and Errington, Carolyn S. 1937. Experimental tagging of young muskrats for purposes of study. *J. Wildl. Manage.* 1(1–2):49–61.

Fall, Michael W. 1971. Seasonal variations in the food consumption of woodchucks, *Marmota monax*. *J. Mamm.* 52(2):370–75.

Faulkner, Clarence E., and Dodge, Wendell E. 1962. Control of the porcupine in New England. *New Hampshire's Conserv. Mag.* 72(Spring):9–10, 18.

Fish, Paul G., and Whitaker, John O. 1971. *Microtus pinetorum* with grooved incisors. *J. Mamm.* 52(4):827.

Fisher, Edward W., and Perry, Alfred E. 1970. Estimating ages of gray squirrels by lens-weights. *J. Wildl. Manage.* 34(4):825–28.

Fitch, H. S. 1958. Home ranges, territories, and seasonal movements of vertebrates of the Natural History Reservations. *Univ. Kansas Publ. Mus. Nat. Hist.* 1(3):157.

Fitzwater, William D., Jr., and Frank, William J. 1944. Leaf nests of gray squirrels in Connecticut. *J. Mamm.* 25(2):160–70.

Fivush, Barbara; Parker, Renee; and Tamarin, Robert H. 1975. Karyotype of the beach vole, *Microtus breweri*, an endemic island species. *J. Mamm.* 56(1):272–73.

Flyger, Vagn F. 1955. Implications of social behavior in gray squirrel management. *Trans. N. Amer. Wildl. Conf.* 20:381–89.

———. 1960. Movements and home range of the gray squirrel. *Sciurus carolinensis*, in two Maryland woodlots. *Ecology* 41(2):365–69.

Forbes, Richard B. 1966. Fall accumulation of fat in chipmunks. *J. Mamm.* 47(4):715–16.

Fraleigh, Lucy B. 1929. Habits of mammals at an Adirondack camp. *New York State Mus. Handbook* 8:119–53.

Friley, Charles E., Jr. 1949. Use of the baculum in age determination of Michigan beaver. *J. Mamm.* 30(3):261–67.

Fulk, George W. 1972. The effect of shrews on the space utilization of voles. *J. Mamm.* 53(3):461–78.

Furrer, Robert K. 1973. Homing of *Peromyscus maniculatus* in the channelled scablands of east-central Washington. *J. Mamm.* 54(2):466–82.

Gentry, John B., and Odum, Eugene P. 1957. The effect of weather on the winter activity of old-field rodents. *J. Mamm.* 38(1):72–77.

Getz, Lowell L. 1959. Activity of *Peromyscus leucopus. J. Mamm.* 40(3):449–50.

———. 1961a. Notes on the local distribution of *Peromyscus leucopus* and *Zapus hudsonius. Amer. Midland Nat.* 65(4):486–500.

———. 1961b. Responses of small mammals to live-trap and weather conditions. *Amer. Midland Nat.* 66(1):160–70.

———. 1961c. Home ranges, territoriality, and movement of the meadow vole. *J. Mamm.* 42(1):24–36.

———. 1968. Influence of weather on the activity of the red-backed vole. *J. Mamm.* 49(3):565–70.

Getz, Lowell L., and Ginsberg, Victoria. 1968. Arboreal behavior of the red-backed vole, *Clethrionomys gapperi. Animal Behaviour* 16(4):418–24.

Goehring, Harry H. 1971. Two rhino mice (*Mus musculus*) from Minnesota. *J. Mamm.* 52(4):834–35.

Goertz, John W. 1965. Late summer breeding of flying squirrels. *J. Mamm.* 46(3):510.

———. 1971. An ecological study of *Microtus pinetorum* in Oklahoma. *Amer. Midland Nat.* 86(1):1–12.

Goertz, John W.; Dawson, Robert M.; and Mowbray, Elmer E. 1975. Response to nest boxes and reproduction by *Glaucomys volans* in northern Louisiana. *J. Mamm.* 56(4):933–39.

Goodwin, George G. 1932. New Records and some observations on Connecticut mammals. *J. Mamm.* 13(1):36–40.

———. 1935. *The mammals of Connecticut.* Bulletin no. 35. Hartford: Connecticut Geological and Natural History Survey.

Gordon, David C. 1962. Adirondack record of flying squirrel above timber line. *J. Mamm.* 43(2):262.

Gould, James, and Morgan, Clifford T. 1941. Hearing in the rat at high frequencies. *Science* 94(2433):168.

Grant, P. R. 1971. The habitat preference of *Microtus pennsylvanicus*, and its revelance to the distribution of this species on islands. *J. Mamm.* 52(2):251–61.

Green, Morris M. 1936. The black rat in Vermont. *J. Mamm.* 17(2):173.

Grizzell, Roy A., Jr. 1955. A study of the southern woodchuck *Marmota monax monax. Amer. Midland Nat.* 53(2):257–93.

Gunson, J. R. 1967. *Reproduction and productivity of the beaver in Saskatchewan.* Quebec: Second Fur Workshop Paper.

Habeck, James R. 1960. Tree-caching behavior in the gray squirrel. *J. Mamm.* 41(1):125–26.

Hall, E. Raymond, and Cockrum, E. Lendell. 1953. A synopsis of the North American microtine rodents. *Univ. Kansas Publ. Mus. Nat. Hist.* 5:373–498.

Hall, E. Raymond, and Kelson, Keith R. 1959. *The mammals of North America.* 2 vols. New York: Ronald Press.

Hall, E. R. S. 1930. Groundhog active in winter. *Canadian Field-Nat.* 44(8):198.

Hamilton, William J., Jr. 1934. The life history of the rufescent woodchuck (*Marmota monax rufescens*) Howell. *Annals Carnegie Mus.* 23:85–178.

———. 1935. Habits of jumping mice. *Amer. Midland Nat.* 16(1):187–200.

———. 1937a. Activity and home range of the field mouse, *Microtus pennsylvanicus pennsylvanicus* (Ord). *Ecology* 18(2):255–63.

———. 1937b. The biology of microtine cycles. *J. Agr. Res.* 54(10):779–90.

———. 1938. Life history of the northern pine mouse. *J. Mamm.* 19(2):163–70.

———. 1939. Observations on the life history of the red squirrel in New York. *Amer. Midland Nat.* 22(3):732–45.

———. 1941a. Reproduction of the field mouse, *Microtus pennsylvanicus* (Ord). *Cornell Univ. Agric. Exper. Sta. Mem.* 237:1–23.

———. 1941b. The food of small forest mammals in eastern United States. *J. Mamm.* 22(3):250–63.

———. 1953. Reproduction and young of the Florida wood rat, *Neotoma f. floridana* (Ord). *J. Mamm.* 24(2):180–89.

Hammer, Donald A. 1969. Changes in the activity

patterns of deer mice, *Peromyscus maniculatus,* caused by wind. *J. Mamm.* 50(4):811–15.

Hansen, Richard M. 1957. Communal litters of (*Peromyscus maniculatus*). *J. Mamm.* 38(4):523.

Hatt, Robert T. 1929. The red squirrel: Its life history and habits. *Bull. New York College Forestry. Roosevelt Wild Life Annals* 2(1):1–140.

Hauser, Doris C. 1964. Anting by gray squirrels. *J. Mamm.* 45(1):136–38.

Hediger, R. 1970. The breeding behaviour of the Canadian beaver (*Castor fiber canadensis*). *Forma et Functio* 2(4):336–51.

Heinrich, Gerd H. 1953. *Microsorex, Sorex palustris,* and *Microtus chrotorrhinus* from Mt. Katahdin, Maine, *J. Mamm.* 34(3):382.

Henry, Dale B., and Bookhout, Theodore A. 1969. Productivity of beavers in northeastern Ohio. *J. Wildl. Manage.* 33(4):927–32.

Hill, Richard W. 1972. The amount of maternal care in *Peromyscus leucopus* and its thermal significance for the young. *J. Mamm.* 53(4):774–90.

Hiner, Laurence E. 1938. Observations on the foraging habits of beavers. *J. Mamm.* 19(3):317–19.

Hirth, Harold F. 1959. Small mammals in old field succession. *Ecology* 40(3):417–25.

Hodgdon, Harry E., and Larson, Joseph S. 1973. Some sexual differences in behaviour within a colony of marked beavers (*Castor canadensis*). *Animal Behaviour* 21(1):147–52.

Hodgdon, Kenneth W. 1949. Productivity data from placental scars in beaver. *J. Wildl. Manage.* 13(4):412–14.

Hodgdon, Kenneth W., and Hunt, John H. 1953. *Beaver management in Maine.* Augusta: Bulletin of the Maine Department of Inland Fisheries and Game.

Hopkins, Milton, Jr. 1953. Distance perception in (*Mus musculus*). *J. Mamm.* 24(3):393.

Horner, B. Elizabeth. 1947. Paternal care of young mice of the genus *Peromyscus. J. Mamm.* 29(1):31–36.

Howard, Walter E., and Cole, Ronald E. 1967. Olfaction in seed detection by deer mice. *J. Mamm.* 48(1):147–50.

Hoyt, Southgate Y., and Hoyt, Sally F. 1950. Gestation period of the woodchuck (*Marmota monax*). *J. Mamm.* 31(4):454.

Huey, W. C. 1956. New Mexico beaver management. *New Mexico Dept. Game and Fish Bull.* 4:1–49.

Hungerford, K. E., and Wilder, N. G. 1941. Observations on the homing behavior of the gray squirrel (*Sciurus carolinensis*). *J. Wildl. Manage.* 5(4):458–60.

Irving, L., and Orr, M.D. 1935. The diving habits of the beaver. *Science* 82:569.

Jackson, Hartley H. T. 1961. *Mammals of Wisconsin.* Madison: University of Wisconsin Press.

Jarvis, F. N. 1927. Rats are good swimmers. *J. Mamm.* 8(3):249.

Jordan, James S. 1948. A midsummer study of the southern flying squirrel. *J. Mamm.* 29(1):44–48.

——. 1956. Notes on a population of eastern flying squirrel. *J. Mamm.* 37(2):294–95.

Kemp, Gerald A., and Keith, Lloyd B. 1970. Dynamics and regulation of red squirrel (*Tamiasciurus hudsonicus*) populations. *Ecology* 51(5):763–70.

King, John A. 1961. Swimming and reaction to electric shock in two subspecies of deermice, *Peromyscus maniculatus,* during development. *Animal Behaviour* 9(1):142–50.

Kirkpatrick, R. L., and Valentine, G. L. 1970. Reproduction in captive pine voles, *Microtus pinetorum. J. Mamm.* 51(4):779–85.

Klein, Harold G. 1957. Inducement of torpidity in the woodland jumping mouse. *J. Mamm.* 38(2):272–74.

Klingener, David. 1971. The question of cheek pouches in *Zapus. J. Mamm.* 52(2):463–64.

Klugh, A. Brooker. 1923. Notes on the habits of the chipmunk, *Tamias striatus lysteri. J. Mamm.* 4(1):29–31.

——. 1924. The flying squirrel. *Nature Mag.* 3(4):205–7.

——. 1927. Ecology of the red squirrel. *J. Mamm.* 8(1):1–32.

Krutzsch, Philip H. 1954. North American jumping mice (genus *Zapus*). *Univ. Kansas Publ. Mus. Nat. Hist.* 7(4):349–472.

Larson, Joseph S. 1962. Notes on a recent squirrel emigration in New England. *J. Mamm.* 43(2):272–73.

——. 1967. Age structure and sexual maturity within a western Maryland beaver (*Castor canadensis*) population. *J. Mamm.* 48(3):408–13.

Larson, Joseph S., and Knapp, Stephen J. 1971. Sexual dimorphism in beaver neutrophils. *J. Mamm.* 52(1):212–15.

Larson, Joseph S., and Van Nostrand, F. C. 1968. An evaluation of beaver aging techniques. *J. Wildl. Manage.* 32(1):99–103.

Layne, James N. 1954. The biology of the red squirrel, *Tamiasciurus hudsonicus loquax* Bangs, in central New York. *Ecol. Monogr.* 24(3):227–67.

——. 1957. Homing behavior of chipmunks in central New York. *J. Mamm.* 38(4):519–20.

——. 1968. Ontogeny. In *Biology of* Peromyscus (*Rodentia*) ed. John A. King, 2:148–248. *Amer. Soc. Mamm. Spec. Publ.*

Layne, James N., and Benton, Allen H. 1954. Some speeds of small mammals. *J. Mamm.* 35(1):103–4.

Layne, James N., and Hamilton, W. J., Jr. 1954. The young of the woodland jumping mouse, *Napaeozapus insignis insignis* (Miller). *Amer. Midland Nat.* 52(1):242–47.

Lidicker, W. Z. 1966. Ecological observations on a feral house mouse population declining to extinction. *Ecol. Monogr.* 36(1):27–50.

Linduska, Joseph P. 1942. Cycles of Virginia voles. *J. Mamm.* 23(2):210.

——. 1950. *Ecology and land-use relationships of small mammals on a Michigan farm.* Lansing. Michigan Department of Conservation, Game Division.

Lloyd, James A., and Christian, John J. 1967. Relationship of activity and aggression to density in two confined populations of house mice, *Mus musculus*. *J. Mamm.* 48(2):262–69.

Longley, William H., and Moyle, John B. 1963. The beaver in Minnesota. *Minnesota Dept. Conserv. Tech. Bull.* 6:1–97.

Lord, Elizabeth M., and Gates, William H. 1929. Shaker, a new mutation of the house mouse, *Mus musculus*. *Amer. Nat.* 63(688):435–42.

McCabe, Robert A. 1947. Homing of flying squirrels. *J. Mamm.* 28(4):404.

M'Closkey, Robert T., and Fieldwick, Bian. 1975. Ecological separation of sympatric rodents (*Peromyscus* and *Microtus*). *J. Mamm.* 56(1):119–29.

McLaughlin, Charles A. 1967. Aplondontoid, sciuroid, geomyoid, castoroid, and anomaluroid rodents. In *Recent mammals of the world* ed. S. Anderson and J. K. Jones, pp. 210–25. New York: Ronald Press.

McLeod, J. A., and Bondar, G. F. 1952. Studies on the biology of the muskrat in Manitoba. Part 1. Oestrus cycle and breeding season. *Canadian J. Zool.* 30:243–53.

McManus, John J. 1974. Bioenergetics and water requirements of the redback vole, *Clethrionomys gapperi*. *J. Mamm.* 55(1):30–44.

Madden, Jacalyn R. 1974. Female territoriality in a Suffolk County, Long Island, population of *Glaucomys volans*. *J. Mamm.* 55(3):647–52.

Manville, Richard H. 1949. A study of small mammal populations in northern Michigan. *Univ. Michigan Misc. Publ. Mus. Zool.* 73:1–83.

——. 1956. Hibernation of meadow jumping mouse. *J. Mamm.* 37(1):122.

Marshall, William H.; Gullion, Gordon W.; and Schwab, Robert G. 1962. Early summer activities of porcupines as determined by radio-positioning techniques. *J. Wildl. Manage.* 26(1):75–79.

Marten, Gerald G. 1973. Time patterns of *Peromyscus* activity and their correlations with weather. *J. Mamm.* 54(1):169–88.

Martin, Robert L. 1965. The yellow nosed vole. *Mount Washington Obs. News Bull.* 6:5–6.

——. 1971. The natural history and taxonomy of the rock vole, *Microtus chrotorrhinus*. *Diss. Abstr. Intn.* 32(5):1–2.

——. 1973. Molting in the rock vole, *Microtus chrotorrhinus*. *Mammalia* 37(2):342–47.

——. 1975. The yellow nosed vole in Maine. *Maine Geogr.* 8:9–12.

Merriam, C. Hart. 1884. *The mammals of the Adirondack region, northeastern New York.* New York: Privately published.

——. 1888. Description of a new species of field-mouse (*Arvicola pallidus*) from Dakota. *Amer. Nat.* 22:702–5.

Merriam, H. Gray. 1966. Temporal distribution of woodchuck interburrow movements. *J. Mamm.* 47(1):103–10.

——. 1971. Woodchuck burrow distribution and related movement patterns. *J. Mamm.* 52(4):732–46.

Miller, Donald H., and Getz, Lowell L. 1969. Life-history notes on *Microtus pinetorum* in central Connecticut. *J. Mamm.* 50(4):777–84.

Miller, Gerrit S., Jr. 1896. The beach mouse of Muskeget Island. *Proc. Boston Soc. Nat. Hist.* 27:75–87.

Mills, Ernest M. 1947. House mouse control with poisoned baits. U.S. Fish and Wildlife Service (mimeographed).

Murie, Martin. 1963. Homing and orientation of deermice. *J. Mamm.* 44(3):338–49.

Murie, Olaus J., and Murie, Adolph. 1931. Travels of *Peromyscus*. *J. Mamm.* 12(3):200–209.

Muul, Illar. 1968. Behavioral and physiological influence on the distribution of the flying squirrel *Glaucomys volans*. *Misc. Publ. Mus. Zool. Univ. Michigan* 134:1–66.

——. 1969a. Mating behavior, gestation period, and development of *Glaucomys sabrinus*. *J. Mamm.* 50(1):121.

——. 1969b. Photoperiod and reproduction in flying squirrels, *Glaucomys volans*. *J. Mamm.* 50(3):542–49.

——. 1974. Geographic variation in the nesting habits of *Glaucomys volans*. *J. Mamm.* 55(4):840–44.

Muul, Illar, and Carlson, Fred W. 1963. Red-back vole in trees. *J. Mamm.* 44(3):415–16.

Nakamura, Mitsuru. 1950. Tularemia in the jumping mouse. *J. Mamm.* 31(2):194.

Nichols, John T. 1958. Food habits and behavior of the gray squirrel. *J. Mamm.* 39(3):376–80.

Nicholson, Arnold J. 1937. A hibernating jumping mouse. *J. Mamm.* 18(1):103.

——. 1941. The homes and social habits of the wood mouse, *Peromyscus leucopus noveboracensis*, in southern Michigan. *Amer. Midland Nat.* 25(1):196–97.

Novak, Melinda A., and Getz, Lowell L. 1969. Aggressive behavior of meadow voles and pine voles. *J. Mamm.* 50(3):637–39.

Novakowski, N. S. 1967. The winter bioenergetics of a beaver population in northern latitudes. *Canadian J. Zool.* 45:1107–18.

Olsen, Peter F. 1959. Dental patterns as age indicators in muskrats. *J. Wildl. Manage.* 23(2):228–31.

O'Neil, Ted. 1949. *The muskrat in the Louisiana Coastal marshes.* New Orleans: Louisiana Department of Wildlife and Fisheries.

Orr, Howard D. 1959. Activity of white-footed mice in relation to environment. *J. Mamm.* 40(2):213–21.

Osborn, Dale J. 1955. Techniques of sexing beaver (*Castor canadensis*). *J. Mamm.* 36(1):141–42.

Osgood, Frederick L., Jr. 1938. The mammals of Vermont. *J. Mamm.* 19(4):435–41.

Pack, James C.; Mosby, Henry S.; and Siegal, Paul B. 1967. Influence of social hierarchy on gray squirrel behavior. *J. Wildl. Manage.* 31(4):720–28.

Panuska, Joseph Allan. 1959. Weight patterns and hibernation in *Tamias striatus. J. Mamm.* 40(4):554–66.

Panuska, Joseph Allan, and Wade, Nelson J. 1960. Captive colonies of *Tamias striatus. J. Mamm.* 37(1):23–31.

Patric, Earl F., and Webb, William L. 1960. An evaluation of three age determination criteria in live beavers. *J. Wildl. Manage.* 24(1):37–44.

Paul, John Robert. 1966. Observations on the ecology, populations and reproductive biology of the pine vole, *Pitymys p. pinetorum*, in North Carolina. Ph.D. diss., North Carolina State University.

Pearson, Oliver P. 1947. The rate of metabolism of some small mammals. *Ecology* 28(2):127–45.

Pearson, Paul G. 1952. Observations concerning the life history and ecology of the woodrat, *Neotoma floridana floridana* (Ord). *J. Mamm.* 33(4):459–63.

Peterson, Arthur Ward. 1950. Backward swimming of the muskrat. *J. Mamm.* 31(4):453.

Petrides, George A. 1951. Notes on age determination in squirrels. *J. Mamm.* 32(1):111–12.

Po-Chedley, Donald S., and Shadle, Albert R. 1955. Pelage of the porcupine, *Erethizon dorsatum dorsatum. J. Mamm.* 36(1):84–95.

Poole, Earl L. 1940. A life history sketch of the Allegheny woodrat. *J. Mamm.* 21(3):249–70.

Pope, Alton S. 1924. Swimming red squirrels. *J. Mamm.* 5(2):134.

Preble, Norman A. 1956. Notes on the life history of *Napaeozapus. J. Mamm.* 37(2):196–200.

Priddy, Ralph R. 1949. Jumping mouse, underwater swimmer. *J. Mamm.* 30(1):74.

Pruitt, William O., Jr. 1958. Winter activity of red squirrels in interior Alaska. *J. Mamm.* 39(3):443–44.

Quadagno, David M. 1968. Home range size in feral house mice. *J. Mamm.* 49(1):149–51.

Quimby, Don C. 1951. The life history and ecology of the jumping mouse, *Zapus hudsonius. Ecol. Monogr.* 21(1):61–95.

Rainey, Dennis G. 1956. Eastern woodrat, *Neotoma floridana*: Life history and ecology. *Univ. Kansas Publ. Mus. Nat. Hist.* 8(10):535–646.

Rawson, K. S., and Hartline, P. H. 1964. Telemetry of homing behavior by the deermouse, *Peromyscus. Science* 146:1596–98.

Richmond, Neil D., and Roslund, Harry R. 1949. *Mammal survey of northeastern Pennsylvania.* Harrisburg: Pennsylvania Game Commission.

Richter, C. P. 1958. The phenomenon of unexplained sudden death in animals and man. In *Physiological bases of psychiatry*, ed. W. H. Gannt, pp. 112–25. Springfield, Ill.: Charles C. Thomas.

Richter, C. P., and Clisby, Kathryn H. 1941. Phenythio-carbamide taste thresholds of rats and human beings. *Amer. J. Physiol.* 134:157–64.

Riley, D. A., and Rosenzweig, M. R. 1957. Echolocation in rats. *J. Compar. Physiol.* 50:323–28.

Robinson, D. J., and Cowan, I. McT. 1953. An introduced population of the gray squirrel, *Sciurus carolinensis* Gmelin, in British Columbia. *Canadian J. Zool.* 32:261–82.

Rowsemitt, Carol; Kunz, Thomas H.; and Tamarin, Robert H. 1975. The timing and patterns of molt in *Microtus breweri. Occ. Paps. Mus. Nat. Hist. Univ. Kansas.* 34:1–11.

Ruffer, David G. 1961. Effect of flooding on a population of mice. *J. Mamm.* 42(4):494–502.

Rzoska, J. 1953. Bait shyness: A study in rat behavior. *British J. Animal Behaviour* 1:128–35.

Salinger, Herbert E. 1950. Mating of muskrats. *J. Mamm.* 31(1):97.

Sather, J. Henry. 1954. The dentition method of aging muskrats. *Chicago Acad. Sci. Nat. Hist. Misc.* 130:1–3.

Saunders, W. E. 1921. Notes on *Napaeozapus. J. Mamm.* 2(4):237–38.

Schooley, J. P. 1934. A summer breeding season in the eastern chipmunk, *Tamias striatus. J. Mamm.* 15(3):194–96.

Schoonmaker, Walter J. 1938. Notes on the home range of the porcupine. *J. Mamm.* 19(3):378.

Schorger, A. W. 1951. *Zapus* with white tail-tip. *J. Mamm.* 32(3):362.

Schwartz, Charles W., and Schwartz, Elizabeth R. 1959. *The wild mammals of Missouri.* Columbia: University of Missouri Press.

Scott, G. W., and Fisher, K. C. 1972. Hibernation of eastern chipmunk (*Tamias striatus*) maintained under controlled conditions. *Canadian J. Zool.* 50:95–105.

Sealander, John A., Jr. 1952. Food consumption in *Peromyscus* in relation to air temperature and previous thermal experience. *J. Mamm.* 33(2):206–18.

Seidel, D. R. 1961. Homing in the eastern chipmunk. *J. Mamm.* 42(2):256–57.

Seton, Ernest Thompson. 1909. *Life histories of northern animals.* 2 vols. New York: Charles Scribner's Sons.

Shadle, Albert R. 1946. Copulation in the porcupine. *J. Wildl. Manage.* 10(2):159–62.

——. 1947. Porcupine spine penetration. *J. Mamm.* 28(2):180–81.

——. 1948. Gestation period in the porcupine, *Erethizon dorsatum dorsatum*. *J. Mamm.* 29(2):162–64.

——. 1951. Laboratory copulation and gestations of porcupines, *Erethizon dorsatum*. *J. Mamm.* 32(2):219–21.

——. 1952. Sexual maturity and first recorded copulation of a 16-month male porcupine, *Erethizon dorsatum dorsatum*. *J. Mamm.* 33(2):239–41.

Shadle, Albert R., and Ploss, Wm. R. 1943. An unusual porcupine parturition and development of the young. *J. Mamm.* 24(4):492–96.

Shadle, Albert R.; Smelzer, Marilyn; and Metz, Margery. 1946. The sex reactions of porcupines, *Erethizon d. dorsatum*, before and after copulation. *J. Mamm.* 27(2):116–21.

Shanks, Charles E. 1948. The pelt-primeness method of aging muskrats. *Amer. Midland Nat.* 39(1):179–87.

Shanks, Charles E., and Arthur, George C. 1952. Muskrat movements and populations dynamics in Missouri ponds and streams. *J. Wildl. Manage.* 16(2):138–48.

Shapiro, Jacob. 1949. Ecological and life history notes on the porcupine in the Adirondacks. *J. Mamm.* 30(3):247–57.

Sharp, Ward M. 1958. Aging gray squirrels by use of tail-pelage characteristics. *J. Wildl. Manage.* 22(1):29–34.

——. 1960. A commentary on the behavior of free-ranging gray squirrels. Pennsylvania Coop. Wildl. Res. Unit. Paper 101:1–13 (mimeographed).

Sheldon, Carolyn. 1934. Studies on the life histories of *Zapus* and *Napaeozapus* in Nova Scotia. *J. Mamm.* 15(4):290–300.

——. 1938a. Vermont jumping mice of the genus *Zapus*. *J. Mamm.* 19(3):324–32.

——. 1938b. Vermont jumping mice of the genus *Napaeozapus*. *J. Mamm.* 19(4):444–53.

Sheldon, W. G. 1971. Alopecia of captive flying squirrels. *J. Wildl. Diseases* 7(2):111–14.

Sheppe, Walter. 1965. Dispersal by swimming in *Peromyscus leucopus*. *J. Mamm.* 46(2):336–37.

Shufeldt, R. W. 1920. Four-footed foresters—the squirrels. *Amer. Forestry* 26:37–44.

Smith, N. B., and Barkalow, F. S., Jr. 1967. Precocious breeding in the gray squirrel. *J. Mamm.* 48(2):328–30.

Smith, Wendell P. 1931. Calendar of disappearance and emergence of some hibernating mammals at Wells River, Vermont. *J. Mamm.* 12(1):78–79.

Snyder, Dana P. 1956. Survival rates, longevity, and population fluctuations in the white-footed mouse, *Peromyscus leucopus*, in southeastern Michigan. *Misc. Pub. Mus. Zool. University of Michigan* 95:1–33.

Snyder, Robert L., and Christian, John J. 1960. Reproductive cycle and litter size of the woodchuck. *Ecology* 41(4):647–56.

Snyder, Robert L.; Davis, David E.; and Christian, John J. 1961. Seasonal changes in the weights of woodchucks. *J. Mamm.* 42(3):297–312.

Sollberger, Dwight E. 1940. Notes on the life history of the small eastern flying squirrel. *J. Mamm.* 21(3):282–93.

——. 1943. Notes on the breeding habits of the eastern flying squirrel (*Glaucomys volans volans*). *J. Mamm.* 24(2):163–73.

Southwick, Charles H. 1955. Regulatory mechanisms of house mouse populations: Social behavior affecting litter survival. *Ecology* 36(4):625–34.

Spencer, Donald A. 1946. A forest mammal moves to the farm—the porcupine. *Trans. N. Amer. Wildl. Conf.* 11:105–99.

Stack, J. W. 1925. Courage shown by flying squirrel (*Glaucomys volans*). *J. Mamm.* 6(2):128–29.

Starrett, Andrew. 1958. Insular variation in mice of the *Microtus pennsylvanicus* group in southeastern Massachusetts. Ph.D. diss., University of Michigan.

——. 1967. Hystricoid, erethizontoid, cavioid, and chinchilloid rodents. In *Recent mammals of the world*, ed. S. Anderson and J. K. Jones., Jr., pp. 254–72. New York: Ronald Press.

Starrett, Andrew, and Starrett, P. 1952. *Peromyscus leucopus* on peak of Mt. Washington, Coos County, New Hampshire. *J. Mamm.* 33(3):398.

Stebbins, H. E. 1932. Mice that waltz. *Nature Mag.* 19(4):237–38.

Steen, Inger, and Steen, J. B. 1965. Thermoregulatory importance of the beaver's tail. *Compar. Biochem. Physiol.* 15:267–70.

Stegeman, LeRoy C. 1954. The production of aspen and its utilization by beaver on the Huntington forest. *J. Wildl. Manage.* 18(3):348–58.

Stephenson, A. B. 1969. Temperatures within a beaver lodge in winter. *J. Mamm.* 50(1):134–36.

Stewart, C. K. 1946. Aspect of plague control. *Pest Contr. Sanit.* 1(4):14–15.

Stewart, R. E. A.; Stephen, J. R.; and Brooks, R. J. 1975. Occurrence of muskrat, *Ondatra zibethicus albus*, in the district of Keewatin, Northwest Territories. *J. Mamm.* 56(2):507.

Stickel, Lucille F. 1948. The trap line as a measure of small mammal populations. *J. Wildl. Manage.* 12(2):153–61.

——. 1954. A comparison of certain methods of measuring ranges of small mammals. *J. Mamm.* 35(1):1–15.

Stickel, Lucille F., and Warbach, Oscar. 1960. Populations of a Maryland woodlot, 1949–1954. *Ecology* 41(2):269–86.

Stoddard, H. L. 1920. The flying squirrel as a bird killer. *J. Mamm.* 1(2):95–96.

Storer, T. I.; Evans, F. C.; and Palmer, F. G. 1944. Some rodent populations in the Sierra Nevada of California. *Ecol. Monogr.* 14:165–92.

Strecker, Robert L. 1955. Food consumption of house mice at low temperatures. *J. Mamm.* 36(3):460–62.

Stupka, Arthur. 1934. Woodland jumping mice. *Nature Notes from Acadia* 1(3):6.

Sutton, R. W. 1956. Aquatic tendencies in the jumping mouse. *J. Mamm.* 37(2):299.

Tamarin, Robert H., and Kunz, Thomas H. 1974. *Microtus breweri. Mammalian Species* 45:1–3. Amer. Soc. Mammalogists.

Tevis, Lloyd, Jr. 1950. Summer behavior of a family of beavers in New York State. *J. Mamm.* 31(1):40–65.

Townsend, M. T. 1935. Studies on some of the small mammals of central New York. *Roosevelt Wild Life Annals* 4:1–120.

Uhlig, Hans G. 1955a. *The gray squirrel, its life history, ecology, and population characteristics in West Virginia.* Pittman-Robertson Project 31-R. Charleston: West Virginia Conservation Commission.

———. 1955b. Weights of adult gray squirrels. *J. Mamm.* 36(2):293–96.

———. 1956. *The gray squirrel in West Virginia.* Bulletin no. 3. Charleston: West Virginia Conservation Commission.

Van Nostrand, F. C., and Stephenson, A. B. 1964. Age determination for beavers by tooth development. *J. Wildl. Manage.* 28(3):430–34.

Van Vleck, David B. 1968. Movements of *Microtus pennsylvanicus* in relation to depopulated areas. *J. Mamm.* 49(1):92–103.

Walker, Ernest P.; Warnick, Florence; Lange, Kenneth I.; Uible, Howrad E.; Hamlet, Sybil E.; Davis, Mary A.; and Wright, Patricia F. 1964. *Mammals of the world.* 2 vols. Baltimore: Johns Hopkins University Press.

Warnock, John E. 1964. Waltzing in the genus *Microtus. J. Mamm.* 45(4):650–51.

Welch, Jack F. 1953. Formation of urinating "posts" by house mice, *Mus,* held under restricted condition. *J. Mamm.* 34(4):502–3.

Wetherbee, D. K.; Coppinger, R. P.; and Walsh, R. E. 1972. *Time lapse ecology, Muskeget Island, Nantucket, Massachusetts.* New York: New York MSS Educational Publishing Co.

Wetzel, Ralph H. 1955. Speciation and dispersal of the southern bog lemming, *Synaptomys cooperi* (Baird). *J. Mamm.* 36(1):1–20.

Whitaker, John O., Jr. 1963a. A study of the meadow jumping mouse, *Zapus hudsonius* (Zimmerman), in central New York. *Ecol. Monogr.* 33(3):215–54.

———. 1963b. Food, habitat and parasites of the woodland jumping mouse in central New York. *J. Mamm.* 44(3):316–21.

Wolfe, James L. 1966. Agonistic behavior and dominance relationships of the eastern chipmunk *Tamias striatus. Amer. Midland Nat.* 76(1):190–200.

———. 1968. Average distance between successive captures as a home range index for *Peromyscus leucopus. J. Mamm.* 49(2):342–43.

———. 1969. Observations on alertness and exploratory behavior in the eastern chipmunk. *Amer. Midland Nat.* 81(1):249–53.

Wood, Thomas J., and Tessier, Gaston D. 1974. First records of eastern flying squirrel (*Glaucomys volans*) from Nova Scotia. *Canadian Field-Nat.* 88(1):83–84.

Wrigley, Robert E. 1969. Ecological notes on the mammals of southern Quebec. *Canadian Field-Nat.* 83(1):210–11.

———. 1972. *Systematics and biology of the woodland jumping mouse, Napaeozapus insignis.* Illinois Biological Monographs, no. 47. Urbana: University of Illinois Press.

Wyman, Leland C. 1923. *Microtus chrotorrhinus* in Maine. *J. Mamm.* 4(2):125–26.

Yerger, Ralph W. 1953. Home range, territoriality, and populations of the chipmunk in central New York. *J. Mamm.* 34(4):448–58.

———. 1955. Life history notes on the eastern chipmunk *Tamias striatus lysteri* (Richardson), in central New York. *Amer. Midland Nat.* 53(2):312–23.

Young, Howard; Strecker, Robert L.; and Emlen, John T. 1950. Localization of activity of two indoor populations of house mice. *J. Mamm.* 31(4):403–10.

# Order Cetacea

THE ORDER CETACEA includes the whales, dolphins, and porpoises, specialized mammals that are wholly aquatic. Some species live in rivers and some enter large inland lakes, but most cetaceans inhabit the oceans and seas. This order contains the largest animals in the world.

Cetaceans are distinguished by their fusiform bodies with the forelimbs modified as flippers lacking functional claws, by tails greatly flattened laterally and modified as horizontal flukes, and by the absence of external hind limbs. The flukes immediately distinguish the cetaceans from the larger sharks and fishes, which all have vertical tail fins. Although fishes and most cetaceans possess a dorsal fin, the fin of cetaceans contains no bony elements. There are no sebaceous glands, and the skin is smooth, thick, and nearly hairless. The body is insulated by thick subcutaneous blubber. The eyes lack nictitating membranes, the ears lack pinnae, and the auditory meatus is reduced to a tiny aperture. The internal ears are highly specialized. The trachea connects the lungs not with the mouth but with a nostril, or blowhole, that opens near the vertex of the head—single in toothed whales, double in baleen whales. The skull is modified accordingly. When the whale exhales, or blows, water vapor condenses to form a spout that may rise many feet into the air.

Cetaceans possess a "melon"—a lense-shaped fatty deposit in the facial depression of the skull. Apparently it serves as an acoustical lens for echolocation. The melon may be well developed so that it forms a bulging, globose forehead as in the pilot whale. It is somewhat homologous to the spermaceti organ of the sperm whale. The name spermaceti is derived from the Latin words *sperma ceti*, meaning whale sperm, based on the erroneous belief that the substance was coagulated semen.

The Cetacea is divided into two suborders that are not closely related and that differ markedly in anatomy—the Odontoceti, or toothed whales, dolphins, and porpoises, and the Mysticeti, or baleen or whalebone whales. The suborder Archaeoceti, or Zeugolodonti, is extinct.

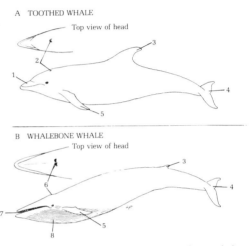

Diagram of a toothed whale, *A*, and a whalebone whale, *B*, illustrating external features. *1*, beak; *2*, single blowhole; *3*, dorsal fin; *4*, flukes; *5*, flipper; *6*, double blowhole; *7*, baleen or whalebone; *8*, pleats.

## SUBORDER ODONTOCETI (TOOTHED WHALES)

The toothed whales have a single blowhole, a relatively small mouth, and from few to many calcified teeth. They have no baleen. These cetaceans catch their relatively large prey one at a time and usually feed on fishes and squid.

The Odontoceti contains six families, of which four have been reported in the waters of New England.

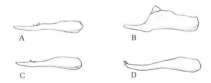

Mandibles of *Mesoplodon*. *A*, Sowerby's beaked whale, *M. bidens*; *B*, tropical beaked whale, *M. densirostris*; *C*, Gervais' beaked whale, *M. europaeus*; *D*, True's beaked whale, *M. mirus*.

### FAMILY ZIPHIIDAE (BEAKED WHALES)

These medium-sized whales are distinguished by a well-developed beak which merges sharply into the bulging forehead in the bottlenosed whales *Hyperoodon* and *Berardius* spp. and gradually into a continuous profile with the head in other genera; by two conspicuous longitudinal grooves on the throat, forming a V shape; by a sickle-shaped dorsal fin situated considerably posterior to the middle of the body; by small flippers set close together; and by flukes that generally lack a medial notch on their posterior edges.

The teeth are more reduced than in any other family of the Odontoceti. Adults have one or two pairs of functional teeth in the lower jaw only, depending on the genus. In females the teeth are much more reduced than in males and sometimes never erupt above the gums.

Little is known of the biology of these oceanic whales. Five species of beaked whales have been recorded in New England waters.

## North Atlantic Beaked Whale

*Mesoplodon bidens* (Sowerby)

*Physeter bidens* Sowerby, 1804. *The British miscellany*, p. 1.
*Mesoplodon bidens*, Fraser and Purves, 1960. *Bull. British Mus. Nat. Hist. Zool.* 7:16, 19, 39, 72.
Type Locality: Coast of Elginshire, Scotland
Names: North Sea beaked whale and Sowerby's beaked whale. The meaning of *Mesoplodon* = floating vessel, and *bidens* = two-toothed.

**Description.** The head of the North Atlantic beaked whale is somewhat flattened laterally. The forehead is moderately high, gradually tapering into a long, slender beak. The mouth cleft is rather large. In adults the lower jaw protrudes somewhat beyond the upper. The crescentic blowhole opens at the middle of the head, and the rostrum of the skull appears sharply pointed when viewed laterally. The relatively large pair of teeth in the lower jaw is situated midway between the tip and the angle of gap of the mouth. Some small teeth may also be present.

The coloration of the animal varies from black or bluish black to dark gray above and usually grayish to whitish below. Some specimens may have many white scars or scratches over the body. The flukes may be dusky or whitish.

Measurements range from 4.5 to 5.5 m (14.8–18.0 ft). One 5.5 m long specimen weighed 450 kg (993 lb) (Tomilin 1957).

**Distribution.** This species is known only in the North Atlantic Ocean and adjacent seas and is well known from European coasts (Tomilin 1957). As far as is known the species has not been encountered in the Pacific Ocean or the Indian Ocean (Moore 1966). In New England the whale is known only from one stranded on Nantucket Island, Massachusetts, in 1867 (True 1910).

**Behavior.** Very little is known about the behavior of the North Atlantic beaked whale. It has been encountered singly and in pairs in the North Atlantic in all seasons. It may feed on cephalopods and deep-sea fishes (Tomilin 1957). One stranded female ate bread and emitted a low sound like the mooing of a cow (Beddard 1900). The breeding season may be projected, based on embryos taken in mid-December (Tomilin 1957) and in early February (Fraser 1937).

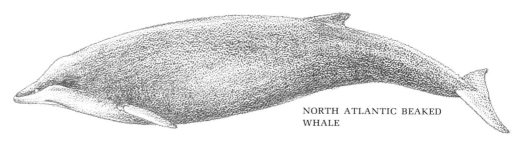

NORTH ATLANTIC BEAKED WHALE

## Tropical Beaked Whale

*Mesoplodon densirostris* (Blainville)

**Description.** The tropical beaked whale is similar to other species of *Mesoplodon* but differs by the relatively enormous mandible bearing a pair of large triangular, tusklike teeth in males only. The teeth are situated well behind the mandibu-

lar symphisis and project anteriorly above the level of the top of the head, somewhat like a pair of horns. Besharse (1971) found sexual dimorphism in this species by comparing the skull, mandible, and teeth of an adult female (the only

*Delphinus densirostris* Blainville,
  1817. *Nouv. Dict. Hist. Nat.* 9:178.
*Mesoplodon densirostris,* Harmer,
  1924. *Proc. Zool. Soc. London,*
  p. 575.
Type Locality: Unknown
Names: Dense-beaked whale, Blainville beaked whale, and Atlantic
  beaked whale. The meaning of
  *Mesoplodon* = floating vessel, and
  *densirostris* = dense beak.

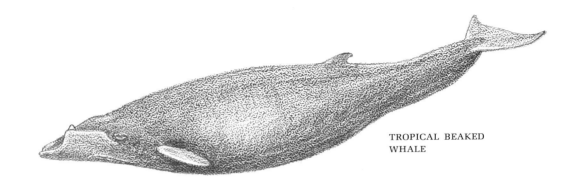

TROPICAL BEAKED
WHALE

reported adult female specimen) with those of males. The rostrum is narrow at the base, and the maxillary tuberosities are absent. The animal is nearly all black with a very slightly paler belly.

Measurements of three males ranged from 4.4 to 4.5 m (14.4–14.8 ft), and the weight of the 4.5 m specimen was 809 kg (1,784 lb) (Raven 1942). A female washed ashore at Annisquam, Massachusetts, in 1898 was 3.7 m (12.1 ft) long (True 1910). A critically ill tropical beaked whale was stranded at Beach Haven, New Jersey, on 3 January 1973 and died several days later. It was 3.8 m (12.5 ft) long and weighed 454 kg (1,000 lb).

**Distribution.** This warm-water species appears to be widely distributed in tropical and subtropical waters but to be very rare throughout its known range. Sick and injured individuals may drift to colder waters on the northwesterly current of the North Atlantic circulation (Moore 1966). A tropical beaked whale was taken at Peggy's Cove, near Halifax, Nova Scotia, in February 1940 (Raven 1942). The only record of this species in the waters of New England was the female taken at Annisquam, Massachusetts, in 1898 (True 1910).

**Behavior.** The behavior of this species is very poorly understood.

# True's Beaked Whale

*Mesoplodon mirus* True

*Mesoplodon mirus* True, 1913.
  *Smithsonian Misc. Coll.* 60(25):1.
*Mesoplodon mirus,* Ulmer, 1941.*Proc.
  Acad. Nat. Sci. Philadelphia*
  83:107.
Type Locality: Bird Island shoal,
  Beaufort Harbor, North Carolina
Names: True's whale and northern
  beaked whale. The meaning of
  *Mesoplodon* = floating vessel, and
  *mirus* = wonderful.

**Description.** True's beaked whale is similar to other species of *Mesoplodon* except that the pair of small teeth in the lower jaw is situated at the tip of the lower jaw; in females they are concealed in the gums. The ventral profile of the rostrum is concave distally, convex proximally. The maxillary foramina are situated behind the premaxillary foramina, and the basirostral grooves are absent.

In dead specimens the coloration of the upperparts is slate black, fading to slate gray on the sides and underparts. The lower sides and underparts may have small elongated spots of light purple, yellow, or pink, and occasionally there is a dark line in the center of the belly. There may be a dark gray area on the thorax. The median of the belly is somewhat darker and grayish in front of the anal vent.

Measurements of five females ranged from 4.9 to 5.2 m (16.1–17.1 ft), and a male measured 4.4 m (14.4 ft) (Moore and Woods 1957). The estimated weight of a 5.2 m female was between

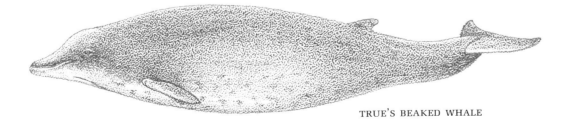

TRUE'S BEAKED WHALE

1,135 and 1,360 kg (2,500–3,000 lb). Her full-term fetus was 2.2 m (7.2 ft) long and estimated at 79 to 91 kg (175–200 lb) (Brimley 1943).

**Distribution.** True's beaked whale occurs principally in the northwestern North Atlantic Ocean and tends to maintain this region despite the sweep of the Gulf Stream (Moore 1966). In New England, a dead specimen was washed ashore on Wells Beach, Maine (1) in 1906 (Raven 1937), and another specimen was washed ashore on Mason Island, Connecticut (2), in 1937 (Thorpe 1938).

**Behavior.** The behavior of this pelagic, or oceanic, whale is poorly understood. The stomachs of beached specimens have contained remains of squid and fish.

## Goose-beaked Whale

*Ziphius cavirostris* G. Cuvier

*Ziphius cavirostris* G. Cuvier, 1823. *Recherches sur les ossements fossiles* 2(5):350, 352.
Type Locality: Between Fos and the mouth of the Gategeon, Département des Bouches-du-Rhône, France
Names: Two-toothed beaked whale and Cuvier's beaked whale. The meaning of *Ziphius* = swordfish, and *cavirostris* = hollow beak or snout.

**Description.** The name goose-beaked whale is based on the shape of the forehead and rostrum of this species. The beak is short and rounded, tapering anteriorly, and appears triangular on the dorsal surface. The rostrum is relatively narrow and the cranial crest steeply elevated. The lower jaw protrudes beyond the tip of the maxilla. The dorsal fin is situated approximately two-thirds posteriorly on the body.

The single pair of teeth is set at the tip of the lower jaw. They project forward some 51 mm (2 in) beyond the gums in males, almost horizontally from the maxilla. In females they are rudimentary and generally concealed in the gums.

The coloration of the animal is dark brown or grayish black, frequently pale white in the facial region and as far back as the dorsal fin. Adults generally have numerous blotches, particularly on the flanks. Some individuals have whitish scars or scratches.

This species may reach a length of 9.2 m (30 ft) (Fraser 1942). Among 51 males and 34 females taken off Japan, males ranged from 5.5 to 6.7 m (18–22 ft) and females from 5.8 to 7.0 m (19–23 ft) (Omura, Fujin, and Kinura 1955). The weight of a 5.8 m (19 ft) long male was 2,535 kg (5,590 lb) (Backus and Schevill 1961).

**Distribution.** This pelagic species inhabits most of the oceans but seems not to be common in any. Strandings have been reported in Massachusetts near Cape Cod (1) (True 1910) and Falmouth (2), 1910, and in Rhode Island near Newport (3), in 1901 (True 1910) and in March 1961 (4) (Backus and Schevill 1961).

**Behavior.** The behavior of goose-beaked whales is poorly known. Some authors have reported that they travel in groups of 30 to 40 individuals and feed together in fairly close associations, while others assert that the animals are usually solitary. These whales are alleged to remain underwater for more than 30 minutes. Squid apparently is their preferred food.

The calving period is protracted, though most births seem to occur in autumn. The gestation period is about 12 months. At birth the calf is about one-third the length of the female (Tomilin 1957). Females attain sexual maturity at about 5.5 m (18 ft) long, and males at about 5.0 m (16 ft) (Omura, Fujin, and Kinura 1955).

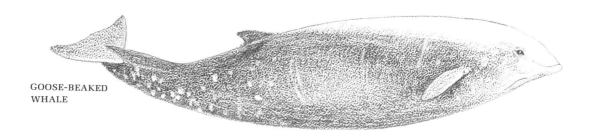

GOOSE-BEAKED
WHALE

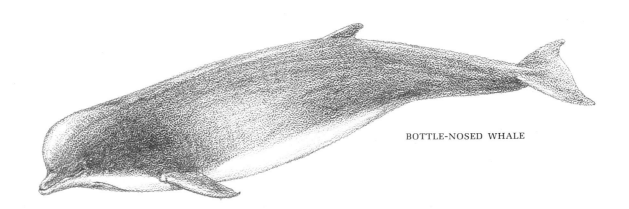

BOTTLE-NOSED WHALE

## Bottle-nosed Whale

*Hyperoodon ampullatus* (Forster)

*Balaena ampullata* Forster, 1770. In Kalm, *Travels into North America*, 1:18, footnote (based on "beaked whale" of Pennant). *The British Zool.* 3:43, 59.
*Hyperoodon ampullatus*, Rhoads, 1902. *Science*, n.s., 15:756.
Type Locality: Maldon, Essex, England
Names: Northern bottle-nose whale, pighead whale, and bottlehead. The meaning of *Hyperoodon* = palate tooth, from the rough papillae of the palate, which Lacépède mistook for teeth, and *ampullatus* = bottle-shaped beak.

**Description.** The body is thickest between the dorsal fin and flippers. Large adults, especially males, have a large, high forehead which descends toward the short beak at almost a right angle, imparting the "bottle-nosed" appearance. The mouth is cleft and slightly S-shaped in lateral view. This species has one or two pairs of teeth at the tip of the lower jaw only. The teeth of adult females are much smaller than those of adult males.

The coloration varies with age. Young whales apparently are brownish and often spotted yellowish white on the flanks and belly. Adults are dark gray above and lighter below. Some individuals, especially adult males, have cream-colored heads. The flippers and flukes are darker than the body. Old individuals may be completely cream colored.

Males are larger than females. Males range from 7.6 to 9.0 m (25–30 ft) long, and females 6 to 7 m (20–23 ft). A female 6 m long weighed 2,500 kg (5,513 lb) (Walker et al. 1964).

**Distribution.** The bottle-nosed whale occurs in all oceans of the northern hemisphere. It is encountered in the North Atlantic in summer and sometimes migrates to the Mediterranean in winter. In New England waters a specimen stranded at North Dennis, Massachusetts (1), in January 1869, and another stranded at Newport, Rhode Island (2), in 1869 (True 1910). A specimen stranded at Stonington, Connecticut (4) (no date; Linsley 1842), two stranded at Tiverton, Rhode Island (3) (no date; Cronan and Brooks 1968), and one stranded at Wells Beach, Maine, in 1906 (Katona, Richardson, and Hazard 1975).

**Behavior.** Little is known about the behavior of this species. Observers have reported that bottle-nosed whales travel in small schools of 20 or more and that they may remain underwater up to 20 minutes when feeding, probably on squid and fish. The gestation period is about 12 months. The calves are born in spring and summer (Tomilin 1957).

### FAMILY PHYSETERIDAE (SPERM WHALES)

This family contains the sperm whales and includes two genera, *Physeter* and *Kogia*, and three species—the huge sperm whale, the pygmy sperm whale, and the dwarf sperm whale. In both genera the skull is markedly asymmetrical, particularly in the rostral part. The upper jaw lacks functional teeth and the lower jaw is much narrower than the contour of the head. The lower jaw of the sperm whale has about 25 teeth on each side and all the cervical vertebrae are fused while the atlas is free. In the pygmy and dwarf sperm whales the lower jaw has from 8 to 16 teeth on each side and all the cervicals are fused.

The physeterids occur in all oceans. The sperm whale and pygmy sperm whale are encountered in New England waters.

# Sperm Whale

*Physeter catodon* Linnaeus

[*Physeter*] *catodon* Linnaeus, 1758,
    *Systema naturae*, 10th ed., 1:76.
Type Locality: Kairston, Orkney Is-
    lands, Scotland (by restriction
    Thomas, 1911. *Proc. Zool. Soc.
    London*, 1911, p. 157).
Names: Cachalot, giant sperm whale,
    spermaceti whale, and pot whale.
    The name sperm is an abbreviation
    of spermaceti, referring to the oil
    contained in the melon of the ani-
    mal's forehead. The meaning of
    *Physeter* = a blower, and *cato-
    don* = having teeth in the lower
    jaw.

**Description.** The sperm whale is the largest of the toothed whales. It is distinguished by its distinctive barrel-shaped head and narrow log-like lower jaw with rows of teeth. The head is high, thick, and blunt, truncated anteriorly with a vertical forehead. The form and proportion of the head vary greatly with the individual and with age. In adult males the head averages approximately one-third of the total body length, and in adult females it averages one-fourth.

The sperm whale is the only odontocete in which the blowhole is situated on the left front end of the head. The bushy spout is directed diagonally to the left and forward at an angle of about 45 degrees. The small eyes are situated just above and behind the angle of the mouth. There are several short grooves on the throat. The dorsal fin is low and thick, compressed into a hump, and some additional humps follow the fin. There is a thick ventral keel on the caudal peduncle. The skin surface is bumpy on the back and flanks. The flippers are small and situated a little behind and below the eyes. The flukes are large and triangular with a deep medial notch.

The facial depression of the skull is greatly developed, obliterating the longitudinal ridge behind the naris. The zygomata are complete. The lower jaw has from 40 to 50 conical teeth that fit into sockets of the palate when the mouth is shut. There are no functional teeth in the upper jaw.

The coloration of the animal is usually brown, or sometimes slate gray, except for occasional patches of white on the lower jaw and venter. Adult males sometimes become paler or piebald, and old bulls often become gray on the snout and the crown of the head. Nishiwaki (1962) reported that all-white sperm whales have been taken in the Pacific Ocean.

Adult males are larger than females. Males may reach a length of 18.0 m (60 ft) and females reach 12.2 m (40 ft). An 18 m specimen may weigh 50 tons.

**Distribution.** Sperm whales occur throughout all oceans; they prefer deep water but may approach the edges of continental shelves. Although they appear to occur most commonly in the temperate and tropical latitudes, adult males may venture near the ice packs in summer.

Females and calves generally remain within the 40° latitude in both hemispheres (Slijper 1962).

Sperm whales are rare in New England waters. A specimen was washed ashore between Cape Porpoise and Biddeford Pool, Maine (2), on 17 July 1668 (Norton 1930); an adult female stranded on Mount Desert Island, Maine (1), in October 1968 (Coman 1972). In New Hampshire, an adult female stranded at Wallace Sands (3) on 16 February 1942 (University of New Hampshire Department of Biology, card catalog no. 100.4.1). Strandings were reported in Massachusetts at West Yarmouth (4) (no date) and at Nantucket Island (5) (no date). A specimen was recorded at Stonington, Connecticut (7) (no date; Linsley 1842), and a calf 4.4 m (14.4 ft) long was washed ashore at Charlestown, Rhode Island (6), on 20 February 1967 (Cronan and Brooks 1968).

**Behavior.** Sperm whales are social animals and may assemble in large groups containing some 1,000 individuals. There are three recognized social associations in sperm whales: (1) large, often tightly packed schools of about 12 to 50 whales, consisting of immature females, pregnant females, lactating females with calves, and immature males; (2) loose associations of about a dozen young adult males; and (3) solitary old males (Caldwell, Caldwell, and Rice 1966).

The succoring behavior of sperm whales is variable. Adult males may approach an injured female; adult females frequently remain near an injured adult female and may come to an injured schoolmate from distances of several miles; adult males do not stand by other injured adult males; females usually flee when a male is harpooned; and juvenile males always flee when a congener is injured (Caldwell and Caldwell 1966).

Sperm whales can leap completely out of the water, and a large sperm whale may throw its flukes 32 feet or more out of the water (Ashley 1942). These whales lobtail by placing themselves perpendicularly in the water, head downward, and slapping the flat of the flukes against the water. Lobtailing may be a signal to congregate, since other sperm whales in the vicinity swim toward lobtailing whales (Gambell 1968).

These mammals can hear the distress call of a congener at distances up to 6 or 7 miles (Ashley

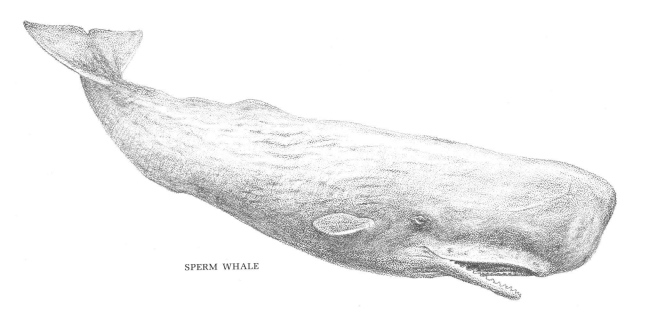

SPERM WHALE

1942; Gilmore 1961). Sperm whales produce low-pitched groans, muffled swashing noises, and a series of sharp clicks (Worthington and Schevill 1957). An alarmed sperm whale may raise its head in the air with the eyes clear of the water and then slowly look around in a complete circle (Ashley 1942). These cetaceans are timid, but when provoked or injured they may be dangerous, especially large males. Occasionally large males will fight.

Sperm whales are deep divers and regularly descend to about 3,000 feet; they have been recorded as entangled in deep-sea cables laid at depths of 3,757 feet (Heezen 1957). The limit of the duration of a dive is determined by the size of the whale—the larger the animal the longer the period of time (Caldwell, Caldwell, and Rice 1966). A large sperm whale may remain underwater from 20 minutes to more than an hour, then surface and blow approximately once for each minute spent underwater (Ashley 1942). A sperm whale that was followed for an entire day submerged for about 57 minutes during each dive (Scholander 1940).

When sperm whales surface the top of the head emerges first, then the blowhole clears the water and the animal blows a forceful spout 10 to 12 feet high (Ashley 1942). The sound of the blow may be heard for nearly 820 feet (Clarke 1954a). These whales execute two different dives. Before a long dive they arch the forward part of the body a few feet into the air, blow a strong spout, round the back in a high arch, and descend steeply. When they intend to make a shallow dive, they raise the hump and simply sink without raising the flukes.

Sperm whales normally swim at speeds of 2 to 4 knots (Davis 1926), rarely more than 5 knots (Murphy 1947). Frightened sperm whales can swim from 6 to 12 knots (Andrews 1916b; Davis 1926). A harpooned sperm whale can briefly pull a whaleboat full of men at speeds from 10 to 21 knots (Clarke 1954a).

These whales feed at great depths. Squid, especially the larger species, are their main food items. Sperm whales also feed on octopuses and on sharks, skates, and other fishes. Clarke (1954b) reported that the stomach of a sperm whale contained 32 squid which averaged 7.2 feet long and weighed a total of 578 pounds. Clarke (1955) found an intact squid 34.5 feet long and weighing 405 pounds in the stomach of a 45-foot sperm whale. Caldwell, Caldwell, and Rice (1966) reported 19 squid, weighing a total of 661 pounds, in the stomach of one sperm whale. Clarke (1956b) suggested that the throat grooves of sperm whales may allow the throat to distend to receive large food items. The regular occurrence of bottom-dwelling fish such as groper and ling in the stomachs of sperm whales suggests that these whales feed on the bottom (Gaskin 1972). Clarke (1956b) suggested that damage to the anterior mandibular teeth of sperm whales indicated bottom feeding since the long, narrow lower jaw may be used to probe among rocks and agitate the bottom, making fish swim upward so that they can be taken.

Sperm whales are polygamous, and males form harems during the breeding season, which lasts from January to July, peaking from March through May in the northern hemisphere (Clarke 1956a). In the southern hemisphere, the breeding season very likely occurs from August through December with a peak in October (Matthews 1938a). Sperm whales may copulate in both the vertical and the horizontal positions. The gestation period is 16 to 17 months (Mizue and Jimbo 1950; Clarke 1956a). In the Atlantic Ocean, calving occurs from May to November in the Azores, peaking from July to September (Clarke 1956a). In the southern hemisphere calving occurs from December through April, peaking in February (Matthews 1938a).

At birth the mean length of fetuses in the Azores is 3.9 m (12.8 ft) (Clarke 1956a). Lactation may last about 12 months (Bennett 1840); at weaning the calf is about 6.7 m (22 ft) long (Clarke 1956a). The female has a strong attachment to her calf and will remain near her wounded young until it dies or recovers. The male is not protective of the young.

Females become sexually mature at about 9 years (Ohsumi, Kasuya, and Nishiwaki 1963), or when they reach a length of 9 m (30 ft) (Mat-

thews 1938a). Males reach sexual maturity at about 9.5 m (31 ft) in the Azores (Clarke 1956a), or at 12 m (39 ft) in the southern hemisphere (Matthews 1938a). Longevity may be up to 50 years (Gaskin 1972).

**Remarks.** Like other great whales, sperm whales have been slaughtered indiscriminately. From the eighteenth to the nineteenth century, thousands of sperm whales were killed annually by whalers. Fortunately the whales have received a respite as petroleum and electricity have supplanted whale oil for lubrication and fuel. Today the whales are afforded protection in the United States by the National Marine Mammal Protection Act of 1972.

Ambergris, an abnormal waxy concretion of the digestive tract that is often expelled by sperm whales and may be found floating or washed ashore, is a uniquely valuable product. Ambergris was used as a medicine by ancient civilizations and today is used as a fragrance-absorbing fixative in perfumes. The color varies from blackish or yellowish brown to whitish, becoming lighter after exposure to air. Fresh ambergris has a disagreeable smell, but after exposure it hardens and gives off a sweet odor. It is found in firm lumps. Tonnessen (1962) reported that the largest lump of ambergris recorded weighed 1,003 pounds. Ambergris brings from $10 to $50 a pound depending on its color. Andrews (1916b) reported that $60,000 worth of ambergris was removed from the intestines of one sperm whale.

## Pygmy Sperm Whale

*Kogia breviceps* (Blainville)

**Description.** The pygmy sperm whale is closely related to the sperm whale in having a spermaceti organ, a strikingly asymmetrical skull, and an inferior, sharklike mouth with functional teeth restricted to the lower jaw. However, it differs conspicuously from the sperm whale by its much smaller size; its porpoiselike shape and anteriorly exaggerated head proportions; the location of its blowhole above the eyes rather than far forward at the end of the snout; and the presence of a well-formed and falcate dorsal fin rather than a rudimentary dorsal fin.

The rostrum is broad and very short. The maxillae dominate the dorsal aspect of the skull, while the lower jaw is fragile, with from 8 to 16 teeth.

The coloration of the animal is dark gray above and light gray below. The lower jaw may be whitish, sometimes onto the upper jaw.

Males are larger than females. Measurements range from 2 to 4 m (6.6–13.1 ft). Weights vary from 318 to 417 kg (700–920 lb).

**Distribution.** The pelagic pygmy sperm whale is widely distributed in the temperate and tropical oceans of the world, though perhaps it is not

PYGMY SPERM WHALE

*Physeter breviceps* Blainville, 1833. *Ann. françaises étrangères ant. phys.* 2:337, pl. 10 (skull).
*Kogia breviceps,* Gray, 1846. In *The Zoology of the Voyage of H.M.S. Erebus and Terror.* Vol. 1, *Mammalia,* p. 22.
Type Locality: Cape of Good Hope, South Africa
Names: Lesser sperm whale, short-headed sperm whale, and lesser cachalot. The meaning of *Kogia* = codger, and *breviceps* = short-headed.

equally abundant in all. In the Atlantic Ocean the species has been reported stranded from the Gulf of Mexico, along the Atlantic coast, north to Nova Scotia. In New England, Brown (1913) recorded a stranded specimen from Nahant, Massachusetts (1), in October 1910. A specimen was recorded at New Bedford, Massachusetts (2), in 1909 (AMNH 34867).

**Behavior.** Very little is known about the behavior of this species. Pygmy sperm whales may migrate toward the poles in summer and return to warmer waters in autumn and winter (Gunter,

Hubbs, and Beal 1955). These whales feed chiefly on crabs, squid, and shrimp, but they do take fish as well (Handley 1966).

The breeding season probably occurs in late summer, and the calves may be born the following spring after a gestation period of only 9 months. Females produce a single calf at each pregnancy. Pregnant and lactating females have been found with small calves, indicating that they may produce young in successive years (Allen 1941). Calves are weaned at about 1 year old, and sexual maturity is probably attained at only 2 years of age (Rice 1967).

## FAMILY MONODONTIDAE (WHITE WHALES AND NARWHALS)

In the male narwhal, the left upper tooth usually grows into a straight, twisted tusk which may reach a length of 3 m (9.8 ft). The right tooth rarely erupts. The function of this tusk is unknown, and female narwhals rarely have tusks.

In both species the dorsal fin is absent or rudimentary, and the flippers are short, broad, and rounded. The flukes have a conspicuous

medial notch; the forehead is high and globose; and grooves are absent on the throat. In the white whale the snout is broad and short; this feature is absent in the narwhal.

These medium-sized cetaceans are gregarious and occur chiefly in arctic seas as well as in most of the larger rivers of Siberia, the Yukon and Kuskokwin rivers in Alaska, and the Saint Lawrence River in eastern Canada (Rice 1967). The white whale occurs in New England waters.

# White Whale

*Delphinapterus leucas* (Pallas)

**Description.** The white whale is recognized by its whitish coloration, high globose forehead, short, broad snout, and absence of a dorsal fin and throat grooves. The skull is asymmetrical.

Both jaws contain sharp, conical teeth on both sides.

The coloration of the animal becomes lighter with age as pigment cells in the skin are lost.

*Delphinus leucas* Pallas, 1776. *Reise durch verschiedene Provinzen des russichen Reichs,* 3(1):85.
*Delphinapterus leucas,* Flower, 1885. *List of Cetacea in the British Museum,* pp. 14–15.
Type Locality: Mouth of Obi [Ob] River, northeastern Siberia, USSR
Names: Beluga, white porpoise, and white fish. The meaning of *Delphinapterus* = dolphin without fin, and *leucas* = white.

Calves are born dark brown, then turn gray, and finally lighten to white or ivory color. Probably the white of adults aids as a camouflage, since it resembles the color of snow, ice, and the white-caps of waves.

Males are larger than females. Males may grow to 5.2 m (17 ft) long and females to 4.6 m (15 ft) (Sergeant 1962*a*). Weights vary from 225 to 675 kg (496–1,488 lb) (Walker et al. 1964).

**Distribution.** White whales occur throughout the arctic and subarctic seas; the farthest south they are resident animals is in the subarctic waters of the Saint Lawrence River estuary, the White Sea, and the Sea of Okhotsk off southeast Alaska (Sergeant 1962*a*), but migrations may take them as far south as 50°N (Nishiwaki 1972). In severe winter weather, white whales may straggle as far south as New Jersey.

In New England waters, white whales occasionally are seen as far south as Long Island Sound. Mairs and Scattergood (1958) reported a white whale in the Penobscot River at the Bangor Salmon Pool in Maine (1). Townsend (1929) observed a white whale off Ipswich Beach, Rockport, and Cape Ann, Massachusetts (2). In early June 1972, a white whale was seen in Cape Cod (3) and Buzzards Bay (4), Massachusetts. The animal remained in the area for about 2 weeks.

The bones of a fossil white whale, *Delphinapterus vermontanus,* were unearthed in the township of Charlotte, about 12 miles south of Burlington, in Chittenden County, Vermont, in August 1849. The discovery site is a little more than a mile east from the shore of Lake Champlain. The skeleton of the fossil whale was displayed in the state museum at Montpellier and is now at the University of Vermont. The total length of the animal was believed to have been at least 4 m (13 ft) (Tracy 1909).

**Behavior.** White whales are gregarious and usually travel in small family groups, but they have been encountered in concentrations of between 5,000 and 10,000 animals. They migrate as far north as ice conditions permit and return south in autumn. Their distribution seems to be closely linked with shifting ice rather than with the movements of their food items (Sergeant 1973). Groups of adult males have been taken far from the rest of the herds (Dorofejev and Klamov 1936), and all-male schools of up to 1,000 animals have been recorded (Lono and Oynes 1961). Solitary individuals occasionally occur.

White whales have well-developed senses of hearing and sight. They can chirp, click, growl, squeal, and creak, and they often emit low sounds like a pig's grunting, produce modulating trill sounds resembling those of a songbird, or sometimes quietly whistle together.

These animals normally swim at 3 to 4 knots, and when frightened can attain a speed of 6 to 8 knots. A normal dive lasts about 2 minutes, but white whales can remain underwater for up to 15 minutes. They can travel considerable distances under the ice, surfacing at air holes in the continuous ice fields. They use their backs to break thin ice and keep the air holes open. White whales can scull with the tail and thus swim backward (Tomilin 1957).

White whales feed mainly on capelin, but they may also eat American sand launce, Atlantic cod, Arctic cod, Atlantic tomcod, sculpins, paddleworms, and squid. They often swallow mud, sand, and stones while feeding at the bottom.

White whales are polygamous. The breeding season takes place in the spring—earlier in the subarctic, later in the arctic. Birth of the calves

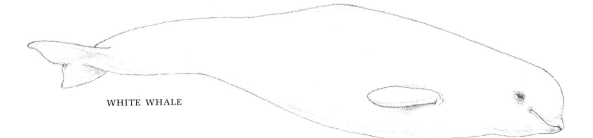

WHITE WHALE

varies from region to region, beginning in late March in west Greenland and ending in July or August in Hudson Bay (Sergeant 1962*a*). The gestation period lasts 14 months and lactation lasts about 20 months. At birth the calf is between 1.2 and 1.8 m (4–6 ft) long. Calving occurs about once in three years. Longevity is about 25 years in both sexes (Sergeant 1973).

## FAMILY DELPHINIDAE (DOLPHINS AND PORPOISES)

The family Delphinidae contains the largest and most diversified of cetaceans. Its members are characterized by small to medium body size, a usually well-developed dorsal fin; flukes with a medial notch; sickle-shaped flippers, and lack of throat grooves. The name dolphin is usually applied to members of this family that have a beaklike snout and a slender body form, whereas the name porpoise generally refers to delphinids having a blunt snout and stocky body.

Males are usually larger than females. The coloration is variable; some species are nearly of one color, some are elaborately patterned with stripes, bands, or spots, and others have contrasting patterns of black and white.

Delphinids are swift swimmers, and many species make synchronized leaps clear out of the water in groups or ride the bow waves of ships. They are shallow divers and surface frequently. Most delphinids are gregarious. They sometimes kill sharks by ramming them.

These cetaceans occur in all oceans, coastal waters, bays, estuaries, and large rivers, and some of them inhabit fresh water. Nine species have been recorded in the waters of New England.

## Striped Dolphin

*Stenella caeruleoalba* (Meyen)

*Delphinus caeruleo-albus* Meyen, 1833. *Nova Acta Acad. Cesarea Nat. Curios* 16(2):609–10, pl. 43, Fig. 2.
*Stenella caeruleoalba*, Scheffer and Rice, 1963. *U.S. Fish and Wildlife Serv. Spec. Rept.* 431:6.
Type Locality: Mouth of the River Plate, South America
Names: Blue dolphin, blue-white dolphin, and euphrosyne dolphin. *Stenella* is a diminutive of Steno, after Nikolaus Steno, a Danish anatomist and geologist, and *caeruleoalba* = of the color sky blue and white, pertaining to animals' blue-white coloration.

**Description.** The striped dolphin is distinguished from the common dolphin by a heavier beak and different color pattern. The best identification mark of the striped dolphin is the "bilge stripe," a narrow black stripe which runs laterally from the eye to the anus, expanding at the posterior end. The forehead and back are dark steel blue, and the chin, throat, and lower half of the flanks and belly are white. A dark blue stripe starts from the dorsal fin and, passing forward, ends abruptly in front. A narrow stripe extends from the eye, which is completely encircled with blue, to the base of the dark blue flippers. The dorsal fin and flukes are dark blue, but the ventral surface of the flukes is white.

In dorsal aspect, the preorbital groove of the skull is moderately deep. The tips of the teeth are sharp and curve slightly inward.

Males are somewhat larger than females. Measurements range from 2.4 to 2.7 m (7.9–8.9 ft) long (Nishiwaki 1972).

**Distribution.** This pelagic species occurs in the warmer waters of the Atlantic Ocean and Pacific Ocean, but it has been reported as far north as Greenland. It infrequently comes near shore and rarely occurs in New England waters. In Massachusetts specimens have been recorded from Woods Hole (2) (no year) (Miller and Kellogg 1955); from Hog Island, Buzzards Bay (1) on 22

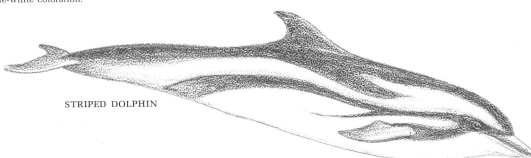

STRIPED DOLPHIN

January 1960 (skull HMCZ 49633); and from Marion (3) on 10 January 1963 (skeleton HMCZ 51763). An immature female was reported from Beverly on 31 January 1975, and in Rhode Island an adult stranded at Narragansett town beach (4) on 5 December 1966 (Cronan and Brooks 1968). In Maine, a juvenile male stranded on the beach at Falmouth on 13 February 1975.

**Behavior.** Very little is known about the behavior of striped dolphins. Schevill and Watkins (1962) reported that a group of about 200 was seen in deep water over the lower Hudson Canyon about 140 miles off New York in August (year not stated). Nishiwaki (1975) reported that a school of 2,838 striped dolphins was captured by drive fisheries as the animals migrated along the coast of Japan. Squid is reported to be the main food of striped dolphins.

These animals have two distinct breeding seasons, in spring and in autumn, and each season seems to be protracted. Calving may occur in spring or autumn, and lactation lasts from 6 to 12 months. Sexual maturity occurs at a length of about 1.8 m (6 ft) and at an age of 4 years (Nishiwaki 1972).

# Common Dolphin

*Delphinus delphis* Linnaeus

*Delphinus delphis* Linnaeus, 1758.
  *Systema naturae*, 10th ed., 1:77.
Type Locality: European seas
Names: Atlantic dolphin, sea dolphin, crisscross dolphin, ring-eyed dolphin, and saddleback dolphin. The meaning of *Delphinus* = a dolphin, and *delphis* = a kind of dolphin.

**Description.** The common dolphin is recognized by its distinctive coloration—dark gray to brownish black above and white below. The blackish color is widest at the top on each side of the dorsal fin, forming a marking shaped like a saddle or a reflection of the dorsal fin. The sides and flanks are marked by varying bands of yellowish gray mixed with long elliptical bands or stripes of gray, white, or yellow. Across the forehead there is a whitish band with a narrow black stripe in the center which unites with the black eye rings. Often there is a narrow black band from the snout to the leading edge of the flippers. The flippers are somewhat lighter below than above. The flukes are blackish on both sides.

The rostrum is elongate, twice as long as the braincase, and split at the anterior end. The premaxillaries tend to fuse together. The lower jaw protrudes slightly ahead of the upper with a short symphysis.

Measurements range from 1.5 to 2.5 m (5.0–8.2 ft), rarely to 2.6 m (8.5 ft) long. The weight is up to 75 kg (165 lb) (Walker et al. 1964).

**Distribution.** The common dolphin is widely distributed in temperate and warm waters and occasionally occurs in northern waters of the Atlantic Ocean. The species sometimes ascends rivers. In New England waters the common dolphin has been recorded for Massachusetts at Denis (1) in 1950 and at South Wellfleet (2) in 1949, and for Connecticut in Long Island Sound (6) (Goodwin 1935). There are three records from Rhode Island: a specimen stranded at Block Island (3) (no date); a specimen stranded at Nannaquaket Pond, Tiverton (4), on 2 December 1965; and an individual captured alive at Point Judith Pond (5) on 12 August 1966 (Cronan and Brooks 1968).

**Behavior.** Little is known about the behavior of this species. Common dolphins are gregarious and may travel in schools of up to several hundred. Rice (1967) wrote that 100,000 saddlebacked dolphins have been observed in a single aggregation. These delphinids are sociable and playful and like to frolic and leap about the bows of moving ships. They are among the swiftest of cetaceans and can attain a speed of up to 22 knots, but usually swim 3 to 5 knots. Common dolphins feed on fish and cephalopods, mostly those found in shoals and near the surface. Flying fish are included in their diet (Walker et al. 1964).

The breeding season is protracted, from May until November, peaking in July through September. The gestation period is 11 months. The calf is born from midwinter to summer and is from 80 to 95 cm (32–37 in) long (Tomilin 1957). Females bear calves about three times in four years because of successive annual displacement of the time of parturition. The calves are weaned in 4 to 5 months. Females apparently reach sexual maturity at an age of 3 years, males at 4 years. Longevity is more than 20 years (Rice 1967).

COMMON DOLPHIN

# Bottle-nosed Dolphin

*Tursiops truncatus* (Montague)

*Delphinus truncatus* Montague, 1821. *Mem. Wernerian Nat. Hist. Soc.* 3:75.
*Tursiops truncatus*, True. 1903. *Proc. Acad. Nat. Sci. Philadelphia* 55:314.
Type Locality: Totness, Devonshire, England
Names: Bottle-nosed porpoise. The meaning of *Tursiops* = a kind of dolphin, and *truncatus* = truncated, referring to the relatively short beak compared with that of some other delphinids.

**Description.** The bottle-nosed dolphin is recognized by its short, depressed beak, clearly divided from the sloping forehead by a furrow. The lower jaw is slightly longer than the upper. The dorsal fin is high, falcately concave posteriorly, and situated midway on the back. The flippers are broad at the base and obtusely rounded. The rostrum is relatively long, and the posterior margins of the orbital plates of the maxillaries are rounded.

The coloration is bluish gray or slate gray above, blending to white or grayish white below. There is a dark stripe from the blowhole to the base of the beak, and often one or two dark lines run from the eye to the beak.

Males are somewhat larger than females. Measurements range from 2.5 to 3.7 m (8–12 ft) long. Weights vary from 150 to 200 kg (330–440 lb).

**Distribution.** Bottle-nosed dolphins are widely distributed in the coastal temperate and tropical waters of the world and occur along the Atlantic coast of North America. They are usually encountered in bays and inshore waters but rarely ascend rivers. They are uncommon in New England waters.

As far as is known, bottle-nosed dolphins have not been recorded from Connecticut. However, Connor (1971) mentioned bottle-nosed dolphins in Long Island Sound, and a specimen was taken from Pelham Bay (5), New York (about 3 to 4 miles from Connecticut), in August 1906. In Rhode Island, Cronan and Brooks (1968) reported that a specimen was seen off Newport (4) (date unknown), and a 3.2 m (10.4 ft) male was found dead on Sand Hill Cove Beach, town

of Narragansett (3), on 17 September 1967. Miller and Kellogg (1955) mentioned that the species was recorded in the waters of Massachusetts (2). In Maine, Norton (1930) reported that bottle-nosed dolphins were frequently seen in schools about Eagle Island in Casco Bay (1) and along the Cape Elizabeth shore from Trundy's Reef to the hook of Cape Elizabeth during autumn.

**Behavior.** Bottle-nosed dolphins are friendly, gregarious, and remarkably intelligent. They are the best-known cetaceans, primarily from specimens in aquariums. They generally form schools of a few to several dozen individuals. They like to gambol about the bows of moving ships.

These delphinids are known to swim at up to 10 knots, but may be capable of speeds up to 20 knots (Walker et al. 1964). At high speeds, the dolphins can leap vertically from the water for up to 16 feet and catch fish held in the air as bait (Kellogg and Rice 1966). Bottle-nosed dolphins have been seen singly and in groups of 2 to 4 animals surf-riding the tops of waves, with the long axis of their bodies oriented at right angles to the beach (Caldwell and Fields 1959).

Female bottle-nosed dolphins show strong attachment to their own kind, as well as toward other species. The male may demonstrate attachment toward the female. Siebenaler and Caldwell (1956) told of an instance in which a dolphin that had just been placed in a holding pen struck its head and immediately sank beneath the water. Two of the three other dolphins that had been previously placed into the pen immediately came to the assistance of the injured animal, raising it to the surface several times until it was able to swim unassisted. Lilly

(1963) reported that a sick female dolphin was given intermittent support for four days by two other bottle-nosed dolphins. Brown and Norris (1956) related how a female bottle-nosed dolphin attempted to raise a Pacific white-sided dolphin to the surface after it had been killed. Norris and Prescott (1961) reported an unusual instance in which a female bottle-nosed dolphin killed a 1.5 m (5 ft) leopard shark by repeatedly forcing it to the surface with her snout. For the next eight days the dolphin carried the shark to the surface on her snout, released it, then retrieved it as soon as it sank. Divers were unable to take the carcass of the shark from the female until it decomposed.

Bottle-nosed dolphins show a hierarchical social structure. On the basis of a captive school of five males and seven females, Tavolga (1966) noted that the adult male was dominant over the others, followed in turn by a group of adult females, a group of juvenile males, and a group of calves. The females kept their calves close to them at all times during the first weeks of life and only gradually allowed them to stray farther away.

Hoese (1971) described the unique behavior of two bottle-nosed dolphins feeding out of water

several times in a saltwater marsh. The dolphins swam together up a tidal creek, usually undulating in a more or less S-fashion pattern several feet apart, then moved close to each other and suddenly rushed together up the slope of a bank, riding a large bow wave that broke on the bank immediately before them. Usually the dolphins whole bodies came out of the water. They immediately slid back because of the slick mud, but just before they slipped back they fed on the small fish that became stranded as the wave broke on the bank.

Bottle-nosed dolphins are vocal. They produce pure-toned sounds or complicated whistles or chirps for communication and pulsed sounds or modified clicks for echolocation.

These dolphins feed on a variety of fishes such as mackerel, mullet, cod, kingfish, tarpon, spotted trout, weakfish, roballo, sheepshead, and other species as available. Other food items include shrimp, squid, and eagle rays.

Reproduction of bottle-nosed dolphins in the wild has been little studied. Walker et al. (1964) reported that most births occur from March through May, after a gestation period of 11 to 12 months. At birth the calf is about 1 m (3.3 ft) long and weighs about 12 kg (26 lb). Essapian (1953) noted that captives begin to eat solid food when 6 months old, and Tavolga (1966) reported that calves may continue to nurse until nearly 18 months old. Rice (1967) found that a captive female born in captivity attained sexual maturity in 6 years and that captives bred only in alternate years. Sergeant, Caldwell, and Caldwell (1973) reported longevity to be about 25 years.

Numerous instances of females supporting stillborn or dead calves at the surface are on record (Moore 1953; Moore 1955). Captive male calves appear precocious sexually and will attempt intromission with their mothers. Captive adult males make sexual advances to females year round, and sexually excited males often bite one another (Caldwell and Caldwell 1966).

BOTTLE-NOSED DOLPHIN

# White-beaked Dolphin

*Lagenorhynchus albirostris* Gray

**Description.** The white-beaked dolphin is recognized by the white color of the upper beak. The coloration of the upperparts is dusky, and the flanks sometimes have small gray spots. The ventral side of the body is white or creamy.

There may be light-colored patches on the flippers, below the dorsal fin, behind the blowhole, and near the flukes. The flippers and flukes are dusky above, paler below.

The rostrum of the white-beaked dolphin is

**WHITE-BEAKED DOLPHIN**

*Lagenorhynchus albirostris* Gray, 1846. *Ann Mag. Nat. Hist.* 17:84.
Type Locality: Great Yarmouth, England
Names: The meaning of *Lagenorhynchus* = flask-shaped beak, and *albirostris* = white beak.

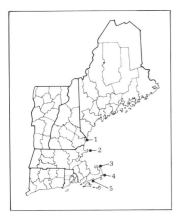

relatively shorter and wider than that of the white-sided dolphin and reaches about one-half of the condylobasal length.

Measurements range from 2.7 to 3.0 m (8.9–9.8 ft) (Tomilin 1957).

**Distribution.** White-beaked dolphins occur in coastal waters of the North Atlantic from the Barents Sea to the coast of France and from Davis Strait to Massachusetts (Tomilin 1957). In New England waters the species has been recorded from Little Boar's Head, Rockingham, New Hampshire (1), in December 1926 (skull HMCZ 51088); from Massachusetts at Ipswich, 1926 (2) (skeleton HMCZ 5322), and at Nauset

Beach (4) on 29 April 1961 (skeleton HMCZ 51051); off Provincetown (3) on 26 April 1961 (HMCZ 51061); and from Falmouth Harbor (5), one-fourth mile off Vineyard Sound on 20 June 1966 (skeleton HMCZ 51859).

**Behavior.** Very little is known about the behavior of this species. White-beaked dolphins have been encountered in groups of 1,000 to 1,500. They eat capelin, mackerel, herring, anchovies, crustaceans, squid, and whelks (Walker et al. 1964). Calving may occur in summer, and juveniles about 1.2 m (4 ft) long have been reported stranded in the latter half of the year (Nishiwaki 1972).

# Atlantic White-sided Dolphin

*Lagenorhynchus acutus* (Gray)

*Delphinus (Grampus) acutus* Gray, 1828. *Spicilegia zoologica* 1:2.
*Lagenorhynchus acutus*, Gray, 1846. In *Zoology of the voyage of H.M.S. Erebus and Terror*, Vol. 1, *Mammalis*, p. 36.
Type Locality: Unknown
Names: Atlantic striped dolphin, short-beaked dolphin, and short-head dolphin. The meaning of *Lagenorhynchus* = flask-shaped beak, and *acutus* = pointed.

**Description.** The Atlantic white-sided dolphin has a relatively small, rounded head with a sloping forehead and a very short beak. The coloration of the back, dorsal fin, flippers, and flukes is black. Between the black back and the white belly, the sides are yellowish gray, except for a wide white band running posteriorly. A streak of yellow or tan runs immediately above the white band up toward, but not over, the ridge of the flukes. A narrow blackish stripe runs from the eye to the base of the flipper. There are black circles around the eyes. The lower lip is either white or fringed with black.

In the skull, the temporal fossae are elongate, and the pterygoid bones are frequently in contact.

Males are larger than females. Measurements range from 2 to 3 m (6.6–9.8 ft). Weights vary from 154 to 227 kg (340–500 lb).

**Distribution.** Atlantic white-sided dolphins are fairly common along North Atlantic coasts. In

the waters of New England, specimens were taken in the nineteenth century. Cronan and Brooks (1968) reported that a specimen was found dead at Monahan's Dock in the town of Narragansett, Rhode Island (4), on 22 July 1967. Twelve stranded dolphins were recovered on the beaches of Orleans (3) and Wellfleet (2), Massachusetts, in early May 1973. On 8 September 1974 up to 1,000 white-sided dolphins were reported to be chasing herring in Longley Cove, near Dennysville, Maine (1). About 100 animals, including many pregnant and nursing females, were stranded on the ebb tide and died. This is the largest sighting and stranding for this species in New England waters.

**Behavior.** Little is known about the behavior of this species. Atlantic white-sided dolphins are gregarious and may be encountered in groups of 1,000 or more. Schevill (1956) reported a group of 12 dolphins that swam about briskly ahead of the ship and did not play at the bow and rarely

approached within 20 yards. When one of the animals was harpooned, the others did not stay with it, although they loitered near the ship for nearly 10 minutes.

These mammals feed on herring, mackerel, striped bass, salmonids, and certain benthic invertebrates.

Calving occurs from spring to midsummer after a gestation period of about 10 months (Guldberg and Nansen 1894). Millais (1906) ob-

served full-term fetuses in June. S. A. Testaverde (personal communication) estimated that at birth calves are from 108 to 114 cm (3.7–3.9 ft) long, based on nine fetuses collected in the Wellfleet, Massachusetts, stranding. Millais (1906) reported that juveniles reach a length of 140 cm (4.7 ft) by November, and Sergeant and Fisher (1957) considered a male 180 cm (6.0 ft) long as immature.

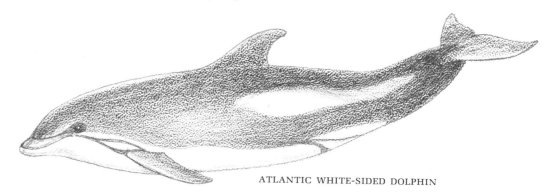

ATLANTIC WHITE-SIDED DOLPHIN

# Killer Whale

*Orcinus orca* (Linnaeus)

[*Delphinus*] *orca* Linnaeus, 1758.
  *Systema naturae*, 10th ed., 1:77.
*Orcinus orca*, Fitziner, 1860.
  *Wissenschaftlich-populare*
  *Naturgeschichte der Säugethiere*,
  6:204.
Type Locality: European seas
Names: White-bellied whale, orca
  whale, thresher, and grampus. The
  meaning of *Orcinus* = a kind of
  whale, and *orca* = a kind of whale.

**Description.** The killer whale is the largest of the delphinids. It is early identified by its prominent erect dorsal fin, especially in males, and by its distinctive color pattern. The color basically is black above and white below with a sharp demarcation between the two areas. The pattern of white extends upward in two or three lobes on each side of the body—from the front of the flippers covering the lower jaw, chin, and throat, narrowing behind the flippers, and expanding behind the navel to the sides of the caudal peduncle and ending behind the anus but far forward of the base of the flukes. When present, the third lobe extends as far as the flukes, expanding to their undersides, which may be partly or entirely white below. There is a white elliptical patch just above or behind the eyes. Just behind the dorsal fin there is a grayish patch that sometimes merges to form a saddle across the back. The saddle is absent in some individuals. The flippers are black on both sides, and the flukes are black above and commonly white fringed with black below.

The rostrum is broad, flattened, and somewhat concave in front of the naris. The premaxil-

laries are much narrower than the maxillaries. The teeth are conical, large, and sharply pointed.

Males are much larger than females. Measurements of males range up to 9 m (30 ft), whereas females range up to 6 m (20 ft) (Walker et al. 1964).

**Distribution.** Killer whales are cosmopolitan but are most common in polar waters. They are rare in the waters of New England. Strandings in Maine include two individuals at Eastport (1) in March 1902 (True 1904b); Tenants Harbor (2) in December 1957 (Mairs and Scattergood 1958); and Harpswell (3) in November 1904 (Norton 1930). Cronan and Brooks (1968) reported a record of a killer whale washed ashore in Rhode Island (4). The species has not been reported in the waters of Connecticut, Massachusetts, or New Hampshire as far as is known.

**Behavior.** Killer whales travel extensively in search of food. They are gregarious, bold, powerful, and ferocious. They swim in close associations, sometimes traveling close together in a

line or a column. Their erect black dorsal fins, rising like poles, almost always remain visible above water.

These cetaceans hunt in packs of 3 to 50 or more, swimming to and fro in all directions, sometimes touching one another. They attack large prey simultaneously, leaping out of the water and tearing the flesh with their powerful teeth. They do not hesitate to grapple with the largest whales. Killer whales often attack birds and seals basking on thin ice. They dash from the deep, smashing or upsetting the ice with their backs and dislodging the hapless prey into the water. They sometimes cruise inshore waters to surprise any unwary seals and may even chase them up rivers.

Killer whales normally swim 6 knots but can swim 17 knots without apparent effort. They sound from 1- to 6-minute intervals (Tomilin 1957). After sounding, killer whales swim for about 120 feet on the surface, performing three to five shallow dives at 5-second intervals (Scheffer and Slipp 1948). They are capable of jumping 40 to 45 feet horizontally and clear the water by about 5 feet (Walker et al. 1964). Occasionally a killer whale holds its head above water and seemingly looks around for a few seconds.

Killer whales show a tightly knit social organization and do not leave injured schoolmates. One female remained three days in the area where her calf was killed (Scheffer and Slipp 1948). These mammals produce two distinct types of high-frequency sounds—whistles and squeaks, probably for communication, and clicks, probably for echolocation (Griffin 1966).

Killer whales feed on fishes, cephalopods, sea turtles, sea birds, seals, sea otters, and whales and other cetaceans. Eschricht (1862) reported fragments of skin and bones of some 14 seals and dolphins in the stomach of a killer whale. This reference is frequently cited as evidence of the size of a single meal, but this is erroneous; Tonnesen (1962) showed that the stomach of a

KILLER WHALE

killer whale could never contain so much food. Although several authors have reported that killer whales prey on larger whales, Jonsgård (1968a) pointed out that there was no proof that larger whales in good health and under normal conditions are attacked by killer whales. Further, Jonsgård (1968b) stated that larger whales may possibly escape by diving so deep that killer whales are not able to follow them. Nishiwaki and Handa (1958) and Rice (1968) reported that older or adult killer whales feed chiefly on mammals. Griffin (1966) found that a captive killer whale ate 400 pounds of salmon per day.

The breeding season may last several months (Collett 1912); pairing occurs in winter (Fraser 1937); and the peak period might be in May to July (Jonsgård and Lyshoel 1970). The gestation period may last from 12 to 16 months. The newborn calf may be about 2.7 m (9 ft) long (Nishiwaki and Handa 1958). Females become sexually mature when about 4.9 m (16 ft) long, and males when about 5.8 m (19 ft) (Jonsgård and Lyshoel 1970).

# Grampus

*Grampus griseus* (G. Cuvier)

*Delphinus griseus* G. Cuvier, 1812. *Ann. Mus. Nat. Hist., Paris* 19:13–14.
*Grampus griseus*, Hamilton, 1837. *Jardine's naturalist's library, Mammalia,* vol. 6, *Whales,* p. 233.
Type Locality: Brest, France
Names: Risso's dolphin. *Grampus* is a corruption of the French *grand poisson* (big fish), and *griseus* = gray.

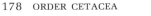

**Description.** The grampus is identified by its prominent rounded forehead, which rises almost vertically from the top of the upper jaw owing to the melon situated over the broad premaxillary and maxillary bones. This species, unlike other dolphins, has only 3 to 7 pairs of teeth in the front part of the lower jaw, and the beak is lacking. The dorsal fin is high, pointed, and strongly recurved, and the flippers are relatively long.

The skull is relatively short and wide, with the rostrum expanded anterior to the antorbital nodes. The premaxillary is almost twice as wide as the maxillary at the middle of the rostrum.

The coloration is grayish black to gray blue or gray tinged with purple above, lighter on the sides and flanks, and paler below. Some individuals have a white anchor-shaped area between the flippers to as far back as the anus. The dorsal fin, flippers, and flukes are colored like the back or still darker. The forehead is almost white in juveniles, and the entire head is almost white in old animals. The whitish streaks on the body are probably healed scars.

The sexes appear to be equal in size. Measurements range from 3.5 to 4.3 m (11.5–14.1 ft) long.

**Distribution.** Grampuses are widely distributed in the temperate and tropical oceans. They are mainly pelagic and probably are most common in the higher latitudes during the summer months. Paul (1968) reported grampuses in the western North Atlantic, though rarely, from Massachusetts to New Jersey and off both the east and Gulf coasts of Florida. Schevill (1954)

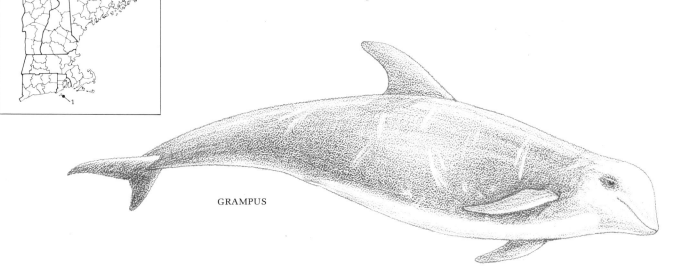

GRAMPUS

reported more than 60 grampuses roughly 70 nautical miles south of Block Island, Rhode Island (1).

**Behavior.** Little is known about the behavior of this species. Schevill (1954) reported that grampuses consorted in small groups of 4 to 6 within the large group of 60 or more animals. They were playful and often leaped out of the water. These delphinids like to follow ships. An interesting story of this behavior of accompanying ships tells about the famous "Pelorus Jack." This grampus received lifelong protection from the Order in Council of New Zealand because of its habit of playing about ships and guiding them into Pelorus Sound from about the years 1896 to 1916. Grampuses feed on cephalopods and fishes.

Full-term fetuses have been found in December, and in late March a very young suckling female was caught in the English Channel. Newborn calves are from 1.4 to 1.7 m (4.6–5.6 ft) long (Tomilin 1957). Interbreeding may occur between grampuses and bottle-nosed dolphins.

# Common Pilot Whale

*Globicephala melaena* (Traill)

*Delphinus melas* Traill, 1809. *J. Nat. Philos. Chem. Arts* 22:81.
*Globicephala melaena*, Thomas, 1898. *The Zoologist*, ser. 4, 2:99.
Type Locality: Scapay Bay, Pomona, Orkney Islands, Scotland
Names: North Atlantic pilot whale, blackfish, driving or herding whale, social whale, pothead whale, howling whale, and caa'ing whale. The name pilot whale comes from European fishermen, who believed that there were herring schools under these whales and used them to guide their boats. The meaning of *Globicephala* = globe or rounded head, and *melaena* = black.

**Description.** The pilot whale is recognized by its high bulging forehead, triangular and high, recurved dorsal fin situated anterior to the center of the body, and very long, slender, and pointed flippers, which are about one-fifth of the total body length. The mouth is rather large and cleft obliquely forward. The lips are protruding, with the upper lip extending farther forward than the lower one. The flippers are quite attenuated, with a prominent bend near the point.

The rostrum is relatively short and very wide. The premaxillaries are large and flat and do not project over the lateral and anterior margin of the maxillaries when viewed from above.

Adult males are distinguished from females by their very large dorsal fin; by the pronounced, laterally compressed dorsal and ventral crests of the caudal peduncle; by a steeper hump situated between the globular head and the dorsal fin; and by a swollen supramaxillary "pot" extending beyond the snout.

The coloration of this species is generally black or charcoal gray, sometimes with a narrow white stripe extending from the throat to the belly. There is a dark gray blaze or teardrop-shaped streak behind the eyes which extends for several inches obliquely upward and backward. It is indistinct in young animals. Immatures are paler than adults.

Males are much larger than females. Meas-

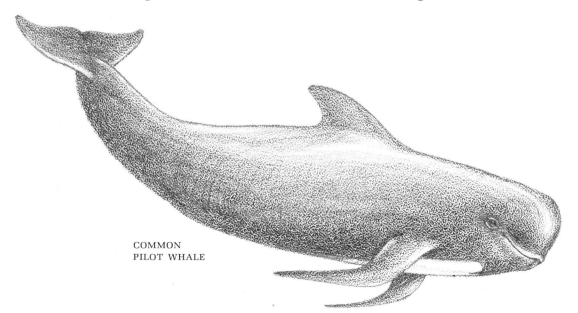

COMMON
PILOT WHALE

urements range from 3.6 to 8.5 m (11.8–27.9 ft) long. A 4 m (13 ft) long specimen was estimated to weigh 680 kg (1,500 lb).

**Distribution.** Pilot whales occur in all oceans except the polar seas. There are two species of pilot whales in the Atlantic Ocean, whose ranges overlap along the middle Atlantic coast of the United States: the common pilot whale *Globicephala melaena* of the North Atlantic, and the short-finned pilot whale *Globicephala macrorhyncha* of the tropical Atlantic. The short-finned pilot whale is difficult to distinguish from the common pilot whale and ranges as far north as New Jersey. It could probably stray into New England waters.

Common pilot whales have stranded along the coast of New England, occasionally in large numbers. About 3,000 beached on Cape Cod in 1874 (Clark 1887), and a herd of 1,975 animals stranded at Wellfleet, Massachusetts, in 1895 (Katona, Richardson, and Hazard 1975). In Maine, a juvenile male was beached in Milbridge (1) on 28 March 1973, and an old adult male beached on Bar Island (2) off Corea on 14 December 1973. In Massachusetts, Brown (1913) recorded that pilot whales have been stranded in Beverly Harbor and Monument Beach in Salem (3) (no dates). In Rhode Island, Cronan and Brooks (1968) reported that a specimen was washed ashore at Middletown (4) in September 1959, an immature pilot whale was taken approximately 30 miles south of Narragansett Bay in March 1961, and another whale was taken in the upper Providence River (5) in July 1962. Goodwin (1935) reported that pilot whales have been seen at Bridgeport and Fairfield Connecticut (6) (no dates).

**Behavior.** Pilot whales are probably the most gregarious of cetaceans and have a strong herding or schooling instinct. They often travel in tight formation in herds of from several hundred to a thousand or more. They seem to derive some sort of security in body contact with other members of a herd, especially when sleeping. These animals seem to follow a leader persistently. They often panic when a leader is lanced and will follow it in face of danger; if the leader swims toward shore, the entire herd follows and may become stranded.

Pilot whales produce various sounds, from a high-pitched squeal or whistle to a loud, prolonged belch (Caldwell and Caldwell 1971). Their bushy spouts may reach 5 feet or more (Olsen et al. 1969).

These animals migrate between warm and cold waters with the changing seasons. Around Newfoundland, pilot whales approach the coast in July, sometimes remaining in inshore waters until late October, when they move back to deep water (Sergeant 1961), and bachelor herds composed of mature males are encountered during late summer (Sergeant 1962b). These delphinids travel up to 6 knots and dive for up to 5 minutes, remaining 2 minutes on the surface. They may thrash the water with their flukes when feeding (Sergeant 1961).

Pilot whales feed mainly on squid but will eat some fish and small invertebrates. Sergeant (1962b), reported that a pilot whale eats about 30 pounds of food at a meal.

In eastern Canadian waters the breeding season is from May to November. The gestation period is between 15 and 16 months, and at birth the calves are approximately 1.8 m (6 ft) long. Males become sexually mature when 4.8 m (15.7 ft) long or about 12 years of age, and females at 3.6 m (11.8 ft) or about 6 or 7 years of age. This disparity between males and females is somewhat compensated for by their polygamous behavior. The calves are weaned at 22 months. Longevity is estimated at 50 years (Sergeant 1962b).

## Harbor Porpoise

*Phocoena phocoena* (Linnaeus)

**Description.** The harbor porpoise is a rather stocky delphinid. It has a comparatively small conical head which is about one-fifth to one-sixth of its body length. The snout is blunt and rounded, without the distinct beak characteristic of dolphins. The dorsal fin is low and triangular, situated somewhat posterior to the middle of the body. The flippers are moderately long, ovate, and pointed obtusely, with tubercles on their anterior margins. The flukes are relatively broad.

The rostrum is short and broad; the symphysis

[Delphinus] phocoena Linnaeus, 1758. Systema naturae, 10th ed., 1:77.

Phocoena phocoena, Norris and McFarland, 1958. J. Mamm. 39(1):24.

Type Locality: Swedish Seas
Names: Common porpoise, bay porpoise, snuffer, puffing pig, lesser grampus, and herring hog. The meaning of Phocoena = a porpoise.

of the mandible is short; and the proximal ends of the premaxillae have raised, irregular bosses in front of external nares. The teeth are two- or three-lobed crowns and are spade-shaped.

The back, dorsal fin, flippers, and flukes are dark gray, greenish brown, or nearly black. The underside from the head as far as the flukes is white. There is usually a zone of light gray between the blackish back and the white belly. There may be a black streak from the corners of the mouth to the flippers. Some specimens are all black, but albinos are rare.

The harbor porpoise is one of the smallest cetaceans. Measurements range from 1.2 to 1.8 m (4–6 ft). Weights vary from 50 to 75 kg (110–65 lb).

**Distribution.** Harbor porpoises are common in the North Atlantic and in the eastern North Pacific, but are rarely encountered in tropical waters. They enter bays, inlets, and rivers, where they may become stranded.

These cetaceans are uncommon in the waters of New England. Cronan and Brooks (1968) cited two reports of harbor porpoises in Mount Hope Bay, Massachusetts (1), in July 1931 and September 1934. Goodwin (1935) mentioned harbor porpoises along the coast of Connecticut and possibly up the Connecticut River (2). Connor (1971) cited reports of harbor porpoises in Long Island Sound (3).

**Behavior.** Harbor porpoises are gregarious and may travel in pairs or in herds of 100 or more. They are less playful than most dolphins—they seldom jump out of the water and usually ignore passing ships. These mammals usually swim just below the surface and rise to breathe about four times a minute.

Harbor porpoises produce sounds singly or in bursts at repetition rates as great as 1,000 per second that consist of low intensity, narrow-band clicks usually near two kilohertz (Schevill, Watkins, and Ray 1969).

These animals feed chiefly on fish but also eat cephalopods and crustaceans. Harbor porpoises have been trapped in nets at depths of 165 feet while feeding near the bottom (Scheffer and Slipp 1948).

In the North Atlantic the breeding season is from late June to early August, and calving occurs mainly in June and July. The gestation period is about 11 months. At birth the calf is from 80 to 90 cm (32–36 in) long. Sexual maturity is reached at a length of about 133 cm (52 in) in males and 145 cm (57 in) in females, probably in their 3rd or 4th year. Females do not become pregnant every year, and calves have been found with milk in their stomachs in early August to late September (Fisher and Harrison 1970). Weaning probably takes place when the calves are about 100 cm (39 in) long (Smith and Gaskin 1974). Old males tend to gather in separate groups, and females associate in mixed-sex groups (Tomilin 1957).

HARBOR PORPOISE

## SUBORDER MYSTICETI (BALEEN WHALES)

The baleen or whalebone whales have paired nostril openings, a relatively large mouth, and no functional teeth when adult. These whales possess a series of flat, flexible horny plates called baleen or whalebone, which hang down from the roof of the mouth, their edges frayed out into long bristles that are used to filter food from the water. As baleen whales move through water rich with crustaceans, they draw water and food into the mouth and with the large, fleshy tongue force the water back through the fringes of the baleen plates, then swallow the food. These whales are shallow divers.

The Mysticeti contains three families, of which two have been reported in New England waters.

### FAMILY BALAENOPTERIDAE (RORQUALS)

This family contains the rorquals or fin-backed whales and includes the largest whale, the blue whale, which may reach a length of 32 m (105 ft). Rorquals are long, slender, and streamlined, with a pointed snout; a falcate dorsal fin usually far posterior on the back; many parallel longitudinal grooves or pleats on the throat and belly; and short baleen plates with bristly fringes. It is generally held that the grooves stretch out when the whale is feeding so that the capacity of the mouth is thereby much enlarged to accommodate the largest possible volume of water and food.

Rorquals feed on a variety of plankton, on dense shoals of "krill," or small shrimplike crustaceans, and on small fishes. The diet varies with the species and geographic region. During summer, most species of rorquals inhabit the cold currents, where they feed heavily and accumulate a thick layer of blubber. In autumn they migrate toward equatorial waters; during the winter breeding season the animals fast for several months, living off stored fat.

Rorquals are exclusively oceanic. Five species have been recorded in New England waters.

# Fin Whale

*Balaenoptera physalus* (Linnaeus)

[*Balaena*] *physalus* Linnaeus, 1758. *Systema naturae*, 10th ed., 1:75.
*Balaenoptera physalus*, Schlegel, 1862. *De Dieren van Nederland, Zoogdieren*, p. 101.
Type Locality: Spitzbergen Sea
Names: Fin-backed whale, common rorqual, razor-backed whale, finner whale, pike whale, herring whale, and sprat whale. The meaning of *Balaenoptera* = whale with wings and fin, and *physalus* = a bellows, as producing a stream of air under pressure, probably referring to the blowing habits of this species.

**Description.** The fin whale is the second largest and the slenderest of the rorquals. It is distinguished from the blue whale by its smaller size, coloration, longer and narrower head and less curved and more slender flippers, and taller dorsal fin. The fin whale differs from the sei whale in size, coloration, slender body form, and relatively smaller dorsal fin, which is more posterior than that of the sei whale.

The baleen of the fin whale can easily be distinguished from that of any other rorqual by its asymmetrical coloration; the front one-half to one-third of the right baleen plates is bright yellowish white, while the plates on the left side and the greater part of the right are dark gray blue. All the bristles are brownish gray. The upper jaw is symmetrical in color except on the baleen plates. The lower jaw is white on the right side and blackish on the left.

The number of baleen plates varies from 350 to 360 on one side of the jaw. The skull is distinguished from that of other rorquals by the relatively smaller size of the anteriorly emarginate nasal bones, relatively long frontonasal processes of the maxillaries and a distinct posterior expansion of the vomer. The skull is not as flat as that of the blue whale. The number of grooves or pleats varies from 50 to 80, and they extend to the navel. The grooves begin at the chin and are longest along the middle of the belly.

The coloration of the upperparts is brownish black, fading into white on the belly. Some animals appear yellow to green because of diatoms which cling to the skin. There are two dark areas passing from the vicinity of the flippers toward the pleats. Another black stripe runs from the middle of the flanks to the anus. There are two pale stripes called chevrons which originate behind the blowhole and run backward forming a broad V along the back and upper sides. Small white or gray spots are scattered laterally from the region of the belly to the end of the caudal peduncle. The flippers and flukes are gray above and white below. The right side of the head is noticeably more whitish than the left side.

Females are slightly larger than males. The following are measurements of specimens taken off Nova Scotia in 1964: 19 males ranged from 15.3 to 18.9 m (50–62 ft), averaging 16.8 m (55 ft), and 32 females ranged from 15.9 to 19.2 m (52–63 ft), averaging 17.4 m (57 ft) (Sergeant 1966). A fin whale measuring 18 m (59 ft) long weighs about 50 tons (Walker et al. 1964).

**Distribution.** Fin whales are cosmopolitan but are rarely seen in tropical waters or among pack ice. They are less abundant in the northern hemisphere than in the southern hemisphere. The northernmost catch of fin whales in the North Atlantic was at 80°42′ N, 11° E in June

(Jonsgård 1966b). The major population of fin whales of northwestern North Atlantic is between latitudes 41°21′ N and 57°00′ N in waters between the coast and the 1,000-fathom line. This region can be considered the summer feeding grounds of the species (Mitchell 1972). On the North American coast the southward limit is indicated by four strandings in North Carolina (Caldwell and Golley 1965) and one stranding in Florida (Moore 1953).

In the waters of New England, fin whales were once plentiful and were regularly hunted on the coast (Allen 1916). Fin whales are the most common big whales in the Gulf of Maine (Katona, Richardson, and Hazard 1975). In Maine, strandings were recorded at Ragged Island (1) in 1927; at Popham Beach (2) in 1905; and Old Orchard Beach (3) (no date). A stranding was also reported at New Haven, Connecticut (4), in 1843.

**Behavior.** Fin whales travel singly, in pairs, or in small groups of 10 to 20 or more. They occasionally congregate in schools of 100 or more at feeding grounds and during migration. They may be seen along the coast and at mouths of rivers where they become stranded while searching for food.

These cetaceans generally migrate north and south during spring and autumn. Their migrations are somewhat irregular and may depend on the local availability of food. On the northward migrations along the coast of North America, fin whales appear in large numbers in the Gulf of Maine early in March, and other schools congregate in May and June. The animals approach the southern coast of Newfoundland in early June, appear off the northern coast of Newfoundland and Labrador in mid-July, and are most numerous in August. The whales leave Newfoundland in October, though some of them winter among the field ice on the Grand Banks off Newfoundland (Kellogg 1929).

Fin whales are the fastest of the Mysticeti (Nishiwaki 1972). They swim a steady 7 to 9 knots, but when alarmed or wounded they can attain bursts of up to 13 knots over short distances. They are very strong, and on several occasions fin whales have been seen towing a whale ship at a rate of 2 to 3 knots for short distances (Tomilin 1957). A fin whale that is being chased by a whaler will not bring its flukes out of the water but may raise its head so far out of water that the ventral grooves are visible to observers on the chase boats (Nishiwaki 1972).

When a fin whale breaks the surface after a deep dive of up to 20 minutes, its body is directed obliquely at a small angle to the surface so that the top of the head and blowhole usually emerge first. After the first surfacing, the animal spouts, then makes four to seven brief and shallow intermediate dives. When the animal breaks the surface after an intermediate dive, its body moves almost parallel to the surface, so that the blowhole and dorsal fin may emerge simultaneously. Before sounding, the fin whale rounds its body into an arch so that the dorsal fin and caudal peduncle are exposed high above the surface. This arch gradually disappears as the

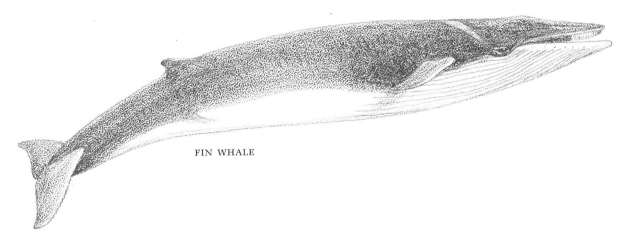

FIN WHALE

inclination of the body increases and the dorsal fin disappears first, followed by the caudal peduncle. The flukes usually do not break the surface (Andrews 1910a).

The height of a spout is variable—the longer the dive, the higher the blow. The spout may reach 10 to 20 feet, and has been described as an inverted cone.

As a rule, fin whales do not breach unless wounded or extremely excited (Tomilin 1957). A large fin whale that had stranded on a rocky shore in Maine succeeded in extricating itself after a few minutes of leaping into the air (Allen 1916). When feeding these animals swim on one side with their mouths open (Andrews 1909a).

Food items of fin whales vary with region and season. In the North Atlantic, they feed chiefly on euphausid crustacea *Calanus* spp., *Meganyctiphanes norvegica*, and *Thysanoessa inermis*. Krill, probably euphausids, are usually eaten in the spring, being replaced by capelin variably between late June and late July off Nova Scotia and Newfoundland (Sergeant 1966). In the Gulf of Maine, the whales depend on seasonal appearance of herring, capelin, squid, and krill (Kellogg 1929). Dogfish is an incidental food item of fin whales in the Gulf of Maine (Allen 1916).

Fin whales are regarded as monogamous, and much affectionate behavior has been observed (Nishiwaki 1972). Breeding may take place throughout the year, although the peak occurs from November and December until March (Tomilin 1957). The gestation period lasts 11 to 12 months, and females usually bear one young every other year. Twin embryos have been found in females off Newfoundland (Allen 1916). Six fetuses have been recovered from a large female (Jonsgård 1953). At birth the calf is from 6.0 to 6.5 m (20–21 ft) long, and it almost doubles its length in the first six months of life. The calf is suckled for about six months until it is between 12.2 and 13.5 m (40–44 ft) long (Mackintosh and Wheeler 1929). Sexual maturity is attained in 3½ to 4 years. Longevity is assumed to be 90 to 100 years (Nishiwaki 1972).

# Sei Whale

*Balaenoptera borealis* Lesson

*Balaenoptera borealis* Lesson, 1828. *Histoire naturelle générale et particulière des mannifères et des oiseaux découverts depuis 1788. Cétacés.* p. 342.
Type Locality: Gromitz, Lubeck Bay, Schleswig-Holstein, Germany
Names: Rorqual, pollack whale, Rudolphi's whale, skimmer whale, coal-fish whale, sardine whale, and Japan finner. The meaning of *Balaenoptera* = whale with wings and fins, and *borealis* = northern.

**Description.** The sei whale is intermediate in size between the larger fin whale and the smaller minke whale. The dorsal fin is relatively large and high. The flippers are rather small compared with those of other rorquals. The number of grooves varies from 30 to 60, and the grooves do not extend much beyond the flippers. The rostrum is triangular with straight edges, as in the fin whale and minke whale. The nasals are almost as wide as they are long. The number of baleen plates on one side of the whalebone varies from 219 to 402.

This species is distinguished from other rorquals by the coloration of the body. The back and flanks are bluish gray, the throat and belly whitish. The whitish coloration on the belly may be expanded on the throat, abruptly narrowed in front of the flippers, and expanded again farther to the rear. The flanks and sometimes the head and back usually have light spots. The flippers and flukes are dark gray above, lighter below. The whalebone is bluish black with white bristles.

Females are slightly larger than males. In the northern hemisphere, 111 males averaged 12.9 m (42.3 ft) long, and 104 females averaged 13.3 m (43.6 ft) (Andrews 1916a). In the southern hemisphere, 65 males averaged 14.5 m (47.6 ft) long, and 155 females averaged 15.5 m (50.9 ft) long (Matthews 1938b).

**Distribution.** Sei whales occur in all oceans. Although they range into the polar seas, sei whales prefer warmer waters than either the fin whale or blue whale (Mackintosh 1965). In the western North Atlantic, sei whales have been taken on the coasts of southern Newfoundland and southern Labrador, mainly in August and September (Sergeant 1961). This species is rare in the waters of New England. In 1948, a stranding was reported at Kingston, Massachusetts (1). Allen (1916) reported a stranded sei whale from Chatham, Massachusetts (2), in August 1910.

**Behavior.** Sei whales migrate northward to temperate and subarctic waters in spring or early summer and southward to subtropical waters in autumn or winter. The two sexes appear to migrate at different times (Mitchell 1972). These whales are usually seen singly or in

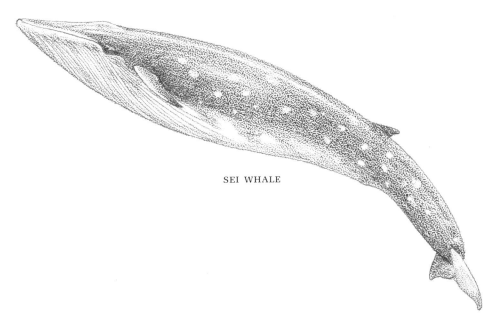

SEI WHALE

pairs but may congregate in loose groups of 50 or more.

Sei whales are among the fastest whales and can attain a speed of 26 knots during an initial rush (Andrews, 1916a). They can breach clear of the water (Tomilin 1957). They breathe two or three times between soundings. When diving, they arch their body much less than fin whales and blue whales do (Collett 1886).

Sei whales are known as "skimmers" because they feed slowly near the surface. In the North Atlantic, they feed chiefly on the minute copepods *Calanus finmarchicus*, other crustaceans such as *Femora longicornis* and *Thysanoessa inermis*, and various species of small fishes.

In the North Atlantic, sei whales breed from November to March, peaking in January. In most instances coitus takes place in winter in warm waters, but may sometimes occur during the summer months in both hemispheres. The gestation period is about 12 months (Tomilin 1957). In the Antarctic Ocean, newborn calves average 4.5 m (15 ft) long and are weaned when about 8 to 9 m (26–30 ft) long (Matthews 1938b). In the northern hemisphere, males reach sexual maturity at a length of 12 m (39 ft) and females at 12.8 to 13.5 m (42–44 ft), or about 8 years of age for both sexes (Tomilin 1957). Longevity is shorter than that of fin whales and blue whales (Nishiwaki 1972).

## Minke Whale

*Balaenoptera acutorostrata*
Lacépède

**Description.** The minke whale is the smallest of the baleen whales. It is easily recognized by the characteristic white band running across the flippers, by the cream white whalebone, and by the ventral grooves that do not extend much beyond the flippers. The body is spindle-shaped and slender, and the snout is sharply pointed. The dorsal fin has a broad base and is high, slightly falcate, and situated in the posterior third of the body. There are approximately 50 grooves. The rostrum is pointed and triangular, shorter and wider at the base than in other rorquals. The baleen plates are yellowish white and are about 20.3 cm (8 in) long.

The coloration is blackish gray above, white below. A broad, light crescent-shaped band runs from near the blowhole down to the base of the flippers. The flippers and flukes are whitish below.

Females are somewhat larger than males. On the basis of 45 minke whales taken off Newfoundland, 23 males ranged from 4.6 to 8.2 m

MINKE WHALE

*Balaenoptera acuto-rostrata*
   Lacépède, 1804. *Histoire naturelle des cétacés*, p. 134.
*Balaenoptera acutorostatra*, Oliver.
   1922. *Proc. Zool. Soc. London*,
   1922, pp. 557–85.
Type Locality: Cherbourg, France
Names: Little-piked whale, pikehead whale, sharp-headed finner whale, little finner, lesser fin back, lesser rorqual, least rorqual, bay whale, and pygmy whale. The meaning of *Balaenoptera* = whale with wings and fin, and *acutorostra* = pointed rostrum, referring to the pointed rostrum or snout.

(15–27 ft) long, averaging 7.1 m (23 ft), and 22 females ranged from 4.6 to 8.8 m (15–29 ft), averaging 7.6 m (25 ft) (Sergeant 1963).

**Distribution.** Minke whales are widely distributed in all oceans from the ice pack to the subtropical region. In the western North Atlantic they are infrequently encountered below the latitude of Long Island, New York. These rorquals are common in the Gulf of Saint Lawrence and along the east coast of Labrador (Allen 1916). There are records of minke whales from Florida (Moore 1953) and the Gulf of Mexico (Moore and Palmer 1955).

True (1940a) reported that minke whales were rare in New England, but Allen (1916) believed them to be more common in the region. He reported stranded or captured specimens from Quoddy Head, Maine, southward to Cape Cod, Massachusetts. Norton (1930) reported that a small female was taken in a net near Portland (1), Maine, in July 1893. Scattergood (1949) remarked that there had been no strandings in New England since 1939, but the animal was seen in the Gulf of Maine off Leonardville, Deer Island (2), in May 1947. Goodwin (1935) reported that a minke whale was killed off Montauk (3), Long Island, New York, in mid-August 1930. In the waters of Rhode Island, Cronan and

Brooks (1968) reported a record of a sighting off Point Judith (4) in May 1849, three sightings in the Sakonnet River (5) during August 1867, and one sighting off Newport (6) in September 1887. These authors also reported two recent Rhode Island records; a young female that drowned in a fishnet off Sakonnet Point, Little Compton, on 11 June 1961, and a calf 4.6 (15 ft) long found dead in the Sakonnet River (7) during July 1967. In the Gulf of Maine and farther south nearly all stranded minke whales have been immatures, whereas strandings from more northerly areas include many adult specimens (Katona, Richardson, and Hazard 1975).

**Behavior.** Minke whales are less sociable than other rorquals. They are seen singly or in pairs, but rarely in large numbers except on feeding grounds. They prefer inshore waters, bays and fjords, where they chase fish and sometimes become entangled in nets. At times minke whales approach and follow ships. They may show their heads obliquely in the air as far as the eyes.

Minke whales normally swim from 8 to 10 knots, but they can swim up to 9 to 13 knots (Nishiwaki 1972). These cetaceans surface more rapidly than do other rorquals. They do not arch the tail as steeply as fin whales and blue whales when sounding, and they do not show their flukes above water (Tomilin 1957). The blow may reach 6 to 7 feet high (Nishiwaki 1972). Minke whales migrate northward to the arctic regions of the eastern North Atlantic Ocean in the summer months and southward in autumn. In their second year of life minke whales may spend the winter in the high northern latitudes. During their feeding migration to the northeastern North Atlantic, the whales segregate; the females usually are found in coastal waters, the males out at sea (Jonsgård 1966b). Minke whales also segregate by age and sex in the inshore waters of Newfoundland (Mitchell and Kozicki 1975).

Minke whales feed chiefly on a variety of fishes, though they eat some squid and crustaceans. In the North Atlantic they eat herring, cod, capelin, pollack, salmon, and other fishes. In the western North Atlantic, capelin is the main food item, and the appearance of minke whales off eastern Newfoundland parallels the spawning concentration of capelin with some cod, herring, and other organisms (Sergeant 1963). Several of these whales have been trapped in fish weirs on the American coast, apparently while chasing fish. Most of the whales trapped were immatures (Allen 1916).

Little is known on the reproduction of this species. The breeding season extends over several months in late winter, and birth occurs in early winter (Sergeant 1963). Females produce one calf each year and a half near Japan (Omura and Sakiura 1956). At birth the calf is about 2.8 to 3.0 m (9.2–9.8 ft) long, and lactation lasts about 6 months (Nishiwaki 1972). Sexual maturity begins at 6.4 m (21 ft) in males and at 7.4 m (24 ft) in females. The largest size for sexually immature whales is 8.2 m (27 ft) in males and 9.5 m (31 ft) in females (Ohsumi and Masaki 1975).

# Blue Whale

*Balaenoptera musculus* (Linnaeus)

[*Balaena*] *musculus* Linnaeus, 1758. *Systema naturae*, 10th ed, 1:76.
*Balaenoptera musculus*, True, 1898. *Proc. U.S. Natl. Mus.* 21:629.
Type Locality: Firth of Forth, Scotland
Names: Sulphur-bottomed whale, Sibbald's whale, blue rorqual, and great northern rorqual. The meaning of *Balaenoptera* = whale with wings and fin, and *musculus* = muscular, possibly a play on *mus*, a mouse.

**Description.** The blue whale is the largest animal that has ever lived. It closely resembles the fin whale but is easily distinguished by the outline of its broad-snouted head, with convex rostral margins, distinct central ridge from the snout to the blowhole, very greatly curved, pointed flippers, and very small dorsal fin. The head is less than one-fourth of the total length of the animal. There are approximately 90 pleats which extend to the navel.

In ventral aspect, the skull of the blue whale differs from other balaenopterids in that the palatine surface of the maxillaries ends abruptly, not forming a left and a right process protruding back at an angle. The whalebone is broader at the base and there are about 320 balleen plates on each side.

The coloration is dark slate blue throughout, except for the tips and undersides of the flippers and flukes. The back, sides, and belly are mottled with light spots. The name sulphur-bottom whale refers to a yellowish film of microscopic diatom algae that occasionally forms on the undersurface of the whale. The baleen is bluish black or black.

Females are larger than males. Measurements range from 15.5 to 26.7 m (51–88 ft) (True 1904a). The largest male killed in the Antarctic Ocean was 32.7 m (107 ft) long and the largest female was 33.3 m (109 ft) (Tomilin 1957). A specimen 24.4 m (80 ft) long weighed 81,000 kg (89.3 tons) (Nishiwaki 1972).

**Distribution.** Blue whales inhabit all oceans, though their northern limit is the edge of the ice pack in the North Atlantic (Jonsgård 1966b). In the western North Atlantic, blue whales frequent the cold current of Nova Scotia and the icy waters that flow from the Saint Lawrence River as well as the Labrador Current (Kellogg 1929).

Blue whales are rare on the east coast of the United States. In the waters of New England, Allen (1916) cited records of blue whales at Popham Beach, Maine (1), in November 1904; at Cape Elizabeth, Maine (2), in August 1912; and at South Coast, Massachusetts, in October 1874. Brown (1913) reported a blue whale stranded on King's Beach, Lynn, Massachusetts (3) (no date). There are no records of sightings or strandings on the coasts of Connecticut, Rhode Island, or New Hampshire.

**Behavior.** Blue whales migrate northward in spring or early summer with an irregular return southward in autumn to breed and calve. These whales are pelagic and travel widely. They are not as gregarious as other rorquals. They are encountered singly, in pairs, or in groups of two or three—rarely more (Mielche 1952). They may gather in herds, but they retain the small family subgroups (Slijper 1962).

These whales seem to have strong succoring behavior. When a pair of blue whales that have become separated by a mile or two are hunted, they will come together very quickly. When one

BLUE WHALE

female blue whale was killed, the male swam slowly around a whaleboat (Lillie 1915). Whalemen have remarked that to capture a pair of blue whales they must kill the female first, for the male will remain with her, whereas the female will not stand by a slain male (Slijper 1962). Caldwell and Caldwell (1966) described an injured blue whale that was protected by two schoolmates from harassment by sharks.

Blue whales can maintain speeds of about 10 knots, but can be pressed to swim 15 knots or more (Walker et al. 1964). They often follow ships, and one blue whale followed a ship for 24 days (Scammon 1874). When a blue whale breaks surface after sounding, it first exposes the top of its head and the blowhole, then the back, the dorsal fin, and finally the caudal peduncle. Immediately before and after sounding, the animal shows a much larger portion of the body than at intermediate dives. It may stay underwater for up to 20 minutes (Mackintosh 1965). The spouts of the blue whale and fin whale are similar. The spout reaches 33 to 50 feet (Nishiwaki 1972).

Blue whales are relatively shallow feeders. Their food consists chiefly of tremendous masses of the small shrimplike crustaceans *Femora longicornis*, *Meganyctiphanes norvegica*, and *Thysanoessa inermis*, called "krill" (Allen 1916). Small fish are sometimes accidentally swallowed.

Breeding takes place in warm waters, mostly in winter. The gestation period is between 10 and 12 months. Females do not breed when suckling calves, so that at least two years intervene between successive pregnancies. At birth the calf is about 7 m (23 ft) long and weighs about 2,495 kg (5,500 lb). The lactation period lasts about 7 months, until the calf is about 16 m (52 ft) long (Mackintosh and Wheeler 1929). Females reach sexual maturity at approximately 23.4 m (76.7 ft) long and males at 22.6 (74.2 ft). Longevity is about 90 years—rarely over 100 years (Nishiwaki, 1972).

## Humpback Whale

*Megaptera novaeangliae* (Borowski)

**Description.** The stocky humpback whale differs markedly from other rorquals by having exceptionally long, narrow flippers, one-fourth to one-third the total body length, knobbed on the anterior edges. The posterior edges of the flukes usually have medial notches. There are wartlike round bumps on the head, forward of the blowhole. The dorsal fin is relatively small. There are from 270 to 400 baleen plates on each side of the mouth. The ventral grooves number about 25, reaching to the navel or behind it. There are usually barnacles on the chin, on the anterior portion of the grooves, along the anterior edges of the flippers, and on the flukes.

The coloration varies with the individual animal but is usually black above and white below. The flippers are usually white above, but may be spotted with black or all black, and are white below. The flukes are normally black with white spots along the posterior edges. The baleen is brownish black or grayish black with a few plates sometimes partially white.

Females are larger than males. Measurements range from 11.5 to 15.0 m (38–49 ft) long. The

*Balaena novae angliae* Borowski, 1781. *Gemeinnuzzige Naturge-schichte des Thierreichs* 2(1):21.
*Megaptera novaengliae*, Kellogg, 1932. *Proc. Biol. Soc. Washington* 45:148.
Type Locality: Coast of New England, United States
Names: Hump whale, hunch-backed whale, and bunch whale. The meaning of *Megaptera* = large winglike, referring to the large and long winglike flippers, and *novaeangliae* = of New England.

average weight was 29 tons for 270 humpbacks that averaged 12.6 m (41.3 ft) long (Walker et al. 1964).

**Distribution.** Humpback whales occur in nearly all the oceans. In the North Atlantic they range from the arctic to the tropic regions. They are seen fairly often in New England waters but rarely become stranded. In 1951 a stranding was reported at Barnstable, Massachusetts (1). Cronan and Brooks (1968) reported that a humpback whale beached at Matunuck, Rhode Island (2), in June 1957.

**Behavior.** In both hemispheres humpback whales migrate to summer feeding grounds in the polar waters and return to winter breeding grounds in tropical waters. There seem to be segregated populations which differ in their migration routes. The overall directional trend during migration of humpbacks is south to north and back in the northern hemisphere, and vice versa in the southern hemisphere (Dawbin 1966). In general, humpbacks follow the coastlines in their migration. Some of them may be deflected inshore by low water depth and bottom topography encountered along the migration routes.

In the western Atlantic Ocean, humpbacks have been reported to pass New England waters mainly in April and May, with some animals remaining up to August during their northward migration. For the most part they arrive off the southern coast of Newfoundland about May, reach southern Labrador by July, and remain until September. On their southward journey they pass New England from October through December (Sergeant 1961).

Humpback whales are playful. They sometimes breach clear of the water, waving their long flippers. They like to roll on the surface of the water, slapping it with their flukes. Sometimes the whales turn somersaults above and under the water. They often lobtail or thrash the surface of the water into foam with their flukes, or sometimes they wave their flukes slowly in the air. They may swim on their backs for short periods.

Humpbacks travel singly, in pairs, or in groups of up to 12 or more (Dakin 1934). They often maintain a speed of 5 to 6 knots for up to 5 hours (Dawbin 1966), though they can swim up to 12 knots in short bursts (Ommanney 1971). Wounded humpbacks can swim up to 17 knots (Zenkovich 1937).

Humpbacks show tightly knit social associations. Whalemen have long recognized that males often stand by injured females. These mammals are known to follow and play about moving ships.

These whales ascend obliquely to the surface and display first the blowhole, then the top of the head, then the posterior portion of the back with the dorsal fin. As they descend under the water the posterior half of the body becomes humped, with the dorsal fin at its highest point. When the whales dive deeply, they take a nearly vertical course and expose a greater portion of the back than usual, curve the body and display the flukes nearly perpendicular to the surface of the water. They may wave the flukes in the air as they go under. Tomilin (1957) reported that the duration of a sounding is usually 6 to 7 minutes but may be 15 to 30 minutes if the whale is wounded.

Humpback whales produce short, loud whistles and wheezing sounds, probably for communication. The wheezing sounds are audible both underwater and in the air and appear to be

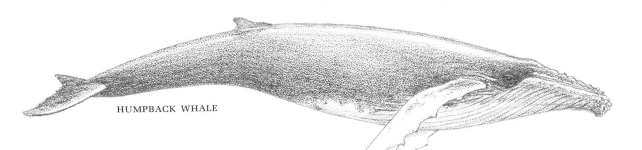

HUMPBACK WHALE

produced through the blowhole during exhalation (Watkins 1967). These whales make a series of varied sounds or "singing" for a period of 7 to 30 minutes, then repeat the same series with considerable precision. Each animal adheres to its own "song type" and the song sessions of several whales may last for several hours. The function of the songs and the sex of the performers are unknown (Payne and McVay 1971).

In the North Atlantic, humpbacks feed mainly on the minute crustaceans *Euphausia inermis*, *Thysanoessa* spp., and *Meganyctiphanes* spp. Occasionally they eat smelt, herring, capelin, cod, pollock, and salmon. They may feed on cephalopod and pteropod mollusks.

The breeding season probably lasts throughout the year. Humpbacks may require tropical coastal conditions with water temperature at about 25°C, for breeding, but they do not require coastal conditions at other stages of the migratory cycle (Dawbin 1966). During the breeding season, humpbacks show much affection toward each other, and the bond between the sexes is very strong. Courtship usually consists of one whale's stroking its partner with its entire body and flippers as it glides past. One partner may playfully slap the other with its long flippers. The maternal instinct is highly developed in humpback whales. In times of danger the female may take her calf under her flippers, push it with her head, or place herself between her calf and danger.

Females generally calve once in two years, but may calve twice in three years, after a gestation period of 11 to 12 months. At birth the calf is about 4 to 5 m (13–16 ft) long. It grows rapidly and is weaned when 5 to 6 months of age or when it attains a length of 8 to 9 m (26–30 ft) (Matthews 1937). Males attain sexual maturity at 11.1 to 11.4 m (36–37 ft) and females at 11.4 to 12.0 m (37–39 ft) (Nishiwaki 1972). Longevity is about 48 years (Gaskin 1972).

### FAMILY BALAENIDAE (RIGHT WHALES)

The members of this family include the bowhead whale, the right whale, and the pygmy right whale. These robust whales are distinguished from the rorquals by their enormous heads, huge and strongly arched mouths, extremely long baleen plates, and lack of throat grooves. The dorsal fin is absent in the bowhead and right whales but present in the pygmy right whale.

These whales are slow swimmers and feed mainly on small crustaceans (copypods and shrimp) and mollusks (pteropods). They are not gregarious and usually occur singly or in small groups. They are moderately migratory, preferring high-latitude waters during winter; bowheads remain near the ice edge all year.

The right whale is the only member of the Balaenidae that has been recorded in the waters of New England.

## Right Whale

*Eubalaena glacialis* (Borowski)

**Description.** The right whale is recognized by its large head, about one-fourth of the total body length, curved mouth, absence of a dorsal fin and throat grooves, and its "bonnet" and other callosities—irregular-shaped horny excrescences or wartlike structures composed of an excessive growth of cornified layers of skin and infested by the parasitic amphipod crustacean *Cyamus* spp. The bonnet is situated on top of the snout, and other callosities are found around the eyes and blowhole and along the lower jaw. The huge mouth is arched markedly. It curves steeply much more backward than forward. The lower lips are very large and fleshy, scalloped along the upper margin, and curve outward and overlap the upper jaw. Some individuals have small horny protuberances along the chin. The flippers are large, broad, and somewhat pointed, and are situated markedly below the level of the eyes. The flukes are very broad and have a medial notch.

The skull is arched more at the base than at the rostrum when viewed from the side. The premaxillaries are wide in dorsal aspect. There are from 250 to 390 olive blackish baleen plates on each side.

The coloration is usually black throughout, but the belly may be white or white spotted.

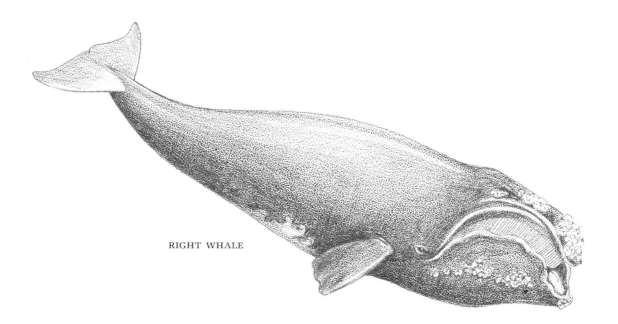

RIGHT WHALE

*Eubalaena glacialis* Borowski, 1781.
*Gemeinnuzzige Naturgeschichte
des Thierreichs* 2(1):18.
*Eubalaena glacialis,* Kükenthal, 1900.
*Fauna Arct.* 1:207.

Type Locality: North Cape, Norway
Names: Atlantic right whale, black
right whale, Biscayan right whale,
North Cape whale, and Nordkaper.
The right whale was called the
"right" whale by whalemen because it was abundant, slow-
swimming, and buoyant when
killed, and it yielded great quantities of oil. The meaning of
*Eubalaena* = a whale, and
*glacialis* = ice, or frozen.

Measurements range from 13.6 to 16.6 m
(35–45 ft) (Walker et al. 1964). An adult female
taken off Amagansett, Long Island, New York, on
22 February 1907 was 16.5 m (54 ft) long (Andrews 1909b). A 16.1 m (53 ft) specimen weighed
65,690 kg (144,846 lb) (Nishiwaki 1972), and a
specimen 17.1 m (56 ft) long weighed 67,197 kg
(148,269 lb) (Rice 1967).

**Distribution.** In the northern hemisphere, right
whales occur in the Atlantic and Pacific oceans
between 20° N and 70° N, and in the southern
hemisphere south of 20° S and 50° S. Some right
whales inhabit the Arctic Sea (Nishiwaki 1972).
Right whales seem to make limited north-south
migrations between the cold temperate and subtropical seas in each hemisphere. In New England waters they migrate northward in spring,
are absent in summer, and reappear in autumn
when they return from the north. They become
less numerous by midwinter, since most of them
migrate farther south and only a few winter in
this region (Allen 1916).

Historical records reveal over 55 strandings of
this species on the coasts of Massachusetts,
Rhode Island, and eastern Long Island between
1620 and 1913. Most of the strandings occurred
between February and May, fewer between June
and September. No strandings occurred in July

or August (Allen 1916). The species was not
common on the Connecticut coast in former
days. It was reported in Long Island Sound at
Stonington, Connecticut (1), and Montauk, Long
Island (Goodwin 1935). In recent years the
species was seen in Cape Cod Canal (2) on 15
June 1957, where it remained for a while and
then returned to sea through the same passage
by which it entered (Clark 1958). Right whales
have recently been seen during summer near
Eastport (3) and Mount Desert Rock (4), Maine
(Katona, Richardson, and Hazard 1975).

**Behavior.** Right whales are seen singly, in pairs,
or in small groups of 25 or more at feeding
grounds. These whales often lie still on the
surface with the bonnet and blowhole protruding high above the water. During migration the
whales generally swim at 2 to 3 knots, but they
can swim up to 6 knots when startled. Right
whales may remain underwater for up to 50
minutes. They do not show their flukes when
diving normally, but they show a wide expanse
of them just before sounding deep. Right whales
make deep dives, but probably not as deep as
other species of baleen whales. They spout 5 or 6
times in succession, then remain underwater for
10 to 20 minutes, partly surfacing but not
breaching. The spout may reach a height of 15

feet and is much thicker than those of other rorquals. The two jets diverge widely sideways like a V (Allen 1916).

Like the sei whale, the right whale is a "skimmer." During feeding it swims slowly along with its mouth open. The main food items are small crustaceans *Calanus finmarchicua*, *Microcalanus* spp., *Oithoun* spp., and *Pseudocalanus* spp. These floating and weak-swimming animals are sometimes called krill.

Not much is known on the reproduction of this species. Breeding has been observed from February to April, and the breeding season probably lasts about 6 months (Nishiwaki 1972). Right whales calve once in two years. Pregnant females tend to stay apart from the other animals. The gestation period is about 12 months. One full-term fetus measured 6 m (19.7 ft) long. The calf is suckled for about 1 year, and females have strong maternal instincts (Allen 1916).

## REFERENCES

Allen, Glover M. 1916. The whalebone whales of New England. *Mem. Boston Soc. Nat. Hist.* 8:108–322.

——. 1941. Pigmy sperm whale in the Atlantic. *Bull. Field Mus. Nat. Hist.* 27:17–36.

Andrews, Roy Chapman. 1909a. Observations on the habits of the finback and humpback whales of the eastern North Pacific. *Bull. Amer. Mus. Nat. Hist.* 26:213–26.

——. 1909b. Further notes on *Eubalaena glacialis*. *Bull. Amer. Mus. Nat. Hist.* 25:272–75.

——. 1916a. Monographs of the Pacific cetacea: II. The sei whale, *Balaenoptera borealis* Lesson. *Mem. Amer. Mus. Nat. Hist.* 1(6):291–502.

——. 1916b. *Whale hunting with gun and camera*. New York and London: D. Appleton.

Ashley, C. W. 1942. *The Yankee whaler*. Garden City, New York: Halcyon House.

Backus, Richard H., and Schevill, William. 1961. The stranding of a Cuvier's beaked whale, *Ziphius cavirostris*, in Rhode Island, U.S.A. *Norwegian Whaling Gaz.* 50:177–81.

Beddard, Frank E. 1900. *A book of whales*. London: John Murray.

Bennett, F. D. 1840. *Narrative of a whaling voyage around the globe from the years 1833 to 1836*. London: Richard Bentley.

Besharse, Joseph C. 1971. Maturity and sexual dimorphism in the skull, mandible, and teeth of the beaked whale, *Mesoplodon densirostris*. *J. Mamm.* 52(2):297–315.

Brimley, H. H. 1943. A second specimen of True's beaked whale, *Mesoplodon mirus* True, from North Carolina. *J. Mamm.* 25(2):199–203.

Brown, C. Emerson. 1913. *Pocket list of mammals of eastern Massachusetts*. Salem: Peabody Museum.

Brown, David H., and Norris, Kenneth S. 1956. Observations of captive and wild cetaceans. *J. Mamm.* 37(3):311–26.

Brown, S. G. 1968. Feeding of sei whales in South Georgia. *Norwegian Whaling Gaz.* 57:118–25.

Caldwell, David K., and Caldwell, Melba C. 1971. Sounds produced by two rare cetaceans stranded in Florida. *Cetology* 4:1–6.

Caldwell, David K.; Caldwell, Melba C.; and Rice, Dale W. 1966. Behavior of the sperm whale, *Physeter catodon* L. In *Whales, dolphins, and porpoises*, ed. Kenneth S. Norris, pp. 677–717. Berkeley and Los Angeles: University of California Press.

Caldwell, David K., and Fields, Hugh M. 1959. Surf-riding by Atlantic bottle-nosed dolphins. *J. Mamm.* 40(3):454–55.

Caldwell, David K., and Golley, Frank B. 1965. Marine mammals from the coast of Georgia to Cape Hatteras. *J. Elisha Mitchell Soc.* 8(1):24–32.

Caldwell, Melba C., and Caldwell, David K. 1966. Epimeletic (care-giving) behavior in Cetacea. In *Whales, dolphins, and porpoises*, ed. Kenneth S. Norris, pp. 755–89. Berkeley and Los Angeles: University of California Press.

Clark, A. H. 1887. The whale fishery: History and present condition of the fishery. In *The fisheries and fishery industries of the United States*, ed. G. B. Goode, 2:3–218. Washington, D.C.: U.S. Commission on Fish and Fisheries.

Clark, Eugene S., Jr. 1958. Right whale, *Balaena glacialis* enters Cape Cod Canal, Massachusetts, U.S.A. *Norwegian Whaling Gaz.* 47:138–43.

Clarke, Robert. 1954a. Open boat whaling in the Azores. *Discovery Repts.* 26:281–354.

——. 1954b. Whales and seals as resources of the sea. *Norwegian Whaling Gaz.* 43:301–17.

——. 1955. A giant squid swallowed by a sperm whale. *Norwegian Whaling Gaz.* 44:589–93.

——. 1956a. The biology of sperm whales captured in the Azores. *Norwegian Whaling Gaz.* 45:439–44.

——. 1956b. Sperm whales of the Azores. *Discovery Repts.* 28:237–98.

Collett, Robert. 1886. On the external characters of Rudolphi's rorqual, *Balaenoptera borealis*. *Proc. Zool. Soc. London*, 1886, pp. 243–65.

——. 1912. *Norges Pattedyr (Norges Hvirveldyr I)*. Kristiania: H. Aschehoug Co.

Coman, Dale Rex. 1972. *The native mammals, reptiles, and amphibians of Mount Desert Island, Maine.* Philadelphia: University of Pennsylvania.

Connor, Paul F. 1971. The mammals of Long Island, New York. *New York State Mus. Sci. Serv. Bull.* 416:1–77.

Cronan, John M., and Brooks, Albert. 1968. The mammals of Rhode Island. Providence, Rhode Island Department of Agriculture and Conservation Division of Fish and Game. *Wildl. Pamphl.* 6:1–133.

Dakin, W. J. 1934. *Whalemen adventurers.* Sydney: Angus and Robertson.

Davis, William Morris. 1926. *Nimrod of the sea; or, The American whalemen.* New York: Harper Bros.

Dawbin, William H. 1966. The seasonal migratory cycle of humpback whales. In *Whales, dolphins, and porpoises,* ed. Kenneth S. Norris, pp. 145–70. Berkeley and Los Angeles: University of California Press.

Dorofejev, S. V., and Klamov, S. K. 1936. On the question of determining the age of the white whale. (In Russian.) *Trudy Vsesoiuznovo N. I. Insti. Rybnovo Khoziaistra Okeanog.* 3:24–34.

Eschricht, D. F. 1866. On the species of the genus *Orca* inhabiting the northern seas. Trans. H. W. Flower. In *Recent memoirs on the Cetacea,* pp. 151–58. London: Royal Society of London. Originally published 1862.

Essapian, Frank S. 1953. The birth and growth of a porpoise. *Nat. Hist.* 62:392–99.

Fisher, H. Dean, and Harrison, Richard J. 1970. Reproduction in the common porpoise (*Phocoena phocoena*) of the North Atlantic. *J. Zool.* 161:471–86.

Fraser, F. C. 1937. Whales and dolphins. In *Giant fishes, whales and dolphins,* ed. J. R. Norman and F. C. Fraser, pp. 203–349. London: Putnam.

——. 1939. Three anomalous dolphins from Blacksod Bay, Ireland. *Proc. Royal Irish Acad.* 45:413–55.

——. 1942. The mesorostral ossification of *Ziphius cavirostris. Proc. Zool. Soc. London,* ser. a, 112:21–30.

Gambell, Ray. 1968. Aerial observations of sperm whale behaviour. *Norwegian Whaling Gaz.* 57:126–38.

Gaskin, D. E. 1972. *Whales, dolphins, and seals with special reference to the New Zealand region.* New York: St. Martin's Press.

Gaskin, D. E.; Smith, G. J. D.; and Watson, A. P. 1975. Preliminary study of movements of harbor porpoises (*Phocoena phocoena*) in the Bay of Fundy using radiotelemetry. *Canadian J. Zool.* 53(10):1466–71.

Gilmore, Raymond M. 1961. Whales, porpoises, and the U.S. Navy. *Norwegian Whaling Gaz.* 50:89–108.

Goodwin, George Gilbert. 1935. The mammals of Connecticut. Hartford. *Connecticut Geol. Nat. Hist. Surv. Bull.* 53:1–221.

Griffin, Edward L. 1966. Making friends with a killer whale. *Natl. Geogr. Mag.* 129:418–46.

Guldberg, Gustav, and Nansen, Fridtjof. 1894. *On the development and structure of the whale.* Bergen: Bergens Museum.

Gunter, Gordon; Hubbs, Carl L.; and Beal, M. Allan. 1955. Records of *Kogia breviceps* from Texas, with remarks on movements and distribution. *J. Mamm.* 36(2):263–70.

Hancock, David. 1965. Killer whales kill and eat a minke whale. *J. Mamm.* 46(2):341–42.

Handley, Charles O., Jr. 1966. A synopsis of the genus *Kogia* (Pygmy sperm whales). In *Whales, dolphins, and porpoises,* ed. Kenneth S. Norris, pp. 62–69. Berkeley and Los Angeles: University of California Press.

Heezen, B. C. 1957. Whales entangled in deep-sea cables. *Deep-sea Res.* 4:105–15.

Hoese, H. D. 1971. Dolphin feeding out of water in a salt marsh. *J. Mamm.* 52(1):222–23.

Jonsgård, Åge. 1951. Studies on the little piked whale or minke whale, *Balaenoptera acutorostrata* Lacépède. *Norwegian Whaling Gaz.* 40:1–55.

——. 1953. Fin whale, *Balaenoptera physalus,* with six foetuses. *Norwegian Whaling Gaz.* 42:685–86.

——. 1955. The stocks of blue whales, *Balaenoptera musculus,* in the northern Atlantic Ocean and adjacent arctic waters. *Norwegian Whaling Gaz.* 44:505–19.

——. 1966a. Biology of the North Atlantic fin whale *Balaenoptera physalus* (1), taxonomy, distribution, migration, and food. *Hvalradets Skr.* 49:1–62.

——. 1966b. The distribution of Balaenopteridae in the North Atlantic Ocean. In *Whales, dolphins, and porpoises,* ed. Kenneth S. Norris, pp. 114–24. Berkeley and Los Angeles: University of California Press.

——. 1968a. A note on the attacking behaviour of the killer whale *Orcinus orca. Norwegian Whaling Gaz.* 57:84–85.

——. 1968b. Another note on the attacking behaviour of killer whale *Orcinus orca. Norwegian Whaling Gaz.* 57:175–76.

Jonsgård, Åge, and Lyshoel, P. B. 1970. A contribution to the knowledge of the biology of the killer whale *Orcinus orca* (L.). *Nytt Mag. Zool.* 18:41–48.

Katona, Steven; Richardson, David; and Hazard, Robin 1975. *A field guide to the whales and seals of the Gulf of Maine.* Rockland, Maine: Maine Coast Printers.

Kellogg, Remington. 1929. What is known of the migrations of some of the whalebone whales. *Ann. Rept. Bd. Regents Smithsonian Inst.* 2981:467–94.

Kellogg, Winthrop N., and Rice, Charles E. 1966. Visual discrimination and problem solving in a bottlenose dolphin. In *Whales, dolphins, and por-*

poises, ed. Kenneth S. Norris, pp. 731–54. Berkeley and Los Angeles: University of California Press.

Lillie, D. G. 1915. Cetacea. British Antarctic (Terra Nova) expedition. Nat. Hist. Rept. Zool., 1910 1(3):85–124.

Lilly, John C. 1963. Distress call of the bottlenose dolphin: Stimuli and evoked behavioral responses. Science 139:116–18.

Linsley, James H. 1842. A catalogue of the Mammalia of Connecticut, arranged according to their natural families. Amer. J. Sci. 43:345–54.

Lono, O., and Oynes, P. 1961. White whale fishery at Spitsbergen. Norwegian Whaling Gaz. 50:267–86.

Mackintosh, N. A. 1965. The stocks of whales. London: Fishing News (Books) Ltd.

Mackintosh, N. A., and Wheeler, J. F. G. 1929. Southern blue and fin whales. Discovery Repts. 1:257–540.

Mairs, Donald F., and Scattergood, Leslie W. 1958. Recent Maine records of the bottlenose porpoise and the beluga. Maine Field Nat. 15(4):78–80.

Matthews, L. Harrison. 1937. The humpback whale, Megaptera nodosa. Discovery Repts. 17:7–91.

——. 1938a. The sperm whale, Physeter catodon. Discovery Repts. 17:95–164.

——. 1938b. The sei whale, Balaenoptera borealis. Discovery Repts. 17:183–290.

Mielche, H. 1952. Thar she blows. London: William Hodge.

Millais, John Guille. 1906. The mammals of Great Britain and Ireland. Vol. 3. London: Longmans, Green.

Miller, Gerrit S., Jr., and Kellogg, Remington. 1955. List of North American Recent mammals. U.S. Natl. Mus. Bull. 205:1–954.

Mitchell, Edward. 1972. Memorandum on northwest Atlantic sei whales. International Comm. on Whaling, 22nd Rept., pp. 119–20.

Mitchell, Edward, and Kozicki, V. Michael. 1975. Supplementary information on minke whale (Balaenoptera acutorostrata) from Newfoundland fishery. J. Fish. Res. Bd. Canada 32(7):985–94.

Mizue, R., and Jimbo, H. 1950. Statistic study of foetuses of whales. Sci. Repts. Whales Res. Inst. (Tokyo) 3:119–31.

Moore, Joseph Curtis. 1953. Distribution of marine mammals to Florida waters. Amer. Midland Nat. 49(1):117–58.

——. 1955. Bottle-nosed dolphins support remains of young. J. Mamm. 36(3):466–67.

——. 1966. Diagnoses and distributions of beaked whales of the genus Mesoplodon known from North American waters. In Whales, dolphins, and porpoises, ed. Kenneth S. Norris, pp. 32–61. Berkeley and Los Angeles: University of California Press.

Moore, Joseph Curtis, and Palmer, Ralph S. 1955. More piked whales from southern North Atlantic. J. Mamm. 36(3):429–33.

Moore, Joseph Curtis, and Woods, F. G., Jr. 1957. Differences between the beaked whales, Mesoplodon mirus and Mesoplodon gervaisi. Amer. Mus. Novitates 1831:1–25.

Murphy, Robert Cushman. 1947. Logbook for Grace. New York: Macmillan Co.

Nishiwaki, Masahura. 1962. Aerial photographs show sperm whales' interesting habits. Norwegian Whaling Gaz. 51:395–98.

——. 1972. General biology. In Mammals of the sea, biology and medicine, ed. Sam H. Ridgway, pp. 3–200. Springfield, Ill.: Charles C. Thomas.

——. 1975. Ecological aspects of smaller cetaceans, with emphasis on the striped dolphin (Stenella caeruleoalba). J. Fish. Res. Bd. Canada 32(7):1069–72.

Nishiwaki, Masahura, and Handa, C. 1958. Killer whales caught in the coastal waters off Japan for recent 10 years. Sci. Rept. Whales Res. Inst. (Tokyo) 13:85–96.

Norris, K. S., and Prescott, J. H. 1961. Observations on Pacific cetaceans of Californian and Mexican waters. Univ. California Pub. Zool. 63:291–402.

Norton, Arthur H. 1930. The mammals of Portland, Maine, and vicinity. Proc. Portland Soc. Nat. Hist. 4:1–151.

Oshumi, S.; Kasuya, T.; and Nishiwaki, M. 1963. Accumulation rate of dentinal growth layers in the maxillary tooth of the sperm whale. Sci. Repts. Whales Res. Inst. (Tokyo) 17:15–35.

Ohsumi, Seiji, and Masaki, Yasuaki. 1975. Biological parameters of the Antarctic minke whale at the virginal population level. J. Fish. Res. Bd. Canada 32(7):995–1004.

Olsen C. Robert; Elsner, Robert; Hale, Frank C.; and Kenney, David W. 1969. "Blow" of the pilot whale. Science 163(3870):953–55.

Ommanney, F. D. 1971. Lost leviathan. New York: Dodd, Mead.

Omura, H., and Sakiura, H. 1956. Studies on the little piked whales from the coast of Japan. Sci. Rept. Whales Res. Inst. (Tokyo), 16:7–18.

Omura, H.; Fujin, K.; and Kinura, S. 1955. Beaked whale Berardius of Japan, with notes on Ziphius cavirostris. Sci. Rept. Whales Inst. (Tokyo) 10:89–132.

Paul, John R. 1968. Risso's dolphin, Grampus griseus, in the Gulf of Mexico. J. Mamm. 49(4):746–48.

Payne, Roger S., and McVay, Scott. 1971. Songs of humpback whales. Science 173(3997):585–97.

Raven, Henry C. 1937. Notes on the taxonomy and osteology of two species of Mesoplodon (M. europaeus Gervais, M. mirus True). Amer. Mus. Novitates 905:1–30.

——. 1942. On the structure of Mesoplodon densirostris, a rare beaked whale. Bull. Amer. Mus. Nat. Hist. 80(2):23–50.

Rice, Dale W. 1967. Cetaceans. In *Recent mammals of the world*, ed. S. Anderson and J. K. Jones, Jr., pp. 291–324. New York: Ronald Press.

———. 1968. Stomach contents and feeding behaviour of killer whales in the eastern North Pacific. *Norwegian Whaling Gaz.* 57:35–38.

Scammon, Charles M. 1874. *The marine mammals of the northwestern coast of North America*. San Francisco and New York: John H. Carmany.

Scattergood, Leslie W. 1949. Notes on the little piked whale. *Murrelet* 39(1):3–16.

Scheffer, Victor B., and Rice, Dale W. 1963. A list of the marine mammals of the world. *U.S. Fish and Wildl. Serv. SSR-Fisheries* 431:1–12.

Scheffer, Victor B., and Slipp, John W. 1948. The whales and dolphins of the west coast of North America. *Amer. Midland Nat.* 39(2):257–337.

Schevill, William E. 1954. Sight records of the gray grampus *Grampus griseus* (Cuvier). *J. Mamm.* 35(1):123–24.

———. 1956. *Lagenorhynchus acutus* off Cape Cod. *J. Mamm.* 37(1):128–29.

Schevill, William E., and Watkins, W. A. 1962. Whale and porpoise voices. *Woods Hole Oceanographic Inst. Contrib.* 1320:1–24.

Schevill, William E.; Watkins, W. A.; and Ray, Carleton. 1969. Click structure in the porpoise, *Phocoena phocoena*. *J. Mamm.* 50(4):721–28.

Scholander, P. F. 1940. Experimental investigations on respiratory function in diving mammals and birds. *Hvalradets Skr.* 22:1–131.

Sergeant, D. E. 1961. Whales and dolphins of the Canadian east coast. *Fish. Res. Bd. Canada Circ.* 7:1–26.

———. 1962a. The biology and hunting of beluga or white whales in the Canadian arctic. *Fish Res. Bd. Canada Circ.* 18:1–13.

———. 1962b. The biology of the pilot or pothead whale *Globicephala melaena* (Traill) in Newfoundland waters. *Fish. Res. Bd. Canada. Bull.* 132:1–84.

———. 1963. Minke whales, *Balaenoptera acutorostrata* Lacépède, of the western North Atlantic. *J. Fish. Res. Bd. Canada* 20(6):1489–1504.

———. 1966. Populations of large whale species in western North Atlantic, with special reference to the fin whale. *Fish Res. Bd. Canada Circ.* 9:1–29.

———. 1973. Biology of white whales (*Delphinapterus leucas*) in western Hudson Bay. *J. Fish. Res. Bd. Canada* 30(8):1065–90.

Sergeant, D. E.; Caldwell, David K.; and Caldwell, Melba C. 1973. Age, growth, and maturity of bottle-nosed dolphin (*Tursiops truncatus*) from northeast Florida. *J. Fish. Res. Bd. Canada* 30(7):1009–11.

Sergeant, D. E., and Fisher, H. D. 1957. The smaller cetacea of eastern Canadian waters. *J. Fish. Res. Bd. Canada* 14:83–115.

Siebenaler, J. B., and Caldwell, David K. 1956. Cooperation among adult dolphins. *J. Mamm.* 37(1):126–28.

Slijper, E. J. 1962. *Whales*. New York: Basic Books.

Smith, G. J. D., and Gaskin, D. E. 1974. The diet of harbor porpoises (*Phocoena phocoena* [L.]) in coastal waters of eastern Canada, with special reference to the Bay of Fundy. *Canadian J. Zool.* 52:777–82.

Starrett, Andrew, and Starrett, Priscilla. 1955. Observations on young blackfish (*Globicephala*). *J. Mamm.* 36(3):424–29.

Tavolga, Margaret C. 1966. Behavior of the bottlenose dolphin (*Tursiops truncatus*): Social interactions in a captive colony. In *Whales, dolphins, and porpoises*, ed. Kenneth S. Norris, pp. 718–30. Berkeley and Los Angeles: University of California Press.

Thorpe, Malcon Rutherford. 1938. Notes on the osteology of a beaked whale. *J. Mamm.* 19(3):354–62.

Tomilin, A. G. 1957. *Mammals of the U.S.S.R. and adjacent countries*. Vol. 9. *Cetacea*. Washington, D.C.: Smithsonian Institution and the National Science Foundation. (Translated by the Israel Program for Scientific Translations, Jerusalem, 1967.)

Tonnessen, J. N. 1962. A great work on whales. *Norwegian Whaling Gaz.* 51:473–80.

Townsend, Charles W. 1929. The white whale at Ipswich, Massachusetts. *J. Mamm.* 10(2):171.

Tracy, James E. 1909. The fossil whale at the state museum. *The Vermonter, The State Magazine*, January 1909, pp. 22–23.

True, Frederick W. 1904a. The whalebone whales of the western North Atlantic compared with those occurring in European waters with some observations on the species of the North Pacific. *Smithsonian Contrib. Knowl.* 33:1–332.

———. 1904b. Notes on a killer whale from the coast of Maine. *Proc. U.S. Natl. Mus.* 27(1375):227–30.

———. 1910. An account of the beaked whales of the family Ziphiidae in the collection of the United States National Museum, with remarks on some specimens in other museums. *Bull. U.S. Natl. Mus.* 73:1–89.

Walker, Ernest P.; Warnick, Florence; Lange, Kenneth I.; Uible, Howard E.; Hamlet, Sybil E.; Davis, Mary A.; and Wright, Patricia F. 1964. *Mammals of the world*. Vol. 2. Baltimore: Johns Hopkins University Press.

Watkins, William A. 1967. Air-borne sounds of the humpback whale (*Megaptera novaeangliae*). *J. Mamm.* 48(4):573–78.

Worthington, L. V., and Schevill, W. E. 1957. Underwater sounds heard from sperm whales. *Nature* 180:291.

Zenkovich, B. A. 1937. Humpback or longarm whale. *Vestnik Dal'nevostochnogo filiala Akad. Nauk USSR* 27:37–62. (In Russian. Original not seen; reference cited from Tomilin 1957).

# Order Carnivora

THE ORDER CARNIVORA, the meat eaters, comprises mammals of great diversity in appearance, size, and behavior. Carnivores are mainly terrestrial and arboreal, although several genera of this order are aquatic or semiaquatic. These mammals are solitary or associate in pairs, family groups, or bands. Food habits vary markedly among the families and among members of a family. For example, in the family Hyaenidae, the hyenas are mainly hunters and meat scavengers, while the aardwolves are mainly insect eaters. Carnivores generally have one or two litters annually. In the family Ursidae and in some species of the family Mustelidae, implantation of the blastocyst or fertilized ovum is delayed, so that the gestation period is considerably longer than the gestation periods of most carnivores. The young are born helpless and are cared for solicitously by the mother, or in some species by both parents.

The dentition of carnivores is especially modified for a meat diet. The small incisors are more or less rounded, but the canines are well developed—elongated, conical, pointed, recurved, strong, and sharp—and project well beyond the other teeth, a specialization for seizing and stabbing prey. The carnassial or sectorial teeth (the fourth upper premolar and first lower molar), usually with opposed trenchant edges like scissors blades, are specialized for cutting and tearing flesh. They are most highly developed in the family Felidae and least developed in the omnivorous families Ursidae and Procyonidae. The premolars are simple and sometimes reduced. The molars have either sharp or blunt cusps and vary in structure.

The order Carnivora is worldwide in distribution except for most oceanic islands (Stains 1967). Five families are represented in New England.

## FAMILY CANIDAE (COYOTES, WOLVES, FOXES, AND ALLIES)

Members of the Canidae are typically doglike and, except for some highly specialized breeds of domestic dogs, generally are lithe, muscular, deep-chested, and long-limbed. They have elongated muzzles, large erect and somewhat pointed ears, and long bushy tails. The incisors are unspecialized; the canines are elongated, strong, sharp, and recurved; the carnassials are well developed and highly sectorial, with sharp cutting edges; and the molars have a grinding surface.

Canids are highly intelligent, alert, sagacious, and cunning. They are active throughout the year, day and night.

The family is nearly cosmopolitan except for Antarctica and some oceanic islands. In New England the family is represented by four species.

## Coyote

*Canis latrans* Say

*Canis latrans* Say, 1823. In Long, *Account of an expedition from Pittsburgh to the Rocky Mountains* . . ., 1:168.
Type Locality: Engineer Cantonment, about 12 miles southeast of the present town of Blair, Washington County, Nebraska, on the west bank of the Missouri River
Dental Formula:
I 3/3, C 1/1, P 4/4, M 2/3 × 2 = 42
Names: Brush wolf, little wolf, prairie wolf, and American jackal. The meaning of *Canis* = a dog, and *latrans* = a barker, referring to the coyote's barking habit.

**Description.** The coyote resembles a small collie dog but is more slender, with erect pointed ears and a bushy, drooping tail. It is recognized by its long, narrow, pointed muzzle, small rounded nose, round pupil and yellow iris of the eye, large ears which are directed forward but are movable, slender legs, and small feet. There are five toes on the forefoot, the first short and rudimentary but bearing a well-developed claw, and the hind foot has four toes. The claws are rather blunt and are nonretractile. There is a prominent callosity near the outside of the posterior surface of the lower part of the forearm, and the top of the base of the tail bears a scent gland. Females have ten mammae—four pectoral, four abdominal, and two inguinal.

The skull is long and slender, with a relatively short, broad rostrum, slightly depressed in the prefrontal region, and moderately spreading zygomatic arches. The sagittal ridge is prominent in adult males, and the auditory bullae are relatively long.

The fur is dense, long, and coarse. The sexes are colored alike and show slight seasonal color variation, but coloration varies greatly among individuals, though it is usually gray to cinnamon gray. The upperparts from the nape to the tail are a mixture of buffy gray and black with more blackish down the midline. The underparts are buff, mixed with deep gray and a few blackish hairs on the throat. The forelegs and forefeet are blackish on the upper surface. The muzzle is brownish, occasionally near cinnamon brown, and the forehead and cheeks are

somewhat paler, much mixed with gray and some black. The crown is more mixed with tawny, becoming ochraceous tawny or brownish on the nape and the back of the ears. The chin is grayish. The tail above is colored like the back, tipped with black at the base and tip of tail, and is paler below, with little or no black.

The annual molt starts about June and is completed in the fall. The pelage is prime from late November into February. Albino and melanistic coyotes are rare.

**Coyote–Wolf Differences.** In addition to being smaller than the wolf, the adult coyote can be distinguished from the adult eastern timber wolf by the following characters.

Coyotes have a relatively large braincase and comparatively more slender rostrum, while the wolf has a relatively small braincase and massive rostrum, presumably a reflection of the large size of the animals on which it preys. The upper canines of the coyote are less than 11 mm (0.4 in) in anteroposterior diameter at the base, and those of the wolf are more than 12 mm (0.5 in). The tips of the upper canines in the coyote extend below a line drawn through the anterior mental foramina when the mandible is in place and the jaws are closed, whereas in the wolf and domestic dog they usually fall well above that line.

The cheek teeth in the coyote are relatively narrower than in the wolf. In most coyotes the posterior part of lower premolar 4 ($P_4$) is relatively long compared with both the length of the tooth and its maximum width, and has two posterior tubercles and a posterior basal cingulum (Lawrence and Bossert 1967).

In Ontario, Kolenosky (1971) mated a female wolf with a male coyote during a series of controlled breeding experiments. The female produced two hybrid litters. In the first litter, the coloration of the three pups resembled the agouti pattern of the coyote more closely than the color of the wolf. In the second hybrid litter the basic gross color pattern of the five pups was similar to that of the first litter but, unlike the former, the hybrids in the second litter displayed little individual variation.

COYOTE

**Coyote–Dog Differences.** The following characters are not wholly reliable. Compared with the domestic dog, the coyote has thicker underfur, a more bottle-shaped tail; straighter and more erect ears; and a proportionately longer and thinner muzzle. Dogs usually have little underfur and ears that often flop over, except in pedigreed German shepherds and a number of other species. The coyote skull is relatively longer and narrower than that of the domestic dog, except for certain slender-headed varieties such as the collie. The canines are relatively longer and narrower (Lawrence and Bossert 1967).

**Coydog.** The coydog is a hybrid between a coyote and a dog and may be a cross between a female coyote and a male dog (Seton 1929; Murie 1936; Dice 1942; Hall 1943) or a female dog and a male coyote (Bee and Hall 1951; Gier 1957; Silver and Silver 1969; Mengel 1971). Coydogs are morphologically intermediate between the parents and vary markedly with the breed of the dog ancestor. For example, Dice (1942) reported that coydogs born of a female coyote and a male hound dog were black-and-tan like the hound rather than gray agouti like the coyote. Hall (1943) compared cranial characters of a 10-month-old coydog and found the resemblance to the female coyote parent was about twice as pronounced as the resemblance to the male wirehaired fox terrier father. Mengel (1971) reported that five of six first-generation coydogs looked like short-haired, short-legged, black coyotes with white-trimmed chins and a white blaze on the chest; the sixth hybrid more closely resembled the dog mother.

Coyotes and dogs have identical karotypes, 2N = 78 (Benirschke 1967). Coyote-dog hybrids are fertile, and two hybrid females artificially inseminated with dog sperm produced pups after a gestation period of 62 to 64 days (Kennelly and Roberts 1969). The gestation period for the coyote is 60 to 65 days, and that for the dog 58 to 62 days (Asdell 1964).

**New England Wild Canid.** Silver and Silver (1969) studied a population of pen-raised coyotelike wild canids and kennel-bred canid hybrids in New Hampshire and concluded: "There was nothing in the findings to indicate

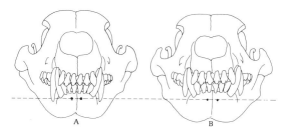

Frontal view of the skull of a coyote, *A*, and the skull of a domestic dog, *B*, showing dentition. In the coyote skull the tips of the upper canine teeth usually fall below a line drawn through the anterior mental foramina in the lower jaw; in the dog the tips of the upper canine teeth usually fall well above this line. Also, the upper canines of the coyote are usually relatively longer and narrower than those of the dog.

recent hybridization in the (wild) New Hampshire animals. Neither did behavioral nor physical data fit entirely those reported for coyotes, dogs, or wolves. The unidentified canids are more nearly like coyotes but possess some wolf- and/or dog-like characters. It is believed that these animals have evolved from coyotes with the introduction of some dog and/or wolf genes sufficiently long ago for the population to have become stabilized. Lawrence and Bossert (1969) concur with this on the basis of their multiple character analysis of skulls.

"Much confusion has arisen from the designation of the wild canids as 'hybrids.' Despite evidence of hybridization at some distant time, their genetic structure has been established to the point where they breed true. Since this is a criterion for recognizing new strains of domestic animals developed through crossbreeding, we feel that the New England canids should be considered a form of coyote. It is suggested that they be designated *Canis latrans* var., and called eastern coyotes."

Males are slightly larger than females. Measurements range from 114 to 132 cm (44.9–52.0 in); tail, 29 to 39 cm (11.4–15.4 in); and hind foot, 17 to 22 cm (6.7–8.7 in). Weights vary from 9.1 to 13.6 kg (20–30 lb). Large males may weigh 20.4 to 22.7 kg (45–50 lb).

**Distribution.** The original distribution of the coyote was the open plains of the United States. It occurs from Alaska south through the western United States to Central America, and within the past fifty years it has extended its range into the northeastern United States and Canada.

In New England, coyotes may have occurred in Vermont and New Hampshire in the 1930s (Silver and Koons 1972) and in Maine about 1936 (Aldous 1939). They were reported in Massachusetts in 1957 and in Connecticut in 1958 (Pringle 1960). Apparently coyotes have not yet been reported in Rhode Island.

**Ecology.** In the East, coyotes mainly inhabit brushy country bordering the edges of second-growth hardwood forests, fields interspersed with thickets, and marshlands. They sleep on the ground in some cover during most of the year but make dens for the pups. The den is found in some concealed spot in a brush-covered slope, in a steep bank with or without cover, on rocky ledges, under a stump, hollow log, log pile, grain bin, or deserted building, or in a dry culvert.

Coyotes may dig dens, but frequently they enlarge an abandoned burrow of a woodchuck, fox, skunk, or rabbit. Dens vary greatly in construction and depth. Some dens extend straight back, while others go straight down 2 to 4 feet, then level off. Some are only a few feet long, others are up to 30 feet long, depending on the soil type. Dens may have one or several entrances which are concealed in high vegetation, and some entrances have a mound 10 inches high and 12 to 30 inches in diameter. The tunnels are 1 to 2 feet in diameter, and several passages may branch from a main tunnel. The den chamber is approximately 3 feet in diameter, situated at the end of the tunnel. It is not lined with nesting material but is kept clean by the adults. After the pups are born, small balls of rolled fur from the female's abdomen may be found at the entrance of the den.

Man and dogs are the chief predators of coyotes. These animals have remarkable recuperative powers. Coyotes have been caught with the lower jaw shot off, with healed bullet wounds, and with peg legs. Their parasites include ticks, mites, lice, cestodes, nematodes, and trematodes, and they are susceptible to distemper, mange, tularemia, rabies, and heartworms.

**Behavior.** Coyotes are active throughout the year. They are chiefly nocturnal but may be seen at dawn and dusk and infrequently during daylight. They are sociable and may be encountered singly, in pairs, or in groups of three or more hunting prey. The senses of hearing, sight, and smell are well developed.

When two or more coyotes hunt together they normally travel single file, often on the well-packed trails of snowshoe hares or deer, though they also follow their own runways (Ozoga and Harger 1966). Coyotes usually establish scent posts along their runways and urinate on stumps, small bushes, and rocks, and at feeding and digging sites. They will defecate on elevated spots such as open ridges, small knolls, and even beaver lodges.

Coyotes seem to howl for pleasure or to call or warn other coyotes. McCarley (1975) reported that adult coyotes produce two basic sounds—the bark and the flat howl. These sounds are given by coyotes in both single and group situations. In addition, the "yip, yip," short howl, warble, laugh, and irregular howl are common group vocalizations. Immature coyotes produce screams and gargles.

The home range may be at least 5 miles in diameter when food is plentiful and may extend 20 to 25 square miles in winter (Ozoga and Harger 1966). They often travel along established lanes that may cover 10 linear miles. They usually trot but may gallop when pressed. Coyotes normally run as fast as 25 to 30 miles per hour, but when being chased they can run 35 miles per hour (Cottam 1945) or even as fast as 43 miles per hour for a short distance (Zimmerman 1943). Coyotes are strong swimmers. Bryant (1920) recorded that a coyote swam a channel 30 feet wide, and Couch (1932) recorded coyotes swimming the Columbia River in Washington at a point where the river was nearly half a mile wide. They will not hesitate to wade in marshes when preying on waterfowl or other animals.

When coyotes stalk rabbits and other small animals they creep up stealthily for some distance, much like a pointer or setter dog, "freeze" momentarily, then pounce. Two or more coyotes may chase an animal in relays. They usually attack from the front and bite the victim at the neck or throat, cutting the jugular vein. Coyotes often bury surplus food. They use the nose to cover the food with dirt, returning later for another meal.

Coyotes are opportunists and eat a wide variety of mammals, including carrion. White-tailed deer carrion is probably the main food of coyotes in New England. In areas where deer are abundant, they doubtless prey on fawns. Coyotes also eat birds, snakes, frogs, lizards, turtles, fishes, crayfish, and insects, as well as fruits, berries, and other plant material.

Very little is known about the reproduction of coyotes in New England. They do not normally pair for life except probably in areas where their populations are low. Some pairs may stay together for several years, but as a rule they pair for only a year. The breeding season varies over a 2-month span owing to the great range of latitude over which coyotes occur and to other environmental factors. In the northern part of the coyote's range breeding occurs mainly during February. The gestation period is 60 to 65 days and the pups are whelped in April or early May. Yearling females breed later than older females, and in Kansas most litters are born in May (Gier 1957). The single annual litter consists of four to eight pups, usually five to seven, though some litters may comprise nine to twelve, and infrequently up to nineteen pups. Occasionally two females, a mother and her daughter of the previous year, share a den with their litters of different sizes (Young and Jackson 1951). These authors suggested that when a den contains two litters of different sizes, "there is usually only one male which would suggest that polygamy occasionally occurs."

At birth the pups are blind and helpless, covered with dark tawny or yellowish brown short, woolly hair, darker on the face, ears, back, and tail. They weigh about 227 g (8 oz) and double their weight by 1 week old. The pups are able to crawl when 2 or 3 days old. They can walk at 8 to 10 days and run fairly well before they are 1 month old. The eyes open at 10 to 14 days. The ears begin to point at about 3 weeks. The pups are nursed for nearly 2 weeks, and at about the time the eyes open they are able to eat partly digested food regurgitated from the stomachs of the parents. The pups are able to eat solid food by the age of 4 weeks. They emerge from the den when 3 to 4 weeks old and are weaned at about 9 weeks of age. About this time the family abandons the den, and the pups are taught to hunt. The family moves about but stays together until early fall (Gier 1957).

Laboratory-raised coyotes kept under good conditions matured in their first year (Gier 1957), and one pair of coyote pups produced a litter the following spring (Snow 1967). Captive New Hampshire coyotes whelped litters when they were 2 years old (Silver and Silver 1969). Coyotes have been known to live up to 10 to 12 years in the wild and 18 years in captivity (Manville 1953).

Bekoff and Jamieson (1975) gave data on the physical development in coyotes, comparing them with the maned wolf, bush dog, cape hunting dog, and fennec.

**Age Determination.** Gier (1957) described a method of determining the age of coyotes from 1 to 8 years old from the wear patterns of the incisors and canines. Rogers (1965) used the wear pattern on the first upper molar and the closure of the presphenoid-basisphenoid suture to determine age, but the suture closure provides determination only up to 22 months of age. Tooth wear and suture closure may be subjective measures because they vary markedly between different geographical areas. However, Linhart and Knowlton (1967) found that the permanent canines erupt at about 4 to 5 months of age and the root canal closes between the 8th and 9th months. At about the 20th month, the first annular layer is formed in the root and one layer is formed each year thereafter.

**Specimens Examined.** MAINE, total 5: Oxford County, Bucksfield 1; Piscataquis County, Lily Bay Township 1; Somerset County, Concord 1, Enchanted Township 1; York County, Limerick Township 1 (USNM).

NEW HAMPSHIRE, total 1: Coos County, Berlin 1 (UVT).

VERMONT, total 1: Chittenden County, Bolton 1 (UVT).

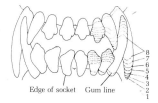

Age determination of coyotes by tooth wear. Frontal view of incisor and canine teeth. Composite drawing made by superimposing pictures of coyotes taken in February in Kansas. Dotted lines show extent of annual wear.

MASSACHUSETTS, total 1: Berkshire County, Otis 1 (HMCZ).

CONNECTICUT, total 14: Hartford County, Farmington 1; Litchfield County, Colebrook 1, Cornwall 1, Goshen 1, New Milford 1, Washington 1; Middlesex County, East Hadden 1; New London County, Stonington 1; Tolland County, Ellington 1, Stafford 2, Union 1, Willington 1; Windham County, Ashford 1 (UCT).

# Red Fox

*Vulpes vulpes* (Desmarest)

*Canis fulvus* Desmarest, 1820. *Mammalogie*, pt. 1, p. 203. In *Encyclopédie méthodique.*
*Vulpes vulpes*, Churcher, 1959. *J. Mammal.* 40(4):513–20.
Type Locality: Virginia
Dental Formula:
  I 3/3, C 1/1, P 4/4, M 2/3 × 2 = 42
Names: American red fox, black fox, cross fox, and silver fox. The meaning of *Vulpes* = a fox.

**Status in North America.** The indigenous red fox was native to North America north of lat. 40° N or 45° N but was either scarce or absent from most of the unbroken mixed hardwood forests where the gray fox was dominant. The European red fox was introduced into the eastern seaboard area about 1750 and either partially displaced the gray fox in the southern portion of the continent or interbred with the scarce population of indigenous red foxes to produce a hybrid population. Holearctic red foxes consist of a single clinal system running from Europe through Asia into North America, with intergradations from population to population. Both the North American and Eurasian red foxes are considered conspecific (Churcher 1959). Churcher proposed that "the North American, Asian and European red foxes be considered to be one species, all the presently described forms being subspecies of this single species. This form must be known as *Vulpes vulpes* Linn. and *Vulpes fulva* (Desmarest) must be considered to be a synonym."

**Description.** The red fox looks like a small collie except for color. The muzzle is elongated and pointed, and the ears are prominent, well-furred, pointed, and erect. The pupil of the eye is vertically elliptical. The pelage is dense and soft, and the tail, which is as long as the head and body, is round and bushy and carried upright. It serves as a balance in walking and helps keep the face and feet warm when the fox sleeps. The legs are moderately long, with small feet. The forefoot has five toes and the hind foot four, with long, blunt, and nonrectractile claws. Females have eight mammae—two pectoral, four abdominal, and two inguinal.

The skull is moderately large. The dorsal surfaces of the postorbital processes are concave and thin, and the indistinct temporal ridges are closely parallel or converge to form a sagittal crest which looks like a narrow V on top of the

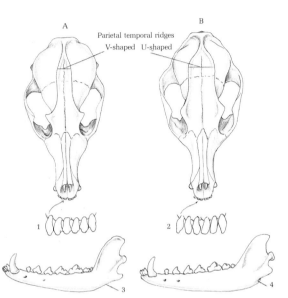

Diagnostic characteristics separating the skull of the red fox, *A*, and the skull of the gray fox, *B*. In the red fox, the parietal temporal ridges form a narrow V-shaped area and may converge to form a weak sagittal crest. In the gray fox, the ridges form a wide U-shaped area and do not form a sagittal crest. In the red fox, the upper incisors are lobed, *1*, and the lower jaw lacks the angular notch on the bottom posterior edge, *2*. In the gray fox, the upper incisors are not notched, *3*, and the lower jaw possesses an angular notch, *4*.

skull. There is a shallow depression below the frontal region of the long, slender rostrum. The auditory bullae are prominent and long. The ventral margin of the mandible lacks a distinct notch or step. The upper incisors are lobed on the cutting edge, and the canines are long and narrow.

The sexes are colored alike, and there is little seasonal color variation except that in winter the pelt is more lustrous. The coloration is bright yellowish red on the sides to deep yellowish brown on the back, darkest on the middle of the back and shoulders. The rump is grizzled to dull white. The face and neck are dusky, the nose, backs of the ears, legs, and feet are black, and the front of the ears, cheeks, and throat and the

RED FOX

midline of the belly are white. The tail is reddish fulvous mixed with black hairs, and the tip is white except in melanistic mutants.

Color phases include the cross fox, silver fox, black or melanistic fox, samson fox, and bastard fox. The cross fox is mostly reddish brown and has much more black on the legs and in the underfur. On the back and shoulders, the darkened guard hairs form a black band down the middle of the back and another across the shoulders, forming the cross from which the fox gets its name. The silver fox is black but is frosted with white-tipped guard hairs that give the silver effect. The black or melanistic fox is all black. The samson fox has a woolly pelage without guard hairs and was named after the biblical Samson and his infamous haircut. The samson pelt is considered commercially worthless. The bastard fox is similar to the red fox but has a smoky appearance and much more black on the belly and legs. The darker color phases are much more common in northern areas of the red fox range. Albinos are extremely rare.

Males are generally larger than females. Measurements range from 90 to 105 cm (35.5–41.4 in); tail, 32 to 40 cm (12.6–15.8 in); and hind foot, 14 to 18 cm (5.5–7.1 in). Weights vary from 3.6 to 5.4 kg (8–12 lb). An unusually heavy red fox may weigh 6.8 to 7.7 kg (15–17 lb).

**Distribution.** The red fox occurs over most of North America from Baffin Island, Canada, and Alaska to the southern United States, except for coastal western Canada, Oregon and California, the great plains, the southwestern desert, and the extreme southeastern United States.

**Ecology.** Red foxes inhabit diverse habitats of broken mixed hardwoods, rolling farmlands, woodlots, sparsely wooded areas, pastures, brushlands, and borders of open areas that are suitable hunting grounds. They avoid open, brushless plains as well as dense forests, preferring sparsely settled and open country with moderate cover, such as farmlands. Red foxes follow regular routes across low rolling hills, valleys, and ravines, across marshes and streams, through swamps, and along waterways.

Red foxes may dig their own dens, but most often they enlarge an abandoned or appropriated woodchuck burrow, a former red-fox den, or the burrow of another mammal. Dens may be situated in wooded areas in banks, gullies, or fencerows, under hollow logs, old sawdust piles, or abandoned buildings, or on level ground in loose, sandy loam or gravelly soils where digging is easy and drainage good. Most dens in deep woods are old, established ones that probably have been used for generations. The chamber is merely a widening of the burrow, usually lined with grass for the pups. Sometimes there are additional chambers where food is stored or where pups can be moved in case of danger. The den itself is kept clean of feces and food remnants but the entrances are not.

The den is at least 4 feet underground in a complex tunnel system that is usually 25 feet long but may extend to 75 feet or more. There may be several entrances to a den, generally not concealed. They are about 8 inches wide and 15 inches high and usually face south. Several trails generally lead to an entrance from different directions.

In New England, man and dogs are the main predators of red foxes, though great horned owls are known to take unwary pups. A fox may die from becoming trapped in a tree cavity or from ingesting porcupine quills. Rabies kills many foxes. Canine distemper, heartworms, tularemia, rickets, Chastek paralysis, salmon disease, coccidiosis, scarcoptic mange, larvae of screwflies, *Wahlfahrtia* spp., and septicemia plague red foxes. Fleas, ticks, cestodes, nematodes, and trematodes parasitize them.

**Behavior.** Red foxes sleep outside, even in zero weather, and use a den mainly during the denning season or as a refuge. They are chiefly nocturnal but may be seen hunting mice at dawn and dusk and occasionally during daylight. They are mild-tempered animals but prefer to be solitary except during the denning season. Their senses of smell, sight, and hearing are well developed. Red foxes are curious and readily approach artificial and recorded distress calls of their prey. In a controlled test, Isley and Gysel (1975) determined that nine red foxes located thirteen different frequencies of pure sounds, ranging from 300 hertz to 34 kilohertz. The foxes located the sound source best from 0.9 to 14 kilohertz, with a slight decrease in accuracy at 8.5 kilohertz.

The voice of the red fox varies from a short, sharp yap or bark followed by a "yap, yap" to a combination of long howls, screeches, and yells.

Red foxes are excellent trotters, and when pursued they can run up to 26 miles per hour for short distances. They usually circle or backtrack to elude dogs rather than stand and fight (Seton 1929).

When hunting, red foxes often trot cautiously back and forth with nose to the ground, or they may crouch low, scarcely moving. When the unwary prey is scented, the fox pauses for several seconds with ears erect and body tense. Then with a spring, rush, or pounce, it seizes its prey and kills it with the powerful jaws. Murie (1936) reported that a red fox pounced 4 feet onto a clump of grass where it probably captured a mouse in 5 to 6 inches of snow. Red foxes are good swimmers and float high in the water. Occasionally they wash themselves by licking their fur. They often roll on strong-smelling carcasses.

We do not always know why certain animals tolerate the presence of other species. Red foxes have been seen making friends and playing with dogs. Seton (1929) cited Charles Sheldon, who observed a red fox walking along a trail with some mountain sheep in Alaska, the fox jumping up and biting the sheep's faces in play, while they butted it gently along in front. When the rams lay down to rest, so did the fox. Instances of friendship between foxes and caribou in Newfoundland have been repeatedly observed. Foxes have been seen running among caribou, leaping and nibbling at their muzzles, brushing and snapping their flanks. Sometimes a fox rests in the middle of a herd of caribou (Dickey 1923; Gianini 1923; Smith 1923). Other unusual interspecific relationships of red foxes include instances of foxes and woodchucks living together. Merriam (1963) observed woodchucks feeding without apparent alarm to within 30 or 40 feet of a litter of playing fox pups.

Red foxes normally restrict their movements to a home range shared by a male-female pair and, seasonally, by their pups (Scott 1943; Sargeant 1972). Vincent (1958), Barash (1974), and Preston (1975) suggested that mutual aggression in early fox encounters progresses to dominance-subordinance polarity as the animals become familiar with each other. Preston (1975) found that resident pairs centered their activities on a den and that the frequency of intruder–resident fox encounters decreased rapidly with increasing distance from the den. Also, the primary home-range defense was continual harassment of male intruders by the resident males through agonistic displays and chases. Physical contact was rare, and resident females seldom interacted with the intruders.

The home range size of nondispersing red foxes is less than 5 miles in diameter (Scott 1943; Storm 1965; Ables 1969; Sargeant 1972). Phillips et al. (1972) found that approximately 70 percent of marked and recovered juvenile male foxes moved more than 5 miles from their natal ranges during their first year, and that about 30 percent of the recovered juvenile females dispersed more than 5 miles in their first year.

Red foxes are opportunists and will eat anything that is readily available, including carrion. They feed on small mammals, birds and their eggs, insects, earthworms, turtles and their eggs, frogs, and snakes. They also eat wild berries, sarsaparilla, grapes, plums, and apples. Nuts, including acorns, and corn and other grains are eaten infrequently. They sometime ingest rope, twine, paper, sticks, and trash. They generally bury their surplus food or cover it with grass or leaves, sprinkling it with urine.

The breeding season occurs from mid-January to late February, occasionally extending into March, and peaks about late January. Vixens have a short estrous period of 2 to 4 days. The gestation period is 51 to 56 days, averaging 53 days. The single annual litter consists of from one to ten, usually four or five, pups born in late March or early April. A litter may occasionally become mixed with other families, suggesting communal denning and a high degree of intraspecific tolerance during the denning season. As many as five litters have been found within 200 acres (Sheldon 1950). Red foxes may be monogamous, remaining paired for life, but unusually large litters may indicate polygamous tendencies (Sheldon 1949).

At birth the pups are blind and helpless, weighing from 71 to 120 g (2.5–4.2 oz), and are covered with dense, fine, dull grayish brown hair. The tail is white-tipped. The eyes are closed until the 9th day of life. In 3 weeks the pups are able to walk. From 5 to 8 weeks of age the coloration is pale tan, except the legs, chest, belly, and tail; the woolly appearance changes between 9 and 14 weeks and the pelage on the body and tail becomes reddish brown and the lower legs black. In September through October the pelage of the molting pups is much glossier than the duller summer coats of the adults. After 15 to 16 weeks the milk teeth are replaced by the permanent teeth (Sheldon 1949).

The pups remain in the den until they are 4 to 5 weeks old. They play about the den entrance with bones, feathers, skins, and leftover food items brought in by the adults. When the adults move the pups they usually also take the play items. Both adults watch the pups closely. The male brings food for the nursing vixen during the first few days after the pups are born; later the female hunts at night and nurses the pups during the day. The parents train the pups to hunt when they are weaned at about 12 weeks old. By this time the pups are nearly three-fourths grown, and they soon leave the den and become almost independent. They remain part of the family unit and may roam within a mile of the den site for the next month or so. By mid-September they are almost fully grown. In autumn the pups begin to shift for themselves fully, the males dispersing first (Sheldon 1949). Sheldon (1950) found that banded male pups dispersed over 15 miles. Ables (1965) reported that a juvenile male fox was taken 245 miles from the site where it was first captured 9 months earlier. Phillips and Mech (1970) reported that a tagged female returned home after traveling 35 miles in 12 days. Red foxes may become sexually mature the winter after their birth.

**Age and Sex Determination.** Red and gray fox pups up to 8 and 9 months old can be distinguished by the distinct epiphyseal cartilage at the wrist end of the radius and ulna, which is not fused with the shaft of the bones when seen on X rays. In January and February the cartilage becomes ossified, and in adults the cartilage is completely ossified and fused with the shaft of the bones. Vixens that have pups can be recognized by their large, dark-colored nipples. Young females have smaller, light-colored nipples less than 1/16 inch in diameter (Sullivan and Haugen 1956).

Churcher (1960) determined that the canine teeth of red foxes erupt throughout life. Cementum deposits progressively form on the roots of the canine teeth, and their points are simultaneously worn away. Hence the tooth changes little in length, and any fixed point on its surface moves away from the alveolus and toward the outer tip. The proximal edge of the enamel forms a determinable fixed point, and the distance between it and a standard point on the edge of the alveolus is related to the age of the individual. Thus the age of the animal can be determined by counting cementum annuli in sectioned canine teeth. Allen (1974) found 100 percent agreement between known ages and ages of red foxes assigned by counting cementum annuli.

**Specimens Examined.** MAINE, total 35: Aroostook County, Sherman 1 (HMCZ); Franklin County, New Sharon 1 (UME); Hancock County, Bucksport 3, West Pembroke 1; Oxford County, Norway 7, Umbago Lake 1, Upton 12 (HMCZ); Piscataquis County, Brownsville 1 (USNM); Somerset County, Enchanted Pond 1, Moose River 1 (AMNH); Washington County, Jonesboro 2, Machias 3 (HMCZ); York County, Eliot 1 (USNM).

NEW HAMPSHIRE, total 6: Carroll County, Ossipee 1 (USNM), Pequaket 1 (HMCZ); Grafton County, Etna 1 (DC); Hillsborough County, Peterboro 1; Merrimack County, Pittsfield 1 (HMCZ); Strafford County, Durham 1 (UNH).

VERMONT, total 40: Addison County, Middlebury 1 (UVT); Bennington County, Readsboro 1 (AMNH); Caledonia County, South Ryegate 1 (USNM); Chittenden County, Burlington 1 (HMCZ), Kirby 2, Richmond 1, Shelburne 3 (UVT); Franklin County, Enosburg Falls 2 (HMCZ); Orleans County, Glover 1 (UVT); Rutland County, Killington Peak 1 (HMCZ), Mendon 1 (UCT), Rutland 1; Windham County, East Dummerston 1, Newfane 20 (AMNH); Windsor County, Norwich 2, Pomfret 1 (DC).

MASSACHUSETTS, total 27: Barnstable County, Brewster 1; Essex County, Danvers 1, Ipswich 1, Wenham 1 (HMCZ); Franklin County, Conway 1; Hampden County, Blanford 1; Hampshire County, Amherst 2, Hadley 1 (UMA); Middlesex County, Concord 1, Hudson 1, Littleton 1, Weston 1 (HMCZ), Wilmington 1; Norfolk County, Canton 1 (USNM); Worcester County, Berlin 2, Boyleston 1, Clinton 1 (UMA), Hardwick 2 (HMCZ), Rutland 5, Shrewsbury 1 (UMA).

CONNECTICUT, total 21: Hartford County, Bloomfield 1, Farmington 2 (UCT); Litchfield County, Salisbury 1 (HMCZ); Middlesex County, East Hampton 1, Portland 5 (UCT); New Haven County, Southbury 1; New London County, Liberty Hill 1 (HMCZ); Tolland County, Columbia 1, Coventry 1, Mansfield 1, Storrs 3, Vernon 1, Willington 1; Windham County, Ashford 1 (UCT).

# Gray Fox

*Urocyon cinereoargenteus*

There are two subspecies of gray fox recognized in New England:
*Urocyon c. cinereoargenteus* (Schreber). *Canis cinereo argenteus* Schreber, 1775. *Die Säugthiere* . . ., part 2, no. 13, pl. 92. *Urocyon cinereo-argenteus* Rhoads, 1894. *Amer. Nat.* 28:524
Type Locality: Eastern North America
*Urocyon c. borealis* Merriam, 1903. *Proc. Biol. Soc. Washington* 16:74.
Type Locality: Marlboro, 7 miles from Monadnock, Cheshire County, New Hampshire (Hall and Kelson 1959)
Dental Formula:
I 3/3, C 1/1, P 4/4, M 2/3 × 2 = 42
Names: New England gray fox, northern gray fox, eastern gray fox, mane-tailed fox, tree fox, wood fox, silvery fox, and gray-backed fox. The meaning of *Urocyon* = tailed dog, and *cinereoargenteus* = ashy silvered.

**Description.** The gray fox is somewhat smaller than the red fox, with a shorter muzzle and legs, and is a different color. The long black-tipped, bushy tail is triangular in cross section and has a mane of short, stiff black hairs along the upper surface. The claws of the forefeet are more curved than those of the red fox, an adaptation for climbing. Females have eight mammae—two pectoral, two abdominal, and four inguinal.

The skull is distinguished from that of the red fox by the widely separated, prominent temporal ridges which form a lyre- or U-shaped pattern but do not unite into a sagittal crest. The incisors are not lobed, and the ventral margin of the mandible has a distinct step. The rostrum is relatively short and the braincase enlarged. The dorsal surface of the postorbital processes has deep pits.

The fur is coarse and dense. The sexes are colored alike, with little seasonal color variation. The coloration of the upperparts is grizzled gray, from a mixture of black and white bands on the black-tipped guard hairs. The sides of the neck, back of the ears, a band across the chest, the inner and back surfaces of the legs, the feet, the sides of the belly, and the undersurface of the tail are reddish brown, with color extent and intensity variable. The cheeks, throat, inner side of the ears and greater part of the belly are white, while the nose pad and chin are black. The muzzle and rims around the eyes are blackish. The sides of the nose and the area above and below the eyes have light brown patches. The tail is black above, reddish brown below.

The only albino gray fox on record is a specimen taken in New York (Shipherd and Stone 1974). Jones (1923) reported black variants. The single annual molt occurs in summer, and the molting line is diffuse.

The sexes are about equal in size. Measurements range from 90 to 104 cm (35.5–41.0 in); tail, 30 to 39 cm (11.8–15.4 in); and hind foot, 13 to 15 cm (5.1–5.9 in). Weights vary from 3.1 to 5.9 kg (7–13 lb). Unusually heavy specimens may weigh as much as 8.6 kg (19 lb).

**Distribution.** The gray fox occurs from southern Canada throughout the United States, except in Montana, Idaho, Wyoming, and most of Washington. It ranges into Mexico and Central America. Although the ranges of gray and red foxes overlap, the gray fox is essentially a southern and western species, whereas the red fox is a northern species. However, the gray fox has

recently extended its range northward to southeastern Canada.

The gray fox is native to New England based on bones exhumed from archaeological sites (Waters 1964). The species occurred in small numbers as far northeast as Cumberland County, Maine, in colonial times (Palmer 1956). The fox was rare in Vermont until 1916, but since then it has been recorded from most of the counties of the state (Foote 1946). In New Hampshire, occasional gray foxes were seen about 1918 in the southwest corner of New Hampshire. Gray fox populations peaked between 1945 and 1950 in Cheshire and Hillsboro counties (Silver 1957). Gray fox populations have increased and reoccupied much of their present range in New England north of Connecticut since 1930 (Palmer 1956). The northward spread of the gray fox has corresponded to that of the cottontail rabbit, which is regarded as an important fox food (Hamilton 1943).

**Ecology.** The gray fox is typically a forest mammal, inhabiting rough, rocky terrain in dense hardwood and mixed coniferous and hardwood forests, swamps, dense thickets and briers, where there is adequate water nearby and cover for hunting and hiding. This fox rarely digs its own den and usually does not use a burrow like the red fox, but has a series of dens concealed in dense cover under logs, in cavities of stumps and hollow trees, crevices in rocks, fissures in cliffs, or in caves, or under abandoned buildings, brush, sawmill piles and even in a discarded ten-gallon can in a dump. The den may be lined with leaves, grass, fur, pieces of bark, or other soft material.

Man and dogs are the chief predators of gray foxes. The bobcat, the larger hawks, and the great horned owl occasionally catch unsuspecting pups. Gray foxes suffer from heartworms, rabies, canine distemper, mange, listeriosis, and tularemia, and are hosts to fleas, lice, ticks, cestodes and nematodes.

**Behavior.** Gray foxes are more secretive and shier, but less cunning and wary and easier to trap, than are red foxes. Gray foxes are mainly nocturnal but may be seen foraging by day.

Gray foxes can attain a speed of about 26 miles per hour but 20 miles per hour is more normal for short distances (Cottam 1937). These foxes are skillful tree climbers. They climb very fast by shinnying up the trunk to a limb, and will jump from branch to branch while ascending. They do not hesitate to climb trees to hunt prey, and when pursued by dogs, gray foxes quickly go up a tree or hide in a den.

The voice of the gray fox is similar to that of the red fox, but louder. Gray foxes are generally quiet and bark or yap less than red foxes.

The home range and movements of gray foxes vary with food supply, denning activity, disturbance by man and dogs, and season. The animals may be confined to an area a mile wide during the denning season and may range an

GRAY FOX

area about 5 miles wide in the fall dispersal. In New York returns of two juvenile vixens which were banded as pups, one female traveled 11 miles in 36 days and the other traveled 52 miles in 28 months (Sheldon 1953).

Like red foxes, gray foxes are opportunists and eat a wide variety of small mammals, birds and their eggs, frogs, turtles, snakes, insects, corn, berries, grapes, apples, acorns, and carrion. In Massachusetts MacGregor (1942) found that one gray fox stomach contained 400 cc (24.4 cu in) of skunk flesh. Foods that are not eaten are generally left in the open.

The breeding season is from mid-January to May, peaking in early March, but latitude affects its timing. Gray foxes in the south breed a month earlier than those in the north (Layne 1958). The gestation period is from 51 to 63 days, averaging 53 days. The single annual litter consists of two to seven pups, usually three or five. At birth the pups are blind and helpless, blackish, and almost hairless. They are born in a den in March or April, depending upon the latitude. The vixen alone tends the pups until they are nearly 3 weeks old, then the male helps by providing some food for the family. The pups are weaned at 8 to 10 weeks of age, and they begin to leave the den and hunt with their parents when they are about 3 months old. The family disperses by autumn, when the pups are nearly full grown. Gray foxes are capable of breeding the first year after their birth. Longevity in the wild may be 5 to 6 years, but some gray foxes have been known to live as long as 14 or 15 years in captivity (Sheldon 1949).

**Age Determination.** As in the red fox, Sullivan and Haugen (1956) could distinguish young of the year from adults by X rays of the wrist end of the radius and ulna. Pups up to 8 or 9 months old can be distinguished by the distinct epiphyseal cartilage, which is not fused with the shaft of the bones. In adults the cartilage is completely ossified and fused.

Wood (1958) classified gray foxes into five distinct age groups by the amount of wear on the conules of the first upper molar. Age can also be determined by the closure of the suture between the undersurface bones of the skull, the basisphenoid and presphenoid. Skulls with open or recently closed sutures are regarded as animals 0 to 12 months old. The vomer is developed and continuous with the presphenoid after 24 months.

Lord (1961) described the technique of weighing dried eye lenses to determine age. He found a positive correlation of .78 between the lens-weight and tooth-wear techniques when both were used to estimate the year of birth, and a positive correlation of .90 when both techniques were used to distinguish between juvenile and adult gray foxes.

**Specimens Examined.** NEW HAMPSHIRE, total 3: Grafton County, Enfield 2, Lebanon 1 (DC).

VERMONT, total 3: Addison County, Shoreham 1 (UCT); Orange County, Fairlee 1 (HMCZ); Washington, County, Plainfield 1 (UVT).

MASSACHUSETTS, total 6: Essex County, West Gloucester 1 (HMCZ); Hampden County, Mt. Wilbraham 1 (AMNH); Hampshire County, Amherst 1, Belchertown 1, Sunderland 2 (UMA).

CONNECTICUT, total 32: Fairfield County, Stamford 1 (AMNH), Trumbull 1; Hartford County, Bristol 1, Canton 1, Farmington 1 (UCT), Hartford 2 (HMCZ); Litchfield County, Bantam 1 (UCT), Colebrook 1, East Canaan 1, Gaylordsville 1, Nepaug 4 (HMCZ), Sharon 1 (AMNH), Winchester 1 (HMCZ); Middlesex County, Portland 2; New London County, East Lyme 1, Sprague 1; Tolland County, Mansfield 4, Merrow 1, North Eagleville 1, Storrs 1; Windham County, Ashford 1, Chaplin 1, Scotland 1, Windham 1 (UCT).

### FAMILY URSIDAE (BEARS)

The bears, or ursids, are the largest land carnivores in North America. They are sluggish and walk with a shuffling, slow plantigrade gait. Yet they are agile and can walk on their hind feet. Bears are usually solitary and wander extensively, especially polar bears.

Bears are usually peaceful animals, but when provoked, they can be dangerous adversaries.

The family is represented in the northern hemisphere and in northern South America. The characteristics of this family are exemplified by the only representative occurring in New England, the black bear.

# Black Bear

*Ursus americanus* Pallas

*Ursus americanus* Pallas, 1780.
  *Spicilegia zoologica* . . . , fasc. 14,
  p. 5.
Type Locality: Eastern North America
Dental Formula:
  I 3/3, C 1/1, P 4/4, M 2/3 ×2 = 42
Names: American black bear, com-
  mon black bear, bruin, and bruno.
  The meaning of *Ursus* = a bear,
  and *americanus* = of America.

**Description.** The black bear is the largest land carnivore in New England. The animal is thickset, with short, rounded, erect ears, well-haired short tail, powerful legs, and a nearly straight facial profile and long muzzle. The broad feet each have five toes with relatively short claws. The front claws are somewhat curved and slightly longer than the hind claws. Females have six mammae—four pectoral and two inguinal.

The skull is relatively wide and heavy, and more or less elongated with the upper profile rounded. The rostrum is short, the nares are large, and the facial profile is straight. The auditory bullae are flat and depressed, and the alisphenoid canal is present. The small orbits are incomplete posteriorly. The palate extends considerably beyond the last molar. The incisors are small, but the canines are elongated, conical, and slightly hooked. The carnassials are weakly developed, and the first three premolars of each jaw usually are rudimentary, single rooted, and frequently lost; the second is rarely present in adults. The flat tubercular crowns of the molars are longer than broad, and the last molar is much larger than the one in front of it.

The pelage is thick, long, and moderately soft. The sexes are colored alike, and seasonal variation is chiefly a change in the length and thickness of the fur. The normal coloration of the entire body is glossy black or brownish black, except for the muzzle, which is tinged with tan. There is usually a small white patch on the throat. Brown color phases occasionally occur. Cinnamon and blue color phases are rare, but individuals of these two phases retain their general color throughout life. These color phases may occur in the same litter. Albinos are extremely rare. Molting usually begins in April or May and is completed about October.

Males are slightly larger than females. Measurements range from 127 to 178 cm (50.0–70.1 in); tail, 8 to 13 cm (3.1–5.1 in); and hind foot, 19 to 28 cm (7.5–11.0 in). Weights vary from 102 to 272 kg (225–600 lb).

**Distribution.** The black bear formerly occurred over most of wooded North America from nearly all of Alsaka to Canada, the United States, and the central northern Mexico. Jonkel and Miller (1970) reported black bears on the barren grounds of Canada, but the species has been extirpated from most of the north central and central United States.

In New England the black bear occurs in the wilder woodland areas. It is fairly common in most of Maine, in northern and central New Hampshire, and in Vermont. It occurs in western Massachusetts and northwestern Connecticut but probably not in Rhode Island.

**Ecology.** This species inhabits forested wilderness and swamps. It prefers mixed stands of conifers and hardwoods, usually with dense, brushy understory, near streams, ponds, and lakes. The winter den may be in a hollow log, cave, or crevice in a rocky ledge, under fallen trees and brush, in a cavity under the roots of a large tree, in the base of a hollow tree, under windfalls, slash, or drainage culverts, in a depression dug in the ground, or in any place that affords protection from cold and snow. The den is sometimes lined with stripped bark, leaves, grasses, and moss. Females are more selective about a den site than males.

Man and dogs are the chief predators of black bears. In Vermont, Willey (1971) found that male bears of all ages were more vulnerable to guns than females because they were less wary and frequented open areas more. Forest fires and automobiles occasionally kill black bears, and some apparently die in their winter dens. Adult males sometimes kill cubs. Erickson (1957) reported that a cub caught in a trap was killed and eaten by an adult black bear, but he believed that large bears preyed on smaller bears only during the spring when they were extremely hungry. Jonkel and Cowan (1971) stated that black bears ate the remains of dead bears when the opportunity arose. Black bears are hosts to fleas, ticks, lice, cestodes, and nematodes, but diseases seem to be uncommon among them.

**Behavior.** Black bears are nocturnal but may sometimes be seen foraging or wandering during the daytime. They are alert and wary, tend to avoid open spaces, and usually dash off at the first sign of danger. They rarely attack men except when provoked, wounded, or cornered, except for females with cubs. Apart from mothers and cubs these carnivores are usually solitary and as a rule are unsociable. When two or more bears meet at a garbage dump they usually avoid contact, though they make various threat displays and dominant-subordinate recognitions. Jonkel and Cowan (1971) indicated that black bears exhibit territorial behavior.

Black bears move with an awkward or shuffling slow gait, but Jackson (1961) reported that they can run up to 32 miles per hour for short distances. Black bears sometime rise on their hind feet to look around long before retreating. They drink freely and enjoy bathing and wallowing in water in hot weather. They swim well, usually in a straight line; they seldom change course and will crawl over an obstacle in their path. Payne (1975) reported that three adult black bears returned to their original capture site after swimming some 0.621 mile in Newman Sound, Newfoundland.

Black bears are expert tree climbers. They may climb a tree just to relax or nap, but most often they climb to get nuts, honey, or other plant food. They climb mainly with the hind feet, which are brought forward under the belly, then extended fully backward. The front feet are used for grasping and balance. They descend rear end first.

Black bears are generally silent and have no regular call notes. On occasion a bear may growl or "woof" at other bears or other animals, and in times of stress it bellows. An injured bear bawls and sobs like a human, and females with cubs often make low grunts, huffs, mumbles, or squeaks. Cubs squall when hungry or frightened and whimper when lost. They purr when content or seeking comfort.

Black bears wander considerably. Their home range is imperfectly known, but undoubtedly it varies with available food, breeding season, harassment by man and dogs, and other factors. Seton (1929) gave the extent of a bear range as 15 miles, and Cahalane (1947) reported a 15-mile radius for adult males, and a 10-mile radius for a female with cubs. Spencer (1961) indicated that the average home range was about 5 miles in radius, or about 78 square miles. Stickley (1961) reported that a tagged 2.5-year-old bear traveled 90 miles. Erickson, Nellor, and Petrides (1964) estimated the minimum summer and annual ranges as approximately 6 and 15 square miles, respectively, and determined that the average movements of recaptured bears were 2.8 miles in summer and 6.7 miles in autumn, with a maximum movement of 19.4 miles. The authors noted that adult males traveled more widely than adult females and that dispersal of young bears was limited. Sauer, Free, and Browne (1969) reported that average movements were 8.6 miles for males and 7.4 miles for females and that a male bear returned to its original capture site after being released 56 miles away. Jonkel and Cowan (1971) determined that the greatest black bear movements averaged 3.9 miles for adult males and 1.6 miles for adult females, and that dispersal of subadults involved mostly young males. The authors indicated that males tend to have more of a seasonal shift within their home ranges than do females. Drahos (1952) reported mass movements of bears during periods of food scarcity, and Harlow (1961) suggested that black bears move into areas when acorns are abundant.

Black bears form well-beaten trails, and along these trails the bear marks any prominent tree by standing on its hind legs, reaching as high as it can, and clawing and biting off a large piece of bark and wood so that distinct marks are made. Such "bear trees" may be associated with territorialism or the breeding season, or perhaps they are simply used to sharpen the claws or to serve as a signpost on which passing bears record their presence. Both sexes mark trees, and the same tree may be marked by several bears.

BLACK BEAR

Black bears do not hibernate. Although they go into a deep sleep, their respiration, heart rate, and body temperature are not drastically reduced. They are relatively easily aroused, although they appear drowsy. The fact that falling snow melts on the back of a sleeping bear at low temperatures indicates that their body temperature does not fall to a great extent. Also, females give birth and nurse their cubs during midwinter. Svihla and Bowman (1954) recorded a respiration rate of two to three times a minute for a sleeping caged bear and noted that its body temperature remained uniform at approximately 95°F during a 14-hour period while the bear was under the influence of Nembutal.

Black bears get very fat in autumn in preparation for the winter dormancy. They ordinarily retire when sufficient fat has been accumulated, without regard for food or temperature (Matson 1954). Bears in poor condition den for shorter periods than bears in good condition (Rausch 1961; Spencer 1961). Black bears den alone, except females with cubs. Females usually enter dormancy earlier than males.

Sleeping bears cannot defecate in winter because the rectum is blocked by a "fecal plug" up to 6 inches long, composed of pine needles, bear hair, dry leaves from fall feeding, and mucus from the intestines. The plug is not expelled until spring. Black bears appear to be in good condition when they awake in spring, usually in March or April, but during the next few weeks they usually lose weight rapidly, become thin and ragged, and spend most of the time looking for food. Rogers (1974) found that for some unknown reason black bears shed their foot pads during denning.

Black bears are opportunists and will eat almost anything available, including the carcass of another bear. They feed on grubs, larvae, insects, seeds, large quantities of a great variety of berries, grapes, apples, acorns and beechnuts, succulent leaves of hardwoods, roots of jack-in-the-pulpit, other vegetable matter, frogs and reptiles, mice, fish, carrion, and garbage. They are fond of sweet items such as cake, pie, candy, and honey and may raid food stores of campers. They often rip open rotten tree stumps and bark to find ants and their eggs.

Females, or sows, usually breed once every two years, but estrus can be induced every year if the cubs are taken from the female before the breeding season, which occurs from early June through mid-July. The season peaks about mid-June, depending on the locality. Several bears of both sexes may congregate in one area during the breeding season. It is not known whether a male, or boar, will mate with more than one sow. Mundy and Flook (1964) saw a pair of black bears copulate twice and reported that copulation lasted a very few minutes. The gestation period is 7 to 7½ months. The cubs are born about 1 February. Smith (1946) found three cubs that were several days old near Calais, Maine, on 27 January, and Spencer (1961) reported cubs born in mid-January in Maine. The litter size varies from locality to locality and from year to year. From one to four or five cubs are born per litter, though two are most frequent. Rowan (1947) reported one litter of six cubs.

At birth the cubs are about 152 to 229 mm (6–9 in) long and weigh 170 to 330 g (6–12 oz). The eyes and ears are closed, and the pelage consists of fine, soft, mottled gray hair, though the skin pigment is still visible. The cubs develop slowly at first and are generally uncoordinated until they are about 46 days old. They begin to walk at about 60 days. The pelage becomes fairly dense by the 14th day; white marking appears on the chest, and the ears open fully by 46 days. Two maxillary incisors erupt at 47 days, and mandibular teeth begin to erupt at about 92 days. The eyes open slightly at about 28 days but continue to develop for about a week, and vision appears weak even at 60 days (Butterworth 1967).

The female is protective of her cubs. Erickson and Martin (1960) tracked a disturbed female that had carried one cub approximately 2¾ miles to a new site and returned to fetch her second cub. The cubs were estimated to weigh 3.18 kg (7 lb) each. The young are weaned in August or September when they are about 7 months old (Erickson, Nellor, and Petrides 1964; Jonkel and Cowan 1971). They leave the den at about 3 months of age and may be self-sufficient at 5½ months (Erickson 1959). The cubs remain with the remale as a family group until the second summer at the onset of the breeding season, at which time they disperse.

Females breed at 3½ years of age, and the first cub is born when the female is approximately 4 years old. Rausch (1961) mentioned that captive black bears mature sexually sooner than wild bears. Stickley (1961) reported that a captive 2.5-year-old female produced cubs. Erickson, Nellor, and Petrides (1964) stated that both sexes reached puberty at 3.5 years of age in the wild, but Jonkel and Cowan (1971) noted that no black bears younger than 4.5 years old were observed in estrus, and no tagged females were successful in raising cubs before 6.5 to 7.5 years of age in the wild.

Black bears have been reported to produce cubs at 24 years of age (Drahos 1952). A captive female produced 34 cubs in 13 litters. Longevity is near 25 years in the wild and near 30 in a zoo (Seton 1929).

**Sex Determination.** Sauer (1966a) found that in New York, the sex of a black bear more than 1 year old could generally be determined by the size of the lower canine teeth. The teeth of males are larger than those of females, based on the maximum root width and thickness.

**Age Determination.** Various criteria have been used to estimate the age of black bears, such as cranial suture closure, tooth replacement and wear, and the rugosity of the masseteric fossae (Stickley 1957); comparison of tooth wear with that of known-age skulls from captives (Black 1958); correlation of annulation on the roots of the upper canines with age and baculum development in relation to age (Rausch 1961); and size and shape of the baculum, skull and body measurements, canine cementum layers, tooth replacement and wear, bone development in the forelimbs, and baculum growth and maturation (Marks and Erickson 1966). There have been unsuccessful attempts to correlate body weight with age (Black 1958; Spencer 1961; Erickson, Nellor, and Petrides 1964; Sauer 1966b).

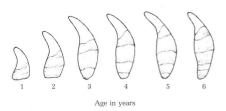

Age in years

Tooth development of black bear in Alaska, taken in November, showing correlation of annulation on the root of upper canine with age. Zones corresponding to periods of activity are laid down annually on the roots of the canines as external, ringlike markings. These annulations demarcate broad zones of dentine laid down in successive seasons of growth.

The most reliable age determination criteria are the external ringlike markings on the roots of the upper canines; the sequence of tooth eruption and root closure (Rausch 1961); and layering of canine cementum (Marks and Erickson 1966; Stoneberg and Jonkel 1966; Willey 1974). The broad zones of dentine are laid down annually on the roots of the canines and represent the boundaries between successive increments of growth.

The annual sequence of the cementum pattern consists of wide light bands alternating with narrow dark bands. A count of these layers provides the age, much like the method of determining the age of a tree by annual rings. The technique involves microscopic examination of decalcified, sectioned, and stained teeth. The wide bands are laid down in the spring and early summer and the narrow bands are added in the fall and winter. The age of a bear can be determined accurately up to 7½ years by examining the cementum layers.

**Specimens Examined.** MAINE, total 2: Hancock County, Aurora 1 (DC); Piscataquis County, Moosehead Lake 1 (UMA).

NEW HAMPSHIRE, total 1: Grafton County, Lyme 1 (DC).

### FAMILY PROCYONIDAE (RACCOONS AND ALLIES)

American procyonids include the raccoons, ringtails, kinkajous, and coatis. These medium-sized carnivores have medium to long bushy tails, usually ringed with alternating light and dark colors. In the kinkajou the tail is prehensile. The ears are small and pointed, and the snout is elongated and flexible, especially in the coatis. Facial markings are usually prominent. The limbs have five flexible toes with semiretractile to nonretractile recurved and compressed claws. The limbs are pentadactyl, and locomotion is bearlike—plantigrade to semiplantigrade.

Procyonids are omnivorous. Most of them are

partially arboreal and are excellent tree climbers. They may be solitary or live in family groups or in bands. Some species den in trees or in rock crevices and caves. These animals are not as vocal as the canids and felids. Their sense of smell is keen, and hearing and vision are well developed.

The procyonids are exclusively American except for the pandas of Southeast Asia (Stains 1967). They are found from southern Canada south through Central America and throughout most of South America. There is only one species of procyonid in New England, and that is the raccoon.

# Raccoon

*Procyon lotor* (Linnaeus)

[*Ursus*] *lotor* Linnaeus, 1758. *Systema naturae*, 10th ed., 1:48.
*Procyon lotor*, Illiger, 1815. *Abhandl. König. Akad. Wissensch., Berlin*, 1804–11, pp. 70, 74.
Type Locality: Pennsylvania
Dental Formula:
I 3/3, C 1/1, P 4/4, M 2/2 × 2 = 40
Names: The meaning of *Procyon* = before the dog or doglike, reflecting its closeness to the primitive stock of the canid-ursid lines, and *lotor* = the washer, from the raccoon's supposed habit of washing its food before eating it. In fact, away from water the raccoon eats its food as it finds it.

**Description.** The raccoon is a robust, medium-sized mammal with a black mask across the eyes and cheeks and a bushy, cylindrical, ringed tail that is shorter than the head and body. The head is broad, with a pointed muzzle, prominent pointed ears, and medium-sized dark eyes. The legs are long and slender; the feet have naked soles, five very long, free digits, the first more than half the length of the second, and nonretractile claws. Females have an os clitoridis and six mammae—two pectoral, two abdominal, and two inguinal. The baculum of the male is well developed.

The skull is broad, with a moderately high, narrow frontal region. The rostrum is short and broad. The braincase is well inflated posteriorly and tapers gradually anteriorly. The sagittal crest is high and trenchant in some old adults, absent in others. The temporal ridges do not unite along the median line. The plate extends well beyond the plane of the last molars. The large auditory bullae are inflated on the inner side, and the postorbital processes of the jugal are well developed. The incisors are more or less grooved. The canines are oval in cross section, and the upper canines are not strongly everted. The molariform teeth are broad, somewhat bunodont, and tuberculate, with weak, rounded cusps.

The pelage is long and thick. The sexes are colored alike and there is little seasonal color variation, though the winter fur is more lustrous, thicker, and longer. The pelage everywhere is dull brown at the base, and the hairs are banded black with lighter rings of gray or yellowish gray and black. The coloration of the upperparts varies considerably, from dull iron gray, to brownish, to blackish, especially on the middle of the back, and more or less buffy or yellowish on the neck and tail. The sides are

grayer and paler. The underparts are thinly overlaid with long grayish or buffy guard hairs that only partly conceal the dense dull brownish or yellowish gray underfur. The top of the head is grizzled. In some individuals, the facial mask is bordered by conspicuous white lines extending from near the middle of the forehead backward under the ears or to the sides of the neck. The sides of the muzzle, lips, and chin and the whiskers are white. The throat is dark gray-brownish black. The ears have short grayish or buffy hairs with black areas at the posterior base. The tops of the feet are yellowish white. The tail is alternately banded with five to seven conspicuous rings, brownish black alternating with broader yellowish gray, and ends in a black tip. The dark rings are more distinct above than below. Immatures are somewhat lighter than adults. Albino, white, melanistic, chestnut, and red raccoons occur infrequently.

The annual molt extends over a long period during the summer, at least in the more northern and more densely haired subspecies. The molt begins in early March and ends in late May. The hairs may be shed from all over the body at the same time or gradually from the head region toward the tail and later on the sides and belly. The fall molt begins in October or November, starting on the belly and moving toward the back and tail. The pelts of most adult raccoons become prime by mid-November and the pelts of juveniles usually become prime in December.

Males tend to be larger than females. Measurements range from 72 to 92 cm (28.4–36.2 in); tail, 22 to 26 cm (8.6–10.2 in); and hind foot, 10 to 13 cm (3.9–5.1 in). Weights vary from 4.1 to 11.8 kg (9–26 lb).

**Distribution.** The raccoon occurs from most of southern Canada through most of the United

RACCOON

States except the high portions of the Rocky Mountains and the arid southwest, and through Mexico to Panama.

**Ecology.** Raccoons inhabit woodlands interspersed with agricultural fields and wetlands and rarely occur in dense evergreen forests. They prefer fairly open mature hardwood areas with hollow trees near streams, rivers, ponds, and lakes. The animals generally den in hollow trees. When necessary they use hollow logs, cavities under stumps, crevices in rocks, caves, abandoned mines, deserted buildings, barn lofts, cornshocks, haystacks, slab piles, brush piles, tile drains, culverts, dry spots in wetlands, abandoned beaver houses, muskrat houses, and abandoned fox dens and woodchuck burrows.

The entrance of a den cavity may be from ground level to a height of 60 feet or more, but most dens are situated at heights between 20 and 40 feet. The den cavity is usually 12 inches in diameter and 24 to 36 inches deep, though some dens may be smaller and deeper. A raccoon may use more than one den and may use the same dens year after year.

Man and dogs are the chief predators of raccoons. Bobcats, coyotes, fishers, and great horned owls capture cubs. Automobiles kill hundreds of raccoons annually, and forest fires kill some. Many young raccoons die from starvation and parasitism during late winter and early spring.

The parasites of raccoons include ticks, lice, fleas, cestodes, nematodes, and trematodes. The larvae of the greenbottle fly, *Phaenicia sericata*, are sometimes present between the skin and the flesh. Raccoons are susceptible to coccidiosis, dog distemper, fox encephalitis, infectious enteritis, listeriosis, pneumonia, trichinosis, tuberculosis, and rabies.

**Behavior.** Raccoons are predominantly nocturnal but may be seen during the daylight. At the onset of cold weather they begin to den for the winter and stay in the den during the cold weather, but they do not hibernate and are active during mild winters. Sharp and Sharp (1956) found that adults were active as long as the ambient temperature remained above 30°F after nightfall, but most of them stayed in their dens when the temperature dropped to 25°F or below. However, the cubs were inclined to be active when the temperature was too cold for adults. Some cubs were even active when temperatures were between 20° and 12°F.

During winter dormancy, raccoons live off fat accumulated in late summer. They generally remain dormant from late November to late March. Raccoons of all ages lose about 50 percent of their body weight when they stop feeding regularly (Stuewer 1943; Mech, Barnes, and Tester 1968), but with renewal of regular feeding the animals gain weight again.

Raccoons tend to avoid each other, except for some pairs and families traveling together, but they may tolerate each other's presence in certain situations. Mech and Turkowski (1966) reported 23 in one winter den. This wintering group may have comprised several generations of the same family.

The home range size of raccoons is related to availability of food, to age and sex of the animals, and to weather conditions. Sunquist, Montgomery, and Storm (1970) found that the home range size was 1 mile by 3 miles. Urban (1970) determined the average size to be 120 square acres, but this varied greatly with the individual raccoon. Schenider, Mech, and Tester (1971) reported that home ranges for individual raccoons were about 2 miles in diameter. Giles (1943) reported that a raccoon traveled 75 miles in 75 days, and Lynch (1967) recorded a raccoon that traveled 165 miles in 164 days.

Raccoons have well-developed senses of hearing, sight, and touch, but taste and smell are less well developed. These animals are curious, and when treed they often look out from behind the branches to see what is going on. They are clever and have good memories. Kitzmiller (1934) found that raccoons that had mastered a puzzle box remembered enough to solve it in just a few seconds on two successive tests given after a 12-month layoff.

Raccoons make numerous sounds, including purrs of contentment, low grunts uttered in warning, loud growls or snarls made during a fight, and a throaty shrill scream, almost like a whistle, given when they are terrorized or caught by dogs. Females with young utter low twittering calls or hissing sounds. Young raccoons emit a plaintive whimper much like a human infant when abandoned or hungry and squeal loudly when handled roughly.

The gaits of raccoons are an ambling pace and a less agile trot and run. They are deliberate rather than swift runners and can travel 10 to 15 miles per hour (Jackson 1961). These cunning animals usually lead hunters and dogs on a long chase and often obscure their trails by taking to water, swimming fairly well. Raccoons are courageous and strong fighters and will fight a dog

their own size. Occasionally raccoons will turn on a dog in water, climb on its back, and drown it. On land they usually defend themselves by lying on their backs and slashing with their teeth.

Raccoons are expert climbers and descend a tree either head or tail first. They are extremely powerful for their size, with extraordinary physical stamina. Whitney and Underwood (1952) reported that an adult raccoon hiding inside a hollow tree, where it had secured a good hold with its legs, supported a 200-pound man hanging onto its tail. Whitney (1933) reported that a pregnant raccoon fell 30 feet or more from a tree and later bore its young. Raccoons may drop their scats anywhere and do not seem to make scent posts.

Raccoons are opportunists, eating both animal and plant foods. In spring and early summer they eat predominately animal matter, and in summer, fall, and winter they feed mainly on fleshy fruits and seeds. Grain, mast, and fleshy fruits may compose the bulk of food eaten throughout the year.

Some plant foods eaten are apples, pears, acorns, beechnuts, hazelnuts, corn, oats, berries, grapes, tomatoes, cantaloupe, plums, watermelon, silky cornel, various species of dogwood, ragweed, smartweed, grass, and tender shoots and buds. Animals eaten include crayfish, frogs, turtles, snails, fish, snakes, earthworms, grasshoppers, crickets, bees, wasps, moths, small birds and their eggs, shrews, mice, muskrats, squirrels, carrion, and garbage.

Raccoons are promiscuous. The breeding season lasts from late January to the middle of March and the gestation period is about 63 days. The young are normally born in late April or early May, but if the female is not fertilized a second breeding cycle may begin 2 to 4 months later (Whitney and Underwood 1952). Lehman (1968) reported that a litter was born between 4 and 15 September.

The single annual litter may consist of three to seven cubs, but the average is four. The cubs are usually born in a hollow tree but may be born in a variety of nest situations, such as in a poultry house, abandoned building, or even a nail keg.

At birth the cubs weigh about 75 g (2.6 oz). They are well furred on the back and sides but scantily haired on the belly and inner sides of the legs. The pelage may be a uniform color or a mixture of yellow and gray, and littermates may be of different colors. The cubs grow rapidly. Their eyes open at about 19 days; they are weaned when about 16 weeks old and may begin eating solid foods at 9 weeks old.

The female usually keeps her cubs in a hollow tree for the first 50 to 60 days, then moves them to a ground bed, most often in a wetland area. She may move them to another nest if disturbed. The cubs begin to travel with the female when they are weaned, and the group beds as a family unit. They become semi-independent in September, but remain within their mother's home range and continue to sleep with her. Later the cubs begin to develop their independence by traveling without the female and then by bedding without her or with siblings. The young do not become fully independent, and temporary liaisons of adult-cub and cub-cub constantly form and dissolve. As winter denning approaches the family bond strengthens and the family beds together more often. During this period the animals select a winter den tree, and the entire family may den for the winter in the same tree or in a group of trees close together (Schneider, Mech, and Tester 1971).

Stuewer (1943) found that approximately 50 percent of the females produced litters as yearlings, but he thought that males probably did not breed as yearlings. Sanderson (1951) indicated that raccoons are not sexually active throughout the year and that they probably are capable of breeding before early January in Missouri. Longevity is about 6 years in the wild and 10 to 14 years in captivity (Haugen 1954).

**Age Determination.** Raccoons taken in the fall must have unworn canines less than 10 mm (0.39 in) long to be considered juveniles (Stuewer 1943). Raccoons can be separated into eight age-classes on the basis of a combination of closure of cranial sutures, wear of mandibular cheek teeth, and cementum ring counts of the mandibular incisor root (Grau, Sanderson, and Rogers 1970). The eye-lens technique is useful only for separating juveniles from adults (Sanderson 1961a). Body weight can be used to separate juveniles from adults taken in the fall and early winter. Parous adult females have longer and usually darker nipples than juveniles

and nulliparous adults (Sanderson 1950). Raccoons in their first winter have distinct epiphyseal cartilage at the distal end of the radius and ulna, and those with the epiphysis joined with diaphysis are considered to be in their second winter or older (Petrides 1959). Raccoons taken during hunting and trapping seasons have a broad area of epiphyseal cartilage during their first winter and either thin or no epiphyseal plates after 15 months of age (Sanderson 1961b). The baculum of males can be used to separate juveniles from adults. The bone has a slender base and truncated tip in juveniles and a massive base, heavier shaft, and ossified, bifurcated tip in adults (Sanderson 1950).

**Specimens Examined.** MAINE, total 25: Franklin County, Farmington 2 (UME); Hancock County, Bucksport 6 (HMCZ), Franklin 7 (UME), Mt. Desert Island 2; Oxford County, Umbagog Lake 1, Upton 4 (HMCZ); Penobscot County, Lincoln 1 (UME), Penobscot River (East Branch) 2 (USNM).

NEW HAMPSHIRE, total 7: Carroll County, Ossipee 1 (USNM); Cheshire County, 1–2 (DC–HMCZ); Coos County, Jefferson 1 (UCT); Grafton County, Etna 2 (DC).

VERMONT, total 21; Addison County, Bridport 1, Ferrisburg 2, New Haven 1 (UVT); Bennington County, Manchester 1 (HMCZ); Chittenden County, Burlington 1, Huntington 2, Richmond 1, Sheburne 2, Williston 3 (UVT); Franklin County, Fairfax 2 (HMCZ), St. Albans 1; Lamoille County, Eden 1; Washington County, Barre 2 (UVT); Windham County, Putney 1 (DC).

MASSACHUSETTS, total 14: Berkshire County, Stockbridge 1; Essex County, Manchester 1 (HMCZ); Hampden County, Montgomerey 1, Westfield 4, West Springfield 1 (UMA); Middlesex County, Pepperell 2; Suffolk County, Brighton 3 (HMCZ); Worcester County, Brookfield 1 (UCT).

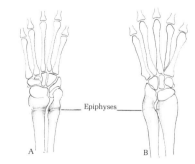

Drawing made from X-ray photograph of radii and ulnae of raccoon showing A, open (immature), and B, closed (adult), epiphyses.

## FAMILY MUSTELIDAE (WEASELS AND ALLIES)

The mustelids vary remarkably in size, color, behavior, and habitat. Some species are fierce and highly predacious, others are gentle. Some mustelids are arboreal, others are terrestrial, partly fossorial, semiaquatic in fresh water, or almost wholly aquatic at sea. These animals are rather short-legged and their movements are quick and graceful. They have great strength and endurance, and their powerful limbs are adapted for climbing, digging, running, or swimming. Locomotion may be digitigrade, semiplantigrade, or plantigrade. Mustelids travel with a loping gait or with a bearlike shuffle. Some species sit on their haunches or stand on their hind feet to get a better view of the surroundings. Mustelids travel alone, in pairs, or in family groups. They may be nocturnal or diurnal in habits. Most members have delayed implantation of the blastocysts, and some of them have induced ovulation. Mustelids possess well-developed anal musk glands.

The family Mustelidae is worldwide in distribution except for Antarctica and most oceanic islands. There are eight species of mustelids in New England.

# Marten

*Martes americana americana*
(Turton)

[*Mustela*] *americanus* Turton, 1806. *A general system of nature . . ., 1:60.*
*Martes americana americana,* Miller, 1912. *Bull. U.S. Natl. Mus.* 79:92.
Type Locality: Eastern North America
Dental Formula:
I 3/3, C 1/1, P 4/4, M 1/2 × 2 = 38
Names: Pine marten, American marten, American sable, and Hudson's Bay sable. The meaning of *Martes* = a marten, and *americana* = of America.

**Description.** The marten is about two-thirds the size of a house cat but is lithe, long, and weasel-like in body form. It has short legs, a comparatively small triangular head with erect prominent, rounded ears, a pointed short nose, and a rather long cylindrical, bushy tail. Each foot has five toes with sharp, slender, semiretractile claws, and the soles are densely furred. The pelage is soft, lustrous, and thick. Females have six mammae—two pectoral and four inguinal.

The skull is elongated and depressed; the facial angle is slight and the rostrum long and slender. The auditory bullae are moderately inflated, more rounded than is usual in the family. The palate extends beyond the last broad molars.

The sexes are colored alike, but there is considerable individual color variation in tone and intensity of browns and yellow oranges. There appears to be marked seasonal color variation, though the summer pelage is thin and coarser. The general coloration of the upperparts is a uniform rich yellowish to light brown mixed with blackish brown hairs on the back, darker on the legs, feet, and tail. The underparts are cinnamon and slightly lighter than the upperparts, darkening posteriorly. The head is grayish brown to almost white, darkest on the nose, and the ears are edged with white. The throat and chest have irregular patches of bright ochraceous orange. Immatures are much like adults. Color phases include light yellow, silver sprinkled, nearly white, slate, and almost all black specimens. Libby (1961) reported a "Samson" condition in a marten from Alaska—the animal had no guard hairs except on the tip of its tail. The single annual molt begins in late summer or in early fall and is completed by middle or late October.

Males are larger than females. Measurements range from 51 to 67 cm (20.1–26.4 in); tail, 15 to 23 cm (5.9–9.0 in); and hind foot, 7 to 10 cm (2.8–3.9 in). Weights vary from 0.7 to 1.6 kg (1.5–3.5 lb).

**Distribution.** The marten previously occurred from Alaska across most of Canada, southward along the higher parts of the mountains of New England and the Alleghenies south to West Virginia and southeastern Ohio, northwest through northern Illinois, most of Wisconsin, and Minnesota, and in the west, in the Rocky Mountains, south to northern New Mexico and the Cascade–Sierra Nevada range, then south to central California.

The range and numbers of marten have decreased in the past several centuries. The present range is much reduced, especially in the south. In New England the species is abundant in northern Aroostook County, Maine (Coulter 1959) and is present in limited numbers in most of Maine and northern New Hampshire. It probably occurs in remote areas of northern Vermont.

The marten was a common inhabitant of the whole of colonial New England, except along the seacoast (Allen 1876; Keay 1901). Allen (1942) reported that marten were common in northern New England, had been much trapped in Maine and the White Mountains of New Hampshire, and were still found there in smaller numbers. Silver (1957) stated that the marten was not abundant at the time of settlement in southern New Hampshire. Emmons (1840) wrote that the species was found in beech woods in the Berkshire Mountains of western Massachusetts during settlement. In Connecticut, Goodwin (1935) reported that formerly the animal was not uncommon in mountains of the northwestern region and possibly occurred in the highlands of northeastern Connecticut. The marten may also have occurred in Rhode Island. Overtrapping, burning, logging, and clearing of land for agriculture were probably the chief reasons for the decline of the species in New England.

**Ecology.** The marten inhabits cool, dense, coniferous forests of northern spruce, balsam, and hemlock in high mountainous regions or rockslides. They may also be found in mixed deciduous-coniferous forests. Dens are usually in hollows of trees, under logs or stumps, and sometimes in rock crevices.

Man is the chief predator of marten, although fishers, lynxes, bobcats, and great horned owls occasionally catch some. Their parasites are fleas, mite, lice, and nematodes, and scabies are known to infest them.

**Behavior.** Marten do not hibernate, though they may hole up during severe winter storms. They may move to lower elevations during winter. Marten wander and seem to have no permanent den site. They are secretive and active both night and day but are more diurnal during the breeding season.

Marten are bold, energetic, fearless, alert, quick, and curious. They are primarily solitary and seldom associate with other marten except during the mating season and when females have young. They are usually silent, but when angry or trapped they snarl, hiss, screech, growl, or scream. They dislike water and rarely enter it. The senses of smell, sight, and hearing are well developed.

Marten climb trees easily. On the ground they diligently investigate every nook and cranny and capture prey by pouncing on it. They seem to enjoy jumping into the snow. One marten twice jumped into the snow from about 15 feet up in a spruce tree; on another occasion a marten jumped from a roof to the snow 7 feet below, and from a slightly leaning cottonwood tree, a marten jumped to land in deep snow 22

MARTEN

feet out from the base of the trunk (Murie 1961). Marten like to plow through soft snow and may tunnel under it.

The home range is 12 to 15 square miles (Marshall 1951a). The average home range size is 0.92 square miles for males and 0.27 square miles for females. Resident marten remain on the same home range as long as 819 days regardless of season. The home range boundaries coincide with various vegetative or topographic features such as the edges of large open meadows and burns, or creeks (Hawley and Newby 1957). Martens make scent posts by rubbing their abdominal scent glands on the ground, branches, and other objects before and during the mating season, apparently to leave a scent trail for mates.

Marten undergo population cycles of approximately 3 years. Adult females and juveniles are most involved in the population decline, which is probably due in part to food shortage. Females in poor physical condition are less able to produce young, and the surviving young apparently are unable to compete with adults and establish home ranges (Hawley and Newby 1957; Weckwerth and Hawley 1962).

Unlike weasels, marten are not known to kill wantonly more than they need for food. They prey mainly on small mammals but will eat birds and their eggs, frogs, toads, fish, reptiles, fish (probably carrion), insects, and fruits when available. They are fond of red squirrels, chipmunks, flying squirrels, field mice, voles, shrews, moles, snowshoe hares, deer carrion, and ruffed grouse.

Marten are polygamous and breed in midsummer, usually in July. Their gestation period is between 259 and 275 days. Jonkel and Weckwerth (1963) determined that implantation of the blastocysts was delayed until 27 days before parturition, which is comparable to the 27 days between implantation and parturition for long-tailed weasels. Enders and Pearson (1943) demonstrated that the gestation period of marten could be shortened considerably by ar-

tificially controlling day length. Marshall and Enders (1942) reported that marten taken as early as January had implanted blastocysts.

Marten usually produce litters from early April (Ashbrook and Hanson 1927; Brassard and Bernard 1939) to mid-May (deVos 1957).

The single annual litter contains from one to five young, averaging three or four. Newborns are blind and helpless, covered with fine, soft yellowish hairs, and weigh between 28 and 40 g (.98–1.4 oz). They grow fast, and the eyes open at 4 to 6 weeks. They are independent at about 3 months of age. Markley and Bassett (1942) reported that some captive females became sexually mature at about 15 months old, though most females reached maturity when 2 or 3 years old. Weckwerth (1957) reported that a marked wild 3-year-old female was suckling young. Jonkel and Weckwerth (1963) found that wild females apparently were not sexually mature as yearlings, but they suggested that wild males may reach sexual maturity then. Longevity of marten may be 6 years in the wild and 11 years in captivity.

**Age Determination.** A sagittal crest of the skull less than 20 mm (0.78 in) long indicates immaturity in males; males with crests over 30 mm (1.17 in) and females with crests over 20 mm are old adults. Some adult females may not have sagittal crests. The nasal sutures are probably unfused in juveniles and fused in adults (Marshall 1951b). Juvenile females reach adult weight at 3 months, juvenile males at about 4 months. The mammae are large and conspicuous in adult females and inconspicuous in juveniles (Newby and Hawley 1954).

**Sex Determination.** The sex of a marten can be determined by palpating for the baculum or looking for the vulva. Also, males are larger and have broader heads (Newby and Hawley 1954).

**Specimens Examined.** MAINE, total 19: Oxford County, Norway 1, Umbagog Lake 1, Upton 15; Piscataquis County, Moosehead Lake 2 (HMCZ).

# Fisher

*Martes pennanti pennanti* (Erxleben)

**Description.** The fisher is twice as large as the marten and a different color. The head is wedge-shaped, with the muzzle somewhat pointed, and the ears are broad, somewhat rounded and less prominent than in the marten. The neck, legs, and feet are stout, with claws strong and curved for climbing. The tail is long, bushy, and tapering. Well developed scent

[Mustela] pennanti Erxleben, 1777.
Systema regni animalis . . ., p. 470.
Martes pennanti pennanti, Miller,
1912. Bull. U.S. Natl. Mus. 79:94.
Type Locality: Eastern Canada, Province of Quebec
Dental Formula:
I 3/3, C 1/1, P 4/4, M 1/2 × 2 = 38
Names: Pennant's marten, big marten, fisher marten, fisher cat, fisher
weasel, pekan, pekan weasel, black
cat, and black fox. The name fisher
is somewhat of a misnomer, since
the animal is not known to fish,
although it readily eats dead fish.
The meaning of Martes = a marten,
and pennanti = for Pennant, a
Welsh naturalist.

glands produce a musky, penetrating odor. Females have four mammae, all inguinal.

The skull is similar to that of the marten but
much more massive, with high crowns over the
occipital region. The sagittal crest is well developed in older animals, especially males, and
the auditory bullae are well inflated.

The pelage is dense, long, and rather glossy.
The coloration is variable between the sexes—
females tend to be darker than males. There are
seasonal changes in color. Older animals, especially males, have many white-tipped hairs on
their shoulders and backs, giving them a grizzled appearance. The coloration of the upperparts is grayish brown or brownish black, lighter
on the sides but darker on the rump. The underparts are brownish. The face, neck, and
shoulders are heavily frosted with gray and pale
brown, the nose, legs, and feet are blackish, and
the tail usually is all black. The ears have pale
linings. There are a few white patches on the
neck, throat, inguinal region and occasionally
the median surface of the legs. Immatures are
somewhat darker than adults. Color variations
include white, melanistic, fawn, and mottled
specimens. A molt occurs in late summer or
early fall.

Adult males are much larger than adult
females and twice as heavy. Measurements
range from 80 to 102 cm (31.5–40.2 in); tail, 30 to
42 cm (11.8–16.5 in); and hind foot, 9 to 14 cm
(3.5–5.5 in). Weights vary from 2.0 to 9.2 kg
(4.4–20.3 lb).

**Distribution.** The fisher occurs from southeastern Alaska and British Columbia to Hudson Bay
and eastern Canada, south in the Adirondack
Mountains of New York, northwestern Maine,
the White Mountains of New Hampshire, the
Green Mountains of Vermont, and probably in
the Berkshires of Massachusetts, south in the
Rocky Mountains of Wyoming and the Sierra
Nevada in California. Originally the range extended into western North Carolina and eastern
Tennessee. It is now much reduced, especially
in the south.

The fisher was common throughout New England during early settlement but soon disappeared in most of New England due to
overtrapping, logging, and clearing of the forest
for agriculture. However, it has greatly expanded its range south and east, and its speed of
dispersal seems to be related to hilly country,
regardless of the kind of forest cover present in
New England.

**Ecology.** Fishers prefer dense forest of mixed
hardwoods and conifers but are also found in
open second-growth stands and occasionally in
areas recently burned over. They rarely dig
burrows. Dens may be in hollows of trees and
logs or holes in rocky ledges, under large boulders, in old porcupine dens, under brush piles,
or in the snow, and may be lined with leaves.
Fishers may change their series of dens occasionally throughout the year.

Man and possibly bobcats, lynxes, and black

FISHER

bears prey on fishers. Ticks, mites, fleas, cestodes, nematodes, and trematodes parasitize them. Sarcoptic mange caused by the mite *Sarcoptes scabei* has been reported to infect fishers.

**Behavior.** Fishers are active night and day throughout the year but remain in a den during winter blizzards. They are solitary except for brief periods during the breeding season and when females have young. Fishers are as agile in trees as on the ground. When hunting, they dart beneath logs or upturned stumps—probably anyplace where prey might be found. Coulter (1966) reported that a fisher regularly jumped up to 9 feet between trees, and that some fishers jumped into the snow from heights of 15 to 20 feet. Fishers descend a tree head first. They swim readily and frequently take to the water.

Fishers are wary and usually irritable. When angry they arch their backs like cats, show their teeth, and emit a growl that ends in a snarl or hiss. During the breeding season, fishers grunt like jackrabbits. They are powerful and fierce fighters and when cornered can fight off a dog or two.

Fishers travel greatly in search of food. They forage along more or less direct routes, denning in suitable places along the way. During the breeding season several fishers may follow one another's tracks. Coulter (1959) stated that fishers prefer to travel along ridges, usually crossing small streams to get to the next ridge, and added that such a "crossing may be used by generations of fisher." Jackson (1961) reported that the normal home range is 8 to 15 miles in diameter, though a fisher may wander 40 to 100 miles in 4 to 12 days during winter. Males tend to range more widely than females.

Although de Vos (1951) noted cyclic fluctuations in fisher populations in Canada, Hamilton and Cook (1955) and Coulter (1960) found no evidence of cyclic tendencies in fishers in New York and in Maine. However, Coulter (1960) stated that it was possible that any cyclic tendency could have been masked by the rapid increase in numbers during the past 10 to 12 years in Maine and might become evident after the fisher population becomes stabilized.

Fishers are opportunists and will eat whatever is available. They hunt food by chance and do not attempt to run down or stalk prey. The porcupine is the only prey that fishers seek out and hunt deliberately. Fishers kill all their prey except porcupines by biting at the back of the head, and consume bones, feathers, and hair. Coulter (1966) noted from interpretation of signs at three kills on good tracking snow that fishers kill porcupines by repeated attacks on the face and head, apparently circling the rodents and making several quick rushes at the head. The porcupines succumb to a succession of wounds. The author stated that fishers may begin to eat a porcupine either on the underside where there are no quills or at the face, head, and neck, where there is minimum exposure to them. The fisher proceeds posteriorly from inside the porcupine skin, which is left flat and remarkably cleaned of most flesh and bones. Both methods permit the fisher to feed with minimum contact with the quills.

Contrary to popular opinion, porcupine quills often penetrate deep into fishers, but they seem to cause no inflammation or festering. Although Daniel (1960) reported finding quills in the adrenal gland, adrenolumbar vein, peritoneal cavity, in the omental fat, and protruding from the stomach, most quills are passed through the fisher's digestive tract.

In addition to porcupines, fishers eat snowshoe hares, carrion, white-tailed deer, shrews, mice, squirrels, small birds, amphibians, fishes, insects, apples, and beechnuts.

In Maine the breeding season lasts from late February to April and reaches a peak in March (Coulter 1966). Breeding occurs a few days after the young are born. This species has an unusually long gestation period of about 51 weeks (Hall 1942), since after breeding the implantation of the blastocysts is delayed for 9 or 10 months. The kits are born in March or early April in a crevice of a ledge or in a hollow tree. The litter size varies from one to four; the usual is three.

Newborns are blind and helpless, partially covered with a growth of fine gray hair over the middorsal area. They utter short, high-pitched cries somewhat similar to those of domestic kittens. They grow rapidly during the first few days. By the 3d or 4th day almost the entire body is covered with fine fawn gray hair. The kits are dependent on the female for 4 months or more. The eyes open about the 53d day, and nursing continues into the 4th month, although the kits eat meat regularly before then (Coulter 1966). In Maine both sexes are sexually mature at less than a year old, and females produce their first litter when two years old (Wright and Coulter 1967).

**Age Determination.** Immature, nonbreeding females can be separated from adults by the presence of the partly open maxillary-palatine suture and by the various degrees of separation of the temporal ridges of the skull. In breeding females, the temporal ridges form a sagittal crest, and the maxillary-palatine sutures are completely fused (Eadie and Hamilton 1958). During early winter immature males can be separated from adult males by the openness of the nasofrontal, nasomaxillary, internasal, and zygomatic-temporal sutures and by the moderately developed sagittal crest. By middle or late winter the skull sutures are closed and the

sagittal crest is well developed. The pelvic girdle and the puboischiae symphysis are completely open in immatures less than a year old and at least partially obliterated in juveniles and adults. The bacula of adult males are more than 100 mm (3.9 in) long and commonly weigh 2 g (.07 oz) or more. However, by February the bacula of immatures overlap with weights of adult bacula (Wright and Coulter 1967).

**Specimens Examined.** MAINE, total 19: Aroostook County (no location) 1, Garfield 2 (UCT); Franklin County, Farmington 1 (UME), Rangely 3 (HMCZ); Oxford County, Bethel 1 (UME), Umbagog Lake 7, Upton 1 (HMCZ); Piscataquis County, Brownsville 1 (USNM), Greenville 2, Moosehead Lake 1 (HMCZ).
NEW HAMPSHIRE, total 2: Hillsborough County, Mt. Monadnock 1 (taken 15 November 1902) (USNM); Grafton County, Warren 1 (HMCZ).
VERMONT, total 2: Orleans County, Glover 1; Windsor County, Royalton 1 (UVT).

# Ermine

*Mustela erminea cicognanii*
Bonaparte

*Mustela cigognanii sic* Bonaparte,
    1838. *Charlesworth's Mag. Nat.
    Hist.* 2:37.
M[ustela] e[rminea] cigognanii, Hall,
    1945. *J. Mamm.* 26(2):180.
Type Locality: Eastern United States
Dental Formula:
    I 3/3, C 1/1, P 3/3, M 1/2 × 2 = 34
Names: Short-tailed weasel,
    Bonaparte's weasel, small brown
    weasel, small weasel, and stoat.
    The meaning of *Mustela* = a
    weasel, *erminea* = an ermine, and
    *cicognanii* = for Felice Cicognani,
    a naturalist.

**Description.** The ermine is a small, slender, long-bodied, long-necked, and short-legged weasel. The head is small and triangular, with a low forehead. The eyes are small and the ears low, rounded, and well haired. There are five small-clawed toes on each foot. The tail is short and bushy, usually less than one-third the total length of the animal. Females have eight mammae—four abdominal and four inguinal. This species can usually be distinguished from the long-tailed weasel by its smaller size and shorter tail, but when using these criteria to separate the two species it is necessary to identify the sex of the individual. Male and female ermines are smaller than male and female long-tailed weasels, but since the males of both species are larger than the females, a large male ermine may be about the same size as a small female long-tailed weasel.

The skull is elongated and flattened, with a slight facial angle, short, narrow precranial portion, and long, smooth, rounded braincase. The rostrum is short, smaller posteriorly, and tapers abruptly from the slender and moderately arched zygomata. The large auditory bullae are greatly inflated and longer than they are wide, the inner sides nearly parallel. The palate ex-

tends behind the upper molars. The teeth are sharp and moderately large.

The pelage is composed of soft, short underfur and long, coarse, hard, glossy guard hairs. The sexes are colored alike and the color shows marked seasonal variation. In summer the coloration of the upperparts is uniformly dark brown, slightly darker on the head and legs, becoming white at the feet. The upper lips are narrowly banded with white, and the chin and throat are whitish, occasionally with one or two brown spots. The underparts are whitish to pale yellowish or creamy buff, the color extending down the insides of the legs and into the feet. The toes are often white-tipped. The terminal third of the brown tail is black, even in winter. The winter pelage is white, tinged with yellow on the back and underparts. The coloration of immatures is similar to that of adults. Albinos rarely occur and melanistics are not known.

The ermine has two molts annually, in spring and in autumn. The autumn molt from brown to white pelage usually begins in October and is completed by late November, but it may be delayed until well into December. During molting the fur appears mixed brown and white, the

ERMINE

white hairs intermixed with brown ones over most of the body. The spring molt usually begins about the middle of March, rarely as early as the middle of February, and is completed by late April. The tail is usually last to molt both in autumn and in spring. The autumn and spring molts usually begin on the belly and spread to the sides and back. Molting may begin on the face.

Bissonette and Bailey (1944) demonstrated that the increased length of daylight is responsible for spring pelage change in ermines. Rust (1962) found that the ambient temperature was not a major factor in the spring pelage change, though it modified the speed and nature of pelage change in captive ermines.

Adult males are larger than females. Measurements range from 19 to 30 cm (7.5–11.8 in); tail, 4 to 8 cm (1.6–3.2 in); and hind foot, 3 to 4 cm (1.2–1.6 in). Weights vary from 45 to 105 g (1.6–3.7 oz).

**Distribution.** The ermine is circumpolar in distribution. In North America it occurs from Baffin Island and Southampton Island, through most of Canada, south to Maryland and Minnesota, and from western Montana south in the Rocky Mountains to New Mexico and from Alaska into northern California.

**Ecology.** In the northern parts of their range ermines inhabit low brush and thickets along waterways in large forested areas. In the south-

ern areas they occur in brushland, open country with hedgerows or stone walls, in old buildings, and occasionally in swampy areas.

The den is a large, loose structure that may be found beneath an old stone wall or rock pile, under a log, tree, or abandoned building, in a pile of hay, or a similar safe retreat. There may be three or four tunnels leading to a den. The natal nest is lined with pieces of grasses, leaves, fur, and feathers.

Man, the large hawks and owls, foxes, domestic cats, and large snakes prey on ermines. Their parasites are fleas, cestodes, nematodes, and trematodes.

**Behavior.** Ermines are active throughout the year; they are mainly nocturnal but may be encountered during daylight. They are intensely alert, very curious, and bold. When surprised they duck into the nearest retreat, but a mouselike squeak will cause them to reappear almost instantly to have a look. They have well-developed senses of sight, smell, and hearing. They produce several calls, such as a rapid "took-took-took," hisses, purrs, low chatters, and soft grunts and screeches. A female utters a high-pitched reedy note when approached or chased by a male. Ermines may stamp their feet when annoyed, and if greatly disturbed they will emit a musky odor from the scent glands.

These animals have clean habits and deposit their long, spiral-shaped, blackish brown feces along trails, often on prominent objects. The scat piles may serve as sign posts. Sometimes they deposit the scats in a separate chamber within a burrow system.

Ermines are fair swimmers and will swim after prey. The normal gait is a series of small bounds or lopes with the back arched. Soper (1919) reported that the normal distance of a bound is about 20 inches, and in open space the animal can leap at least 5 or 6 feet. Jackson (1961) reported that ermines can run about 8 miles per hour for short distances.

In winter and early spring ermines travel extensively in search of food. The normal home range is between 30 and 40 square acres, but when prey is lacking ermines may travel 2 or 3 miles in one night (Jackson 1961).

Ermines are primarily terrestrial but sometime climb trees to a height of about 15 feet (Booth 1946). These mustelids are agile, fearless, cunning, and tireless hunters. They dart in and out of the burrows of small mammals, among crevices in rocks or through brush piles; pausing now and then with neck outstretched and standing on the hind feet, they look about with ears attuned to catch the faintest noise of prey. Ermines persistently stalk their unwary prey by scent and when a few feet away they pounce with lightning quickness. Ermines bite their prey over the back of the skull, while their forelegs encircle the captive as though hugging it and the hind legs scratch it wildly. If the prey is a large animal such as a rat or a rabbit, ermines usually lie on their side while the diminishing struggles of the prey continue, but if the prey is a mouse or a small bird they are apt to crouch over it.

Ermines usually kill more than they can eat. They consume the flesh, fur, feathers, and bones of small prey and generally the flesh of larger animals. They do not suck blood from prey as is commonly believed, but they lick the warm blood from the base of the skull or neck.

Ermines feed mainly on mice, voles, rats, rabbits, chipmunks, shrews, cottontail rabbits, frogs, lizards, small snakes, birds, insects, and earthworms, and they eat carrion when hunting is poor. Hamilton (1933) reported that captive ermines eat food equal to about one-third their weight every 24 hours.

Little is known about the reproduction of ermines. The breeding season occurs in early summer, and the young are born from mid-April to mid-May in an underground nest, after a gestation period of approximately 9 months (Hamilton 1933). Delayed implantation of the blastocyst occurs in this species (Wright 1942). The female may breed at the age of 3 months or possibly earlier, but males do not become sexually mature until late winter or early spring of their second year of life (Hamilton 1958). From four to nine young are born per litter, though six or seven are usual. Newborns are blind; they are flesh-colored but lightly covered with short fine white hairs on the neck, and weigh about 1.7 g (0.60 oz). They are surprisingly strong and can lift their heads and support themselves for an instant or two. The young develop rapidly and at 2 weeks of age have a prominent brown mane on the neck, though the rest of the body is only scantily covered with white fur. They can eat half their weight in 24 hours. The eyes open at 35 days, and at 45 days of age the brown fur on the back obscures the mane. By this time the young are active and play with one another like kittens. Young ermines are rather silent, in sharp contrast to the noisy young of long-tailed weasels. At 7 weeks old the males are larger than their mothers (Hamilton 1933).

The male may help the female care for the young. Longevity may be 5 or 6 years (Jackson 1961).

**Specimens Examined.** MAINE, total 112: Androscoggin County, Auburn 1 (UVT); Aroostook County, T11–R10 3, T12–R15 2, T12–16 1 (UCT), Allegash 1 (USNM), Quimby 1 (HMCZ); Franklin County, Dryden 1–2 (AMNH–UME); Hancock County, Bucksport 20 (HMCZ), Mt. Desert Island 1; Oxford County, Hartford 1 (UME), Norway 2, Umbagog Lake 3, Upton 22 (HMCZ); Penobscot County, Lincoln 1 (UME), Sherman 2 (HMCZ), South Twin Lake 1; Piscataquis County, Barnard 1 (AMNH), Greenville 8, Milo 3, Moosehead Lake 7 (HMCZ), Mt. Katahdin 4 (USNM); Sagadahoc County, East Harpswell 1; Somerset County, Crocker Pond 1 (AMNH); Washington County, Baring 2, East Machias 15, Steuben 1, West Pembroke 4 (HMCZ).

NEW HAMPSHIRE, total 19: Carroll County, Ossipee 6; Cheshire County, Dublin 1 (USNM); Coos County, College Grant (Errol) 2, Dummer 1 (HMCZ); Grafton County, Bridgewater 1, Canaan 1, Hoover 2 (DC); Hillsborough County, Amherst 1 (HMCZ); Rockingham County, Hampton Falls 1 (UCT), Northwood 1; Strafford County, Durham 2 (UNH).

VERMONT, total 36: Bennington County, Bennington

1 (HMCZ); Caledonia County, Hardwick 1 (UCT); Chittenden County, Shelburne 1 (UVT); Grand Isle County, North Hero 1 (UCT); Lamoille County, Mt. Mansfield 1 (USNM); Orange County, Ely 1; Rutland County, Belmont 1, East Wallingford 1 (DC), Mendon 2–4 (AMNH–UCT), Rutland 1–3–13 (AMNH–DC–UCT), Sherburne 1 (DC); Washington County, Montpelier 1 (UVT); Windham County, Brattleboro 1, Londonderry 1 (UCT); Windsor County, Woodstock 1 (USNM).

MASSACHUSETTS, total 27: Essex County, Wenham 1 (HMCZ); Franklin County, Gill 2 (UMA); Hampden County, Feltonville 1 (HMCZ); Hampshire County, Amherst 2, Cunnington 1 (UCT), Easthampton 1 (UMA), Northampton 1, Sunderland 3 (UMA); Middlesex County, Burlington 1–5 (HMCZ–USNM), Cambridge 1, Lexington 1, Malden 2 (HMCZ), Wilmington 2 (USNM); Worcester County, Harvard 2, Lancaster 1 (HMCZ).

CONNECTICUT, total 23: Litchfield County, Canaan 1, Riverton 1; Hartford County, Newington 1; Middlesex County, Portland 3 (UCT); New London County, Liberty Hill 3 (HMCZ); Tolland County, Charter Marsh 1, Coventry 1, Mansfield 2 (UCT), Rockville 6 (UMA), Spring Hill 1, Tolland 1; Windham County, Eastford 1 (UCT), South Woodstock 1 (AMNH), Windham 1 (UCT).

## Long-tailed Weasel

*Mustela frenata*

There are two subspecies of long-tailed weasel recognized in New England:

*Mustela f. noveboracensis* (Emmons). *Putorius noveboracensis* Emmons, 1840. *A report on the quadrupeds of Massachusetts*, p. 45. *Mustela frenata noveboracensis* Hall, 1936. *Carnegie Inst. Washington, Publ.* 473:104.

Type Locality: Williamstown, Berkshire County, Massachusetts

*Mustela f. occisor. Putorius occisor* Bangs, 1899. *Proc. New England Zool. Club* 1:54. *Mustela frenata occisor* Hall, 1936. *Carnegie Inst. Washington, Publ.* 473:1–4.

Type Locality: Bucksport, near mouth of Penobscot River, Hancock County, Maine (Hall and Kelson 1959).

Dental Formula:

I 3/3, C 1/1, P 3/3, M 1/2 × 2 = 34

Names: New York weasel, northern long-tailed weasel, large brown weasel, large ermine, and big stoat. The meaning of *Mustela* = a weasel, and *frenata* = bridled, referring to facial markings in some subspecies occurring in southern parts of the range.

**Description.** The long-tailed weasel is one of the larger weasels in the United States. It is similar in general appearance and coloration to the ermine but is larger and has a longer tail. The feet have five slightly webbed toes, each with a small claw. Females have eight mammae—four abdominal and four inguinal.

The skull is larger and more angular, with a longer precranial portion and more pointed postorbital processes than that of the ermine. The skulls of the long-tailed weasel and the mink are similar, but the former is much smaller and the auditory bullae are relatively much longer.

The pelage is rather short, moderately fine, and not dense. The sexes are colored alike and show marked seasonal variation. In summer the coloration of the upperparts is a uniform dark brown which usually extends down onto the feet and toes. The underparts are yellowish white, with occasional brown spots. The tail is brown and has a black tip both in summer and in winter. It is generally believed that this weasel normally changes to white only where the winters are very cold; however, Hamilton (1933) reported that only males remain brown in winter.

The autumn molt usually begins in mid-October and is completed by early or mid-November. The change from white to brown begins from mid-February to early March and is completed by mid-April. In autumn, molting starts on the belly and moves upward, whereas spring molting starts on the back and moves downward (Hamilton 1933).

Adult males are much larger than females. Measurements range from 30 to 44 cm (11.8–17.3 in); tail, 8 to 16 cm (3.2–6.3 in); and hind foot, 3 to 5 cm (1.2–2.0 in). Weights vary from 72 to 267 g (2.5–9.3 oz).

**Distribution.** This species occurs from southern Canada to Peru. It is absent in the southwestern United States and northwestern Mexico, except for northern Baja California.

**Ecology.** The long-tailed weasel is found from sea level to the timberline. Unlike the ermines, long-tailed weasels prefer open woodlands, brushlands, and fencerows in rocky areas near water. They seldom dig their own dens, but use a modified chipmunk burrow, a hole enlarged under a stump, or a natural crevice, or they may den among rocks or stone walls or around abandoned buildings.

The nest is situated about 6 inches underground and about 2 feet from the entrance of the burrow. It is about 9 inches in diameter and may have four entrances. The nest is made of grass packed in layers and lined with mouse and shrew fur. The burrow may contain bones and skins of prey that were eaten. This mustelid defecates in its nest as well as outside near the nest entrance. It may cover this filth and debris with new grass, giving the nest lining a stratified appearance.

Man preys on long-tailed weasels, and dogs, coyotes, red foxes, large hawks, and owls are said to catch and eat them. Porcupine quills seems to be a problem for weasels, and some suffer from accidents. Fleas, ticks, lice, cestodes, nematodes, and trematodes are known to be parasites of long-tailed weasels.

LONG-TAILED WEASEL

**Behavior.** The behavior of long-tailed weasels is similar to that of ermines. They are active throughout the year and seldom stay in a den for long periods of time regardless of weather conditions. They are mainly nocturnal but may be encountered during daylight.

Long-tailed weasels are good swimmers. They are also adept climbers and will chase a squirrel up a tree. They can climb a tree vertically or make short ascending spirals around the trunk. They are generally solitary, though sometimes two individuals may play together. Long-tailed weasels are persistent and relentless predators and will attack man or another animal that tries to interfere in their hunting. Like ermines, long-tailed weasels are highly efficient and voracious predators, striking the prey at the base of the skull and hugging it with their long body, usually resulting in a quick kill.

The distance long-tailed weasels travel from their home den varies considerably with food availability and cover type. Hamilton (1933) thought that they might travel at least 7 miles in a single night. Polderboer, Kuhn, and Hendrickson (1941) obtained an average of 312 feet and a maximum of 642 feet from tracks made in the snow by four weasels. Glover (1943) found that the distance traveled in a single night averaged 704 feet for 11 males and 346 feet for 10 females in new snow. Quick (1944) determined that the average distance traveled by a large male was 2 miles and the maximum distance in one day or night was 3.43 miles. Criddle and Criddle (1925) noted that a female that lived in their basement often traveled more than half a mile away in search of food. The home range of this species is about 400 square acres in Missouri (Schwartz and Schwartz 1959), and 30 to 40 square acres in Wisconsin (Jackson 1961).

Long-tailed weasels prey on small terrestrial mammals, bats, hares, rabbits, birds and their eggs, frogs, snakes, earthworms, insects, and carrion. They eat the smaller victims entirely. Hamilton (1933) reported that the long-tailed weasel eats about a third of its weight every 24 hours, but growing young weasels eat much more. Sanderson (1949) noted that captive young weasels eat as much as 40 percent of their body weight in one day.

This weasel breeds during July and August, and the young are born the following April or May after a gestation period of 205 and 337 days, averaging 279 days (Wright 1948a). Implantation of the blastocysts is delayed for approximately 7½ months, and development resumes during the last 27 days of pregnancy (Wright 1948b). Young females breed when 3 or or 4 months old, so that all females may produce young in the spring when they are about 1 year old. Males are not sexually mature until almost a year old. The litter may be from one to twelve young, but the usual number is six to eight (Wright 1942).

Newborns are blind, naked, and pink, but covered with a few fine white hairs over the back and head. They are about 65 mm (2.5 in) in total length and weigh about 3.1 g (0.11 oz). They grow rapidly, and by 2 weeks of age the sexes can be readily separated. By 3 weeks, they are well furred over the back, the white natal fur becoming grayish. At 4 weeks the teeth begin to erupt. The eyes open at about 36 days, and the female starts to wean the young at this time. The young continue their rapid growth throughout the summer, and by November they are nearly full-grown. After weaning the young leave the nest and soon disperse. The male usually helps care for the young (Hamilton 1933; Quick 1944). Longevity may be 5 or more years in captivity.

**Specimens Examined.** MAINE, total 32: Aroostook County, Garfield 1 (UCT), Sherman 1; Hancock County, Bucksport 12; Oxford County, Norway 1, Umbagog Lake 1, Upton 9; Piscataquis County, Greenville 1, Milo 2, Moosehead Lake 1 (HMCZ); Somerset County (no location) 1 (AMNH); Washington County, Baring 1, East Machias 1, Steuben 1 (HMCZ).

NEW HAMPSHIRE, total 20: Belknap County, Alton 1 (UCT), Belmont 1 (UNH); Carroll County, Ossipee 1 (USNM); Cheshire County, Dublin 2 (HMCZ); Coos County, College Grant (Errol) 1 (UNH); Grafton County, Franconia Notch 1 (AMNH), Hanover 2 (DC), Rumney Center 2; Merrimack County, Lake Sunapee 1 (AMNH), Pittsfield 1 (UCT), Webster 2 (HMCZ); Rockingham County, East Hampton Falls 1 (UCT), East Nottingham 1, Epping 1 (HMCZ); Strafford County, Barrington 1 (UNH), South Lee 1 (HMCZ).

VERMONT, total 12: Addison County, Hancock 1 (HMCZ), New Haven 1 (DC); Bennington County, Bennington 3 (HMCZ); Rutland County, East Wallingford 1 (DC), Mendon 3 (UCT), Rutland 1, Sherburne 1 (DC); Windsor County, Hartland 1 (AMNH).

MASSACHUSETTS, total 33: Essex County, Andover 1 (HMCZ); Franklin County, Shutesbury 1; Hampden County, Westfield 2; Hampshire County, Belchertown 1 (UMA); Middlesex County, Belmont 1 (HMCZ), Burlington 1–6 (HMCZ–USNM), Lexington 2 (HMCZ), Wakefield 1 (USNM), Wayland 2 (HMCZ), Wilmington 1 (USNM); Norfolk County, Brookline 1, Ponkapoag 1, South Weymouth 1; Plymouth County, Wareham 5; Worcester County, Lancaster 2 (HMCZ), Lunenburg 2 (USNM), Princeton 2 (HMCZ).

CONNECTICUT, total 69: Fairfield County, Greenwich 2 (AMNH), Reading 1 (HMCZ), Shelton 2 (UCT); Hartford County, East Hartford 1–1–2 (HMCZ–UCT–USNM), Farmington 1 (UCT); Litchfield County, Barkhamsted 3 (HMCZ), Gaylordsville 1 (USNM), Norfolk 1 (UCT); Middlesex County, Clinton 1 (AMNH), Portland 29; New Haven County, Durham 4, Haddam 6 (UCT); New London County, Liberty Hill 4 (HMCZ), Stonington 1; Tolland County, Mansfield 1, Staffordville 1, Storrs 2, Union 2, Willington 1 (UCT); Windham County, Plainfield 1 (USNM), Pomfret Center 1 (UCT).

RHODE ISLAND, total 1: Providence County, Chepachet 1 (USNM).

# Mink

*Mustela vison*

**Description.** The mink is much more robust than the other weasels and is nearly as large as a house cat, though much more slender. The head is relatively small, broad, and flattened and tapers to a pointed muzzle. The neck is thick and muscular. The ears are short, low, and rounded. The legs are short and stout, and the broad feet each have five slightly webbed toes. The soles are hairy, but the pads are naked. The tail is moderately long and bushy. The anal glands are well developed. Females have six mammae—two abdominal and four inguinal.

The skull is somewhat flattened, with a long braincase and short, broad rostrum. The audi-tory bullae are flattened. The long palate extends well beyond the last molars. The last upper molar is dumbbell-shaped and smaller than the preceding tooth.

The pelage is thick and dense, adapted for aquatic life. The underfur is dense and oily, overlaid with long, coarse, glistening guard hairs. The sexes are colored alike and show little seasonal variation. The coloration is a nearly uniform rich, dark brown throughout, except for a white patch on the chin and an occasional white spot on the chest and belly. Immatures are paler than adults and lack most of the guard hairs. Apparently color phases do not occur in

There are two subspecies of mink recognized in New England:

*Mustela v. vison* Schreber, 1777. *Die Säugthiere . . .*, part 3, no. 25, pl. 127b. *Mustela v. vison* Miller, 1912 *U.S. Natl. Mus. Bull.* 79:101.

Type Locality: Eastern Canada, Province of Quebec

*Mustela v. mink* Peale and Palisot de Beauvois. *Mustela mink* Peale and Palisot de Beauvois, 1796. *A scientific and descriptive catalogue of Peale's museum, Philadelphia*, p. 39. *Mustela v. mink* Hollister, 1914. *Proc. Biol. Soc. Washington* 27:215 (Hall Kelson 1959).

Type Locality: Maryland

Dental Formula:

I 3/3, C 1/1, P 3/3, M 1/2 × 2 = 34

Names: Common mink, woods mink, large brown mink, vison, water weasel, and least otter. The meaning of *Mustela* = a weasel; the meaning of *vison* is not certainly known, but it probably comes from *visor*, a scout. The name seems to be derived from the French *vison* or *Foutereaux*, meaning a kind of marten that lives near water.

the wild, but mink farms have bred color mutations under controlled conditions. Some of these are jet black, topaz, white, natural dark, pastel, sapphire, platinum, silver-blue, and amber-gold.

Molting occurs twice a year. The spring molt usually begins in March or early April and is completed by mid-July. It starts on the muzzle and feet and moves over the body. The autumn molt takes place about mid-August or early September, and the pelt is prime by late November or early December, at which time the skin is creamy white. Molting starts at the tip of the tail and spreads forward. The summer fur is flat, shorter, and less dense and less lustrous, usually with a reddish brown tint. Rust, Shackelford, and Meyer (1965) indicated that the pituitary gland is necessary for normal pelage cycles in the mink, based on hypophysectomized ranch mink.

Males are much larger than females. Measurements range from 46 to 68 cm (18.1–26.8 in); tail, 15 to 24 cm (6.0–9.5 in); and hind foot, 5 to 8 cm (2.0–3.1 in). Weights vary from 0.5 to 1.1 kg (1.1–2.5 lb).

**Distribution.** The mink occurs throughout Alaska, most of Canada except the tundra region, and in the United States, except for the desert region of the southwest.

**Ecology.** Mink are found along streams, rivers, and lakes and in or near marshes. They prefer forested, log-strewn, and thicketed areas. The den may be in a burrow along or near the banks of a body of water, under a log or stump, or in a rock recess, or a mink may use an abandoned or appropriated muskrat burrow or house or other suitable natural shelter. A mink may dig its own burrow or may take over one dug by some other animal. The burrow may be 8 to 12 or more feet long, 4 to 6 inches in diameter, and 1 to 3 feet below the surface of the ground. The number of entrances varies from two to five. The entrances may be just above the water level or a few yards from water.

The natal den is an enlargement of the burrow and is a few feet back from the water's edge. This chamber is about 12 inches in diameter, lined with a mixture of grass, leaves, fur, and feathers. The mink urinates and defecates outside the den or burrow.

Males may not use the same dens repeatedly but utilize the dens that are most convenient or accessible. Females may use only two dens regularly during winter (Marshall 1936). A family of mink may use many dens before the young disperse (Errington 1943). One mink family used twenty different den sites within a 77-acre area (Schladweiler and Storm 1969).

Man, dogs, red foxes, bobcats, and occasionally great horned owls prey on mink. They are hosts to fleas, ticks, lice, cestodes, nematodes, and trematodes. Gorham (1956) reported that *Wohlfahrtia* maggots are responsible for the death of many young ranch mink.

Although tularemia and abcesses are known in wild mink, little is known of other disease. Diseases of ranch mink include anthrax, abscesses, botulism, tularemia, distemper, enteritis, hydrocephalus or "water head," wet belly, rickets, septicemia, tuberculosis, salmonella and streptococcus infections, sinusitis, Chastek paralysis, steatitis (yellow fat), gastroenteritis, and poisoning from foreign bodies (Gorham and Griffiths 1952).

**Behavior.** Mink are mainly nocturnal but may be seen hunting at dawn and dusk and sometimes during the daytime. They are active throughout the year but may den up during cold, stormy days. They are solitary and unsociable, except during the breeding season and when females have young. These restless animals are curious and bold and may try to steal fish caught by fishermen.

On land, mink are not as agile as weasels. They do not chase their prey into burrows, nor do they habitually prowl about stone heaps and similar recesses. Mink travel by a series of low bounds of 10 to 20 inches long, with the back arched at each step. They lope along slowly for considerable distances at a speed of 6 to 8 miles per hour. They are alert and often rear up on their hind legs to look around.

These animals often climb on logs and stumps and may climb trees to escape danger. Mink like to push through snow and sometimes slide down snow-covered slopes on their bellies like river otters.

Mink are well adapted to water. They swim with most of the body submerged and move underwater with ease. On the surface they normally swim at a speed of 1 mile per hour. They can dive to depths of 15 to 18 feet and can swim underwater for 75 to 100 feet (Jackson 1961).

Mink are tireless wanderers and frequently travel far, often cruising over a wide range from one watershed to another when food is scarce or when their water freezes or dries up. Hibbard and Adams (1957) reported that a tagged mink traveled 20 miles in 4 months in search of new habitat when potholes dried up in North Dakota. Marshall (1936) reported that the home range of females is 20 square acres and the range for males is much larger. Mitchell (1961) found that the maximum home range of 35 tagged juvenile males was 3 square miles and that of 2 adult males was 2.7 miles. The home range of 23 tagged juvenile females did not exceed 1.8 square miles and that of 7 adult females was 0.4 square miles. Movements of 6 tagged juvenile males were 2 to 28 miles, averaging 11.2 miles, and those of 2 tagged juvenile females were 14.5 and 18.5 miles.

Mink are skillful and versatile hunters and will persistently trail a prey by scent. They are courageous fighters who will attack animals larger than themselves, which they kill as a weasel does, by biting behind the skull. They sometimes drag prey heavier than themselves long distances to the den, where they prefer to eat, and unlike weasels they cache their food.

Yeager (1943) reported an instance where a mink stored thirteen freshly killed muskrats, two mallard ducks, and one American coot in a den in a hollow ash stump 6 feet above ground at the edge of a marsh. In winter mink will forage underwater and come up for air through a smooth air hole in the ice.

Mink produce purring grunts, loud growls, snarls, screams, or shrill screeches. The senses of sight and hearing are not well developed, but smell is very keen. When much excited, irritated, or injured, mink spray on stones or rocks an acrid liquid musk from the prominent anal scent glands, perhaps to make mink sign posts.

Mink prey on fishes, frogs, crayfish, clams, turtles, snakes, lizards, earthworms, insects, mice, rats, bats, muskrats, moles, rabbits, and birds. They occasionally eat grass.

Digestion in the mink is rapid. Sibbald et al. (1962) found that the mean time of food passage in captive minks was 142 minutes, ranging from 62 to 215 minutes.

MINK

Mink are promiscuous. The breeding season occurs from late February to early April. The single annual litter is born in April or May, the majority arriving early in May. Estrus in the female lasts approximately 3 weeks, but it varies greatly, being longer in some females, shorter in others.

Copulation in the mink is often furious. It may be preceded by a rough-and-tumble courtship fight, after which the male seizes a firm grip on the back of the female's neck. Prolonged copulation is the rule, consisting of periods of activity followed by periods of rests. The male may drag the female about, or they may rest united with her upright or with both on their sides. Intromission may last as long as 3 hours. If the female does not become pregnant during ovulation, pseudopregnancy usually follows and lasts as long as pregnancy would.

The gestation period is 40 to 75 days, with an average length of 51 days. This variation in length of pregnancy is associated with the age of the female. Older females have slightly shorter pregnancies than yearlings, and females that mate early in the season have a longer gestation period than those mating later. The implantation of the blastocysts is delayed until 30 to 32 days before parturition (Enders 1952). The number of kits per litter may be two to ten, rarely more, but the usual is three or four.

At birth the kits are blind and helpless; they may be naked or partially covered with very fine, short, whitish hairs. They grow very rapidly during the first 8 weeks, then more slowly. Mitchell (1961) reported that young males attain the average weight of adult males sometime between early November and late February, and young females attain the average weight of adult females by mid-August, or at approximately 4 months of age.

The eyes open approximately 25 days after birth, and by then the body is covered with sleek, short, reddish gray fur. The deciduous teeth erupt between 16 and 49 days after birth, and the permanent teeth erupt between 44 and 71 days. Weaning takes place at 5 or 6 weeks, at which time the female brings food to her young. By 8 weeks of age the kits begin to capture prey. A male may help the female care for the young. The family remains together until late August, then disperses. The kits are playful and often practice for hunting by stalking and pouncing and by following their parents on hunting trips. The parents may carry the kits on land by the scruff of the neck. The mink becomes sexually mature at 10 months of age. Ranch mink may continue to produce young until 11 years of age. Wild mink may live to 3 to 6 years, and captives live 8 to 10 years or more.

**Age and Sex Determination.** Petrides (1950) and Elder (1951) have used the baculum to distinguish juvenile and adult mink. They found that both length and weight of the baculum increase each year. Greer (1957) correctly identified the age of 843, or 95.5 percent, of a total of 883 juvenile and adult mink by combining two characters—presence or absence of the tubercle on the femur and the jugal-squamosal suture. Franson, Dahm, and Wing (1975) have demonstrated a method for preparing and sectioning mink mandibles for age determination.

Immature females have very small, unpigmented teats, whereas in adults the teats are dark and raised. In cased skins the sex can be identified by the presence or absence of the penis scar.

**Specimens Examined.** MAINE, total 9: Hancock County, Orland 1; Penobscot County, Lincoln 1 (UME), South Twin Lake 2 (AMNH); Piscataquis County, Greenville 1 (UME), Mt. Katahdin 1; Washington County, Columbia Falls 1; York County, Eliot 2 (USNM).

NEW HAMPSHIRE, total 10: Grafton County, Hanover 1 (DC), Wentworth 1 (AMNH); Hillsborough County, Sharon 5 (UMA); Rockingham County, Derry 1; Strafford County, Durham 2 (UNH).

VERMONT, total 8: Chittenden County, Milton 1, Williston 1 (UVT); Rutland County, Brandon 1, Rutland 1 (DC); Windham County, Newfane 3 (AMNH); Windsor County, Bridgewater 1 (USNM).

MASSACHUSETTS, total 48: Bristol County, Swansea 1 (USNM); Hampden County, Montgomery 1; Hampshire County, Amherst 3, Hadley 1, Haydenville 1, Florence 1, Pelham 28, Southampton 1, Sunderland 9 (UMA); Middlesex County, Wilmington 1 (USNM); Worcester County, Berlin 1 (UMA).

CONNECTICUT, total 58: Litchfield County, Bantam 1, Litchfield 1 (UCT); Middlesex County, Portland 12 (USNM); New Haven County, Middlebury 1; Tolland County, Columbia 3, Mansfield 13, Merrow 1, Storrs 4, Willington 3; Windham County, Ashford 2, Canterbury 2, Chaplin 9, North Windham 1, Thompson 5 (UCT).

# Sea Mink

*Mustela vison macrodon* (Prentiss)

*Lutreola macrodon* Prentiss, 1903.
  *Proc. U.S. Natl. Mus.* 26:886–87.
*Mustela vison macrodon,* Manville,
  1966. *Proc. U.S. Natl. Mus.*
  122(3584):1–12.
Type Locality: Shell heaps at Brooklin, Bluehill Bay, Hancock County,
  Maine
Names: Saltwater mink, seashore
  mink, shell-heap mink, bull mink,
  big or giant mink, and ancient
  mink. The meaning of *mustela* = a
  weasel, and *macrodon* = bigtoothed.

**Description.** The extinct sea mink is known only from skeletal remains. It was first described by Prentiss (1903) from skull fragments found in an Indian shell heap or kitchen midden dating back to pre-Columbian times at Brooklin, Hancock County, Maine, the type locality. Prentiss compared these fragments with the largest mink extant, the Alaska mink, *Mustela vision ingens,* and found that the sea mink was decidedly larger, with a tooth row measurement of 30 mm (1.17 in) compared with 28 mm (1.09 in) in the Alaska mink (Manville 1966).

Hardy (1903), of Brewer, Maine, who traded for some 50,000 mink skins with Indians of the Penobscot and Jericho Bay regions, reported that the sea mink became extinct about 1860 or somewhat later. The author stated that the fur of the sea mink was much coarser and much more reddish than that of the common inland mink, and that the sea mink had a very strong fishy odor. Referring to various accounts of what was known of the sea mink, Seton (1921) called attention to the possibility of obtaining a specimen and stated: "There are traditions along the coasts of Maine, New Brunswick, etc., of a gigantic mink known as the sea-mink, which was commonly trapped there during the early part of the 19th century."

Some years ago an unusually large mounted mink was located in the possession of Clarence H. Clark. According to Mr. Clark, the specimen was captured by a neighbor of his near the Bay of Fundy about 1874. About 1964, James C. Sullivan, of East Winthrop, Maine, acquired the Clark collection and was anxious to sell the mounted specimen to the National Museum. He agreed to let the museum examine it to verify its identification before further negotiations regarding its acquisition. The supposed specimen of sea mink was examined by Richard Manville and his staff in 1965, and by visiting mammalogists. As a result, these scientists concluded that the legendary Clark specimen was merely an unusually large mink, probably an intergrade between *M. v. mink* and *M. macrodon.* Accordingly, the sea mink was reduced to subspecific rank, as *Mustela vision macrodon*

(Prentiss). The specimen was returned to Mr. Sullivan.

Manville (1966) reported that the fur of the Clark specimen was coarse and much faded, light reddish tan in color. It had a large whitish patch between the forelegs about 15 by 50 mm in area, with smaller white spots on the left forearm and medially in the inguinal region.

Measurements of the Clark specimen were approximately as follows: total length 72 cm (28.4 in), tail 21 cm (8.3 in), and hind foot 7 cm (2.8 in).

**Distribution.** The probable distribution of the sea mink was along the north Atlantic seaboard from southwestern Nova Scotia (Anderson 1947) to the coast of Connecticut (Goodwin 1935). In Maine, remains of the large mink have been reported from middens near South Harpswell on Casco Bay (Loomis 1911), between Rouge's Island and Great Diamond Island, Casco Bay (Norton 1930), and in the eastern part of the Gulf of Maine (Allen 1942). In Massachusetts, Waters and Ray (1961) unearthed remains of *M. v. macrodon* at Assawampsett Pond in Middleboro, Plymouth County, and speculated that the animal may have reached there via the Taunton River or Mattapoisett River, or may have been transported by Indians from Narragansett Bay or Buzzards Bay. Waters and Mack (1962b) reported remains of the sea mink from Connant's Hill on the Weweantic River; Cronan and Brooks (1968) remarked that it may have occurred in Rhode Island; and Goodwin (1935) thought that it might have occurred in the salt marshes and rivers along the coast of Connecticut.

Manville (1966) examined 57 fragments of skulls of *M. v. macrodon* and numerous specimens of *M. vision* and reviewed what was known on the sea mink, then summed up the distribution of *M. v. macrodon* as follows: "Known only from skeletal remains; coast of New England from Penobscot and Casco Bays south to Middleboro, Plymouth Co., Mass.; possibly north to Campobello Island, New Brunswick, and south to the saltmarshes and rivers of Connecticut."

# Striped Skunk

*Mephitis mephitis nigra* (Peal and Palisot de Beauvois)

*Viverra nigra* Peale and Palisot de Beauvois, 1796. *A scientific and descriptive catalogue of Peale's Museum, Philadelphia*, p. 37.
*Mephitis mephitis nigra*, A. H. Howell, 1921. *N. Amer. Fauna* 45:39.
Type Locality: Maryland
Dental Formula:
  I 3/3, C 1/1, P 3/3, M 1/2 × 2 = 34
Names: Common skunk, lined skunk, polecat, and wood pussy. The meaning of *Mephitis* = pestilent or bad odor, referring to the characteristic obnoxious odor given off by the scent glands, and *nigra* = black.

**Description.** The striped skunk is about the size of a house cat but has a relatively small head, short legs, robust, wide rear end, and bushy, long-haired tail. The forefeet have long, curved claws for digging; the claws of the hind feet are shorter and straighter. The feet are plantigrade, each with five toes. Females may have from ten to fourteen mammae, though the usual number is twelve.

The skull is long, relatively narrow, highly arched, and deepest in the frontal region, and the rostrum is deep and truncate. The auditory bullae are slightly inflated; the infraorbital foramina are small and sometimes divided by thin septa. The posterior border of short palate is nearly in line with the posterior borders of upper molars. The squarish mandible has a distinct posterior notch. The teeth are relatively heavy, and the upper molars are somewhat dumbbell-shaped. The upper rear molars are larger than the carnassials.

The long, thick pelage is composed of soft, wavy underfur overlaid with long, coarse, shiny guard hairs. The sexes are colored alike and show no noticeable seasonal change. The color is mostly glossy black but a thin white stripe runs from the nose to the back of the forehead and a broad white stripe extends from the crown of the head over the nape, branching at the shoulders, and continues posteriorly along the upper sides to the rump or onto the tip of the tail. A distinct white stripe may also occur on the outside of each front leg.

The amount and distribution of white varies considerably among individual skunks. The fur industry grades skunk pelts according to the amount of white in the pelage. In the best grade, or no. 1, the white may be absent or restricted to the head and neck. The no. 2 pelts, or "short stripe" pelts, have white stripes that do not extend beyond the middle of the body. No. 3 pelts have long narrow stripes, and no. 4 pelts are broad-striped. Northern pelts are more valuable than southern pelts because the fur is finer and the black color more intense. Some skunks have a white patch on the chest. Immatures are striped more or less like the adults. Cream-colored, brown, and all-black specimens occur sometimes, silvering seldom, and albinos rarely.

Molting begins in April and is completed by early September. Molting of the underfur and guard hairs progresses from front to back. Immatures apparently begin to molt when approximately a year old.

Males are somewhat larger than females. Measurements range from 54 to 67 cm (21.3–

STRIPED SKUNK

26.4 in); tail, 18 to 29 cm (7.1–11.4 in); hind foot, 6 to 8 cm (2.4–3.1 in). Weights vary from 1.1 to 3.4 kg (2.5–7.5 lb).

**Distribution.** The striped skunk occurs throughout southern Canada and all of the United States, except for arid areas in the Southwest, and in extreme northern Mexico.

**Ecology.** Striped skunks occur from sea level to timberline. They inhabit a variety of habitat types, from rolling weedy fields, fencerows, wooded ravines, rocky outcrops, and drainage ditches where water is available. They are encountered under porches of houses and vacant buildings, in culverts, and near dumps. Striped skunks may dig their own burrows, but they prefer to use natural cavities among rocks, caves, recesses under stone walls or fences, and holes under fallen logs and old stumps. Most often they retreat to abandoned burrows of woodchucks, red foxes, muskrats, or other mammals. Slopes seem to be the preferred den sites, probably because drainage is good. Scott and Selko (1939) found that most striped skunk dens in Michigan were on slopes of 5 to 10 percent or greater, and Verts (1967) reported a slope of 5.3 percent for skunk dens in Illinois.

Most burrows are from 6 to 20 feet long, but an unusually long burrow may be 50 feet long or more. Burrows generally reach 3 to 4 feet underground, and those less than 20 inches underground are apt to be summer dens without nests. The burrow ends in one, two, or three spherical chambers, 12 to 15 inches in diameter. The chambers may be lined with a bushel of mixed leaves and grass. In cold weather the natal nest material is used to plug the entrances. A den may have from one to five well-hidden entrances with openings about 8 inches in circumference, which are found more or less on the sunny side of slopes.

Man, lynxes, bobcats, foxes, coyotes, and great horned owls prey on skunks. Most dogs avoid skunks after the first encounter. Many skunks are killed by automobiles, especially during the fall. Striped skunks are hosts to fleas, ticks, lice, botfly larvae, cestodes, nematodes, and trematodes.

Diseases of striped skunks are leptospirosis, listeriosis, pulmonary aspergillosis, Q fever, Chagas's disease, murine typhus, pleuritis, tularemia, histoplasmosis, brucellosis, canine distemper, bronchopneumonia, and rabies.

**Behavior.** Striped skunks are well known for their characteristic odor of musk, produced by a yellowish, oily, phosphorescent, nauseating fluid sometimes containing creamy yellow curds. This fluid is secreted by a pair of oval anal glands on each side of the rectum, embedded in a mass of sphincter muscles which can compress one or both glands so forcefully that the secretion can be ejected up to 15 feet or more downwind. Each gland has a single duct leading to a prominent nipplelike papilla that can be protruded from the anus. The pungent fluid is composed of butylmercaptan, a malodorous sulphur-alcohol compound. Contrary to popular belief, striped skunk secretion causes no permanent ill effects when sprayed into the eyes, but it causes an intense burning sensation and causes tears to flow for from a few moments to nearly an hour or so.

Striped skunks emit scent or spray the fluid as a means of defense. When they are provoked, they arch their backs, plume out the tail high above the back, stamp their front feet very rapidly, and shuffle backwards; if pressed further, they eject either a fine foglike spray or a short stream of droplets. They may first face an adversary, then in a flash bend into a U-shape with both head and rear facing the intruder and swiftly scent. They do not scatter the secretion with the tail as is commonly believed. They can scent while suspended by the tail. Striped skunks generally avoid scenting themselves and will refrain from scenting if the tail is held over the anus or if the anus is held tightly against the ground.

Striped skunks are not sociable animals, but several skunks of both sexes may den together in winter. They are not agile and normally walk slowly, and they do not climb trees. They usually avoid water but will swim when necessary. Wilber and Weidenbacher (1961) demonstrated that striped skunks were able to swim for at least 463 minutes (7.7 hours) in water at 23°C.

The senses of sight, smell, and hearing are poor to fair, but touch appears well developed.

Striped skunks are usually silent but occasionally utter low growls, grunts, churrings, squeals, chatters, and hissing noises.

Striped skunks sometimes wander a considerable distance but normally do not roam far. Their nightly movements are between ¼ and ½ a square mile, and during the breeding season males may travel 4 to 5 miles a night (Schwartz and Schwartz 1959). Female skunks generally have a daily range only 700 feet in diameter (Dean 1965). Movements of juvenile skunks average 699 yards in late summer and autumn, and after the breeding season movements of adult males average about ¼ to ½ mile more than those of juvenile males in autumn. Pregnant females range about ¼ mile from their dens (Verts 1967). Males, both adults and juveniles, tend to move greater distances than females during late summer and autumn (Storm 1972). Although the movement of a rabid skunk is somewhat aberrant, it is no more extensive than that of a healthy skunk (Storm and Verts 1966).

Striped skunks are crepuscular or nocturnal but are sometimes seen during daylight hours. Parker (1962) reported that in areas where rabies is prevalent, a skunk encountered in daylight may be rabid. Verts (1967) reported that many newly weaned young skunks appear to spend the daylight hours in cornfields. Storm (1972) found that the animals extensively used hayfields, pastures, fencerows, and waterways as day retreats.

During late summer striped skunks acquire much body fat, and by late autumn and colder weather they spend more time in the den. They do not hibernate but merely go into a winter sleep, since their rate of metabolism does not correspond to lowering of ambient temperatures.

Striped skunks feed on a great variety of insects, earthworms, snails, grains and nuts, sweet corn, grasses, leaves and buds, apples, and a host of wild berries, bird's eggs, frogs, snakes, turtle eggs, mice, voles, moles, shrews, rats, young chipmunks, bats, squirrels, and rabbits, garbage, and carrion.

Striped skunks are polygamous and normally breed once a year. Shadle (1953) reported that a captive female produced two litters, on 16 May and on 28 July. Parks (1967) reported that two captive females each produced second litters after losing their first litters, and a third female produced a litter on 20 July. Although Shadle (1953) suggested that lactation inhibits the next sexual cycle, Parks (1967) stated that lactation did not seem to inhibit copulation or pregnancy. The breeding season begins in February and lasts until late March. By mid-February males are seeking females even at nightly temperatures of 10°F (Hamilton 1963).

The gestation period is between 62 and 68 days, though Parks (1967) reported a gestation period of 59 days in a captive female in New York. The cubs or kits are born between late April and early June. A litter may contain from two to ten young, but usually consists of six or seven. Older females generally produce litters during the first part of May, whereas younger females usually produce litters in early June and generally have fewer young than do older females.

At birth the kits are thinly furred, wrinkled, and blind. They are at least 14.3 cm (5.6 in) long and weigh 31 to 36 g (1.09–1.26 oz). The black-and-white patterns show distinctly on the pink skin, the front claws are well developed, and the sexes can be distinguished. The eyes open at about 3 to 4 weeks of age. About the time the eyes open, the young can assume the defensive posture and can scent a small amount of fluid. Weaning may occur when the young are between 6 and 8 weeks old.

The kits grow rapidly and follow the female on hunting trips when they are nearly 2 months old, often keeping close behind her in a long single file along a trail. Soon thereafter the young become independent and disperse. They are capable of breeding the spring following their birth (Verts 1967). Longevity may be from 2 years (Linduska 1947) to 42 months in the wild (Verts 1967). In captivity the animals may live as long as ten years (Schwartz and Schwartz 1959).

**Age and Sex Determination.** Casey and Webster (1975) described age determination of striped skunks by counting the number of annual layers in the tooth cementum and sex determination by examining the hippocampal neurons for sex chromatin.

**Specimens Examined.** MAINE, total 33: Franklin County, Farmington 1 (UME); Hancock County, Bucksport 14; Oxford County, Norway 3, Umbagog Lake 4, Upton 6 (HMCZ); Penobscot County, Orono 1 (UME), South Twin Lake 2 (AMNH); Piscataquis County, Greenville 1 (HMCZ); Somerset County, Jackman 1 (AMNH).

NEW HAMPSHIRE, total 11: Carroll County, Ossipee 8 (USNM); Coos County, Mt. Washington 1 (AMNH); Merrimack County, Webster 2 (HMCZ).

VERMONT, total 18: Bennington County, Manchester 1–1 (AMNH–HMCZ); Chittenden County, Ferrisburg 1, Williston 1 (UVT); Essex County, Lunenburg 1; Franklin County, Enosburg Falls 2 (HMCZ); Windham County, Newfane 10, Whitingham 1 (AMNH).

MASSACHUSETTS, total 58: Barnstable County, Barnstable 1 (HMCZ); Bristol County, Taunton 1 (USNM); Essex County, Ipswich 2 (HMCZ); Hampden County, Westfield 1; Hampshire County, Amherst 3 (UMA); Middlesex County, Belmont 1 (HMCZ), Burlington 2 (USNM), Cambridge 1, Concord 1, Newton 4, Newtonville 2, Watertown 1 (HMCZ), Wilmington 14 (USNM); Norfolk County, Bellingham 7, Lincoln 2; Plymouth County, Wareham 14; Worcester County, Harvard 1 (HMCZ).

CONNECTICUT, total 37: Fairfield County, Danbury 1, Noroton 1, Redding 2 (HMCZ), Ridgefield 1 (AMNH); Hartford County, East Hartford 1–3 (HMCZ–USNM); Litchfield County, Terryville 1; Middlesex County, Portland 1; New Haven County, Waterbury 1 (UCT); New London County, Liberty Hill 5 (HMCZ), Pachaug 1, Stonington 1; Tolland County, Mansfield 4, Storrs 3, Strafford 1, Tolland 1, Vernon 2 (UCT); Windham County, Plainfield 2 (USNM), South Woodstock 4 (AMNH), Warrenville 1 (UCT).

RHODE ISLAND, total 2: Providence County, Mentaconkanut Hill 1 (AMNH); Washington County, Bradford 1 (UCT).

# River Otter

*Lutra canadensis* (Schreber)

*Mustela lutra canadensis* Schreber, 1776. *Die Säugthiere* . . . , part 3, no. 18, pl. 126b.
*Lutra canadensis*, Sabin, 1823. In Franklin, *Narrative of a journey to the shores of the Polar Sea* . . . 1819–22, p. 653.
Type Locality: Eastern Canada, Province of Quebec
Dental Formula:
I 3/3, C 1/1, P 4/3, M 1/2 × 2 = 36
Names: Canada otter, northeastern otter, and land otter. The meaning of *Lutra* = an otter or land otter, and *canadensis* = of Canada.

**Description.** The river otter is a streamlined semiaquatic mustelid with a long, lithe, and muscular cylindrical body and short, stout legs. The head is relatively small, broad, and flattened. The muzzle is short and the nose pad is prominent and flat. The ears are short, the small eyes are set high and forward on the head, and the whiskers are prominent. The tail is long and rounded, thick at the base and tapering to the tip. The feet are pentadactyl and broad, with webbed toes and strong claws. The dense, sleek pelage is composed of oily short, soft underfur intermixed with long, glistening hard guard hairs which serve as insulators in cold water. The ears and nose can be closed underwater. Females have six mammae—two abdominal and four inguinal.

The skull is broad and strongly flattened, with a broad, short rostrum and large braincase. The palate extends behind the upper molars, which are squarish, cuspidate, and broader than long. The auditory bullae are small and depressed; the infraorbital foramen is large and nearly oval in anterior view; and the parietals are somewhat rounded and swollen. The second lower incisor is large and situated behind the other incisors. The upper carnassials have tricuspid blades and large inner lobes.

The sexes are colored alike and show little or no seasonal color variation. The coloration is a dark rich brown, becoming somewhat paler or grayer below. The muzzle and throat are grayish. Immatures are colored like adults. Variations of black, slate, white, and albino occur rarely. Jackson (1961) indicated that there might be two molts annually, in the spring and in autumn, with molting extending over a long period.

Males are slightly larger than females. Measurements range from 89 to 130 cm (35.1–51.2 in); tail, 30 to 50 cm (11.8–19.7 in); and hind foot, 11 to 15 cm (4.3–5.9 in). Weights vary from 5.4 to 11.4 kg (12.0–25.0 lb), rarely up to 13.6 kg (30 lb).

**Distribution.** The river otter occurs throughout most of Canada and Alaska and the continental United States, except for treeless areas and desert regions. The species has never been abundant in any part of its range, and it has been nearly extirpated in many areas since the period of settlement.

**Ecology.** River otters occur along streams, sloughs, swamps, rivers, and lakes, and not infrequently near brackish water. The well-hidden den may be merely a convenient resting place under roots of large trees, under fallen trees, beneath rocky ledges, in hollow logs, or in thickets near water. River otters may use an abandoned beaver lodge or bank den or enlarge a muskrat house or woodchuck burrow. Dens along a stream or lake have an opening above

RIVER OTTER

water in summer and usually below the ice line in winter. The den may be simple with either short or extensive and complex tunnels. The floor of a den may be bare or covered with some vegetation. The females do not clean the den.

Man is the chief predator of river otters. A coyote, lynx, bobcat, or great horned owl may catch an unwary kit, and sometimes an otter is caught by a fishnet or fishhook. Ticks, nematodes, and trematodes are known to infest river otters.

**Behavior.** River otters are active throughout the year. Although they are principally nocturnal, they can be seen hunting or playing during the daytime. They are intelligent, shy, gregarious, and loyal to each other. They are strong and capable fighters: an otter can usually whip a dog of comparable size. These animals utter low chuckles, chirps, purring grunts, growls, shrills, hissing barks, and screams. Females caterwaul when copulating.

River otters are one of the most aquatic of mustelids, graceful and excellent swimmers. Normally they paddle with the hind feet, but when pressed they swim by twisting and writhing the body and tail. They can outswim and catch a trout in open water. River otters may swim in a nearly horizontal position with the head held high above water, swim completely submerged, or undulate above and below the surface with scarcely a ripple. These animals

may remain underwater for up to 2 minutes and can swim up to 6 or 7 miles per hour on the surface as well as underwater. They spend much time under the ice fishing, obtaining oxygen from open holes in the ice or from air bubbles under the ice. Scott (1939) reported that a drowned river otter was retrieved on a fishhook set in 42 feet of water 500 feet from the shore of Lake Superior. Scheffer (1953) recorded an otter drowned in a fishnet set at a depth of 60 feet at Fish Bay, Alaska, and two otters were later taken from a net at the same depth in Deep Bay, Alaska.

River otters enjoy tumbling, wrestling, chasing each other, following the leader, and playing with rocks. They delight in sliding down steep banks on their chests and bellies with their feet folded out of the way. The slide may be on smooth, steep grass- mud- or snow-covered slopes and may end in deep water or in a snowdrift. A slide may be more than 25 feet long.

River otters are not fast runners; they lope or canter up to 18 miles per hour. But they are great travelers and have been known to wander 100 miles or more seeking new territory. During the warmer seasons they tend to remain near water and constantly move up and down the shore near water, often crossing from one stream to another to find easy food sources. In winter they often travel many miles along shores and overland in search of entrances to water for fishing. They love to roll in water and on land. These mustelids have haul-out places along streams and lakes where they roll on the grass to dry off and frequently defecate and urinate.

River otters feed on fishes, crayfish, frogs, clams, salamanders, snails, turtles, earthworms, snakes, small birds, and some plant matter. They excel at catching fish and exhibit a high degree of cooperation. Sheldon and Toll (1964) saw two otters, swimming on the surface about 10 feet apart, dive and swim rapidly toward a shallow cove, apparently herding fish before them. As they attempted to catch the fish they had driven into shallow water, the otters went through contortions that splashed the water in all directions. Both otters came up with fish.

River otters eat fish on land and may play with their prey before eating it piecemeal. They grub for aquatic animals by standing on their heads while rooting in mud and debris with only the tips of their tails showing at the surface of the water. Liers (1951) reported that it takes about an hour for an otter to digest a crayfish.

Not much is known on the reproduction of river otters. Some biologists believe they are monogamous, while others believe that the bond between pairs is broken shortly after breeding. Liers (1951) reported that tame otters breed in winter and in early spring and that the variable gestation period is between 9 and 12 months. Hamilton and Eadie (1964) assumed that most wild otters breed in March and April and that the blastocyst is implanted in January or early February. Development then proceeds rapidly, and the kits are born in March or April.

River otters are docile during most of the year except during the breeding season, when several males may follow a female in estrus and fight among themselves for her. Copulation takes place in or under water or on land, though wild otters probably breed exclusively in water. The several vigorous copulations may last from 15 to 25 minutes, interspersed with rest periods. Females may breed again soon after birth of the kits (Liers 1951).

The single annual litter consists of one to five kits, or rarely six, but the usual is two or three. The kits are born over a period of 3 to 8 hours, depending on the number being born. The female stands on all four feet during parturition, then curls herself tightly around her young and may put her head over them for protection and for nursing. At birth the kits are blind, and toothless, with ears closed, but are fully furred, with well-formed claws, and are typically otter-like in appearance. They are helpless for 5 to 6 weeks (Liers 1951). Hamilton and Eadie (1964) reported two full-term fetuses whose approximate size was: total length, 27.5 cm (10.8 in); tail, 6.4 cm (2.5 in); hind foot, 2.8 cm (1.1 in); and weight, 132 g (4.6 oz).

The kits grow slowly, and their eyes open at about 35 days. Both parents are attentive, but the female does not allow the male near the kits until they leave the nest when they are 10 to 12 weeks old. The kits are weaned at about 4 months of age. They begin to play with one another and with the female when they are 5 to 6 weeks old. The female trains her kits to swim and hunt. At first the kits may be reluctant to enter water, but they are coaxed to swim. The female may allow the kits to climb on her back in the water. The kits remain with the parents until they are about 1 year old, when a new litter is born (Liers 1951).

Captive females do not normally breed until 2 years of age (Liers 1951). Wild females may breed for the first time in March or April when they are 2 years old, and males may become sexually mature at 2 years (Hamilton and Eadie 1964). Longevity may be 15 to 20 years in captivity and is probably 8 to 10 years in the wild (Scheffer 1958).

**Sex Determination.** The sex of adults can be determined by the relative positions of the anus and urogenital openings. The presence of the penis can be readily determined in males by palpation.

**Specimens Examined.** MAINE, total 17: Hancock County, Bucksport 1, Mt. Desert Island 1; Oxford County, Umbagog Lake 5, Upton 5; Penobscot County, Lincoln 1; Piscataquis County, Moosehead Lake 4 (HMCZ).

NEW HAMPSHIRE, total 3: Cheshire County, Westmoreland 1 (HMCZ); Grafton County, Enfield 1, Lyme 1 (DC).

MASSACHUSETTS, total 14: Essex County, Swampscott 1 (HMCZ); Hampden County, Westfield 1; Hampshire County, Easthampton 1, Hadley 2 (UMA); Middlesex County, North Reading 1, Sudbury 1, Westford 1; Norfolk County, Canton 2; Plymouth County, Kingston 1, Wareham 2 (HMCZ); Worcester County, Petersham 1 (AMNH).

CONNECTICUT, total 12: New London County, Liberty Hill 1 (HMCZ); Tolland County, Columbia 1, Mansfield 2, Willington 2; Windham County, Chaplin 4, Hampton 2 (UCT).

### FAMILY FELIDAE (CATS)

Members of the cat family have teeth highly specialized for stabbing, slashing, and biting. The incisors are small and chisellike, and the canines are elongated, pointed, and slightly recurved. There are large, well-developed carnassials; the upper carnassial tooth has a three-lobed blade, the lower one has a talon or inner cusp. The premolars and molars are reduced in number and size.

Cats are highly efficient predators that stalk their prey with fluid stealth or lie in wait. They are agile climbers and good swimmers. Some species are nocturnal in habits, some diurnal.

Most cats have one or two litters a year, but some species may breed only every two or three years.

The family occurs worldwide, except in Antarctica and some remote oceanic islands. It is represented in New England by three species.

# Lynx

*Lynx canadensis* Kerr

*Lynx canadensis* Kerr, 1792. *The animal kingdom* . . . , 1:157.
Type Locality: Eastern Canada, Province of Quebec
Dental Formula:
I 3/3, C 1/1, P 2/2, M 1/1 × 2 = 28
Names: Canada lynx, lynx cat, and gray wild cat. The meaning of *Lynx* = a glarer, or one who can see well in dim light, and *canadensis* = of Canada.

**Description.** On the average the lynx looks much taller than the bobcat, but it weighs about the same. It is much more than twice as large as a domestic tomcat. The lynx is lean and stout-bodied, with long, muscular legs, large, spreading feet, a very short, blunt tail with a completely black tip, and prominent triangular long tufted ears. The eyes are prominent, and the nose is broad and blunt. The cheeks have a ruff of long hairs forming sideburns. Females have four mammae—two abdominal and two inguinal.

The skull is short, low, and broad, much smaller than that of the mountain lion, and has only two premolars in the upper jaw. In the lynx skull, the postorbital processes of the frontals are small, and the posterior palatine foramen is situated near the rim of the palate. It differs from the bobcat skull in having a wide posterior presphenoid, more than 5 mm (0.20 in) at its greatest width. The anterior condyloid foramen is distinct and separate from the foramen lacerum posterius. The long upper carnassials are more than 16 mm (0.62 in) long, and the condyle of ramus is shorter in exterior diameter. Many lynx skulls are smaller than large bobcat skulls.

The winter pelage is long, thick, silky, and loose, giving the animal a fluffy appearance. The sexes are colored alike and show individual variation as well as marked seasonal color variation. In winter the coloration of the upperparts is pale grizzled gray or brown. The guard hairs are white at the base, darkest in the middle, and black at the tip. The crown of the head is brownish but heavily suffused with white-tipped hairs. The nose and cheeks are grayish. The insides of the ears are grayish white edged with buff, with a central gray spot on the black posterior surface, and the tips of the ear tufts and the lines down the margins of the ears are brownish black. There are some black spots at the corners of the mouth, and the eyelids are

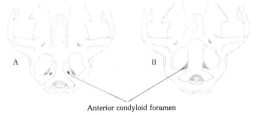

Basicranium of Canada lynx, *A*, with separate anterior condyloid foramen; bobcat, *B*, with anterior condyloid foramen confluent with foramen lacerum posteriorus.

white. The well-developed cheek ruff, chin, and throat are grayish white or light buffy brown mixed with long brownish black bars. The underparts are buffy white sparsely mottled with light brown, particularly on the insides of the legs. In winter the feet are heavily furred above and below; they are so broad that they serve remarkably well to support the cat in deep snow.

By late spring the fur is worn and ragged, paler and more buffy than the winter fur. The summer fur is darker, more grizzled and brownish than the winter fur. Immatures are spotted and streaked with brown and blackish on a light fawn color base. Albinos and melanistics are rare.

There seems to be a single annual molt that begins in late spring, with a continuous growth of new long, silky grayish-tipped hairs being acquired by late fall or early winter.

Males are somewhat larger than females. Measurements range from 83 to 100 cm (32.7–39.4 in); tail, 10 to 13 cm (3.9–5.1 in); and hind foot, 22 to 25 cm (8.7–9.9 in). Weights vary from 6.8 to 11.4 kg (15–25 lb). Exceptionally heavy individuals may weigh up to 15.9 kg (35 lb).

**Distribution.** The original range of the Canada lynx was boreal North America south of the timberline from Alaska to Nova Scotia, south to southern New England and New York, west possibly to the Michigan Upper Peninsula, northern Wisconsin, and southern Saskatchewan to the Pacific Ocean, and south in the Rockies into Colorado to the Uinta Mountains and the central mountains of Utah. It has since been extirpated from much of that range.

The lynx is uncommon to rare in New England, and its distribution is imperfectly known. In Maine, Hunt (1964) reported that this cat is rarely found east of the headwaters of the west branch of the Penobscot River, east of the upper headwaters of the Saint John and Allagash rivers, or south or west of Moosehead Lake. In New Hampshire, Silver (1957) reported that the lynx was restricted to the northern half of the state, particularly in the White Mountain National Forest region of Zeland Notch in Bethlehem and Lincoln, although on occasion a lynx may be taken as far south as Hillsborough or Merrimack counties. Siegler (1971) stated that the lynx has become very rare even in the White Mountain region and that lynx trapping has been banned in the National Forest, New Hampshire. In Vermont, Osgood (1938) stated that the animal was formerly taken occasionally, and Hamilton (1943) reported a lynx taken in Ripton, Vermont,

LYNX

in 1937. Current records of the lynx in Massachusetts and Connecticut are unavailable, but lynxes probably may be encountered in these states. The species may have occurred in Rhode Island during historical times.

**Ecology.** Lynxes inhabit deep, dark, unbroken mountainous boreal forests far from man. They rarely venture into open lands. They den in a hollow tree or under windfalls, in tangle thickets, under logs, stumps, or fallen timber, or in holes in the rocks. Lynxes often hide by flattening themselves on a limb or ledge, or on top of a log, rock, or knoll where they can easily escape an enemy or attack prey.

Man is the chief predator of the lynx. Fleas, cestodes, nematodes, and trematodes are known parasites of this cat.

**Behavior.** Lynxes are extremely shy and elusive and are rarely seen even where they are common. The animals are active throughout the year, mainly at night, but they may be seen prowling during the daytime in search of food. They are solitary except when hunting, during the breeding season, or when females have kittens.

These cats are usually silent, but they can purr, hiss, growl, spit, yowl, caterwaul, howl, and mew to the kittens. They are agile climbers and travel easily among fallen timber and moss-covered logs and boulders. But on the ground lynxes are surprisingly slow and run with an awkward long gallop; a dog could easily run them down. These cats are good jumpers and swim well, high in the water. They are powerful fighters and inquisitive, and they may follow a person for many hours, but they are not known to attack man.

Lynxes may hunt alone, in family groups, or possibly in two or more families, spreading out in a line across the woods instead of single file. Barash (1971) observed cooperative hunting among two adults and a juvenile lynx in Montana. One adult circled around a rockslide and conspicuously trotted in a zigzag pattern down a slope, flushing a Columbian ground squirrel toward the other adult lynx, which had crept up from behind the unsuspecting squirrel. The cat that made the kill then gave a short, low-pitched "meow," and the juvenile joined the adults. All

three cats appeared to share in eating the squirrel.

The population cycles of the lynx seem to correspond closely with those of the snowshoe hare, its basic food in winter. When the hares become scarce, the lynxes reproduce more slowly. This low reproduction of lynxes is well reflected in the correlation between a low snowshoe hare population and the number of lynx pelts that reach the fur market. Elton and Nicholson (1942) found from data of the Hudson Bay Company and other records left by traders over 206 years that during "good years" the company took in an average of 115,000 rabbit skins, compared with about 10,000 in "bad years." The authors correlated the numbers of the two species and found them synchronous. They wrote: "The cycle in lynx furs is very violent and regular and has persisted unchanged for the whole period. Its average period is about 9.6 years." Saunders (1961) also suggested that high lynx populations were generally correlated with high hare populations.

When snowshoe hares are scarce, lynxes migrate extensively. Jackson (1961) reported that the normal home range is less than a 5-mile radius, but lynxes may wander 50 miles or more. Saunders (1963b) found that the animals averaged 2 miles' travel and ranged from zero to 64 miles in winter, and that their home ranges varied from 6 to 8 square miles in Newfoundland. Nellis and Wetmore (1969) found that a male lynx moved 102 miles in 163 days in Alberta, but they believed that retrapping and trailing the animal may have prompted this long-range movement.

Lynxes feed chiefly on snowshoe hares, but when available they also eat rabbits, squirrels, chipmunks, mice, skunks, porcupines, waterfowl, game birds and their eggs, and occasionally fish and carrion. Saunders (1963a) determined that in one year a lynx might consume about 170 hares and a few birds and mice.

Lynxes may cannibalize another lynx. Elsey (1954) reported that a large lynx attacked and killed one of two young lynxes that were hunting snowshoe hares. These cats rarely eat plants.

Little is known about the reproduction of the lynx. The breeding season occurs in late winter, the gestation period is about 60 days, and from one to four kittens are born in a natural cavity on a rocky ledge or in a thicket. At birth the kittens weigh about 340 g (12 oz), and their eyes open in 9 or 10 days. They are weaned at 2 months. The females breed after 1 year, and the family disbands in about 6 to 9 months. Longevity is 10 to 20 years (Walker et al. 1964). The kittens are furred and are blotched and streaked on the upperparts and blotched and spotted on the underparts and legs until they attain the adult pelage at about 9 months of age (Jackson 1961). Yearling females approach the size of the adult females by the end of their second year (Saunders 1964).

**Specimens Examined.** MAINE, total 23: Aroostook County, Ashland 1 (UME), Masardis 1; Oxford County, Umbagog Lake 11, Upton 7 (HMCZ); Penobscot County, Eddington 1 (UME); Piscataquis County, Moosehead Lake 2 (HMCZ).

NEW HAMPSHIRE, total 2: Coos County, Milan 2 (HMCZ).

# Bobcat

*Lynx rufus* (Schreber)

*Felis rufus* Schreber, 1777. *Die Säugthiere . . .* , part 3, no. 95, pl. 109b.
*Lynx rufus,* Rafinesque, 1817. *Amer. Monthly Mag.* 2(1):46.
Type Locality: New York
Dental Formula:
  I 3/3, C 1/1, P 2/2, M 1/1 × 2 = 28
Names: Wildcat, bay lynx, red lynx, and lynx cat. The meaning of *Lynx* = a glarer, or one who can see well in dim light, and *rufus* = reddish, referring to the reddish summer pelage.

**Description.** The bobcat differs from the lynx in having shorter legs, smaller, bare feet, ears slightly or not at all tufted, a longer tail not black all around at the tip, pelage not pale grizzled but brownish and spotted, and shorter fur. Females have four mammae—two abdominal and two inguinal. The skull differs from that of the lynx by the characters previously pointed out in the lynx skull.

The pelage is dense, short, and very soft when in prime condition. The sexes are colored alike and there is marked seasonal color variation. The coloration varies considerably among individuals. In summer the upperparts are grayish, buffy, or reddish, spotted or streaked with black, darkest and most intense along the back from the head to base of the tail and becoming lighter on the sides. The rump and hind legs are buffy. The underparts and the insides of the legs are whitish with black spots, and there are black bars on the front legs. The head is streaked with black, and the buffy neck is heavily streaked or spotted blackish. The backs of the ears are black with a large gray or white patch, and the tufts are black when present. The eyelids are white. The tail above is colored like the back and has three or four indistinct black bars, the last being broadest and blackish at the tip, and the tail below is whitish to the tip. The winter pelage is much paler. Immatures are spotted like the adults. Albinos and melanistics occur occasionally. Molting probably occurs annually, beginning in early summer.

Males are somewhat larger than females. Measurements range from 71 to 120 cm (28.0–47.3 in); tail, 9 to 20 cm (3.5–7.9 in); and hind foot, 15 to 22 cm (6.0–8.7 in). Weights vary from 6.7 to 15.9 kg (15–35 lb). An unusually large bobcat may weigh 20.5 kg (45 lb) or more.

**Distribution.** This species occurs across southern Canada and southward through the United States into most of Mexico.

**Ecology.** Bobcats roam wild broken mountainous country with rocky habitats and adequate brushy timber interspersed with roads, old fields, and semiopen farmlands. They frequently haunt spruce thickets and cedar swamps and may be encountered in open areas of hardwoods mixed with stands of hemlock, white pine, or other conifers. During warmer months bobcats seek shelter and rest concealed in windfalls, under shrubs or low trees, or in rock crevices. They may den in rock crevices, under ledges, in caves, or in hollows of trees, stumps, or logs. The den is lined with dried grass, leaves, moss, or other soft vegetation which is scraped and scratched into the den.

Man and dogs are the chief predators of bobcats. A great horned owl sometimes catches a young bobcat. Bobcats are hosts to fleas, lice, ticks, cestodes, nematodes, and trematodes and are susceptible to mange, rabies, and feline enteritis or cat-scratch fever.

**Behavior.** Bobcats are solitary, elusive, and shy. They are active throughout the year, mainly at night, but may be seen hunting during the day, especially in winter. Bobcats have a fairly good sense of smell but rely chiefly on their keen eyesight and hearing to detect prey or enemies. When stalking, these cats creep stealthily along from cover to cover until they are close enough to pounce, or they may lie motionless in a tree,

BOBCAT

listening and watching for unwary prey, or crouch in ambush on a trail. At the first sign of danger they silently steal away. Bobcats will take to the trees to rest, escape hounds, chase prey, or catch a nesting bird.

Bobcats are quick and very strong and make formidable fighters. A large bobcat can pull down and kill a weakened deer in deep snow. McCord (1974) found in Massachusetts that 96 percent of deer carcasses cached by bobcats were hidden in coniferous woods. In deep snow, bobcats travel on roads, exposed shore and lake ice, logs, and in the tracks of other animals.

These cats usually walk or sometimes trot, but they are not fast runners. At best they run 15 miles per hour (Jackson 1961). Bobcats dislike water and will often go out of their way to find a log bridge rather than wade a stream; but if necessary they take to water and swim well. Young (1958) reported that bobcats can bound out of deep water like a bounding deer, and often a bobcat crowded by hounds will take to the water, make an about face, and with bounding leaps of as much as 12 feet, run unharmed through the pack of hounds. Amundson (1943) saw a bobcat swim a 150-foot wide canal. Bobcats are quiet but may scowl, caterwaul, snarl, and spit with bared teeth at an enemy. They

scream from time to time during the breeding season.

Bobcats are prone to follow established routes, choosing favorite places to defecate and urinate, but will deviate from a route to follow prey. The home range varies greatly with the availability of food; if food is scarce the home range is large. These cats often travel far in search of mates. In Maine, Marston (1942) reported that a bobcat traveled a triangular route, covering 18 miles within a month. In Massachusetts, Pollack (1951) suggested that bobcats travel 2 to 5 miles in a night. Rollins (1945) noted that bobcats in Minnesota traveled from 3 to 7 miles, averaging 5½ miles. Jackson (1961) reported that in Wisconsin the home range was seldom more than 2 square miles.

Bobcats feed chiefly on deer and snowshoe hares but will eat mice, squirrels, porcupines, mink, rabbits, skunks, muskrats, moles, shrews, opossums, chipmunks, birds and their eggs, snakes, fishes, crustaceans, insects, carrion, and some plant matter.

The breeding season usually begins in late February and extends into March, and some females may breed as late as June. The gestation period is about 62 days. As a rule, one litter is produced each year, but a second litter is sometimes born in early August. One to four, but usually two, kittens are born per litter. At birth the kittens are blind and helpless, but well-furred and spotted. They are approximatly 254 mm (10 in) long and weigh about 340 g (12 oz). The eyes open in about 10 days. The kittens are weaned at between 60 and 70 days. They often stay with the female until they are fairly well grown, or until autumn or early winter or later. Females will protect their young (Young 1958), but if unduly disturbed by man and dogs, a female may desert her newborn kittens. The female may tolerate the male with the kittens when they are nearly weaned (Gashwiler, Robinette, and Morris 1961).

Females reach sexual maturity within a year after birth; males become seasonally fertile but are not sexually mature until their second year (Crowe 1975).

There are three known instances of hybridization between the bobcat and the domestic house cat. In each instance, a male bobcat bred with a female house cat. The offspring either were bob-tailed and spotted like the bobcat or had long tails and resembled the house cat (Young 1958; Gashwiler, Robinette, and Morris 1961). Longevity in the wild is approximately 12 years, and bobcats remain sexually active until death (Crowe 1975). A captive male lived 25 years (Carter 1955).

**Age Determination.** Crowe (1972) demonstrated opaque annular layers of the cementum in bobcat canine teeth and suggested that the layers may be used to determine the age of a bobcat. Crowe (1975) found that no annulus is deposited during the first winter of a bobcat's life, because the teeth are growing rapidly at this time and the accelerated rate of cementum deposition may override or obscure those factors that precipitate annual deposits in later life. Annuli apparently are deposited during late winter and early spring.

**Specimens Examined.** MAINE, total 34: Aroostook County, Ashland 1 (UME), Bancroft 2 (HMCZ), Garfield 1 (UCT); Hancock County, Bucksport 7, Ellsworth 1, Hancock 2 (HMCZ); Somerset County (no location) 7 (USNM), Washington County, East Machias 4, Jonesboro 2, Steuben 2, West Pembroke 5 (HMCZ).

NEW HAMPSHIRE, total 75: Carroll County, Conway 4 (UMA); Cheshire County, Dublin 1, Keene 8; Coos County, Berlin 3, Colebrook 4, Lancaster 3 (USNM), Milan 1 (HMCZ); Grafton County, Canaan 5, Franconia 7, Littleton 1, Piermont 2, Warren 4 (USNM), Wentworth 3 (AMNH); Hillsborough County, Antrim 1, Bennington 1, Goffstown 1, Wilton 7; Merrimack County, Canterbury 1 (UCT), Pittsburg 6 (USNM), Sutton 1 (HMCZ); Sullivan County, Guild 12 (USNM).

VERMONT, total 19: Addison County, Hancock 8, Lincoln 1; Bennington County, Arlington 2 (HMCZ); Franklin County, East Fairfield 1 (UVT); Orange County, Fairlee 2 (DC); Rutland County, Clarendon 1 (HMCZ), Mendon 1 (UCT); Washington County, Worcester 1 (UVT); Windsor County, Bridgewater 1–1 (HMCZ–USNM).

MASSACHUSETTS, total 79: Berkshire County, Huntington 2 (UMA), Lenox 1, West Otis 1 (HMCZ); Franklin County, Leverett 2 (AMNH), New Salem 1, North Leverett 5, Wendell 1 (USNM); Hampden County, Blanford 2 (UMA), Sandisford 2 (HMCZ), Westfield 3; Hampshire County, Amherst 1, Easthampton 1 (USNM), Pelham 28 (UMA), Southampton 1 (USNM), Sunderland 9 (UMA), Westhampton 1 (USNM); Middlesex County, Pepperell 1; Worcester County, Barre 1 (HMCZ), Berlin 11 (UMA), Brookfield 1, Hubberston 1 (USNM), Petersham 1–1 (HMCZ–USNM), South Lancaster 1 (HMCZ).

CONNECTICUT, total 22: Fairfield County, Sherman 1 (UCT); Hartford County, Hartland 1; Litchfield County, Barkhamsted 5, Colebrook 1, Cornwall 1, Nepaug 1 (HMCZ), Norfolk 1–5 (AMNH–HMCZ), Washington 1 (UCT), Winchester 2 (HMCZ); Middlesex County, Portland 1; New Haven County, Southbury 1; Tolland County, Columbia 1 (UCT).

## REFERENCES

Ables, Ernest D. 1965. An exceptional fox movement. *J. Mamm.* 46(1):102.

——. 1969. Home-range studies of red foxes, *Vulpes vulpes. J. Mamm.* 50(1):108–20.

Aldous, C. M. 1939. Coyotes in Maine. *J. Mamm.* 20(1):104–6.

Allen, Durward L., and Shapton, Warren W. 1942. An ecological study of winter dens, with special reference to the eastern skunk. *Ecology* 23(1):59–68.

Allen, Glover M. 1942. Extinct and vanishing mammals of the western hemisphere with the marine species of all the oceans. *Amer. Int. Wild Life Proct. Spec. Pub.* 11:1–629.

Allen, Joel A. 1876. Former range of New England mammals. *Amer. Nat.* 10(12):708–15.

Allen, Stephen H. 1974. Modified techniques for aging red fox using canine teeth. *J. Wildl. Manage.* 39(1):152–54.

Amundson, Geno A. 1943. Bobcat swims. *J. Mamm.* 24(3):399–400.

Anderson, Rudolph Martin. 1947. Catalogue of Canadian Recent mammals. *Natl. Mus. Canada Bull.* 102:1–238.

Asdell, S. A. 1964. *Patterns of mammalian reproduction.* Ithaca, N.Y.: Comstock Pub. Co.

Ashbrook, Frank G., and Hanson, Karl B. 1927. Breeding martens in captivity. *J. Heredity* 18(11):498–503.

Barash, David P. 1971. Cooperative hunting in the lynx. *J. Mamm.* 52(2):480.

———. 1974. Neighbor recognition in two "solitary" carnivores: The raccoon (*Procyon lotor*) and red fox (*Vulpes fulva*). *Science* 185(4153):794–96.

Bee, James W., and Hall, E. Raymond. 1951. An instance of coyote-dog hybridization. *Trans. Kansas Acad. Sci.* 54(1):73–77.

Bekoff, Marc, and Jamieson, Robert. 1975. Physical development in coyotes (*Canis latrans*), with a comparison to other canids. *J. Mamm.* 56(3):685–92.

Benirschke, K. 1967. Sterility and fertility of interspecific mammalian hybrids. In *Comparative aspects of reproductive failure,* ed. K. Benirschke, pp. 218–34. New York: Springer-Verlag.

Bissonette, Thomas Hume, and Bailey, Early Elmore. 1944. Experimental modification and control of molts and changes of coat-color in weasels by controlled lighting. *Ann. New York Acad. Sci.* 45(6):221–60.

Black, Hugh C. 1958. Black bear research in New York. *Trans. N. Amer. Wildl. Conf.* 23:443–60.

Booth, Ernest S. 1946. Account of a weasel in a tree. *J. Mamm.* 26(4):439.

Brassard, J. A., and Bernard, R. 1939. Observations on breeding and development of marten, *Martes a. americana* (Kerr). *Canadian Field-Nat.* 53(2):15–21.

Bryant, H. C. 1920. The coyote not afraid of water. *J. Mamm.* 1(1):87–88.

Butterworth, Bernard B. 1967. Postnatal growth and development of *Ursus americanus. J. Mamm.* 50(3):615–16.

Cahalane, Victor H. 1947. *Mammals of North America.* New York: Macmillan Co.

Carter, T. Donald. 1955. Remarkable age attained by a bobcat. *J. Mamm.* 36(2):290.

Casey, G. A., and Webster, W. A. 1975. Age and sex determination of striped skunk (*Mephitis mephitis*) from Ontario, Manitoba, and Quebec. *Canadian J. Zool.* 53(3):223–26.

Churcher, Charles S. 1959. The specific status of the New World red fox. *J. Mamm.* 40(4):513–20.

———. 1960. Cranial variation in the North American red fox. *J. Mamm.* 41(3):349–60.

Cottam, Clarence. 1937. Speed of the gray fox. *J. Mamm.* 18(2):240–41.

———. 1945. Speed and endurance of the coyote. *J. Mamm.* 26(1):94.

Couch, Leo K. 1932. River swimming coyotes. *Murrelet* 13(1):24–25.

Coulter, Malcolm W. 1959. Some recent records of martens in Maine. *Maine Field Nat.* 15(2):50–53.

———. 1960. The status and distribution of fisher in Maine. *J. Mamm.* 41(1):1–9.

———. 1966. Ecology and management of fishers in Maine. PhD. diss., Syracuse University.

Crane, Jocelyn. 1931. Mammals of Hampshire County, Massachusetts. *J. Mamm.* 12(3):267–73.

Criddle, Norman, and Criddle, Stuart. 1925. The weasels of southern Manitoba. *Canadian Field-Nat.* 39:142–48.

Cronan, John M., and Brooks, Albert. 1968. *The mammals of Rhode Island.* Wildlife Pamphlet no. 6. Providence: Rhode Island Department of Agriculture and Conservation, Division of Fish and Game.

Crowe, Douglas M. 1972. The presence of annuli in bobcat tooth cementum layers. *J. Wildl. Manage.* 36(4):1330–32.

———. 1975. Aspects of ageing, growth, and reproduction of bobcats in Wyoming. *J. Mamm.* 56(1):177–98.

Daniel, M. J. 1960. Porcupine quills in viscera of fisher. *J. Mamm.* 41(1):133.

Dean, Frederick C. 1965. Winter and spring habits and density of Maine skunks. *J. Mamm.* 46(4):673–75.

DeVos, Antoon. 1951. Overflow and dispersal of marten and fisher from wildlife refuges. *J. Wildl. Manage.* 15(2):164–75.

———. 1957. Pregnancy and parasites of marten. *J. Mamm.* 38(3):412.

Dice, Lee R. 1942. A family of dog-coyote hybrids. *J. Mamm.* 23(2):186–92.

Dickey, Darold R. 1923. Evidence of interrelation between fox and caribou. *J. Mamm.* 4(2):121–22.

Drahos, Nick. 1952. Notes on bears. *New York State Conserv.* 7(2):14–18.

Eadie, W. R., and Hamilton, W. J., Jr., 1958. Reproduction in the fisher in New York. *New York Fish and Game J.* 5(1):77–83.

Elder, William H. 1951. The baculum as an age criterion in mink. *J. Mamm.* 32(1):43–50.

Elsey, C. A. 1954. Case of cannibalism in Canada lynx, *Lynx canadensis. J. Mamm.* 35(1):129.

Elton, Charles, and Nicholson, Mary. 1942. The ten year cycle in numbers of the lynx in Canada. *J. Animal Ecol.* 11(2):215–44.

Emmons, Ebenezer. 1840. *A report on the quadrupeds of Massachusetts.* Cambridge, Mass.: Folsom, Wells, and Thurston.

Enders, Robert K. 1952. Reproduction in the mink. *Proc. Amer. Philo. Soc.* 96(6):691–775.

Enders, Robert K., and Pearson, O. P. 1943. Shortening gestation by inducing early implantation with increased light in the marten. *Amer. Fur Breeder* 15(7):18.

Erickson, Albert W. 1957. Techniques for live-trapping

and handling black bears. *Trans. N. Amer. Wildl. Conf.* 22:520–43.

———. 1959. The age of self-sufficiency in the black bear. *J. Wildl. Manage.* 23(4):401–5.

Erickson, Albert W., and Martin, Paul. 1960. Black bear carries cubs from den. *J. Mamm.* 41(3):408.

Erickson, Albert W.; Nellor, John; and Petrides, George A. 1964. The black bear in Michigan. *Michigan State Univ. Exper. Sta. Res. Bull.* 4:1–102.

Ernst, Carl H. 1965. Rutting activities in a captive striped skunk. *J. Mamm.* 46(4):702–3.

Errington, Paul L. 1943. An analysis of mink predation upon muskrats in north central United States. *Iowa Agr. Exper. Sta. Res. Bull.* 320:799–924.

Foote, Leonard E. 1946. A history of wild game in Vermont. *Vermont Fish and Game Serv. State Bull.* 11:1–55.

Franson, J. Christian; Dahm, Paul A.; and Wing, Larry D. 1975. A method for preparing and sectioning mink (*Mustela vison*) mandibles for age determination. *Amer. Midland Nat.* 93(2):507–8.

Gashwiler, Jay S.; Robinette, W. Leslie; and Morris, Owen W. 1961. Breeding habits of bobcats in Utah. *J. Mamm.* 42(1):76–84.

Gianini, Charles A. 1923. Caribou and fox. *J. Mamm.* 4(4):253–54.

Gier, H. T. 1957. Coyotes in Kansas. *Bull. Agr. Exper. Sta. Kansas State Coll.* 393:1–97.

Giles, LeRoy W. 1943. Evidences of raccoon mobility obtained by tagging. *J. Wildl. Manage.* 7(2):235.

Glover, Fred A. 1943. A study of the winter activities of the New York weasel. *Pennsylvania Game News* 14(6):8–9.

Goodwin, George Gilbert. 1935. The mammals of Connecticut. *Connecticut Geol. and Nat. Hist. Surv. Bull.* 53:1–221.

———. 1936. Big game animals in the northeastern United States. *J. Mamm.* 17(1):48–50.

Gorham, John R. 1956. Diseases and parasites of minks. In *Animal diseases*, pp. 567–73. The Yearbook of Agriculture 1956. Washington, D.C.: U.S. Department of Agriculture.

Gorham, John R., and Griffiths, H. J. 1952. Diseases and parasites of minks. *U.S. Dept. Agr. Farmer's Bull.* 2050:1–41.

Grau, Gerald A.; Sanderson, Glen C.; and Rogers, John P. 1970. Age determination of raccoons. *J. Wildl. Manage.* 34(2):364–72.

Greer, Kenneth R. 1957. Some osteological characters of known-age ranch minks. *J. Mamm.* 38(3):319–30.

Hagmeier, Edwin M. 1956. Distribution of marten and fisher in North America. *Canadian Field-Nat.* 70(4):149–68.

Hall, E. Raymond. 1942. Gestation period in the fisher with recommendations for the animal's protection in California. *California Fish and Game* 28(3):143–47.

———. 1943. Cranial characters of a dog-coyote hybrid. *Amer. Midland Nat.* 29(2):371–74.

———. 1945. A revised classification of the American ermines with description of a new subspecies from the western Great Lakes region. *J. Mamm.* 26(2):175–82.

Hall, E. Raymond, and Kelson, Keith R. 1959. *The mammals of North America.* 2 vols. New York: Ronald Press.

Hamilton, William J., Jr. 1933. The weasels of New York: Their natural history and economic status. *Amer. Midland Nat.* 14(4):289–344.

———. 1943. *The mammals of eastern United States.* New York and London: Hafner Pub. Co.

———. 1958. Early sexual maturity in the female short-tailed weasel. *Science* 127(3305):1057.

———. 1963. Reproduction of the striped skunk in New York. *J. Mamm.* 44(1):123–24.

Hamilton, William J., Jr., and Cook, Arthur H. 1955. The biology and management of the fisher in New York. *New York Fish and Game J.* 2(1):13–35.

Hamilton, William J., Jr., and Eadie, W. Robert. 1964. Reproduction in the otter (*Lutra canadensis*). *J. Mamm.* 45(2):242–52.

Hamilton, William J., Jr.; Hosley, N. W.; and MacGregor, A. E. 1937. Late summer and early fall foods of the red fox in central Massachusetts. *J. Mamm.* 18(3):366–67.

Hardy, Manly. 1903. The extinct mink from the Maine shell heaps. *Forest and Stream* 61(7):125.

Harlow, R. F. 1961. Characteristics and status of the Florida black bear. *Forest and Streams* 61(7):125.

Haugen, Orland L. 1954. Longevity of the raccon in the wild. *J. Mamm.* 35(3):439.

Hawley, Vernon D., and Newby, Fletcher E. 1957. Marten home ranges and population fluctuations. *J. Mamm.* 38(2):174–84.

Hibbard, E., and Adams, A. 1957. Furbearer investigations in the Coteau. Bismarck: North Dakota Game and Fish Dept. (mimeographed).

Hunt, John H. 1964. The lynx. *Maine Fish and Game Mag.* 6(2):14.

Iljin, N. A. 1941. Wolf-dog genetics. *J. Genetics* 42(3):359–414.

Ingles, Lloyd G. 1965. *Mammals of the Pacific States.* Stanford, Calif.: Stanford University Press.

Isley, Thomas Earl, and Gysel, Leslie W. 1975. Sound-source localization by the red fox. *J. Mamm.* 56(2):397–404.

Jackson, Hartley H. T. 1961. *Mammals of Wisconsin.* Madison: University of Wisconsin Press.

Jeanne, Robert L. 1965. A case of a weasel climbing trees. *J. Mamm.* 46(2):344–45.

Jones, Sarah V. H. 1923. Color variants in wild animals. *J. Mamm.* 4(3):172–77.

Jonkel, Charles J., and Cowan, Ian McT. 1971. The

black bear in the spruce-fir forest. *Wildl. Monogr.* 27:1–57.

Jonkel, Charles J., and Miller, Frank L. 1970. Recent records of black bears, *Ursus americanus*, on the barren grounds of Canada. *J. Mamm.* 51(4):826–28.

Jonkel, Charles, J., and Weckwerth, Richard P. 1963. Sexual maturity and implantation of blastocysts in the wild pine marten. *J. Wildl. Manage.* 27(1):93–98.

Keay, Fred E. 1901. The animals our fathers found in New England. *New England Mag.*, 24 March.

Kennelly, James, and Roberts, Jerry D. 1969. Fertility of coyote-dog hybrids. *J. Mamm.* 50(4):830–31.

Kitzmiller, A. B. 1934. Memory of raccoons. *Amer. J. Psychol.* 46:511–12.

Kolenosky, George B. 1971. Hybridization between wolf and coyote. *J. Mamm.* 52(2):446–49.

Lawrence, B., and Bossert, William H. 1967. Multiple character of *Canis lupus, latrans,* and *familiaris,* with a discussion on the relationships of *Canis niger. Amer. Zool.* 7(2):223–32.

——. 1969. The cranial evidence for hybridization in New England canids. *Breviora* 330:1–13.

Layne, James N. 1958. Reproductive characteristics of the gray fox in southern Illinois. *J. Wildl. Manage.* 22(2):157–63.

Lehman, Larry E. 1968. September birth of raccoons in Indiana. *J. Mamm.* 49(1):126–27.

Libby, Wilbur L. 1961. Occurrence of "samson" condition in marten. *J. Mamm.* 42(1):112.

Liers, Emil E. 1951. Notes on the river otter (*Lutra canadensis*). *J. Mamm.* 32(1):1–9.

——. 1958. Early breeding in the river otter. *J. Mamm.* 39(3):438–39.

Linduska, J. P. 1947. Longevity of some Michigan farm game mammals. *J. Mamm.* 28(2):126–29.

Linhart, Samuel B., and Knowlton, Frederick F. 1967. Determining age of coyotes by cementum layers. *J. Wildl. Manage.* 31(2):362–65.

Linsley, James H. 1842. A catalogue of the mammals of Connecticut, arranged according to their natural families. *Amer. J. Sci.* 43(7):345–54.

Loomis, F. B. 1911. New mink from the shell heaps of Maine. *Amer. J. Sci.* 4(31):227–29.

Lord, Rexford D., Jr. 1961. The lens as an indicator of age in the gray fox. *J. Mamm.* 42(1):109–11.

Lynch, Gerry M. 1967. Long-range movement of a raccoon in Manitoba. *J. Mamm.* 48(4):109–11.

McCarley, Howard. 1975. Long-distance vocalizations of coyotes (*Canis latrans*). *J. Mamm.* 56(4):847–56.

McCord, Chet M. 1974. Selection of winter habitat by bobcats (*Lynx rufus*) on the Quabbin Reservation, Massachusetts. *J. Mamm.* 55(2):428–37.

MacGregor, Arthur E. 1942. Late fall and winter food of foxes in Massachusetts. *J. Wildl. Manage.* 6(3):221–24.

Manville, Richard H. 1953. Longevity of the coyote. *J. Mamm.* 34(3):390.

——. 1966. The extinct sea mink with taxonomic notes. *Proc. Portland Soc. Nat. Hist.* 4(1):1–151.

Markely, Merle H., and Bassett, Charles P. 1942. Habits of captive marten. *Amer. Midland Nat.* 28(3):604–16.

Marks, Stuart, and Erickson, Albert W. 1966. Age determination in the black bear. *J. Wildl. Manage.* 30(2):389–410.

Marshall, William H. 1935. Mink display sliding habits. *J. Mamm.* 16(3):228–29.

——. 1936. A study of the winter activities of the mink. *J. Mamm.* 17(4):382–92.

——. 1951a. Pine marten as a forest product. *J. Forestry* 49(12):899–905.

——. 1951b. An age determination method for the pine marten. *J. Wildl. Manage.* 15(3):276–83.

Marshall, William H., and Enders, R. K. 1942. The blastocyst of the marten (*Martes*). *Anat. Record* 84(9):307–10.

Marston, Merwin A. 1942. Winter relations of bobcats to white-tailed deer in Maine. *J. Wildl. Manage.* 6(4):328–37.

Matson, J. R. 1954. Observations on the dormant phase of a female black bear. *J. Mamm.* 35(1):28–35.

Mech, L. David; Barnes, Donald M.; and Tester, John R. 1968. Seasonal weight changes, mortality, and population structure of raccoons in Minnesota. *J. Mamm.* 49(1):63–73.

Mech, L. David, and Turkowski, Frank J. 1966. Twenty-three raccoons in one winter den. *J. Mamm.* 47(3):529–30.

Mengel, Robert M. 1971. A study of dog-coyote hybrids and implications concerning hybridization in *Canis. J. Mamm.* 52(2):316–36.

Merriam H. Gray. 1963. An unusual fox : woodchuck relationship. *J. Mamm.* 44(1):115–16.

Mitchell, James L. 1961. Mink movements and populations on a Montana River. *J. Wildl. Manage.* 25(1):49–54.

Monson, Ruth A.; Stone, Ward B.; and Webber, Bruce L. 1973. Heartworms in foxes and wild canids in New York. *New York Fish and Game J.* 20(1):48–53.

Montgomery, G. G. 1968. Pelage development of young raccoon. *J. Mamm.* 49(1):142–45.

Mumford, Russell E. 1969. Long-tailed weasel preys on big brown bats. *J. Mamm.* 50(2):360.

Mundy, K. R. D., and Flook, D. R. 1964. Notes on the mating activity of grizzly and black bears. *J. Mamm.* 45(4):637–38.

Murie, Adolph. 1936. Following fox trails. *Univ. Michigan Mus. Zool. Misc. Pub.* 32:7–45.

——. 1961. Some food habits of the marten. *J. Mamm.* 42(4):516–21.

Nellis, Carl H., and Wetmore, Stephens P. 1969. Long-range movement of lynx in Alberta. *J. Mamm.* 50(3):640.

Newby, F. E., and Hawley, V. D. 1954. Progress on a

marten live trapping study. *Trans. N. Amer. Wildl. Conf.* 19:452–62.

Norton, Arthur H. 1930. The mammals of Portland, Maine and vicinity. *Proc. Portland Soc. Nat. Hist.* 4(1):1–151.

Osgood, Frederick L., Jr. 1938. The mammals of Vermont. *J. Mamm.* 19(4):435–41.

Ozoga, John J., and Harger, Elsworth M. 1966. Winter activities and feeding habits of northern Michigan coyotes. *J. Wildl. Manage.* 30(4):808–18.

Palmer, Ralph S. 1956. Gray fox in the northeast. *Maine Field Nat.* 12(3):62–70.

Parker, R. L. 1962. Rabies in skunks in the north-central states. *Proc. U.S. Livestock Sanitation Assoc.* 65:273–80.

Parks, Eugene. 1967. Second litters in the striped skunk. *New York Fish and Game J.* 14(2):208–9.

Payne, Neil F. 1975. Unusual movements of Newfoundland black bears. *J. Wildl. Manage.* 39(4):812–13.

Petrides, George A. 1950. The determination of sex and age ratios in fur animals. *Amer. Midland Nat.* 43:355–82.

——. 1959. Age ratios in raccoons. *J. Mamm.* 40(2):249.

Phillips, Robert L.; Andrews, R. D.; Storm, G. L.; and Bishop, R. A. 1972. Dispersal and mortality of red foxes. *J. Wildl. Manage.* 36(2):237–48.

Phillips, Robert L., and Mech, L. David. 1970. Homing behavior of a red fox. *J. Mamm.* 51(3):621.

Polderboer, Emmett R.; Kuhn, Lee W.; and Hendrickson, George O. 1941. Winter and spring habits of weasels in central Iowa. *J. Wildl. Manage.* 5(1):115–19.

Pollack, E. Michael. 1951. Observations on New England bobcats. *J. Mamm.* 32(3):356–58.

Prentiss, Daniel Webster. 1903. Description of an extinct mink from the shell-heaps of the Maine coast. *Proc. U. S. Natl. Mus.* 26(1336):887–88.

Preston, Eric M. 1975. Home range defense in the red fox, *Vulpes vulpes* L. *J. Mamm.* 56(3):645–52.

Pringle, Laurence P. 1960. Notes on coyotes in southern New England. *J. Mamm.* 41(2):278.

Quick, Horace F. 1944. Habits on economics of the New York weasel in Michigan. *J. Wildl. Manage.* 8(1):71–78.

Rausch, Robert L. 1961. Notes on the black bear, *Ursus americanus* Pallas, in Alaska, with particular reference to dentition and growth. *Zeitschrift für Säugetierkunde* 26(2):65–128.

Richens, Voit B., and Hugie, Roy D. 1974. Distribution, taxonomic status, and characteristics of coyotes in Maine. *J. Wildl. Manage.* 38(3):447–54.

Rogers, J. G. 1965. Analysis of the coyote population of Dona Ana County, New Mexico. M.S. thesis, University of New Mexico.

Rogers, Lynn L. 1974. Shedding of foot pads by black bears during denning. *J. Mamm.* 55(3):672–74.

Rollins, Clair T. 1945. Habits, foods, and parasites of the bobcat in Minnesota. *J. Wildl. Manage.* 9(2):131–45.

Rowan, William. 1947. A case of six cubs in the common black bear. *J. Mamm.* 28(4):404–5.

Rust, Charles Chapin. 1962. Temperature as a modifying factor in the spring pelage of short-tailed weasels. *J. Mamm.* 43(3):323–28.

Rust, Charles Chapin; Shackelford, Richard M.; and Meyer, Roland K. 1965. Hormonal control of pelage cycles in the mink. *J. Mamm.* 46(4):549–65.

Sanderson, Glen C. 1949. Growth and behavior of a litter of captive long-tailed weasels. *J. Mamm.* 30(4):412–15.

——. 1950. Methods of measuring productivity in raccoons. *J. Wildl. Manage.* 14(4):389–402.

——. 1951. Breeding habits and a history of the Missouri raccoon population from 1941 to 1948. *Trans. N. Amer. Wildl. Conf.* 16:445–61.

——. 1961a. The lens as an indicator of age in the raccoon. *Amer. Midland Nat.* 65(2):481–85.

——. 1961b. Techniques for determining age of raccoons. *Illinois Nat. Hist. Surv. Biol. Notes* 45:1–16.

Sargeant, Alan B. 1972. Red fox spatial characteristics in relation to waterfowl predation. *J. Wildl. Manage.* 36(2):225–36.

Sauer, Peggy R. 1966a. Determining sex of black bears from the size of the lower canine tooth. *New York Fish and Game J.* 13(2):140–45.

——. 1966b. Growth of black bears from the Adirondacks. *Proc. N. E. Sec. Wildl. Soc.* 22:1–31.

Sauer, Peggy R.; Free, Stuart L.; and Browne, Stephen D. 1969. Movement of tagged black bears in the Adirondacks. *New York Fish and Game J.* 16(2):205–23.

Saunders, Jack, Jr. 1961. The biology of the Newfoundland lynx (*Lynx canadensis subsolanus* Bangs). Ph.D. diss., Cornell University.

——. 1963a. Food habits of the lynx in Newfoundland. *J. Wildl. Manage.* 27(3):384–90.

——. 1963b. Movements and activities of the lynx in Newfoundland. *J. Wildl. Manage.* 27(3):390–400.

——. 1964. Physical characteristics of the Newfoundland lynx. *J. Mamm.* 45(1):36–47.

Scheffer, Victor B. 1953. Otter diving to a depth of sixty feet. *J. Mamm.* 34(2):255.

——. 1958. Long life of a river otter. *J. Mamm.* 39(4):591.

Schladweiler, J. L., and Storm, G. L. 1969. Den-use by mink. *J. Wildl. Manage.* 33(4):1025–26.

Schneider, Dean G.; Mech, L. David; and Tester, John R. 1971. Movements of female raccoons and their young as determined by radio-tracking. *Animal Behav. Monogr.* 4(1):1–43.

Schwartz, Charles W., and Schwartz, Elizabeth R. 1959. *The wild mammals of Missouri.* Columbia: University of Missouri Press.

Scott, Thomas G. 1943. Some food coactions of the northern plains red fox. *Ecol. Monogr.* 13(4):427–79.

Scott, Thomas G., and Selko, L. F. 1939. A census of red foxes and striped skunks in Clay and Boone Counties, Iowa. *J. Wildl. Manage.* 3(2):92–98.

Scott, Walter E. 1939. Swimming power of the Canadian otter. *J. Mamm.* 20(3):371.

Seton, Ernest Thompson. 1921. The sea mink, *Mustela macrodon* (Prentiss). *J. Mamm.* 2(3):168.

———. 1929. *Lives of game animals.* Vols. 1–4. Garden City, N.Y.: Doubleday, Doran and Co.

Shadle, Albert R. 1953. Captive striped skunk produces two litters. *J. Wildl. Manage.* 17(3):388–89.

Sharp, Ward M., and Sharp, Louise H. 1956. Nocturnal movements and behavior of wild raccoons at a winter feeding station. *J. Mamm.* 37(2):170–77.

Sheldon, William G. 1949. Reproductive behavior of foxes in New York State. *J. Wildl. Manage.* 14(1):236–46.

———. 1950. Denning habits and home range of red foxes in New York State. *J. Wildl. Manage.* 14(1):236–46.

———. 1953. Returns on banded red and gray foxes in New York State. *J. Mamm.* 34(1):125.

Sheldon, William G., and Toll, William G. 1964. Feeding habits of the river otter in a reservoir in central Massachusetts. *J. Mamm.* 45(3):449–55.

Shipherd, Susan V., and Stone, Ward B. 1974. An albino gray fox. *New York Fish and Game J.* 21(2):184–85.

Sibbald, I. R.; Sinclair, D. G.; Evans, E. V.; and Smith, D. L. T. 1962. The rate of passage of feed through the digestive tract of the mink. *Canadian J. Biochem. Phys.* 40(10):1391–94.

Siegler, Hilbert R. 1971. The status of wildcats in New Hampshire. In *Proceedings of a Symposium on the Native Cats of North America,* held in conjunction with the 36th North American Wildlife National Resources Conference.

Silver, Helenette. 1957. A history of New Hampshire game and furbearers. *New Hampshire Fish and Game Dept. Surv. Rept.* 6:1–466.

Silver, Helenette, and Koons, G. 1972. Those New Hampshire coyotes. *New Hampshire Profiles,* pp. 40–47.

Silver, Helenette, and Silver, Walter T. 1969. Growth and behavior of the coyote-like canid of northern New England with observations of canid hybrids. *Wildl. Monogr.* 17:1–14.

Smith, Bertrand E. 1946. Bear facts. *J. Mamm.* 27(1):31–37.

Smith, H. A. P. 1923. Friendship between fox and caribou. *J. Mamm.* 4(2):122–23.

Snow, Carol J. 1967. Some observations on the behavioral and morphological development of coyote pups. *Amer. Zool.* 7(2):353–55.

Soper, J. D. 1919. Notes on Canadian weasels. *Canadian Field-Nat.* 33(1):43–47.

Spencer, Howard E., Jr. 1961. *The black bear and its status in Maine.* Bulletin no. 4. Augusta: Maine Department of Inland Fisheries and Game, pp. 1–55.

Stains, Howard J. 1967. Carnivores and pinnipeds. In *Recent mammals of the world,* ed. S. Anderson and J. K. Jones, Jr., pp. 325–26. New York: Ronald Press.

Stanton, Don C. 1963. A history of the white-tailed deer in Maine. Bulletin no. 8. Augusta: Maine Department of Inland Fisheries and Game, pp. 1–75.

Stickley, A. R., Jr. 1957. The status and characteristics of the black bear in Virginia. M.S. thesis, Virginia Polytechnic Institute.

———. 1961. A black bear tagging study in Virginia. *Proc. Ann. Conf. S. E. Assoc. Game and Fish Comm.* 15:43–54.

Stoneberg, Ronald P., and Jonkel, Charles J. 1966. Age determination of black bears in cementum layers. *J. Wildl. Manage.* 30(2):411–14.

Storm, Gerald L. 1965. Movements and activities of foxes as determined by radio-tracking. *J. Wildl. Manage.* 29(1):1–13.

———. 1972. Daytime retreats and movements of skunks on farmlands in Illinois. *J. Wildl. Manage.* 36(1):31–45.

Storm, Gerald L., and Ables, Ernest D. 1966. Notes on newborn and full-term wild red foxes. *J. Mamm.* 47(1):116–18.

Storm, Gerald L., and Verts, B. J. 1966. Movements of a striped skunk infected with rabies. *J. Mamm.* 47(4):705–8.

Stuewer, Frederick W. 1943. Raccoons: Their habits and management in Michigan. *Ecol. Monogr.* 13(2):203–58.

Sullivan, Edward G., and Haugen, Arnold O. 1956. Age determination of foxes by x-ray of forefeet. *J. Wildl. Manage.* 20(2):210–12.

Sunquist, M. E.; Montgomery, G. G.; and Storm, G. L. 1969. Movements of a blind raccoon. *J. Mamm.* 50(1):145–47.

Svihla, Arthur, and Bowman, Howard S. 1954. Hibernation in the American black bear. *Amer. Midland Nat.* 52(1):248–52.

Urban, David. 1970. Raccoon populations, movement patterns, and predation on a managed waterfowl marsh. *J. Wildl. Manage.* 34(2):372–82.

Verts, B. J. 1967. *The biology of the striped skunk.* Urbana: University of Illinois Press.

Vincent, R. E. 1958. Observations of red fox behavior. *Ecology* 39:755–57.

Walker, Ernest P.; Warnick, Florence; Lang, Kenneth I.; Uible, Howard E.; Hamlet, Sybil E.; Davis, Mary A.; and Wright, Patricia F. 1964. *Mammals of the world.* 2 vols. Baltimore: Johns Hopkins University Press.

Waters, Joseph H. 1964. Red fox and gray fox from New England archeological sites. *J. Mamm.* 45(2):307–8.

——. 1967. Foxes on Martha's Vineyard, Massachusetts. *J. Mamm.* 48(1):137–38.

Waters, Joseph H., and Mack, Charles W. 1962. Second find of sea mink in southeastern Massachusetts. *J. Mamm.* 43(3):429–30.

Waters, Joseph H., and Ray, Clayton E. 1961. Former range of the sea mink. *J. Mamm.* 42(3):380–83.

Weckwerth, Richard P. 1957. The relationship between the marten population and the abundance of small mammals in Glacier National Park. M.S. thesis, Montana State University.

Weckwerth, Richard P., and Hawley, Vernon D. 1962. Marten food habits and population fluctuations in Montana. *J. Wildl. Manage.* 26(1):55–74.

Whitney, Leon F. 1933. The raccoon—some mental attributes. *J. Mamm.* 14(2):108–14.

Whitney, Leon F., and Underwood, Acil B. 1952. *The raccoon.* Orange, Conn.: Practical Science Pub. Co.

Wilber, Charles G., and Weidenbacher, George H. 1961. Swimming capacity of some wild mammals. *J. Mamm.* 42(3):428–29.

Willey, Charles H. 1971. Vulnerability in Vermont's bear population. *Proc. N. E. Fish and Wildl. Conf.* 28:109–40.

——. 1974. Aging black bears from first premolar tooth sections. *J. Wildl. Manage.* 38(1):97–100.

Wood, John E. 1958. Age structure and productivity of a gray fox population. *J. Mamm.* 39(1):74–86.

Wright, Philip L. 1942. Delayed implantation in the long-tailed weasel (*Mustela frenata*), the short-tailed weasel (*Mustela cicognani*), and the marten (*Martes americana*). *Anat. Record* 83(3):341–53.

——. 1948*a*. Breeding habits of captive long-tailed weasels (*Mustela frenata*). *Amer. Midland Nat.* 39(2):338–44.

——. 1948*b*. Preimplantation stages in the long-tailed weasel (*Mustela frenata*). *Anat. Record* 100(4):593–608.

Wright, Philip L., and Coulter, Malcolm W. 1967. Reproduction and growth in Maine fishers. *J. Wildl. Manage.* 31(1):70–86.

Yeager, Lee E. 1943. Storing of muskrats' and other food by minks. *J. Mamm.* 24(1):100–101.

Young, Stanley P. 1958. *The bobcat in North America.* Harrisburg, Penn.: Stackpole Co.

Young, Stanley P., and Jackson, Hartley H. T. 1951. *The clever coyote.* Harrisburg, Penn.: Stackpole Co.

Zimmerman, R. Scott. 1943. A coyote's speed and endurance. *J. Mamm.* 24(3):400.

# Order Pinnipedia

THE ORDER PINNIPEDIA comprises the walruses, seals, and sea lions. These animals have adapted to both aquatic and terrestrial existence. Their body form is markedly streamlined. They are characterized by having the limbs modified for swimming; the elbows and knees are deeply folded within the body and the humerus is relatively massive, while the limbs are short with the extremities flattened into flippers (Pinnipedia means "wing-footed"). The front flippers are usually tapered and pointed, used mainly for balancing and maneuvering in the water, and the hind flippers are wide and paddlelike, used for propulsion. The five digits of each limb are covered with a thick web of skin. The tail is very short and flattened, and the neck is thick, muscular, and flexible. External ears are rudimentary or absent. The eyes may be small or large. The whiskers are well developed for touch, and the slitlike nostrils and ears close underwater. The skin is thick and tough, while the blubber provides insulation against cold, buoyancy, padding, and reserve energy when food is scarce.

The canines are conical, and the cheek teeth are homodont, two-rooted, and usually simple and conical. There are one or two lower incisors on each side of the jawbone. The carnassials are not developed. The teeth are adapted for grasping and tearing rather than for chewing flesh.

In adults, the color pattern of the pelage may be uniform, flecked, spotted, streaked, or sharply banded. The fetal pelage is shed shortly before or shortly after birth, depending on the species, and the pelage of the precocious pups is often dense and woolly, unlike that of adults.

Pinnipeds are social animals. They congregate on breeding grounds in large colonies called rookeries. Some species, such as the northern fur seal, spend at least eight months of the year at sea, whereas the harbor seal remains on land much of the time.

Pinnipeds occur along most of the shores and

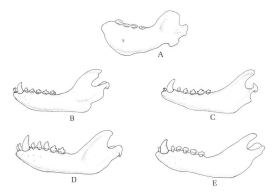

Lateral views of left lower jaws of A, walrus; B, harbor seal; C, harp seal; D, gray seal; E, hooded seal.

ice packs of the world and in some rivers. A few have been isolated in lakes. Most species are found in polar and temperate waters, except for monk seals, which inhabit the warm waters of the Mediterranean Sea, the Carribean Sea, and Hawaiian regions (Stains 1967). The order is represented in New England by two families—the Odobenidae (walrus) and Phocidae (earless seals or hair seals).

## FAMILY PHOCIDAE (EARLESS SEALS OR HAIR SEALS)

Seals of the family Phocidae have no external ears, and their hind flippers cannot be turned forward alongside the body. On land they move forward by a hunching movement like an inchworm or by repeatedly rolling over sideways. The pelage is stiff, with no woolly underfur in adults. The fur of the young, at least in several genera, is long and woolly, but is shed either in utero or shortly after birth.

The phocids occur along most coastlines north of 30° north latitude and south of 50° south latitude, and there is a scattered distribution in intermediate tropical and subtropical areas (Stains 1967). Four species occur in the waters of New England.

## Harbor Seal

*Phoca vitulina concolor* De Kay

**Description.** The harbor seal has a short, blunt doglike muzzle with V-shaped nostrils; large eyes; ears with no pinnae; a short neck; and hairy flippers. The skull is rather thick, with a large, deep lower jaw and short mandibular symphysis. The cheek teeth are broad, large, and set obliquely in the lower jaw, especially in immatures. The oblique position of the cheek teeth is characteristic of this species.

The sexes are colored alike, but the color

*Phoca concolor* De Kay, 1842. Zoology of New York . . . , *Mammalia* 1(1):53.

*Phoca vitulina concolor*, Brown, 1913. *Pocket list of mammals of eastern Massachusetts*, p. 30.

Type Locality: Long Island Sound, near Sands Point, Nassau County, New York

Dental Formula:

I 3/2, C 1/1, P 4/4, M 1/1 × 2 = 34

Names: Common seal, hair seal, spotted seal, leopard seal, river seal, bay seal, land seal, fresh water seal, sea calf, and seal dog. The meaning of *Phoca* = a seal, *vitulina* = sea calf, and *concolor* = of the same color, referring to the all-gray color of the type speciman.

Common

Scattered

pattern is variable—essentially dark yellow gray, mottled with irregular-sized dark brown or black spots and a whitish network of broken small rings and loops above, paler below. When the spots are far apart the seal appears light-colored; when they run together, it looks dark-colored. The dry pelage appears silvery or whitish.

Males are slightly larger than females. Measurements range from 1.2 to 1.8 m (4–6 ft). Weights vary from 45 to 136 kg (100–300 lb) (King 1964).

**Distribution.** Harbor seals are widely distributed along the coasts of the oceans of the northern hemisphere. They are found in the western Atlantic Ocean, on both coasts of Greenland, on Ellesmere Island 79° N, and south to Florida, but they are more common from southern Greenland to Maine.

These animals ascend the Saint Lawrence River to Montreal, and one individual has been found in Lake Ontario. Occasionally they find their way into Lake Champlain (Osgood 1938). In Vermont, Thompson (1842) reported harbor seals from Burlington (1) in 1810, and Kirk (1916) reported them from Otter Creek (2) in 1846 and from Weybridge (3) in 1876. Katona, Richardson, and Hazard (1975) reported that there are at least 6,000 harbor seals in Maine waters, with great concentrations occurring in lower Penobscot Bay, Machias Bay, and off Swans Island and Mount Desert Island. Probably several hundred individuals occur in waters south of Maine.

They seem to be only winter residents of bays south of Cape Cod.

**Ecology.** Harbor seals live mainly in coastal waters, harbors, bays, estuaries, and rivers and haul out on islands, mudbanks, sandbanks, rocky shores, and ice. They have a strong liking for fresh water and actively ascend rivers and lakes, sometimes far from the sea. Some of them live in lakes throughout the year where open water is available. They are essentially animals of open water and do not frequent the fast ice.

Man, killer whales, polar bears, and sharks prey on harbor seals. Some pups die of starvation, while others are deserted by their mothers or become separated from them. Captives suffer from pneumonia, conjunctivitis, cirrhosis of the liver, respiratory distress, loss of hair, and other ailments. These seals are hosts to lice, cestodes, nematodes, and trematodes.

**Behavior.** Harbor seals do not seem to make definite migrations, but unusually heavy ice and lack of food may force them to move. They appear not to keep breathing holes open in fast ice. They form loosely organized colonies. Adults are quite active at high-tide hours and generally haul out day and night to rest and sleep at low tide, especially during warm weather. These animals are shy and difficult to approach but are curious and may gather around to look at a boat from a safe distance. Adults

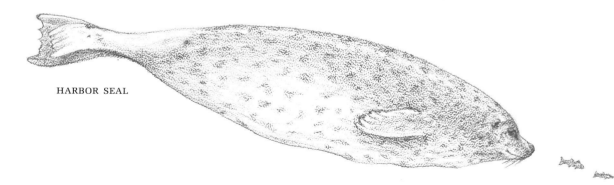

HARBOR SEAL

make loud snorts, growls, barks, or grunts, and pups emit bleets or mewing cries when lost. On land they move slowly and laboriously. In water they play, roll, leap, and often swim in circles. Maxwell (1967) reported that harbor seals can attain speeds of up to 13 knots in short bursts and can remain underwater for at least 15 minutes.

Harbor seals eat a wide variety of fishes, shrimps, squids, crustaceans, and mollusks.

These seals do not form harems, but adult bulls often fight among themselves during the breeding season. Copulation takes place in the water in August. After a delay in implantation of the blastocyst, true gestation begins in late October and lasts for 7½ months (King 1964). Large regional variations and clines occur in the timing of the whelping season along the east coast of North America. The pups tend to be born later as one goes north from New England to southern Baffin Island (Bigg 1969). Birth occurs from late March to early June along the coast of New England (Allen 1942); from late March to mid-April in Massachusetts to southern Maine and from late April to mid-June in Maine (Bigg 1969); and from early May to early June, peaking during the last week in May, on Sable Island, Nova Scotia (Boulva 1971).

The pups are born on land, sandbanks, mudflats, or in river estuaries. Male pups are longer than females at birth. The males average 80.5 cm (31.7 in) long and females 78.5 cm (30.9 in), and they grow 11 cm (4.3 in) and 8 cm (3.2 in) respectively during the first 3 weeks of life. At birth both males and females average 11 kg (24 lb), but after 3 weeks of age the males more than double their weight, reaching an average of 24 kg (53 lb), whereas females weigh only 19.5 kg (43 lb) (Boulva 1971).

The pups are precocious and are obliged to swim before the next tide covers their birthplace. The embryonic coat (the lanugo) of long, soft white hair is shed in utero or immediately after birth. The second coat consists of short, stiff hairs, bluish gray above and silvery white below, mottled with faint spots. The pups are nursed for 3 or 4 weeks. After weaning they eat crustaceans and then fish. Katona, Richardson, and Hazard (1975) reported that sexual maturity is attained at 3 to 5 years and full growth at about 10 years, and that captives may live up to 35 years.

# Harp Seal

*Pagophilus groenlandicus* (Erxleben)

[*Phoca*] *groenlandica* Erxleben, 1777.
  *Systema regni animalis* . . . , 1:588.
*Pagophilus groenlandicus*, Gray,
  1844. In *The zoology of the voyage
  of H.M.S. Erebus and Terror.* . . . ,
  vol. 1, pt. 1, p. 3.
Type Locality: Greenland and Newfoundland
Dental Formula:
  I 3/2, C 1/1, P 4/4, M 1/1 × 2 = 34
Names: Greenland seal, saddle seal,
  and saddleback seal. Newborn
  pups are called whitecoats; fully
  molted pups, beaters; one-year-
  olds, rusty; and two-year-olds to
  adults, bedlamers. The meaning of
  *Pagophilus* = ice-loving, since this
  species keeps to the edge of drifting ice for at least half the year, and
  *groenlandicus* = of Greenland.

**Description.** Harp seals are somewhat larger than harbor seals. Harp seals and ribbon seals are the only phocids having an irregular banded pelage, and there are sexual differences in pelage coloration, which is usually bright tawny gray or yellowish white above and silvery below. Harp seals get their names from a deep brown harp-shaped band or saddle that crosses the back along the shoulders and runs along the flanks to near the tail in adult males. In females the band may be broken, indistinct, or absent. Adults of both sexes have a dark brown or blackish head to just behind the eyes, but this is often somewhat lighter or broken into spots in females. Newborn pups have long, silky white fur which they begin to molt in about a week, starting from the head and moving along the back. This pelage is replaced in 3 to 4 weeks by a coarser, shorter silvery fur with small dusky spots. As the male matures the spots grow larger and darker and form the saddle. Young females may retain the spotted coat for some years after reaching sexual maturity.

The nasals of the skull are long and narrow, tapering gradually from front to back. The posterior margin of the palate is U-shaped. In the lower jaw the cheek teeth are small with accessory cusps set in a straight line and appear three-pointed rather than four-pointed as in the harbor seal. The third cheek tooth is the largest in both jaws.

Adults are about 1.8 m (6 ft) long and weigh about 182 kg (400 lb).

**Distribution.** Harp seals occur from the arctic coast of Russia to Greenland and on the northern coasts of Canada and Newfoundland. In this distribution, there are three stocks of harp seals with distinct breeding grounds. These are the eastern, in the White Sea; the central, in the Greenland Sea between Iceland and Spitsbergen; and the western, between northwest

Greenland and the Canadian arctic archipelago in summer and Newfoundland and the Gulf of Saint Lawrence in winter. Tagging experiments have so far not shown any mixing between the three stocks. The western stock is divided into two breeding and molting populations; "Gulf," which inhabits the Gulf of Saint Lawrence, and the "Front," off the east coast of Newfoundland (Sergeant 1965).

In New England, harp seals have been seen in Penobscot Bay and Casco Bay, Maine (no date) (Katona, Richardson, and Hazard 1975). A specimen was taken at Nahant, Massachusetts (1) (no date) (Brown 1913), and one animal was seen at Stonington, Connecticut (2), during the winter of 1841–42 (Linsley 1842).

Stragglers have been reported from Coney Island, Long Island, New York in winter (no date) (Kieran 1959) and from New Jersey (no date) (Rhoads 1903); and one individual was seen at Cape Henry, Virginia, in March 1945 (Scheffer 1958).

**Ecology.** Harp seals haul out on ice to molt and breed. Man is the chief predator of harp seals. Lice, cestodes, nematodes, trematodes, and acanthocephalids are known to parasitize them, and they are susceptible to gastroenteritis, peritonitis, and pleuritis.

**Behavior.** Harp seals are gregarious and migratory; the herd moves south in winter when the ice begins to close their feeding grounds and returns in spring after the pups leave the ice for the sea. These animals have a complex annual cycle of migration and reproduction. Harp seals of the western stock leave the Arctic in September and November and move slowly south along the coast of Labrador, reaching Belle Isle by January. Most immatures lag behind and may remain in midwest Greenland as late as February. In January the seals at Belle Isle split up, part of the population going through the Strait of Belle Isle into the Gulf of Saint Lawrence, the remainder moving close to the shore in northeast Newfoundland. In mid-February, the breeding females haul out on the ice fields, and they whelp in early to mid-March. In late March they ovulate and breed with the males, which have waited near the edge of the whelping places. During lactation females feed sporadically near the ice pack. By early April the pups begin feeding on euphausid shrimp and capelin. During April the adults and immatures molt; in the northern Gulf of Saint Lawrence, herds of molting seals leave the ice fields for several days at a time, presumably to feed (Sergeant 1963).

Molting begins near the top of the head and the area of the flippers and moves back over the rest of the body. In captives it is completed in from 10 to 90 days (Ronald et al. 1970). At the time of molting, the seals haul out on the ice, especially on sunny days. The Gulf seals spend the later part of the molting period in the water around the Magdalen Islands in April and May. During the peak of molting, the Front seals remain off the coast of Newfoundland and Labrador, drifting along the moving ice and swimming north by stages. In May and June the Gulf seals move northward through the Strait of Belle Isle and follow the Front seals up the Labrador coast. The pups are solitary but gather in herds and move independently somewhat behind the adults; then, becoming highly social, they join the adults in Greenland and arctic Canadian waters for the summer, completing the annual migratory cycle (Sergeant 1963).

Harp seals can swim up to 17 knots for short

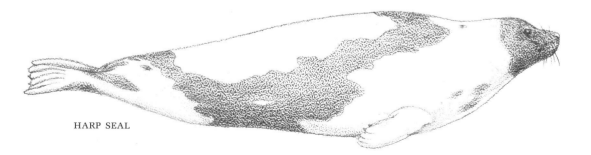

HARP SEAL

periods and can remain submerged for about 20 minutes (Bartlett 1927). They can dive to depths of 100 fathoms or more (Sergeant 1963).

These seals feed chiefly on capelin and on pelagic and benthic crustaceans (decapods, euphausids, mysids, and amphipods), with smaller quantities of herring, cod, polar cod, flatfish, redfish, skate, and barracuda.

Harp seals appear to be promiscuous. In the northwestern Atlantic, populations breed and conceive in mid- to late March, with immediate growth of the new corpus luteum, but implantation of the blastocyst is delayed some 3 months. The gestation period is approximately 12 months. At birth the pup is about 91.4 cm (36 in) long and weighs about 6.8 kg (15 lb). Twins are rare. The milk teeth are resorbed before birth, and the permanent teeth erupt at or shortly after birth. The pup is weaned 10 to 12 days after birth and is independent of the female about 2 weeks later. During the next 2 weeks the pup lives off its body fat reserve (Sergeant 1963). Females reach sexual maturity at 4 to 7 years, averaging 5½ years of age (Sergeant 1965). Males probably do not breed until their 6th or 7th year (Fisher 1954). Females can produce pups up to 16 years of age and probably up to 20 years. Longevity is 20 years, rarely up to 28 years (Fisher 1952).

# Gray Seal

*Halichoerus grypus* (Fabricius)

*Phoca grypus* Fabricius, 1791. *Skriver of Naturn.—Selskabet Kjbenhaon,* 1(2):167.
*Halichoerus grypus,* Nilsson, 1841. *Arch. Naturg., Jahrg.* 7(1):318.
Type Locality: Greenland
Dental Formula:
 I 3/2, C 1/1, P 4/4, M 1/1 × 2 = 34
Names: Horsehead, hump-nosed seal, bristled seal, big seal, and Atlantic seal. Pups in their first molt are called moulters. The meaning of *Halichoerus* = sea swine, and *grypus* = hump-nosed, referring to the high roman nose of the animal, which is more pronounced in males than in females.

**Description.** The gray seal is distinguished from the harbor seal by the straight and rather long profile of the head, uniformly dark gray color of the body, and larger size. Adult males have several conspicuous wrinkles on the stout neck. In adult females the profile is almost straight, and in adult males it is somewhat convex and like a roman nose, which hangs slightly in old individuals. The muzzle is broad and donkeylike, which gives the animal the name horsehead. The nostrils are well separated and look like a W when viewed from the front.

The skull is arched dorsally, with the facial portion longer than the braincase; the sagittal crest is well developed in old males, and the profile of parietals, frontals, and nasals forms a straight line. The cheek teeth are large and are single rooted except for the two posteriormost teeth in the upper jaw and the single posteriormost tooth in the lower jaw.

The coloration of both sexes varies greatly with the individual; some seals are uniformly colored, while others are blotched with irregular dark gray spots. Wet gray seals appear dark gray or grayish black, but when dry they are gray or brown. Both sexes have dark backs and paler bellies, with varying degrees of irregular spotting. The sexes can be identified by the background color. Adult males have a background of either dark gray, dark brown, or nearly black above, paler below, upon which are lighter spots or patches. Adult females have a paler background of smoky gray above, blending to silver gray, cream-colored, or tan on the sides and belly, which are overlaid with darker spots and patches. Immatures have less mottling than adults but are difficult to distinguish.

Males are larger than females. Adult males measure from 1.8 to 3.1 m (6–10 ft) long and weigh 160 to 290 kg (353–640 lb); females measure from 1.7 to 2.3 m (5.6–7.5 ft) and weigh 120 to 250 kg (265–551 lb) (Walker et al. 1964).

**Distribution.** Gray seals occur on both sides of the northern Atlantic Ocean and adjacent seas, with major populations in northwestern Europe, Iceland, and eastern Canada. Some individuals stray as far south as New Jersey.

There are three distinct populations of gray seals, one in the Gulf of Saint Lawrence, a second in the British Isles, the Faroes, Iceland, Norway, and the White Sea region, and a third in the Baltic Sea. The eastern Atlantic and western Atlantic groups apparently have been separated by a distance of 1,500 miles (in Iceland and Newfoundland) for approximately 100,000 years, and they have different breeding seasons: the eastern Atlantic group breeds in autumn, whereas the Baltic and western Atlantic seals breed in winter. Records are not available of gray seals breeding in Greenland, the type locality (Davies 1957). Van Bree (1972) thought that the species was originally very common and had a wider distribution that it does today, and that formerly gray seals migrated to and from the American continent by way of south Greenland during summer, and that a migration occurred between Iceland and western Europe,

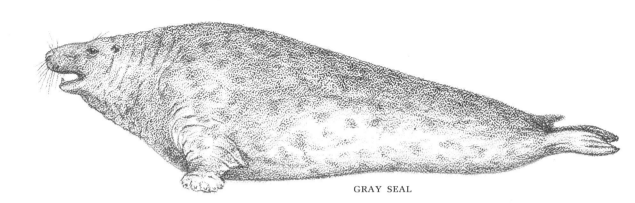

GRAY SEAL

so that no discontinuous distribution existed at that time. The author believed that the apparent rarity of gray seals and its present distribution are related to human influences during the last 3,000 years. Fisher (1950) reported that the West Atlantic population is centered in the Gulf of Saint Lawrence, with a relatively large breeding center in the Northumberland Strait and breeding colonies near the Basque Islands, Magdalen Islands, and Sable Island.

In New England, as many as 100 gray seals occur in Maine waters, mainly in the approaches to Swans Island and Mount Desert Island and lower Penobscot Bay (Katona, Richardson, and Hazard 1975). In Massachusetts waters a colony of about fifteen or more seals occurs at Nantucket Island, south of Cape Cod. Some gray seals have been seen around Muskeget Island and Tuckernuck Island, and their remains have been found at Martha's Vineyard and at Block Island, Rhode Island.

Man is the main predator of gray seals. They are hosts to lice, cestodes, nematodes, and trematodes. They are susceptible to an epiflora called "green gray seal," which makes the pelage look greenish because of an infestation of the green algae *Enteromorpha groenlandica* (Mackenzie 1954). The pups are susceptible to pneumonia.

**Behavior.** Gray seals usually haul out on secluded rocky ledges. They are elusive and wary and spend much time in the water, usually hovering in one spot. On land, they are ungainly and waddle along on their chests. They are gregarious, but live in loose colonies.

The seals sleep underwater, on the surface, or on land. In the water, the most common sleeping position is semifloating; the nostrils are at the surface, usually just clear, and the rear flippers occasionally make swimming movements. At other times the animals bob up and down slowly, taking several breaths at the surface and then sinking below for a minute or so before rising to breathe again. In a more extreme version of this behavior, one gray seal was seen to take 10 to 20 breaths at the surface and sink for as long as 4 minutes (Ridgway, Harrison, and Joyce 1975).

These seals disperse from the rookeries at the end of the breeding season, the young dispersing far more widely than the adults, and some pups wander as far as 600 miles from their birthplaces in the eastern Atlantic populations (Lockley 1954; Smith 1962). Some tagged pups have been recovered 550 miles away from the place where they were tagged 2 months earlier (Hickling 1956). Young gray seals appear to disperse radially, whereas dispersal in adults tends to be organized and deliberate, probably relating to availability of food and the size of a colony of seals (Davies 1957).

Gray seals feed on a variety of fishes and occasionally on crustaceans, mollusks, and squid.

In eastern Canadian waters the breeding season begins in late December, when mature bulls and pregnant cows congregate on reefs and islets near the breeding islands. In early January the seals in the vanguard of breeders haul out on the breeding islands and establish territories (Mansfield 1967a). Resident or established bulls often are challenged by bulls who have no territories, and fighting occurs. The resident bull often rolls over after routing an intruding bull, perhaps as a demonstration of victory (Cameron 1967).

Gray seals are polygamous; the older bulls maintain harems of up to six breeding females (Mansfield 1967b), and some bulls make an effort to keep the cows from wandering out of their territories (Cameron 1967). Pregnant cows haul out and wander onto the breeding ground to whelp. Pupping occurs from late December through mid-February and peaks about mid-January. Lactation lasts 2 to 3 weeks, and breeding occurs before the pups are weaned (Mansfield 1966). Copulation may occur in the water, though it is more frequently performed on land or ice, where disturbance is minimal (Mansfield 1967b). Delayed implantation of the blastocyst occurs in this species (Backhouse and Hewer 1956).

Newborn pups average 89 cm (35 in) long and weigh about 14 kg (30 lb), increasing to about 107 cm (42 in) and 45 kg (100 lb) at weaning, when they are 3 weeks old. The long, silky, cream-white natal fur (lanugo) is shed after 3 or 4 weeks and replaced by a coat of short stiff hair of a distinctive pattern in each sex. The male moulter is recognized by its large nose and black coat with silver-gray reticulations; the female moulter is largely silver gray with scattered dark spots (Mansfield 1967b).

The pups do not go out to sea before they are 3 weeks old and the females desert them for the sea. The pups probably remain ashore for at least a month. The amount of attention a pup receives probably depends on the temperament of the female. Most females defend their young against other adults (Cameron 1967).

Females become sexually mature when 3 years old and males when 6 or 7 years old. Longevity is about 40 years in captivity and about 35 years in the wild (King 1964). In captivity, several interspecific breedings between gray seals and ringed seals have occurred, but all the pups have been born dead (Scheffer 1958). Attempted mating between a male gray seal and a female harbor seal on Sable Island, Nova Scotia, has been described by Wilson (1975).

# Hooded Seal

*Cystophora cristata* (Erxleben)

[*Phoca*] *cristata* Erxleben, 1777.
    *Systema regni animalis* . . ., 1:590.
*Cystophora cristata*, True, 1884. *Proc. U.S. Natl. Mus.* 7(29):608.
Type Locality: Southern Greenland and Newfoundland
Dental Formula:
    I, 2/1, C 1/1, P 4/4, M 1/1 × 2 = 30
Names: Hood seal, crested seal, and bladdernose. The pups are called bedlamers, and young in their first hair coat are called bluebacks. The meaning of *Cystophora* = bladdernose, and *cristata* = a crest. Both scientific names refer to the inflatable nasal appendage.

**Description.** The hooded seal is larger than the harp seal. Its most striking feature is the remarkable enlarged skin of the snout, which forms a hood on top of the head from the snout to just past the level of the eyes in adult males only. In young males the hood becomes progressively larger with increasing age and body size. It is covered with short, rather stiff black hairs and is about 30 cm (12 in) long and 18 cm (7 in) in diameter when inflated. When deflated, the structure hangs limply down in front of the mouth like a loose proboscis. The hood can be inflated when the male is calm, excited, or angered. In addition to the hood, males have an inflatable nasal sac that can be blown out through the nostrils like a red balloon. This sac is formed from the very extensible membranous part of the internasal septum. The balloon is extruded through one nostril by internal pressure while the other nostril is kept closed. It is about 16 cm (6 in) long and 15 cm (6 in) in diameter and can be extruded several times in quick succession as the male shakes its head up and down. It is not extruded at the same time as the hood is inflated. Like the hood, the balloon can be inflated when the animal is calm, excited, or angered. The hood is a secondary sexual characteristic used to impress or frighten competitive males and intruders; it is not used as an air reservoir during diving, but it may act as a cushion in lessening the impact of blows on the forehead during fighting. The balloon might also be an extension of the hood's secondary sexual characteristics and may play a role in heat regulation (Berland 1966).

In the skull, the anterior third of the nasal bones protrudes freely forward. The palatine bones are nearly square, the auditory bullae are square in front, and the superior notch of the mandible is deeply U-shaped. The upper canines are about twice the size of the adjacent incisors. The first and second premolars often have single cusps; premolars and molars usually are well separated and appear peglike, with narrow necks and fine, wrinkly grooves in the enamel.

HOODED SEAL

Adults are slate blue or brownish above, mixed with a pattern of irregular dark spots on the back, shading to silver gray on the sides and belly. The muzzle and face of males are blackish. The face of females and young is blackish to just behind the eyes. Pups shed their light gray fetal hair before birth, and it is swallowed by the fetus and excreted into the amniotic fluid as compact felted disks. At birth the pups have a lustrous fur, bluish black above and silvery white below.

Males are larger than females. Adult males measure from 2.1 to 3.5 m (7–11 ft) long and weigh about 410 kg (904 lb). Females measure up to nearly 3.1 m (10 ft) long and weigh about 270 kg (595 lb) (Walker et al. 1964).

**Distribution.** Hooded seals are associated with the ice pack in deep waters of the Arctic and Atlantic Oceans from Northern Siberia and Barents Sea to Jan Mayen Island, Iceland, Denmark Strait, southern Greenland, Baffin Island, Newfoundland, and the Gulf of Saint Lawrence. There seem to be two main breeding areas; one to the north of Jan Mayen Island, where the seals breed in the spring and then move northward toward Svalbard, and another south of Greenland near Newfoundland, where the seals breed in the spring and then migrate to the ice off the coast of Greenland (Scheffer 1958).

In New England, stragglers have been recorded in Maine at North Harpswell (2) on 25 March 1928 (Norton 1930) and South Brooksville (1) on 10 April 1974 (Richardson 1975); in Massachusetts, at Newburyport (3) in May 1882 (Brown 1913); in Rhode Island, in the Providence River (4) (no date) (Cronan and Brooks 1968); and in Long Island Sound (5) (no date) (Goodwin 1935). There are also records of strays from New Jersey (Rhoads 1903), Maryland (Mansueti 1950), North Carolina (Goodwin 1954), and as far south as Cape Canaveral, Florida (Miller 1917).

**Ecology.** Hooded seals haul out on large, heavy floes of arctic ice to molt and reproduce. Man and polar bears are the main predators of hooded seals, but accidents also take their toll. Parasites are lice, cestodes, nematodes, trematodes, and acanthocephalids.

**Behavior.** Little is known about the behavior of hooded seals. They are rather quarrelsome, much less gregarious than harp seals, and are solitary for most of the year, except during the molting and breeding seasons. Adults fiercely defend their pups. These animals are nomadic and prefer thick ice floes on the high seas, rarely hauling out on land. They swim slowly with the top of the head just above water.

Hooded seals travel with harp seals, but the two species tend to keep apart, the hooded seals staying farther out at sea. Although the two species breed in the same areas, hooded seals usually form separate family groups consisting of a bull, a cow, and their pup, sometimes staying at considerable distances from the nearest breathing hole or open water. The adults appear to fast during the breeding season. These animals are known to eat fish, squid, shrimp, octopus, and mussels.

The gestation period is unknown, but the pups are born on ice in late February to late March. They are about 1.1 m (3.6 ft) long and weigh from 11 to 15 kg (24–33 lbs). They grow rapidly and are nursed from 3 to 4 weeks. During this time the bulls remain nearby, and just after the pups are weaned the adults breed and return to sea. The pups remain on the drifting ice for the first 20 to 22 days of life, then disperse at sea (Bartlett 1927). Nothing is known of the feeding habits of pups at sea. Young hooded seals tend to keep to themselves until sexual maturity. Most hooded seals may breed at about 4 years of age.

## REFERENCES

Allen, Glover M. 1942. The harbor seal. *New England Nat.* 15:38.

Backhouse, K. M., and Hewer, R. R. 1956. Delayed implantation in the grey seal, *Halichoerus grypus* (Fab.). *Nature* (London) 178:550.

Bartlett, Robert A. 1927. Newfoundland seals. *J. Mamm.* 8(3):207–12.

Berland, Bjørn. 1966. The hood and its extrusible balloon in the hooded seal, *Cystophora cristata* Erxl. Reprint from *Årbok 1965.* Oslo: Norsk Polarinstitutt, pp. 95–102.

Bigg, Michael A. 1969. Clines in the pupping season of the harbour seals, *Phoca vitulina. J. Fish. Res. Bd. Canada* 26:449–55.

Boulva, Jean. 1971. Observations on a colony of whelping harbour seals, *Phoca vitulina concolor*, on Sable Island, Nova Scotia. *J. Fish. Res. Bd. Canada* 28:755–59.

Brown, C. Emerson. 1913. *Pocket list of mammals of eastern Massachusetts.* Salem, Mass.: Peabody Museum.

Caldwell, David K., and Caldwell, Melba C. 1969. The harbor seal, *Phoca vitulina concolor*, in Florida. *J. Mamm.* 50(2):379–80.

Cameron, Austin W. 1967. Breeding behavior in a colony of western Atlantic gray seals. *Canadian J. Zool.* 45(2):161–73.

Cronan, John M., and Brooks, Albert. 1968. *The mammals of Rhode Island.* Wildlife Pamphlet no. 6 Providence: Rhode Island Department of Agriculture and Conservation, Division of Fish and Game.

Davies, J. L. 1957. The geography of the gray seal. *J. Mamm.* 38(3):297–310.

Fisher, H. D. 1950. Seals of the Canadian east coast. *Fish Res. Bd. Canada, Circ.* 18:1–4.

———. 1952. Harp seals of the northwest Atlantic. *Fish. Res. Bd. Canada, Circ.* 20:1–4.

———. 1954. Studies on reproduction in the harp seal, *Phoca groenlandica* Erxleben in the northwest Atlantic. *Fish. Res. Bd. Canada, MS Rept. (Biol.).* 588:1–109.

Gallacher, J. B., and Waters, W. E. 1964. Pneumonia in grey seal pups at St. Kilda. *Proc. Zool. Soc. London* 142(1):177–80.

Goodwin, George Gilbert. 1933. Occurrence of a gray seal at Atlantic City, N.J. *J. Mamm.* 14(1):73.

———. 1935. The mammals of Connecticut. *Connecticut Geol. and Nat. Hist. Surv. Bull.* 53:1–221.

———. 1954. Southern records of arctic mammals and a northern record for Alfaro's rice rat. *J. Mamm.* 35(2):258.

Hewer, H. R. 1957. The Hebridean breeding colony of grey seals (*Halichoerus grypus*) (Fabr.) with comparative notes on the grey seals of Ramsey Island, Pembrokeshire. *Proc. Zool. Soc. London* 128:23–66.

Hickling, Grace. 1956. The grey seals of the Farne Islands. *Trans. Nat. Hist. Soc. Northumberland, Durham and Tyne,* n.s., 11:230–44.

Katona, Steven; Richardson, David; and Hazard, Robin. 1975. *A field guide to the whales and seals of the Gulf of Maine.* Rockland, Maine: Maine Coast Printers.

Kieran, J. 1959. *An natural history of New York City.* Boston: Houghton Mifflin Co.

King, Judith E. 1964. *Seals of the world.* London: British Museum of Natural History.

Kirk, George L. 1916. The mammals of Vermont. *Joint Bull., Vermont Botanical and Bird Clubs* 2:28–34.

Linsley, James H. 1842. A catalogue of the Mammalia of Connecticut, arranged according to their natural families. *Amer. J. Sci.* 43:345–54.

Lockley, R. M. 1954. The Atlantic grey seal. *Oryx* 2:384–87.

Mackenzie, B. A. 1954. Green algal growth on gray seals. *J. Mamm.* 35(4):595–96.

Mansfield, A. W. 1966. The grey seal in eastern Canadian waters. *Canadian Audubon* 28(4):161–66.

——. 1967a. The mammals of Sable Island. *Canadian Field-Nat.* 81(1):40–49.

——. 1967b. Seals of arctic and eastern Canada. *Fish. Res. Bd. Canada, Bull.* 137:1–35.

Mansueti, Romeo. 1950. Extinct and vanishing mammals of Maryland and District of Columbia. *Maryland Nat.* 20:1–48.

Maxwell, Gavin. 1967. *Seals of the world.* World Wildlife Series no. 2. Boston: Houghton Mifflin Co.

Miller, Gerrit S., Jr. 1917. A hooded seal in Florida. *Proc. Biol. Soc. Washington* 30:121.

Norton, Arthur H. 1930. The mammals of Portland, Maine, and vicinity. *Proc. Portland Soc. Nat. Hist.* 4:1–151.

Osgood, Frederick L., Jr. 1938. The mammals of Vermont. *J. Mamm.* 19(4):435–41.

Ray, Carleton. 1963. Locomotion in pinnipeds. *Nat. Hist.* 67(3):11–21.

Rhoads, Samuel N. 1903. *The mammals of Pennsylvania and New Jersey.* Philadelphia: Privately published.

Richardson, David T. 1975. Hooded seal whelps at South Brooksville, Maine. *J. Mamm.* 46(3):698–99.

Ridgway, S. R.; Harrison, R. J.; and Joyce, P. L. 1975. Sleep and cardiac rhythm in the gray seal. *Science* 187(4176):553–55.

Ronald, K.; Johnson, E.; Foster, M.; and Vander Pol, D. 1970. The harp seal, *Pagophilus groenlandicus* (Erxleben, 1777). I. Methods of handling, molt, and diseases in captivity. *Canadian J. Zool.* 48(5):1035–40.

Scheffer, Victor B. 1958. *Seals, sea lions and walruses: A review of the Pinnepedia.* Stanford, Calif.: Stanford University Press.

Sergeant, David. 1963. Harp seals and the sealing industry. *Canadian Audubon* 25(2):29–35.

——. 1965. Migrations of harp seals *Pagophilus groenlandicus* (Erxleben) in the northwest Atlantic. *J. Fish. Res. Bd. Canada* 22:433–64.

Smith, E. A. 1962. Tracking the gray seals. *Nat. Hist.* 71(3):18–27.

Stains, Howard J. 1967. Carnivores and pinnipeds. In *Recent mammals of the world,* ed. S. Anderson and J. K. Jones, Jr., pp. 325–54. New York: Ronald Press.

Stegman, LeRoy C. 1938. The European wild boar in the Cherokee National Forest, Tennessee. *J. Mamm.* 19(3):279–90.

Thompson, Zadock, 1842. *History of Vermont: Nattural, civil, and statistical.* Part 1:1–224. Burlington.

Van Bree, P. J. H. 1972. On a luxation of the skull-atlas joint and consecutive ankylosis in a grey seal. *Halichoerus grypus* (Fabricius, 1791), with notes on other grey seals from the Netherlands. *Instituut voor Taxonomische Zool (Zoologisch Mus.), Amsterdam Zool. Med. Dell.* 47:330–36.

Walker, Ernest P.; Warnick, Florence; Lange, Kenneth I.; Uible, Howard E.; Hamlet, Sybil E.; Davis, Mary A.; and Wright, Patricia F. 1964. *Mammals of the world.* Vol. 2. Baltimore: Johns Hopkins Press.

Wilson, Susan C. 1975. Attempted mating between a male grey seal and female harbor seal. *J. Mamm.* 56(2):531–34.

# Order Artiodactyla

THE ORDER ARTIODACTYLA comprises those of the ungulates, or hoofed mammals, that have an even number of toes. They are characterized by having the main axis of the foot passing between the third and fourth digits (paraxonic) so that the weight of the animal is borne mainly by these digits. The lateral, or second and fifth digits, are reduced or rudimentary, and the third and fourth metapodials are usually fused to form the cannon bone. In the order Perissodactyla, or ungulates with an odd number of toes, the main axis of the foot passes down through the third digit (mesaxonic), which is the longest on all four feet. The Perissodactyla includes modern horses (*Equus*), tapirs (*Tapirus*), and rhinoceros (*Rhinoceros*, *Didermocerus*, *Diceros*, and *Ceratoherium*).

In the artiodactyls the premolars are simpler than the molars, the upper incisors are reduced or absent, and the canines are usually reduced or absent, except in the hogs. Most artiodactyls are strictly herbivorous and browse or graze, and most are also ruminants and chew the cud. The stomach is simple to complex, and the cecum is usually smooth.

Artiodactyls occur worldwide as native wild animals except in the Australian region, on unpopulated islands, and in Antarctica (Koopman 1967), and include such diverse animals as hippopotamuses, camels, giraffes, deer, sheep, cattle, peccaries, pronghorn antelopes, and bison. The order is represented in New England by two families.

## FAMILY SUIDAE (HOGS)

The family Suidae contains the hogs, pigs, or swine. These animals have simple stomachs and do not chew the cud. The canine teeth grow continuously and may form tusks. The cheek teeth are bunodont and are usually distinguishable as premolars and molars. Hogs are Old World species but are almost cosmopolitan by introduction, except for Antarctica and some oceanic islands. The wild boar has been introduced into New Hampshire and may occur in the wild in New Hampshire and Vermont.

## European Wild Boar

*Sus scrofa* Linnaeus

*Sus scrofa* Linnaeus, 1758. *Systema naturae*, 10th ed., 1:49.
Type Locality: Germany
Dental Formula:
  I 3/3, C 1/1, P 4/4, M 3/3 × 2 = 44
Names: Wild boar and wild hog. The word "boar" is used for both males and females of the European and Asiatic wild pig. The meaning of *Sus* = a swine, and *scrofa* = a breeding sow.

**Description.** The European wild boar differs from the domestic swine chiefly in its longer legs, its tail, and its erect ears. The head is large, long, and pointed, with a long, flexible snout. The ears are rather small, pointed, and densely haired. The tail is lightly covered with hair and has a large tuft of long, bristly hair at the tip. The front feet are considerably larger than the hind feet. The body is narrower and higher from the ground than that of domestic swine. Large European boars have a thick layer of skin over the shoulders which forms a heavy protective shield. Females have twelve mammae.

One conspicuous characteristic of the wild boar is the four remarkably long canine tusks, two in each jaw. The upper canines curve outward and upward instead of downward; the lower canines curve upward and rub against the upper canines, forming sharp edges. These are much more prominent in males and vary from 5 to 30 cm (2–12 in) long. The tusks are used to dig and rip roots and to gouge and rub trees and other vegetation; they can rip open a dog with one sweep. The tusks continue to grow throughout life.

The profile of the skull is long and more or less straight. It is distinguished by the posteriorly inclined slope of the very high occipital crest, the long, narrow nasals, and the long, narrow palate that extends back beyond the plane of the last molars. The canines are prominent, particularly so in males, with persistent roots. The incisors are rooted; the lower incisors are long and narrow, set close together and in an almost horizontal position, while the upper incisors decrease in size from first to third. The cheek teeth are cuspidate, gradually becoming larger and more complex from the front to the back of the jaws. The upper premolars are simpler in structure than the bunodont molars; the third molars are very large, nearly as long as the first and second molars combined, and seldom wear with age.

The pelage consists of long, bristlelike guard hairs over a usually woolly underfur. Many individuals have side whiskers and a mane on the nape. The coloration is grizzled dark gray or rusty brown. Some individuals are piebald or blackish with silver gray heads or are nearly all black, tawny, or brownish. The young are

striped longitudinally, with six light brown and five black stripes on each side of the body. The stripes gradually disappear when they are about 6 months old.

Males are larger than females. Silver (1957) reported that the largest wild boar killed in New Hampshire was a male which weighed 123.5 kg (272 lb), and she stated that the average animal weighs from 36.3 to 40.9 kg (80–90 lb).

**Distribution.** The European wild boar is a native of Europe, Asia Minor, and North Africa. It was introduced into California, Texas, New York, Georgia, North Carolina, Tennessee, and New Hampshire. The introduction into New Hampshire occurred in 1899 when Austin Corbin imported a herd of eleven wild boars from Germany to stock his 25,000-acre preserve a few miles north of Newport, Sullivan County, New Hampshire. The preserve was renamed the Blue Mountain Forest Association. Spears (1893) reported that there seemed to be two lots of introduced wild boar in the Corbin Preserve; one of them—the larger, darker race—probably came from Russia.

Interbreeding of the two lots obscured any racial differences. From the start the animals took naturally to the preserve, and by late 1896 it was estimated that there were at least five hundred of them. In that year there was a failure of the beechnut crop; the boars were not artificially fed during the winter of 1896–97, and all but fifty perished. By 1903 there again was an estimated population of five hundred boars (Manville 1964).

An unknown number of boars escaped from the preserve after the hurricane of 1938, and a small population existed for some years within a radius of about 15 miles of the preserve. A few boars have spread as far north as Alexandria, Grafton County, New Hampshire (Silver 1957). According to Stewart (personal communication) two boars have been shot in Vermont: one in Hartland, Windsor County, in 1954, the other in Barton, Orleans County, in 1956.

In Corbin's day the European wild boar was hunted from horseback with javelins and tracked with Austrian boar setters. The main population control in the Corbin Preserve is hunting, with prescribed bag limits and a four-month open season. From 1948 to 1955 a total of 317 boars were taken, the largest number (79) in the 1951–52 season (Silver 1957).

In 1949 the New Hampshire legislature passed an act encouraging the capture and killing of free-ranging wild boars. Outside the preserve boars may be hunted at any season without limit, although a license is required and hunting at night is prohibited. The reported kills outside the preserve are 25 in 1958, 20 in 1959, 12 in 1960, and 3 in 1961. Probably there are no wild boars free ranging in New England at present.

**Ecology.** Wild boars prefer thick forested areas in rugged mountainous terrain but may be found along the edges of woods or in farmland. In warmer seasons they rest in elongated depressions which they gouge in sandy soil. In colder seasons they build crude beds of boughs or grass in dense thickets resembling huge bird's nests, where several boars may sleep together or where the young may be born.

Man is the chief predator of wild boars in New England. Black bears may kill large and small boars, and bobcats and lynx may kill piglets. Ticks, lice, nematodes, ascarids, and whipworms are known parasites of wild boars.

**Behavior.** Wild boars are active throughout the year. They are elusive, active mainly at night, but may be seen during daytime. Their senses of smell and hearing are keen, but vision is poor. These animals enjoy wallowing in mud for hours throughout the year. Stegeman (1938) reported that wild boars may break thin ice to wallow in water, and when being chased during winter they often wallow in each stream they cross. They are good swimmers; Bourlière (1954) reported that several boars have been seen swimming across at least a half mile of salt water to reach the island of Oleron in France. Wild boars are sure-footed and jump well, and they will leap over logs that domestic pigs would go around. They frequently cross streams by walking across logs, whereas domestic pigs would go through the water.

Wild boars are fond of rubbing against small evergreen trees near wallowing places and along trails. Rubbing trees are useful for determining the presence of boars in an area, because the

EUROPEAN WILD BOAR

ground around the trees has wild boar hairs trampled into it. They often gouge trees with their tusks.

Wild boars often travel in bands of three to several or more and feed together at night. They drift from place to place grubbing or rooting for food and may travel several miles during a feeding period. Although little is known about the home range and movements of these animals, Lewis (1966) found that, when freed, young boars traveled 2.6 miles in 24 hours and had a home range of 1.5 to 2 square miles.

These animals are opportunists and will eat almost any kind of plant or animal matter that is available, including nuts, grains, berries, leaves, fungi, roots, small mammals, carrion, birds and their eggs, snails, amphibians, reptiles, and large quantities of insects and worms.

European wild boars breed throughout the year. Adult males are solitary or may travel in small groups. They join the females during their estrous period and leave them when they are no longer in estrus. The estrous period lasts about 48 hours; the normal length of the estrous cycle is 21 to 23 days, but irregular cycles are common. The gestation period is 108 to 120 days, usually 115 days. The number of young per litter is three to twelve, usually four to six (Henry 1968).

The piglets are born in some secluded spot in a dense thicket. At birth they are fully haired, with eyes open, and weigh about 907 g (32 oz). Wild boars become sexually mature at about 1½ years of age and are fully grown at 5 to 6 years old. Longevity is 15 to 20 years and occasionally up to 27 years (Henry 1968).

### FAMILY CERVIDAE (DEER AND ALLIES)

The family Cervidae is characterized by paired antlers, primarily in males but in both sexes of caribou. The antlers are absent in the musk deer and Chinese water deer. The gallbladder is absent, except in the musk deer (Koopman 1967). There are no upper incisors and, in most species, no upper canines. The cheek teeth are selenodont.

Antlers are not bones, but osseous structures

that are shed and regenerated annually. They begin to grow in early spring with the swelling of buds from the bony pedicels of the frontal bone. During the fast growing period the antlers are soft and tender and covered with a soft, hairy skin, the velvet, which is richly supplied with blood vessels that carry nutriments and minerals to the growing structures. The velvet probably also provides protection to the growing antlers.

Ossification of the antlers begins at the base and keeps pace with the growing tip so that a section at any level is harder than that above it and less ossified than that below, until the entire antler becomes ossified and rigid. By early fall the antlers reach full development; the blood vessels then constrict so that the blood flow becomes sluggish and is shut off, and the velvet begins to dry up. The dried velvet is rubbed off on trees, shrubs, or rocks. The well-defined grooves in hardened antlers result from the former channels of dried-up blood vessels. By late winter antlers begin to loosen and are usually shed one at a time.

The family is widely distributed over the world except for Antarctica, southern South America, and some oceanic islands (Koopman 1967). There are five species of cervids in New England.

# White-tailed Deer

*Odocoileus virginianus borealis* Miller

*Odocoileus virginianus borealis* Miller, 1900. *Bull. New York State Mus. Nat. Hist.* 8:23.
Type Locality: Bucksport, Hancock County, Maine
Dental Formula:
 I 0/3, C 0/1, P 3/3, M 3/3 × 2 = 32
Names: Northern white-tailed deer, whitetail, and bannertail. The meaning of *Odocoileus* = hollow tooth (the type tooth was found in a cave), *virginianus* = of Virginia, and *borealis* = boreal, or northern.

**Description.** The graceful white-tailed deer is distinguished by its long legs, conspicuous ears, long, bushy tail, naked nose pad, narrow, painted hooves, and, in males, antlers. Females have four mammae, all inguinal. White-tailed deer have four prominent cutaneous glands—the preorbitals, at the inner corners of the eyes; the metatarsals, on the outsides of the hind legs between the ankles and hooves; the tarsals, on the insides of the hind legs at the hocks or heel joints; and the interdigitals, between the toes on each foot.

The skull of the white-tailed deer may be distinguished from skulls of other cervids by the vomer dividing the nares into two separate chambers posteriorly and by the width of the slightly inflated auditory bullae, which is equal to the length of the bony tube leading to the meatus. The lacrimal vacuity is very large, and the lacrimal fossae are small and shallow. The upper incisors are absent. The lower canines are modified to incisors in form and function, while the first premolars take on the form and function of canines.

The sexes are colored alike and show marked seasonal variation. The summer pelage is called the red coat, and hunters consider the deer to be in poor condition when in the "red"; the winter pelage is called the blue coat, being a grayish color, and the animal is worth killing when in the "blue." The pelage is short, thin, straight, and somewhat wiry in summer; in winter it is long and thick and may be slightly crinkled. In summer the color of the upperparts is reddish brown to tan, sometimes paler on the sides, and the underparts and insides of the legs are whitish. The top of the head and the tip of the ears are often tipped with ochraceous tawny hairs. There is a white patch on the throat and white across the blackish nose, around the eyes, and inside the ears. There is a black spot on each side of the white chin. The tail is dusky above, white around the edges and below. When the deer bounds away it erects its tail and shows its white "flag." In winter the upperparts of the pelage are grayish brown or grayish and the underparts are whitish.

Fawns are reddish brown with white dorsal spots. They usually begin to lose their spots when 3 to 4 months old. Fawns born in May normally molt in September, but spotted fawns may be seen in late October and even into early December. The new coat is like the winter gray of the adult in texture and color.

There are two molts each year. The fall molt begins from August to mid-September and is completed from September to early October. The spring molt begins in May or June. Albino and melanistic white-tailed deer are rare.

Males have moderately spreading, large, and branching antlers; the large beams, or main shafts, grow slightly backward and outward at first and then directly forward, slightly inward, and almost horizontal. Each beam bears several unbranched sharp tines which appear to rise vertically. The brow tines are absent. Unlike the antlers of most cervids, the antlers of white-tailed deer rarely exceed the length of the head.

Deer antlers are grown and shed annually. In Connecticut they are generally shed from mid-December to mid-January (Behrend and McDowell 1967); in Massachusetts, usually from late December through January and February (Shaw and McLaughlin 1951); and in New Hampshire, about mid-December, though some deer may retain their antlers until April (Siegler et al. 1968).

Antlers begin to grow in April or May, and they become hard by September when the velvet dries up. The annual growth and shedding of antlers appears related to the annual sexual cycle of the species. Wislocki (1943) proposed that the pituitary gland may secrete a hormone that stimulates growth of the antlers, then releases another hormone that stimulates growth and activity of the testes during late summer. The testes then secrete testosterone into the bloodstream, which results in drying of the velvet. The author demonstrated that castrating bucks with antlers in velvet resulted in permanent retention of the velvet and failure of the antlers to be shed. Castration in the presence of antlers that had lost their velvet resulted in immediate shedding of the antlers and their subsequent renewal and permanent retention in the following year. Wislocki, Aub, and Waldo (1947) showed that drying of the velvet was precipitated by a rise in the testosterone content of the blood, and that antlers dropped off as a result of a decline in the production of the hormone as activity of the testes diminished near the end of the rutting season.

The color of a mature antler varies from white to deep brown. Some people believe this is due to staining by the juices of the shrubs and plants on which deer like to rub their antlers. Others suspect that dried blood is the principal source of the color.

Antlers are formidable weapons among fighting bucks during the breeding season. The question often asked is whether "dehorning," or prematurely sawing off the antlers, can affect a buck's attitude and mating ability. Severinghaus and Cheatum (1956) reported that dehorning a buck usually moderated its aggressiveness and lessened its interest in mating. However, Siegler et al. (1968) reported incidents in which two belligerent bucks were dehorned shortly after the velvet had dried up. Their aggressiveness did not lessen, and both successfully bred does. In another instance a captive dominant buck was dehorned. For the next two weeks it lacked aggressiveness around the younger bucks, but thereafter it regained dominance and bred two does in one day.

Although the age of a buck cannot be determined by the size of its antlers or the number of tines, for a given buck the diameter of the antler beam near its base tends to enlarge with each new set of antlers. In young bucks, especially yearlings, the antlers do not develop tines but consist of a single slender, spikelike tine on each side called a "spikehorn." Bucks in the prime of life usually have broader and stouter antlers with more tines than bucks in advanced age. Bucks in poor health may have antlers resembling spike horns. Severinghaus et al. (1950) have shown that antler beam diameters vary with the quality and quantity of food available and that the best antler development is found on the best deer range. French et al. (1956) demonstrated that "young" deer maintained on an unbalanced or quantitatively deficient ration produced antlers inferior to those of deer on an adequate ration. Banasiak (1961) used antler beam diameters of yearlings to estimate range quality in Maine. Does seldom have antlers, but when they do they are usually sexually normal and fertile and can bear and rear fawns. Does' antlers are usually small and irregular and may remain in the velvet.

Bucks are larger than does. The total length of adult bucks averages 180 cm (71 in); tail, 29 cm (11 in); hind foot, 52 cm (20 in); and height at shoulders, 100 cm (39 in) (Goodwin 1935). Weights vary considerably with age and sex, physical condition of the individual, range condition, quality and quantity of food available, season, time of year of the kill, duration of snow cover before time of kill, and other environmental factors.

Seigler et al. (1968) found that the average hog-dressed weight of all bucks was 55.4 kg (122 lb), and that of all does was 42.1 kg (93 lb) (heart, lungs, liver, and paunch removed), based on some 5,000 New Hampshire deer of all ages examined between 1 November and 31 December during the period 1951–54. Weights did not increase substantially with age after deer reached 4½ years old.

WHITE-TAILED DEER

It is difficult to calculate the live weight of a deer accurately from the hog-dressed weight because of the considerable differences in dressing techniques and amount of viscera removed. Also, some deer are badly shot, with varying amounts of flesh and bones blown off, and some carcasses are weighed several days after shooting, with no allowances made for shrinkage. However, hunters can estimate the approximate live weight of their dressed deer by adding 5 ciphers to the dressed weight and dividing by 78,612, or by dividing the dressed weight by 4 and adding the result to the dressed weight.

**Distribution.** White-tailed deer are found over most of southern Canada and the United States except for most of California, Nevada, and Utah, and south to Coiba Island, Panama.

**Ecology.** This species is essentially a forest-edge animal, preferring thickets alternating with open, sunny glades and abandoned fields. It is seldom found in remote, extensive, dense forests with closed canopy. During early summer, the white-tailed deer can be seen feeding near lakes, ponds, and streams where there are abundant succulent grasses, pondweeds, and other water plants.

White-tailed deer have bed spots and loitering grounds. The bed spots are located in a little hollow where the ground is level, grassy, or dry, often with one or several trees at the back and on the sides, and they may be found in hardwoods and tall meadow grasses open to the sun at a vantage point from which the deer can see around them. Deer may use a bed spot only once.

The loitering grounds are usually close to the feeding grounds, in open places exposed to direct sun. Here the deer stand about and chew the cud for a while before going farther back into the brush.

During summer, autumn, and winter, white-tailed deer are widely distributed and may be encountered in all wooded areas that provide a great variety of food and cover. During mild winters with little snow, the animals move freely over their range, but when temperatures drop and the snow gets deeper they congregate and restrict their movements to a "yard" within their normal range. A yard consists of a network of packed deer trails formed by deer constantly plowing through the snow in search of food and shelter. The trails sometimes radiate from a central resting place.

In Massachusetts, Hosley and Ziebarth (1935) found that stands of pines were favored sites for winter yarding. In Maine, Banasiak (1961) reported that 85 percent of 350 yards were along streams or near lakeshores and ponds surrounded by stands of conifers or conifers and hardwoods.

Topography seems to influence the distribution of yards. In Vermont, Seamans (1947) reported that white-tailed deer preferred the protected valleys with southern or southeastern exposures that protected them from winter winds.

Man and dogs are the chief predators of the white-tailed deer, and coyotes and bobcats may kill some. Deer may be killed by automobiles or trains, or by fighting, falling through ice and drowning, falling over snowy banks and cliffs, becoming entangled in fences, impaling themselves on branches while running, getting caught in tree crotches while reaching for high browse, being mired in muck, and occasionally by being trapped in forest fires. Many deer are killed or crippled during the hunting season but not retrieved. Severe winters and lack of food will cause starvation or "die-off" in herds.

Parasites of white-tailed deer are lice, mites, ticks, cestodes, nematodes, trematodes, nasal bots, and footworms. There is evidence that the meningeal roundworm *Pneumostrongylus tenuis* of the white-tailed deer is deadly to young moose, elk, and caribou (Anderson and Strelive 1967, 1968). The parasite has little pathological effect on white-tailed deer, but where these cervids coexist, deer populations could conceivably control moose, elk, and caribou populations by transmission of this parasite.

White-tailed deer are susceptible to skin tumors and warts, malignant edema or algae, hemorrhagic septicemia, dermatophilosis (cutaneous streptothricosis), lumpy jaw or mandibular and dental anomalies, and impaction of hair balls in the esophagus. Helmboldt and McDowell (1958) discovered granular venereal disease in a female white-tailed deer in Connecticut. This genital disease (infectious or nodular vaginitis) affects the mucosa of the vagina, vulva, or prepuce and resembles that described in the domestic cow.

**Behavior.** White-tailed deer are cautious, alert, curious, and retiring animals. They have good eyesight for perceiving movement but apparently cannot detect a stationary object well. However, the least motion or the faintest scent or sound carried by the breeze will send them crashing through the brush. These animals have a keen sense of smell, which may be their chief method of detecting danger. The sense of hearing is also very keen, and the slightest noise will cause the deer to raise its head and peer intently in the direction of the disturbance.

Bucks appear to be more wary than does. White-tailed deer emit bleats, whistles, whining sounds, loud snorts, squawks, or "whiews." The doe utters a murmur of low whining sound to her fawn, and it gives a bleat to call the doe.

White-tailed deer normally walk slowly and quietly or trot at a leisurely pace, but when startled or frightened they may canter for a few paces, then take a few graceful jumps. Severinghaus and Cheatum (1956) reported that white-tailed deer can run up to 25 miles per hour for a long distance.

White-tailed deer are somewhat social, living in loosely organized family groups after the rutting season. A group consists of an adult doe with her fawns and yearlings. The old doe leads the family, and any bucks with antlers in this group are likely to be her yearlings. Older bucks tend to form more or less definite groups of their own. Incidental groupings of mature bucks, yearlings of both sexes, does, and fawns may be seen at favorite feeding grounds during summer.

There seem to be aggressive-submissive interactions among white-tailed deer within each sex and age group when food becomes scarce. Collias (1950) found that when white-tailed deer congregated around bales of hay distributed in an attempt to keep them from starving, the adult bucks generally dominated all the other deer, whereas with a few exceptions does and yearlings dominated the fawns.

White-tailed deer are very strong swimmers. Severinghaus and Cheatum (1956) recorded that a deer swam the width of Lake Cayuga, New York, covering a distance of 3 to 3½ miles at about 13 miles per hour, then slowing to a steady pace of 11 miles per hour for a little more than ½ mile. Schemnitz (1975) reported that two deer swam 2½ miles in open ocean water off the coast of Maine.

Deer activity depends on local factors, season of the year, weather conditions, and wind as well as on individuals. In localities where deer are disturbed by man, the animals keep to their beds during most of the day and feed mostly at night; in remote areas the reverse tends to be true. Normally, white-tailed deer forage from dawn till a few hours later, then wander to a brushy hillside or the top of a ridge and lie down for the afternoon. At dusk they begin to feed until night grows black, then they retire and doze or chew the cud till dawn. The animals are more active before rain and snow. They are likely to bed in a protected spot and let a storm pass.

White-tailed deer follow much the same routine day after day, visiting different parts of their home range at certain times each day. The routine changes slowly. In moving about the animals take the path of least resistance, going around obstacles when possible, until well-beaten paths or runways are made. They prefer to go around a fallen tree rather than to leap over or crawl under. Deer paths or runways vary greatly in length; some are very short owing to obstacles such as windfalls, brush heaps, large rocks, swamps, bogs, and often snowdrifts. Short runways often break up into individual tracks. Long runways usually parallel lakes, bogs, swamps, ponds, and rivers to and from good feeding grounds. From these long runways, side trails often branch off into short individual trails.

White-tailed deer are not naturally inclined to roam and may remain for weeks in a definite range of 2 or 3 square miles, following the same routine day after day, except during the rutting season or when adequate food, water, and cover are not available, or when they are disturbed by cold weather, deep snow, and other environmental factors. A deer may leave its home range when wounded, tracked by hunters, or trailed by dogs, but barring accident it usually returns.

White-tailed deer probably have a homing instinct, since they tend to return regularly to the same home ranges in summer and in winter. McBeath (1941) reported that a tagged deer returned to the original site of capture after it was released 10 miles away, and that another tagged deer returned to its original site of capture after it was released 16 miles away in another direction. Dahlberg and Guettinger (1956) reported that a deer returned to the original site of capture after it was released 13 miles away.

White-tailed deer are chiefly browsers but may graze occasionally on grasses, lichens, mushrooms, and other nonwoody plants.

The variety of foods eaten by white-tailed deer is enormous. The following is a composite list of woody plants browsed by deer during winter in Massachusetts (Hosley and Ziebarth 1935), in Maine (Banasiak 1961), and in New Hampshire (Siegler et. al. 1968). Preferred plants include: American yew or ground hemlock, apple, basswood, black cherry, dwarf raspberry, eastern hemlock, hazel, hobblebush, mountain ash, mountain maple, northern white cedar, red maple, red oak, smooth sumac, staghorn sumac, striped maple or moosewood, and wintergreen. Species commonly browsed are: American cranberry bush, American elm, American hazelnut, alternate-leaf dogwood, balsam fir, balsam poplar, beaked hazelnut, beech, bigtooth aspen, black alder, black ash, berries, common elder, common juniper, creeping juniper, dogwood, fungi, hawthorn, hickory, honeysuckles, maple-leaf viburnum, mountain holly, mountain laurel, paper birch, pitch pine, poplar, quaking aspen, red-berried elder, redosier dogwood, red pine, roses, shadbush, sorrel, spiny-shield fern, spirea, sugar maple, sweet fern, sweetgale, white ash, white cedar, white pine, white oak, willows, wild raisin, wintergreen, witch hazel, witherod, and yellow birth. Species infrequently browsed are: bear oak, black chockberry, black spruce, bog rosemary, eastern hop hornbean or ironwood, gray birch, hardhack, Labrador tea, lambkill, leather-leaf, meadow sweet, red oak, red spruce, speckled alder, tamarack, and white spruce.

White-tailed deer occasionally grub for roots, and they frequently chew at old rotten stumps and wood, probably to get roots that have penetrated the soft wood. They are fond of lichen and will stand on their hind legs to reach it on trees.

In New Hampshire, deer commonly strip the bark of small hemlocks, fir, elderberry, alder, apple, cherry, hobblebush, striped maple, red maple, mountain ash, shadbush, and willow. Usually only smooth, tender bark is stripped. Severe stripping normally occurs in March or April and varies from year to year. Some trees are stripped repeatedly, while others nearby go untouched (Siegler et al. 1968).

The basic daily browse requirement of white-tailed deer varies with the size of the animal, time of year, and type of activity, such as the rut and parturition and suckling of young. Gerstell (1936) determined that in winter Pennsylvania deer required slightly over 2 pounds of food per 100 pounds of deer in 24 hours to maintain fair physical condition. In Michigan, Taylor (1947) found that yarded deer needed an average of 5 pounds of browse per day. Swift (1946) demonstrated that in Wisconsin adult deer needed 5 to 7 pounds daily of first-class browse per 100 pounds of deer.

The rutting urge develops earlier in the buck than in the doe. In northern woods, during late September through mid-November, the bucks' necks swell, they have shed their velvet, and they begin to spar with bushes and small trees with their hardened antlers.

During the rutting season bucks move about extensively, seeking receptive does by smell. Sometimes two or more bucks will follow the same doe; usually the dominant buck is in front, trailed at a safe distance by a lesser buck, and sometimes by a small buck. Some bucks may establish territories during this prebreeding season. Although the bucks do not collect harems, a buck is an opportunist and may service several does.

In Massachusetts, Shaw and McLaughlin (1951) determined that 77 percent of breeding activity among adults took place between 10 and 25 November and that about 90 percent of the adults bred before 1 December and none after 15 December. Approximately 70 percent of the fawns that bred bred in December and January. In New Hampshire, Siegler et al. (1968) reported that the earliest breeding occurred between 9 November and 16 December, with 22 November the day of the greatest breeding activity; the mean breeding date of fawns was 13 December. In Maine, Banasiak (1961) indicated that breeding dates ranged from 20 October to 28 December, with mid-November the height of the breeding season.

In the northern latitudes the gestation period is approximately 201 days. The fawns are usually born in May and June. Since there are extreme early and late breeding dates for white-tailed deer, there is also an extreme spread of dates of fawning. Severinghaus and Cheatum (1956) reported that fawns have been seen as early as March in New York.

Most does become sexually mature at 18 months of age, although some mature at 6 months and perhaps some not until 2½ years in poor habitat. Bucks may reach puberty when they are about 18 months old (Cheatum and Norton 1946). Silver (1965) reported successful breeding of two penned fawns in New Hampshire. The precocious buck, which had polished antler buttons, was 202 days old when he mated with a doe 214 days old. Cheatum (1949) reported that the estrous period lasts 24 to 36 hours. If a doe is not bred during this time, she comes into estrus again approximately 28 days later. In cases of persistent breeding failure, a doe may have three estrous cycles.

Reproduction rate depends on age of the doe and range quality. The number of fawns varies from one to four annually, but quadruplets are rare. In Massachusetts, Shaw and McLaughlin (1951) reported that all pregnant fawns, with one exception, produced a single fawn. The one exception produced twins. Among the yearlings, 28 percent produced single fawns, 63 percent twins, and 9 percent triplets. Among the adults, 14 percent produced single fawns, 68 percent twins, and 18 percent triplets. In New Hampshire, Siegler et al. (1968) reported that does, including fawns, produced an average of 1.39 fawns annually. Does 4½ years old produced an average of 2.0 fawns. In Maine,

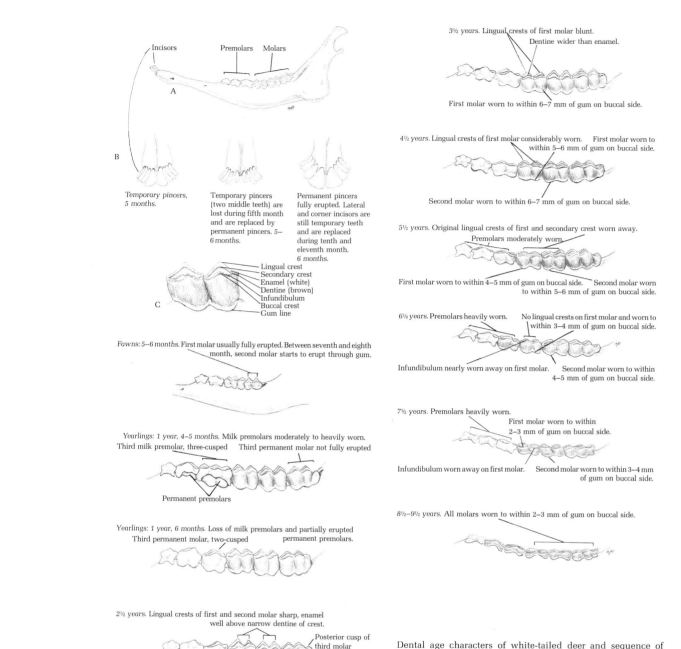

Incisors    Premolars  Molars

A

B

*Temporary pincers, 5 months.*

Temporary pincers (two middle teeth) are lost during fifth month and are replaced by permanent pincers. 5–6 months.

Permanent pincers fully erupted. Lateral and corner incisors are still temporary teeth and are replaced during tenth and eleventh month. 6 months.

Lingual crest
Secondary crest
Enamel (white)
Dentine (brown)
Infundibulum
Buccal crest
Gum line

C

*Fawns: 5–6 months.* First molar usually fully erupted. Between seventh and eighth month, second molar starts to erupt through gum.

*Yearlings: 1 year, 4–5 months.* Milk premolars moderately to heavily worn.
Third milk premolar, three-cusped    Third permanent molar not fully erupted

Permanent premolars

*Yearlings: 1 year, 6 months.* Loss of milk premolars and partially erupted
Third permanent molar, two-cusped    permanent premolars.

*2½ years.* Lingual crests of first and second molar sharp, enamel well above narrow dentine of crest.
Posterior cusp of third molar slightly worn.

*3½ years.* Lingual crests of first molar blunt.
Dentine wider than enamel.
First molar worn to within 6–7 mm of gum on buccal side.

*4½ years.* Lingual crests of first molar considerably worn.    First molar worn to within 5–6 mm of gum on buccal side.
Second molar worn to within 6–7 mm of gum on buccal side.

*5½ years.* Original lingual crests of first and secondary crest worn away.
Premolars moderately worn.
First molar worn to within 4–5 mm of gum on buccal side.    Second molar worn to within 5–6 mm of gum on buccal side.

*6½ years.* Premolars heavily worn.    No lingual crests on first molar and worn to within 3–4 mm of gum on buccal side.
Infundibulum nearly worn away on first molar.    Second molar worn to within 4–5 mm of gum on buccal side.

*7½ years.* Premolars heavily worn.    First molar worn to within 2–3 mm of gum on buccal side.
Infundibulum worn away on first molar.    Second molar worn to within 3–4 mm of gum on buccal side.

*8½–9½ years.* All molars worn to within 2–3 mm of gum on buccal side.

Dental age characters of white-tailed deer and sequence of tooth eruption. *A*, lower jaw; *B*, top view of incisors or pincer teeth; *C*, lateral view of lower molar.

Banasiak (1961) reported that 21 percent of does had single embryos, 74 percent twins, and 5 percent triplets. The average was 1.8 embryos per doe.

As fawning time approaches the pregnant doe leaves the herd and seeks seclusion in some well-concealed spot to give birth to her young. The doe makes no natal bed. The place of birth may be in a thicket or similar cover, in an open grassy field, near a stream or lake, or in a bog or swamp.

The birth of a fawn is usually quick and

almost bloodless. At birth the fawn is weak, wobbly, and helpless and remains hidden beneath clumps of shrubs or similar cover. It instinctively flattens itself against the ground and remains still even though a predator may pass within a few yards. When a fawn is motionless like this, the reddish brown coat spotted with white is almost perfect camouflage and blends into sunlit foliage. Very young fawns are scentless.

The weights of fawns vary considerably, from 1.5 to 6.7 kg (3.3–14.8 lb) (Haugen and Davenport 1950). At first the fawns are awkward, but they grow rapidly on the does' rich milk. Fawns will stand up when about 12 hours old to nurse while the doe is standing. They do not wander far during the first week, and for the most part they remain hidden.

Fawns begin to nibble on green vegetation when 2 or 3 weeks old. By the end of a month they browse on seedlings and similar foods. Although they are weaned at about 4 months, some does let their fawns nurse occasionally up to the next breeding season. During the summer the fawns forage with the doe and trail her rather closely.

Most fawns begin to lose their spots at 3 months old. The spotted coat is gradually replaced by the winter coat of the adult during September in the northern United States. Occasionally spotted fawns are seen during the hunting season. These fawns probably are born in July or August or later. Captive white-tailed deer may live up to 15 to 20 years.

**Age Determination.** Haugen and Speake (1958) gave a method for determining the age of white-tailed deer fawns up to 30 days, based on growth of the front hooves. At birth the hooves are quite soft and grayish, but they harden and darken within the first day. New growth proceeds from a ringed "groove." Growth is like that of a fingernail, from the base outward. Proximal to this groove there are longitudinal striations on the hoof, whereas distal to it the striations are radial. Measurement is taken in millimeters from the base of the distal hairs on the outside half of a front hoof to the ring described above, along the front edge of the hoof. The growth rate is relatively consistent. The measurement is multiplied by 2.2 and 0.66 is added to this product to give the age of the fawn in days.

Severinghaus (1949) described criteria for determining the age of white-tailed deer in New York on the basis of sequence of tooth growth and wear of the cheek teeth in the lower jaw. There are slight differences in the schedule of eruption and wear of teeth between regional areas of the deer's range.

# Moose

*Alces alces americana* (Clinton)

*Cervus americanus* [Clinton], 1822. *Letters on the natural history and internal resources of the State of New York*, p. 193.
*Alces alces americana*, Lydekker, 1915. *Catalogue of the ungulate mammals in the . . . British Museum*, 4:234.
Type Locality: "Country north of Whitestown" [probably in the western Adirondack region], New York.
Dental Formula:
I 0/3, C 0/1, P 3/3, M 3/3 × 2 = 32
Names: Common moose, black moose, and flat-horned elk. The meaning of *Alces* = an elk or moose, and *americana* = of America.

**Description.** The ungainly moose is the largest of all the deer. It has long legs, short neck, very short tail, a high hump on the shoulders and relatively small rump. There is a slight mane on the neck and shoulders. The head is relatively large, but narrow, with large ears and small eyes. The muzzle is broad, inflated, and pendulous, the anterior end of the upper lip overhanging the lower lip. The large nasal pad is haired except for the extreme lower portion. Both sexes have a dewlap of skin and long hair called the "bell" which hangs from the underside of the throat. It is more developed in males than in females, but its function is unknown. The front legs are longer than the hind legs, which makes the gait awkward but helps the moose jump over fallen trees and other obstacles. The hooves are long, narrow, and pointed, with well-developed lateral hooves, and the front hooves are larger than the hind ones. The metatarsal glands are absent.

Adult males have massive and broadly palmate antlers with many tines. The antlers often have a secondary palmation of the brow prongs at right angles to the main palmation. Although yearling males usually grow their first set of spikes, which vary greatly in shape and size, male calves may develop buttons or knobs, and some calves may even grow a set of spikes.

New antlers begin to grow in April. Growth is slow at first but becomes much faster as summer advances. By August or September full development is reached when the velvet dries and is rubbed off on shrubs and trees. The antlers are usually shed in December or January, but may be shed as late as early March.

The skull has a long premaxillary region with very short nasals and a very large nasal opening; the distance from the front of the nasal to the front of premaxilla is about equal to that from the back of the nasal or occiput. The vomer is low posteriorly and does not divide the posterior nares into two chambers. The lacrimal vacuity is widely open and the pit well developed. The lower canines are incisiform, incisors are little differentiated, and premolars and molars are broad and low-crowned.

The fur is thick, brittle, and very coarse, approximately 152 mm (6 in) long on the neck and shoulders. Both sexes are colored alike and show slight seasonal color variation. The new spring pelage is blackish brown and grayish brown, paler on the head and sometimes gray on the face. The underparts are the same color as above except that the lower belly and lower legs are distinctly lighter or pale brownish gray, sometimes almost whitish. The pelage in summer, autumn, and winter is somewhat lighter in color, almost grayish. The calf is reddish brown and not spotted. Albinos and melanistics have not been recorded as far as is known.

The single annual molt occurs in early spring with the shedding of the faded and ragged hairs. Adult bulls molt first, the cows and yearlings shortly after. Molting starts on the shoulder hump region and proceeds along the sides of the neck and back of the ears and backward over the body (Peterson 1955).

Females average about three-quarters the size of males. Measurements of a typical eastern moose from Quebec are: total length, 290 cm (114 in); tail, 6 cm (2 in); hind foot, 79 cm (31 in); height at shoulder, 183 cm (72 in); and spread of antlers, 146 cm (58 in) (Seton 1929). Weights

MOOSE

vary from 408 to 635 kg (900–1,400 lb) (Anthony 1928). A large moose killed in Maine weighed more than 635 kg (1,400 lb).

**Distribution.** The moose occurs below the timberline and on the tundra between the Alaska-Yukon boundary and Hudson Bay, and in parts of the northern United States into New England south to Pennsylvania, westward across southern Ontario and northern Wisconsin and Minnesota to the Rocky Mountain range, with a southern extension into Wyoming and northeastern Utah. The present distribution in the United States is much reduced.

The eastern moose occurs from eastern Canada, except Prince Edward Island, from Nova Scotia and New Brunswick west through Quebec to eastern Ontario, south into Maine, northern New Hampshire, and Vermont. Some individuals straggle into Massachusetts and Connecticut.

The moose apparently was common as far south as Massachusetts during early settlement. Crane (1931) reported that the species was practically extinct in Massachusetts about the beginning of the nineteenth century, though occasional stragglers were sighted in the mountainous and unsettled regions of the state. Merrill (1920) reported that about 1900 William C. Whitney imported three pairs of moose from Manitoba for his wire-fenced 1,200-acre October Mountain preserve in Lee, Berkshire County, Massachusetts.

In Connecticut, Goodwin (1935) stated that there were no records of moose in Connecticut before colonial times, but that at the beginning of the sixteenth century moose were seen in the state. In Rhode Island, Cronan and Brooks (1968) reported that an occasional moose may have been encountered before colonial times.

During colonial times moose were plentiful in Vermont, where the settlers depended upon their flesh for subsistence. The animals were such conspicuous and easy targets that they were nearly exterminated by 1828 (Foote 1946). Moose were seen in northern Vermont until about 1900 (Goodwin 1936). Stragglers have been seen at Saint Johnsbury, Caledonia County, Randolph, Orange County, Brattleboro, Windham County, and other places (Osgood 1938). They have been reported from Shaftsbury, Ben-

nington County, and Marshfield, Washington County, Vermont (Warfel 1937). A moose was killed in Westminster, Windham County, in 1942, and two moose were seen near Bridgewater, Windsor County, and Rockingham, Windham County, Vermont, in 1945 (Foote 1946).

Moose were also common in northern New Hampshire in the early days of settlement but were so widely killed for food that they were nearly extirpated there by 1900 (Goodwin 1936). Silver (1957) wrote: "Some idea of their numbers may be gained from the record of Nathan Caswell, who killed 99 in one winter near East Lancaster. A man by the name of Hilliard—one of a family of noted hunters—destroyed 80 in one season. By 1810 the total score for the year around Lancaster was 90. By 1820 moose were getting scarce in southern Coos."

Austin Corbin imported 16 moose from Minnesota for his park in Sullivan County, New Hampshire, in 1889. Later shipments, plus births, increased the number to 60 animals. Moose persisted in the park until about 1940, when the last of them died of starvation (Manville 1964). Carpenter and Siegler (1958) listed the moose as uncommon and wrote that about 40 moose existed in New Hampshire at that time.

In Maine, moose were abundant during colonial days. St. Michael-Podmore (1904) wrote that Maine was "the sporting paradise for moose." Hornaday (1904) estimated that some 3,000 moose were killed in Maine between 1894 and 1902. Aldous and Mendall (1941) reported that by 1904 the animals were found only in the northern counties, but for some unknown reason they then appeared in the southern half of Maine while in the extreme northern and western parts of the state they were reported to be rare.

**Ecology.** Moose inhabit the northern forests, thickets, swamps, and swales that provide plants to eat and good cover for concealment. In summer they frequent streams, ponds, and shallow lakes that provide important aquatic plant food items and some relief from swarming flies and mosquitoes.

Man is the chief predator of moose. Black bears may kill some calves, but coyotes apparently are ineffectual in killing moose. Some moose drown when they fall through thin ice or die from becoming mired in deep marshes and bogs, from forest fires, or from collisions with trains and automobiles.

Parasites of moose include ticks, the moose fly, blackfly, cestodes, nematodes, trematodes, and larvae of the botfly *Cephenemyia jellisoni*. Thomas and Cahn (1932) described a "moose disease" which includes as symptoms partial or complete blindness, emaciation, head held to one side, dropping of one ear, partial paralysis of the legs, and inability to rise or stand; the victim becomes tame and stays close to certain places or open land and sometimes travels in circles, then dies. Although Cahn, Wallace, and Thomas (1932) believed that the outbreak of the disease was coincidental with the life cycle of the tick *Dermacentor albinipictus* infesting the moose, Fenstermacher (1937) showed that the tick did not necessarily cause moose disease, since he found sick and dead moose at every season of the year except late summer. Benson (1952) suggested that the disease may be correlated with cobalt deficiency. Lamson (1941) found that captive sick moose became healthy after being fed a diet supplemented by foods high in vitamins.

Anderson and Strelive (1967) reported that the meningeal roundworm *Pneumostrongylus tenuis* of the white-tailed deer is highly pathogenic to young moose. Other diseases of moose are arthritis, brucellosis, encephalitis, sarcocystis, tuberculosis, and necrotic stomatitis.

**Behavior.** The moose is alert and its senses of hearing and smell are very keen, but its vision is poor. Lamson (1941) reported that a moose can detect the footsteps of an approaching person 1½ to 3 minutes before man first becomes aware of the sound. Peterson (1955) reported that it was more difficult to approach moose from downwind than from upwind, even when the animals were seen across a lake as far away as half a mile.

Moose are not very sociable, but for most of the year they seem to enjoy some companionship, regardless of sex. During summer several moose may congregate in a large body of shallow water containing abundant aquatic food, but often they pay little attention to each other when feeding (deVos 1958). They may stand or lie in shallow water as a refuge from flies (Flook 1959).

When the snow is deep as many as twelve to fifteen moose may congregate or "yard" together in a small area. Probably such associations in winter yards are a result of some common condition such as food availability or restriction of movement rather than social instinct (Lönnberg 1923). Toward spring the loosely knit associations begin to disintegrate, and the bulls remain alone or in small groups called "satellites" with younger bulls who follow the leading bull (Altmann 1956).

Moose are chiefly active at dawn and dusk, less often at midday. Skunke (1949) reported that the animals will lie up in dense forest cover on foggy, rainy, and windy days when they have difficulty in hearing or picking up scent.

These animals ordinarily remain within a radius of from 2 to 10 miles in areas that provide palatable food the year round (Peterson 1955). Moose in British Columbia have altitudinal migrations (Edwards and Ritcey 1956). Movements of moose are more extensive during the breeding season, and bulls will investigate the bellow of a cow several miles away. Some moose may wander extensively at other seasons.

Moose walk casually, trot gracefully, and gallop clumsily. They can be heard moving through the brush several yards away when undisturbed, but they move silently when alerted or suspicious. When frightened they often crash headlong through dense brush. Findley (1951) determined that moose normally run 20 to 25 miles per hour, but they can travel 30 to 35 miles per hour for short distances. Moose can walk easily through deep bogs and muskegs. Stanwell-Fletcher and Stanwell-Fletcher (1943) reported that moose use their forelegs as snowshoes and move in a kneeling position in deep snow. Merrill (1920) stated that moose are strong swimmers and can swim slowly for up to 12 miles.

Moose are silent animals except during the breeding season and except for cows with calves. Both sexes call during the rut. The call of the moose is subdued but clear and may be heard up to 200 yards away (Hosley 1949). The bull moose utters either a low, moo-like plea, broken off short with an upward inflection at the end, or a throaty gulp. The call of the cow moose is longer and more like that of the domestic cow, but not nearly as loud, and that of the calf is similar to that of the domestic calf with a moo-like beginning, ending in a petulant bawl. The call of the cow to her calf is a single grunt.

Moose are chiefly browsers but graze occasionally on grass, moss, lichen, mushrooms, and herbaceous and leafy plants. The food items and methods of feeding of moose are more varied than those of most deer. Moose commonly are forced to bend or spread their front legs or drop on their knees to feed from the ground because of their short necks. They often rear up on their hind legs to reach tender terminal twigs, stripping them through their mouths. They may straddle or "ride down" small saplings by walking over them to bring the twigs within easy reach.

In spring, summer, and early fall moose eat ground plants, chew tender leaves, twigs, bark of deciduous trees, and feed on semiaquatic and aquatic plants such as pondweeds, arrowheads, yellow pond lily, bur reed, eelgrass, spikerush, sedges, water horsetail, pickerelweed, and other water plants. In winter they browse on conifers and hardwood trees. The preferred plants at all seasons are leaves and twigs of willows, balsam poplar, beaked hazelnut, gray birch, white birch, mountain alder, white ash, quaking aspen, Juneberry, pin cherry, maples, dogwoods, wild raisin, viburnums, American yew, balsam fir, eastern hemlock, and to a lesser extent white cedar, white spruce, and white pine.

These animals consume large amounts, probably because of the low nutritional value of the foods they eat. Palmer (1944) estimated that 35 pounds (dry weight) of forage is the daily food requirement for a 545 kg (1,200 lb) moose in Alaska. In Michigan, Kellum (1941) found that captive moose consumed (wet weight) 40 to 50 pounds of food per day in summer and 50 to 60 pounds in winter.

Moose eagerly utilize mineral springs or "salt licks," particularly during the spring. Surber (1940) reported that one well-used natural lick was 75 to 100 feet in area and 3 feet deep, with well-worn trails leading to it. Although Murie (1934) reported that moose commonly frequented "artificial licks" on Isle Royale, Lake Superior, Peterson (1955) demonstrated that the animals showed no interest in salt blocks placed near natural licks in Algonquin Provincial Park, Canada.

The first indication of rutting behavior of moose in Newfoundland begins from the middle of August to the middle of September when adult, 2-year-old, and yearling bulls congregate in small groups of two, three, or four. Before the rut, the bulls establish a breeding status in three stages—grouping of bulls, belligerent action between two or more bulls within a group, and acceptance of status. In each group there is some interplay: an adult bull challenges a younger bull, and if the younger bull accepts the challenge, both animals carefully place their velvet-covered antlers together and engage in a shoving match that may last for 50 minutes. Sometimes two yearlings may challenge a large adult bull, or an adult may challenge a bull of equal size. By early to mid-September the rutting status becomes established; the large bulls show interest in the cows and force the younger bulls to keep their distance. By this time the bulls have white-tipped and polished antlers. They become excited and travel widely in search of drifting cows. The bulls are extremely curious about any low, blatting love call of a cow or the short continuous grunts of a rival bull, and they frequently respond to these calls (Dodds 1958). Some hunters take advantage of this behavior and imitate the call of a cow to lure the bulls within gun range.

Bull moose paw out depressions or wallows and urinate in them. A wallow in use has a musky odor. Woodin (1956) found four unused wallows averaging 8 feet across and 1 foot deep, situated about 30 to 40 feet apart on the southern side of a low hill. The exact function of the wallow is not known, but it may be related to courtship and behavior of cows during the breeding season. Thompson (1949) reported that a bull moose was seen repeatedly pawing and urinating into a depression in a forested area. As a cow approached, the bull rolled in the mud. The cow drove the bull from the wallow and wallowed herself. The bull returned and drove the cow out, then wallowed again. Later the bull smelled the cow's genital region and rubbed his neck and belly along her body. The cow moved on and the bull followed. Altmann (1959) reported that when a bull locates a cow, he will display, then drive and follow her.

The breeding season begins in early September and extends through late October, with a few exceptions in December. Moisan (1956) reported the sighting on 15 August, in Quebec, of a cow that was accompanied by a calf estimated to be only a few hours old. Since the gestation period is between 240 and 246 days the cow was believed to have become pregnant about 15 December.

A bull generally stays with one cow at a time for 7 to 12 days, then leaves to take another mate, repeats the pattern of search, display, defense, and drive, and breeds with successive cows until he is spent by the end of the breeding season. The bull usually returns to his first mate, and in such cases the cow usually waits for him near the wallow. By the end of the breeding season the bulls spend long hours recuperating. A bull may be seen with two or more cows, but no harem is maintained. Serious combats over a cow may occur between bulls, and the two animals may lock antlers (Altmann 1959).

Not much is known about when moose attain sexual maturity. Peterson (1955) remarked that a few cows may breed successfully at the age of 16 months and produce young in their second year. Although the author found no evidence that 16-month-old bulls bred successfully, he felt that they might have been capable of doing so were they not commonly refused the opportunity by larger, dominant bulls. Skunke (1949) concluded that although a few yearlings in Sweden do breed, some cows breed for the first time when they are 3½ years old. Edwards and Ritcey (1956) and Pimlott (1959) indicated that breeding of yearlings is highly variable.

One or two calves are produced per litter; frequently there is only one. Triplets are rare. Hosley (1949) noted a wide variation in the occurrence of twins in different continental areas and suggested that less frequent twinning was correlated with poorer range. Variation in the fecundity of cow moose in different regions may be caused by the supply of particular nutritional elements essential for ovulation and breeding. Pimlott (1959) stated that the nutritional level was the most important factor for a yearling female during her first breeding season. Simkin (1965) reported pregnancy rates of 17 calves per 100 yearlings and 113 calves per 100 older cows, based on 210 pairs of ovaries examined from moose in Ontario.

Edwards and Ritcey (1958) found that the incidence of twins in pregnancies increased with the onset of winter and varied from year to year as a result of moose migrating to lowland areas; that is, "in years when hunters encounter few migrating moose, fewer twins are encountered than in years when early migration brings moose from high elevations to the hunted lowlands." The authors found that 85 percent of pregnant moose were bred in a 10-day period in late September. The interval between periods of estrus is almost 30 days, and a few cows become pregnant in one previous or one of two subsequent periods of estrus.

In May or June pregnant cows, still accompanied by their yearlings, seek out secluded spots favorable for parturition, but about 2 weeks before the birth the yearlings are usually driven away. However, some cows may be seen with both calf and yearling (Altmann 1959). At birth the calf has short, woolly pelage, dull light reddish brown with a dark dorsal stripe, dark muzzle, and dark areas about the eyes. It lacks the whitish spotting found in fawns of white-tailed deer and wapiti. The ungainly calf has a short body, high, hunched shoulders, very short neck, relatively long ears, and long, feeble legs.

At birth the calf weighs between 11.3 and 15.9 kg (25–35 lb). It grows rapidly and gains 0.45 to 0.91 kg (1–2 lb) per day during the first month and 0.91 to 2.3 kg (2–5 lb) per day during the second month (Kellum 1941). According to Dodds (1959), a captive calf gained 34.1 kg (75 lb) in 9 weeks.

The newborn calf is awkward and helpless and is likely to trip over logs or get caught in brush. It is often kept hidden by the cow. The calf is able to swim with its mother while still very young, and it can support itself by placing its forelegs on the cow's back and thus be convoyed across wide expanses of water. The calf can be easily approached by man and will often follow after being petted. Couey (1948) reported that the cow moose has a stronger maternal instinct than the cow elk, and that the cow moose with a small calf not old enough to travel will stand her ground and will attack a man if he approaches within 20 yards of the calf. DeVos (1958) saw a cow playing with her calf.

Weaning occurs about the time the next calf is born (Merrill 1920; Murie 1934; Cahalane 1947) or in December (McDowell and Moy 1942;

Cooney 1943). The prime of life for moose is between 6 and 10 years (Dufresne 1946), but a 21-year-old cow moose in Sweden gave birth to calves. Moose may live to 18 to 23 years old (Merrill 1920).

**Age Determination.** Passmore, Peterson, and Cringan (1955) provided estimates of age of moose on the basis of mandibular tooth wear. Sergeant and Pimlott (1959) determined the age of moose from growth layers in the cementum of longitudinally sectioned incisors. The authors found that cementum layering continues at a fairly constant rate almost throughout life, and

the cementum layers are detectable as an alternation of opaque and translucent zones. Simkin (1967) compared tooth wear, cementum-layer counts, and eye-lens weight as methods of determining age in moose and concluded that both counting cementum layers and the eye-lens weight method were more reliable than tooth wear. Wolfe (1969) described a method for determining age by counting cementum layers of cross-sectioned first molar teeth. Roussel (1975) found that 4-month-old calves, regardless of sex, exhibited a paler patch in the anovulvar region but that female calves were easily recognized by the well-defined white patch surrounding the vulva.

# Fallow Deer

*Cervus dama* Linnaeus

*Cervus dama* Linnaeus, 1758.
  *Systema naturae*, 10th ed, 1:67.
Type Locality: Sweden (introduced)
Dental Formula:
  I 0/3, C 0/1, P 3/3, M 3/3 × 2 = 32
Names: The meaning of *Cervus* = a stag, and *dama* = a kind of gazelle of the Sudan.

**Description.** The fallow deer differs markedly from the white-tailed deer in the impressive coloration of the pelage and the shape of the antlers. From May to October the coloration of both sexes may be various shades of bright fawn or yellowish brown (hence the name "fallow") on the head and neck and dappled with white or yellowish white spots on the back and flanks. The throat, underparts, innersides of the legs, undersurface of the tail, and the rump are whitish. A black line runs down the middle of the back from the nape of the neck to the end of the tail. There is often a clear white stripe separating the colors of the upperparts and underparts. In winter the pelage become darker and the spots indistinct. The winter coat is dark grayish brown on the upper half of the body and paler and grayer on the underparts. The rump patch is whitish throughout the year.

There is considerable range in color variations. At times the entire summer coat may be a uniform glossy blackish brown without spots above and sooty gray below, and the winter coat may be similar but slightly lighter and duller. Other color variations include white, albino, black, and blue. The bluish phase is grayish but appears blue-tinted. These color phases are valued for ornamental purposes. Fawns are slightly darker than adults and may or may not be spotted.

Male fallow deer are also distinguished by

their antlers. In adult males the antlers are shaped somewhat like those of moose. The antlers are divergently flattened and broadly palmate, with numerous small snags on the outer edge and with brow and trez tines but no bez tine. In December and during the early spring following its birth, a buck fawn develops bumps from which pedicles grow. From these pedicles grow unbranched antlers, or spikes, the length varying considerably with the animal. The velvet is usually shed in late August and September. Each succeeding year the antlers are larger, with more tines, varying in size and complexity among individuals. As in other deer, antler growth is closely associated with the quality of the habitat.

In the skull the lacrimal vacuities and orbitals are large. The vomer is low behind and does not divide the posterior nares. The upper canines are absent. The cheek teeth have very short, broad crowns, and there is a marked difference in the size of the crowns of the three pairs of lower incisors.

Males are larger than females. Measurements range from 130 to 160 cm (51–63 in); tail, 17 to 23 cm (6.7–9.1 in); hind foot, 38 to 43 cm (15–17 in); and height at shoulder, 81 to 91 cm (32–36 in). The weight of adult males varies from 59 to 145 kg (130–320 lb), and adult females weigh from 23 to 26 kg (51–57 lb) less than males (Harris and Duff 1970). Shaw and McLaughlin (1951) re-

FALLOW DEER

ported that in Massachusetts the fallow deer is from 10 to 20 cm (4–8 in) smaller in length and height than the white-tailed deer.

**Distribution.** The fallow deer is a native of Asia Minor and countries bordering the Mediterranean Sea. It was successfully introduced into central Europe. It is supposed that the ancient Egyptians transported it to Egypt and that the Romans brought it to the British Isles. Today these beautiful animals are prized in the zoos, parks, and private estates of the world.

In New England the fallow deer was released on the islands of Nantucket and Martha's Vineyard, Massachusetts. The records are incomplete as to when the animals were released on Nan-

tucket, but Shaw and McLaughlin (1951) reported that 3 or 5 fallow deer were released as game animals on Martha's Vineyard in 1932, and 8 more females were released in the autumn of 1938. It is believed that these animals were obtained from a park in Worcester, Massachusetts. There were an estimated 150 fallow deer on Martha's Vineyard when the first deer-hunting season opened in 1949. During the one-week hunting season, 15 fallow deer and 35 white-tailed deer were taken on the island. In the 1950 hunting season approximately the same number of deer were taken. There seem to be no data on the fallow deer on Martha's Vineyard since 1950, except that 3 animals were killed during the 1971 deer season.

**Ecology.** Fallow deer occur in the wild on Martha's Vineyard as a game animal; man and dogs seem to be their main predators there. Information and data are lacking on the parasites and diseases of fallow deer in the United States.

**Behavior.** Fallow deer have highly developed senses of hearing and smell; sight is the least developed, but the animals are quick to spot movement. Fallow deer are rather silent. Does and fawns utter low wickering notes when communicating with each other, and does may make a plaintive cry to call a fawn. Does make the same note during the rut. When suspicious or curious the animals utter a short, deep bark as a warning of danger, but when fully alarmed they run off silently. During the rut the bucks make grunting, or sometimes belching, sounds.

Fallow deer are gregarious, but the males remain apart from the does except during the breeding season and early winter. These animals are more active in early morning and in the evening in summer, but they feed throughout the day during winter. When they are disturbed, fallow deer feed during the evening and night, especially when there is moonlight. They also feed at dawn. When not disturbed the deer may be found during the daytime browsing in semiopen forested areas. They usually face the wind when bedding down. These animals move about quietly when feeding, walking at a steady pace and stopping often to look and listen. When they spot danger, the deer curl their tails upward, squirrellike, and show the white underside. Then they take three, four, or five bounds on rigid legs, drumming their feet on the ground, which acts as a warning to other deer. The animals may pause for a moment after bounding, then run off very fast. Although fallow deer prefer to crawl under a fence or pass through a gap rather than jump over, they can jump high fences, often from a stationary position (Cadman 1966).

Fallow deer browse on the twigs, shoots, leaves, shrubs, and bark of most hardwood trees. They also eat a great variety of grasses, herbs, fruits, berries, acorns, and fungi (Cadman 1966).

There appears to be no information available on the reproductive behavior of the fallow deer on Martha's Vineyard.

## REFERENCES

Aldous, Clarence M., and Mendall, Howard L. 1941. *The status of big game and fur animals in Maine.* Orono: University of Maine, Maine Cooperative Wildlife Research Unit.

Altmann, Margaret. 1956. Patterns of social behavior in big game. *Trans. N. Amer. Wildl. Conf.* 21:538–45.

———. 1959. Group dynamics in Wyoming moose during the rutting season. *J. Mamm.* 40(3):420–24.

Anderson, Roy C., and Strelive, Uta R. 1967. The penetration of *Pneumostrongylus tenuis* into the tissues of white-tailed deer. *Canadian J. Zool.* 45(3):258–89.

———. 1968. The experimental transmission of *Pneumostrongylus tenuis* to caribou (*Rangifer tarandus terraenovae*). *Canadian J. Zool.* 46(3):501–10.

Anthony, H. E. 1928. *Field book of North American mammals.* New York and London: G. P. Putnam's Sons.

Banasiak, Chester F. 1961. *Deer in Maine.* Bulletin no. 6. Augusta: Maine Dept. Inland Fisheries and Game, pp. 1–159.

Behrend, Donald F., and McDowell, Robert D. 1967. Antler shedding among white-tailed deer in Connecticut. *J. Wildl. Manage.* 31(3):588–90.

Benson, D. A. 1952. Treatment of a sick moose with cobaltous chloride. *J. Wildl. Manage.* 16(1):110–11.

Bourlière, François. 1954. *The natural history of mammals.* New York: Alfred A. Knopf.

Cadman, W. A. 1966. *Fallow deer.* Forestry Commission Leaflet no. 52. London: F. Mildner and Sons.

Cahalane, Victor H. 1939. Deer of the world. *Natl. Geogr. Mag.* 76(4):463–510.

———. 1947. *Mammals of North America.* New York: Macmillan Co.

Cahn, Alvin R.; Wallace, G. I.; and Thomas, L. J. 1932. A new disease of moose. *Science* 76(1974):385.

Carpenter, Ralph G., II, and Siegler, Hilbert R. 1958. A list of New Hampshire mammals and their distribution. *New Hampshire Fish and Game Dept. Bull.*, pp. 1–15 (mimeographed).

Cheatum, E. L. 1949. The use of corpora lutea for determining ovulation incidence and variations in the fertility of white-tailed deer. *Cornell Vet.* 39(3):282–91.

Cheatum, E. L., and Morton, Glenn H. 1946. Breeding

season of white-tailed deer in New York. *J. Wildl. Manage.* 19(3):249–63.

Collias, N. E. 1950. Some variations in grouping and dominance patterns among birds and mammals. *Zoologica* 25(2):97–119.

Cooney, Robert M. 1943. Montana moose survey. Hellroaring-Buffalo-Slough Creek Unit. Montana Fish and Game Dept. Wildl. Rest. Div. Rept. (typed).

Couey, Faye M. 1948. Montana wildlife survey and management. *P-R Quart., U.S. Dept. of the Interior, Fish and Wildl. Serv.* 8(3):306–11.

Crane, Jocelyn. 1931. Mammals of Hampshire County, Massachusetts. *J. Mamm.* 12(3):267–73.

Cronan, John M., and Brooks, Albert. 1968. *The mammals of Rhode Island.* Wildlife Pamphlet no. 6. Providence: Rhode Island Department of Agriculture and Conservation, Division of Fish and Game.

Dahlberg, Burton L., and Guettinger, Ralph C. 1956. The white-tailed deer in Wisconsin. *Wisconsin Conserv. Dept. Tech. Wildl. Bull.* 14:1–282.

DeVos, Antoon. 1958. Summer observation on moose behavior in Ontario. *J. Mamm.* 39(1):128–39.

Dodds, Donald G. 1958. Observations of pre-rutting behavior in Newfoundland moose. *J. Mamm.* 39(3):412–16.

Dufresne, Frank. 1946. *Alaska's animals and fishes.* New York: A. S. Barnes.

Edwards, R. Yorke, and Ritcey, Ralph W. 1956. The migrations of a moose herd. *J. Mamm.* 37(4):486–94.

Fenstermacher, R. 1937. Further studies of diseases affecting moose. *Cornell Vet.* 27(1):25–37.

Findley, James S. 1951. A record of moose speed. *J. Mamm.* 32(1):116.

Flook, Donald R. 1959. Moose using water as refuge from flies. *J. Mamm.* 40(3):455.

Foote, Leonard E. 1946. A history of wild game in Vermont. *Vermont Fish and Game Serv. State Bull.* 11:1–55.

French, C. E.; McEwen, L. C.; Magruder, N. D.; Ingram, R. H.; and Swift, R. W. 1956. Nutrient requirements for growth and antler development in the white-tailed deer. *J. Wildl. Manage.* 20(3):221–32.

Gerstell, Richard. 1936. Breeding experiments with the whitetail deer. *Pennsylvania Game News* 6(12):4, 20.

Goodwin, George Gilbert. 1935. The mammals of Connecticut. *Connecticut Geol. Nat. Hist. Surv. Bull.* 53:1–221.

———. 1936. Big game animals in the northeastern United States. *J. Mamm.* 17(1):48–50.

Harris, Roy A., and Duff, K. R. 1970. *Wild deer in Britain.* New York: Taplinger Pub. Co.

Haugen, Arnold O., and Davenport, L. A. 1950. Breeding records of white-tailed deer in the Upper Peninsula of Michigan. *J. Wildl. Manage.* 14(3):290–95.

Haugen, Arnold O., and Speake, Daniel W. 1958.

Determining age of young fawn white-tailed deer. *J. Wildl. Manage.* 22(3):319–21.

Helmboldt, C. F., and McDowell, R. D. 1958. Granular venereal disease in a white-tailed deer. *J. Wildl. Manage.* 22(2):203.

Henry, Vernon G. 1968. Length of estrous cycle and gestation in European wild hogs. *J. Wildl. Manage.* 32(2):406–8.

Hinkson, Henry R. 1951. How fast can deer swim? *New York State Conservationist* 6(2):36.

Hornaday, William T. 1904. *The American natural history.* New York: Charles Scribner's Sons.

Hosley, Neil W. 1949. *The moose and its ecology.* Wildlife Leaflet no. 312. Washington, D.C.: U.S. Dept. of the Interior, Fish and Wildlife Service.

Hosley, Neil W., and Ziebarth, R. K. 1935. Some winter relations of the white-tailed deer to the forests in north central Massachusetts. *Ecology* 16(4):535–53.

Kellum, Ford. 1941. Cusino's captive moose. *Michigan Conservationist* 19(7):4–5.

Koopman, Karl F. 1967. Artiodactyls. In *Recent mammals of the world,* ed. S. Anderson and J. K. Jones, Jr., pp. 385–406. New York: Ronald Press.

Lamson, Arroll L. 1941. Maine moose disease studies. M.S. thesis, University of Maine, Orono.

Lewis, James C. 1966. Observations of pen-reared European hogs released for stocking. *J. Wildl. Manage.* 30(4):832–35.

Lönnberg, Einar. 1923. *Severges jaktbara djur.* Stockholm: Svenska Jordbrakets bok. Albert Bonnier.

McBeath, Donald Y. 1941. Whitetail traps and tags. *Michigan Conservationist* 10(11):6–7, 11; 10(12):6–7.

McDowell, Lloyd, and Moy, Marshall. 1942. Montana moose survey. Hellroaring-Buffalo-Slough Creek Unit. Montana Fish and Game Dept. Wildl. Rest. Div. Rept. (typed).

Manville, Richard H. 1964. History of the Corbin Preserve. *Smithsonian Inst. Pub.* 4581:427–46.

Merrill, Samuel. 1920. *The moose book.* New York: E. P. Dutton.

Moisan, Gaston. 1956. Late breeding in moose, *Alces alces. J. Mamm.* 37(2):300.

Murie, Olaus J. 1934. The moose of Isle Royal. *Univ. Michigan Mus. Zool. Misc. Pub.* 25:1–44.

Newsom, William M. 1926. *White-tailed deer.* New York: Charles Scribner's Sons.

Osgood, Frederick L., Jr. 1938. The mammals of Vermont. *J. Mamm.* 19(4):435–41.

Palmer, L. J. 1944. Food requirements of some Alaskan game animals. *J. Mamm.* 25(1):49–54.

Palmer, Ralph. 1951. The whitetail deer of Tomhegan camps, Maine, with added notes on fecundity. *J. Mamm.* 32(3):267–80.

Passmore, R. C.; Peterson, R. L.; and Cringan, A. T. 1955. A study of mandibular tooth wear as an index to age of moose. In *North American moose,* ed.

Randolph L. Peterson, pp. 223–38. Toronto: University of Toronto Press.

Peterson, Randolph L. 1955. *North American moose.* Toronto: University of Toronto Press.

Pimlott, Douglas H. 1959. Reproduction and productivity of Newfoundland moose. *J. Wildl. Manage.* 23(4):381–401.

Roussel, Yvon E. 1975. Aerial sexing of antlerless moose by white vulval patch. *J. Wildl. Manage.* 39(2):450–51.

St. Michael-Podmore, P. 1904. *A sporting paradise, with stories of adventure in America and the backwoods of Muskoka.* London: Hutchinson.

Schemnitz, Sanford D. 1975. Marine Island–mainland movements of white-tailed deer. *J. Mamm.* 56(2):535–37.

Seamans, Roger A. 1947. The time is now! *Vermont Fish and Game Serv. State Bull.* 15:1–48.

Sergeant, David E., and Pimlott, Douglas H. 1959. Age determination in moose from sectioned incisor teeth. *J. Wildl. Manage.* 23(3):315–21.

Seton, Ernest Thompson. 1929. *Lives of game animals.* Vols 1–4. Garden City, N.Y.: Doubleday, Doran.

Severinghaus, C. W. 1949. Tooth development and wear as criteria of age in white-tailed deer. *J. Wildl. Manage.* 13(2):195–216.

Severinghaus, C. W., and Cheatum, E. L. 1956. Life and times of the white-tailed deer. In *The deer of North America*, ed. Walter P. Taylor, pp. 57–186. Harrisburg, Penn.: Stackpole Co.

Severinghaus, C. W.; Maguire, H. F.; Cookingham, R. A.; and Tanek, J. E. 1950. Variations by age class in the antler beam diameters of white-tailed deer related to range conditions. *Trans. N. Amer. Wildl. Conf.* 14:551–70.

Shaw, Samuel P., and McLaughlin, Charles L. 1951. The management of white-tailed deer in Massachusetts. *Massachusetts Div. Fish and Game Res. Bull.* 13:1–59.

Siegler, Hilbert R.; Silver, Helenette; White, David L.; and Laramie, Henry A., Jr. 1968. *The white-tailed deer of New Hampshire.* Survey Report no. 10. Concord: New Hampshire Fish and Game Department.

Silver, Helenette. 1957. *A history of New Hampshire game and furbearers.* Survey Report no. 6. Concord: New Hampshire Fish and Game Department.

——. 1965. An instance of fertility in a white-tailed buck fawn. *J. Wildl. Manage.* 29(3):634–36.

Simkin, D. W. 1965. Reproduction and productivity of moose in northwestern Ontario. *J. Wildl. Manage.* 29(4):740–50.

——. 1967. A comparison of three methods used to age moose. *Proc. N.E. Sect. Wildl. Soc.*, pp. 1–16 (multilith).

Skunke, Folke. 1949. *Älgen. Studier jakt och vård.* Stockholm: P. A. Norstedt.

Spears, John R. 1893. The Corbin Game Park. *Ann. Rept. Smithsonian Inst. for 1891*, pp. 417–23.

Stanwell-Fletcher, John F., and Stanwell-Fletcher, Theodora C. 1943. Some accounts of the flora and fauna of the Driftwood Valley region of north central British Columbia. *Occ. Pap. British Columbia Prov. Mus.* 4:1–97.

Sugden, Lawson G. 1964. An antlered calf moose. *J. Mamm.* 45(3):490.

Surber, Thaddeus. 1940. Minnesota moose—then and now. *Minnesota Conserv. Volunteer* 1(1):41–44.

Swift, Ernest. 1946. A history of Wisconsin deer. *Wisconsin Conserv. Dept. Pub.* 323:1–96.

Taylor, Walter P. 1947. Some new techniques—hoofed mammals. *Trans. N. Amer. Wildl. Conf.* 12:293–324.

Thomas, Jack Ward; Robinson, R. M.; and Marburger, R. G. 1965. Social behavior in a white-tailed deer herd containing hypogonadal males. *J. Mamm.* 46(2):314–27.

Thomas, L. J., and Cahn, A. R. 1932. A new disease of moose. 1. Preliminary report. *J. Parasitol.* 18:219–31.

Thompson, W. K. 1949. Observations of moose courting behavior. *J. Wildl. Manage.* 13(3):313.

Warfel, H. E. 1937. Moose records for Vermont. *J. Mamm.* 18(4):519.

Wislocki, George B. 1943. Studies on growth of deer antlers. 11. Seasonal changes in the male reproductive tract of the Virginia deer (*Odocoileus virginianus borealis*) with a discussion of the factors controlling the antler-gonad periodicity. In *Essays in biology in honor of Herbert M. Evans*, pp. 629–53. Berkeley and Los Angeles: University of California Press.

Wislocki, George B.; Aub, Joseph C.; and Waldo, Charles M. 1947. The effects of gonadectomy and the administration of testosterone propionate on the growth of antlers in male and female deer. *Endocrinology* 40(3):202–24.

Wolfe, Michael L. 1969. Age determination in moose from cemental layers of molar teeth. *J. Wildl. Manage.* 33(2):428–31.

Woodin, Howard E. 1956. Appearance of a moose rutting ground. *J. Mamm.* 37(3):458–59.

# Extirpated Species

EARLY MAN had nothing to do with the disappearance of mammoths, mastodons, bisons, and some other animals that once roamed New England. But the white man has, regrettably, had a hand either directly or indirectly in the extirpation of the eastern wolf, wolverine, eastern mountain lion, wapiti, and woodland caribou. The Indiana myotis is on the verge of extirpation in New England, and the Canada lynx and walrus doubtless were once far more numerous here than at present and may well be threatened with extirpation.

## Eastern Timber Wolf

*Canis lupus lycaon* Schreber

*Canis lycaon* Schreber, 1775. *Die Säugthiere . . .*, vol. 2, part 13, pl. 89.
*Canis lupus lycaon*, Goldman, 1937. *J. Mamm.* 18(1):45.
Type Locality: Vicinity of Quebec, Quebec, Canada (fixed by Goldman, *J. Mamm.* 18[1]:38)
Dental Formula:
I 3/3, C 1/1, P 4/4, M 2/3 × 2 = 42
Names: Gray wolf, timber wolf, and black wolf. The meaning of *lupus* = a wolf, and *lycaon* = referring to an Arcadian king who, when Zeus came in disguise, set before the god a dish of human flesh to test his divinity. Zeus transformed the king into a wolf.

**Description.** The wolf bears a strong resemblance to the largest of the canids, a huge German shepherd dog. It is strong, robust, and wild, with a pointed muzzle, prominent, erect ears, bushy tail, and moderately long legs. It has a distinctive face as a result of wide tufts of hair that project outward and downward from below the ears. Females have ten mammae—four pectoral, four abdominal, and two inguinal.

The eastern timber wolf is one of the smallest subspecies of North American wolves. Compared with those of other wolves, the skull of *Canis lupus lycaon* possesses a relatively slender rostrum, with nasals usually more deeply emarginate anteriorly and the supraoccipital shield projecting less posteriorly over the foramen magnum. The upper carnassials are smaller than in most other wolves; the protocones are prominent in some specimens and absent in others.

There is much color variation in wolves, from white, cream-colored, dusky, rufous, and gray to black. The sexes are colored alike, and there is little seasonal variation except that the summer pelage is somewhat paler. In winter the pelage is often shaggy. Old wolves tend to be much grayer.

**Wolf–Dog Differences.** The anatomy of the wolf differs somewhat from that of the domestic dog. The chest of the wolf is much shallower and keel-like, and the shoulders are narrower and more firmly set with the elbows turned well inward. This allows the fore- and hind legs on the same side to swing in the same line (Young and Goldman 1944). In contrast, the dog places its hind feet between the tracks of the forefeet rather than in the same tracks. In the wolf the orbital angle of the skull measures 40 to 45 degrees, whereas in dogs it generally ranges from 50 to 53 degrees. (The orbital angle is the angle between a line drawn through the upper and lower edges of the eye socket and a line drawn across the top of the skull.) The auditory bullae are always large, convex, and nearly spherical in the wolf; they are small and compressed and present a crumpled appearance in most dogs. The wolf generally carries its tail low, whereas the dog carries its tail high (Iljin 1941).

Males are larger than females. Measurements range from 142 to 162 cm (56.0–64.0 in); tail, 36 to 42 cm (14.2–16.5 in); and hind foot, 25 to 27 cm (10.0–10.6 in). Weights vary from 29.5 to 45.4 kg (65–100 lbs) (Young and Goldman 1944).

**Distribution.** The original distribution of the wolf in North America was the extreme Arctic region from the Atlantic to the Pacific, and south in the west to the tropical latitude of middle Mexico. The wolf apparently did not occur in the desert regions or in much of the southeastern United States. The former range of the eastern timber wolf was eastern Quebec to western and northern Ontario, eastern Minnesota, Michigan, Ohio, and the northeastern and middle Atlantic states; the southern limits of the range are not known, but it was apparently absent in much of the southeastern United States. Although still present in southern Quebec, Ontario, and parts of northern Minnesota and northern Michigan, the eastern timber wolf is an endangered species in the United States.

The eastern timber wolf has been extirpated in New England, and records on its history here are sketchy. The first colonists in New England found wolves numerous everywhere and considered them a threat to the precious livestock they had brought from England. Morton (1637) wrote: "The wolfes are of divers coloures: some sandy coloured; some griselled, and some black, their foode is fish which they catch when they pass up the rivers, into the ponds to spawne, at the spring time. The Deare are also their pray,

EASTERN TIMBER WOLF

and at summer, when they have whelpes, the bitch will fetch a puppy dogg from our dores, to feede their whelpes with. They are fearefull Curres, and will runne away from a man (that meeteth them by chance at a banke end) as fast as any fearefull dogge. These pray upon the Deare very much. The skinnes are used by the Savages, especially the skinne of the black wolfe, which is esteemed a present for a prince there." The author stated that an Indian would gladly exchange forty beaver skins for one black wolf skin.

As early as 1631 the Plymouth and Massachusetts Bay colonies waged war on the wolves. In 1698 the town of Lynn, Massachusetts, voted a bounty of 20 shillings for every wolf killed in the town. Wolves were still common in or near the settled parts of New England down to the time of the American Revolution (Allen 1942). A wolf was killed in New Marlborough, Massachusetts, on 2 December 1918 (Seton 1929).

In the past the wolf ranged throughout Connecticut (Goodwin 1935). Linsley (1842) wrote: "Only three years since, a very large wolf was killed near Bridgeport by Mr. Moses Buckley; and one over 10 years since, in Newton, by Aaron Glover."

In Rhode Island a number of old Providence records refer to wolves. Cronan and Brooks

(1968) wrote: "In January, 1659 a wolf bounty was initiated by the government in Providence and the bounty was continued well into the 1700's. Newport also had wolf problems as the general court in Newport on September 19, 1642 requested that Roger Williams consult with the Narragansett Indians concerning the control of wolves on Aquidneck Island."

In New Hampshire, Silver (1957) wrote that wolves "were hunted and trapped in all manners of ways. Great hunts were sometimes organized by 100 or more men, who surrounded an area, driving the wolves before them toward the center. Occasionally as many as 500 or 600 men assembled for these drives, which were often effective in cleaning up a whole township." She stated that the wolf disappeared in New Hampshire about 1895.

Norton (1930) wrote that "wolves were numerous in the Portland (Maine) region, and existed at least down to 1740 in the immediate vicinity of the present city." Stanton (1963) reported that wolves were common in eastern Maine about 1845–50 and persisted there for some time.

In Vermont, Green (1822) recounted an observation a few years before 1821 when a wolf received a fatal thrust from a free-ranging domestic swine aided by a herd of swine. Thompson (1853) reported that a wolf was killed in Addison County, Vermont, about 1830 and that the species probably was occasionally found well after that time.

Young and Goldman (1944) stated that the wolves had practically disappeared from most of New England by the end of the eighteenth century, but that they remained fairly common in the southern parts of Maine, Vermont, and New Hampshire as well as in the mountainous sections of western Massachusetts.

# Wolverine

*Gulo luscus luscus* (Linnaeus)

[*Ursus*] *luscus* Linnaeus, 1758.
    *Systema naturae*, 10th ed, 10(1):57.
*Gulo luscus luscus*, Anderson, 1947.
    *Nat. Mus. Canada Bull.*
    102(1946):68.
Type Locality: Hudson Bay
Dental Formula:
    I 3/3, C 1/1, P 4/4, M 1/2 × 2 = 38
Names: Skunk bear, devil bear, wood devil, and glutton. The name glutton refers to the wolverine's supposedly greedy feeding habits; the name skunk bear arose because of the animal's disagreeable scent and because it somewhat resembles those two animals. The meaning of *Gulo* = a gullet, on account of its supposed gluttony or gormandizing, and *luscus* = half-blind, referring to the wolverine's poor eyesight.

**Description.** The wolverine is the largest and most formidable of terrestrial mustelids. It is a powerful, low-built, burly mammal with a bear-like body form. The head is broad, with beady eyes, short and slightly pointed ears, and a broad, truncated nose. The legs are short and strong, and the feet subplantigrade, each with five toes and large sharp, partially retractile claws. The soles of the feet have thick bristly hairs, except for the pads of the toes. The tail is fairly long, heavy, and bushy. The pelage is long, coarse, thick, durable, and glossy. Females have eight mammae—four abdominal and four inguinal.

The facial angle of the massive skull is steep, and the orbits are small and rounded. The zygomatic arches are very massive, particularly posteriorly. The paroccipital processes protrude markedly behind the moderately inflated auditory bullae, and the prominent sagittal crest ends at the cranial vertex. The infraorbital foramina are large and elliptical in outline. The palate extends behind the molars, and the palatine foramina are small and roundly elongated. The mandible is massive. The tooth crowns are robust and the lower carnassials large, with the upper molars much smaller than the carnassials. The third upper incisors are large and caninelike. According to Shufeldt (1924), the lower jaw is so firmly hinged to the cranium that it cannot be removed without breaking it.

The sexes are colored alike and show slight seasonal color variation. The predominant color is blackish brown, with forehead and cheeks grayish and muzzle dark brown. Two bands extend from the sides of the neck over the shoulders, expanding posteriorly, continue low on the sides of the body, and join across the back and base of tail, forming a horseshoe pattern. The color of this band varies markedly, from yellowish white to orange cinnamon and dark brown. The throat and chest are spotted with yellowish white, and the belly occasionally has whitish blotches. The limbs and feet are dark brown and the claws horn-colored. The tail may

WOLVERINE

be cinnamon brown or entirely dark brown. Immatures resemble adults. Albinos and nearly all-black individuals are rare.

Males are larger than females. Measurements range from 90 to 113 cm (35.1–44.0 in); tail, 19 to 26 cm (7.4–10.0 in); and hind foot, 18 to 19 cm (7.0–7.5 in). Weights vary from 11.0 to 18.2 kg (24–40 lb) (Hall and Kelson 1959).

**Distribution.** This species occurs from the shores of the Arctic Ocean, straggling north to northern Baffin Island, Ellesmere Island, and Melville Island, east to the coast of Labrador, and west to Alaska; formerly the species extended south to the extreme northeastern United States, Michigan, Wisconsin, Minnesota, and North Dakota and southward in the Rocky Mountains into Colorado.

The wolverine was not very plentiful in early New England. Allen (1876) wrote: "The wol-verine (*Gulo luscus*), now rarely recognized as an animal that was ever found in New England, seems to have been formerly of frequent occurrence in the northern parts of Vermont, New Hampshire, and Maine, and probably once inhabited the highlands of western Massachusetts." Jackson (1922) reported that a pair of young wolverines were taken in the Diamond region, east of the Connecticut Lakes, New Hampshire, in 1918 and thought that the species was breeding to a certain extent within this wild region. In Vermont, Thompson (1853) wrote: "This animal was occasionally found when the country was new, in all parts of the state, but was never very plentiful. For many years past, however, it has been known only in the most woody and unsettled districts, and in such places it is now extremely rare, none having been met with to my knowledge for several years." The wolverine has been extirpated in New England.

# Eastern Mountain Lion

*Felis concolor couguar* Kerr

*Felis couguar* Kerr, 1792. *The animal kingdom . . . ,* p. 151.
*Felis concolor couguar,* Nelson and Goldman, 1929. *J. Mamm.* 10(4):347.
Type Locality: Pennsylvania
Dental Formula:
I 3/3, C 1/1, P 3/2, M 1/1 × 2 = 30
Names: Puma, cougar, panther, catamount, painter, and mountain screamer. The meaning of *Felis* = a cat, *concolor* = uniform color, and *couguar* = false deer (so-called because it resembles the tawny brown color of a deer).

**Description.** Next to the jaguar, the mountain lion is the largest native cat in North America. It is a long, slender-bodied felid with a small, broad, rounded head, erect, short, rounded ears without elongated terminal tufts, and a long, round, thick, rather long-haired tail. The fur is rather short and fairly soft. Females have eight mammae—four pectoral (the anterior two are nonfunctional), two abdominal, and two inguinal.

The skull is short, broad, and rounded. The sagittal crest is convex in profile, and the lambdoidal crest has a deep lateral concavity. The speculelike bregmatic processes of the parietal extend diagonally inward and upward over the frontals, approaching or reaching the temporal ridge and blending with the frontals in old specimens. The palate is wide and the nasals are expanded posteriorly. The mandible is deep and short, with four cheek teeth on each side, the first and last simply tiny.

The sexes are colored alike and show considerable individual color variation but no noticeable seasonal variation. The coloration above is dull, dark reddish brown, somewhat darker from the top of the head to the base of the tail, intermixed with black hairs. The shoulders and flanks are paler, and the underparts are dull whitish or reddish white, overlaid with buff and dusky spots, especially on the flanks and inner sides of the limbs. The feet are dark brownish. The muzzle is whitish, with black marking just back of the white at the base of the vibrissae and in front of the eyes. The median portion of the upper lip adjoining the lower part of the rhinarium, the lower lip, and the chin are nearly pure white. The underside of the neck and throat is dusky, and the ears are light-colored within, blackish behind, with or without grayish median patches. The tail above is usually similar to the back but paler below, with the tip generally black above and below. A gray phase probably occurs, but albinos and melanistics are not known in North America. Immatures are paler brown than adults and are plainly marked with large dark brown or blackish spots on the body and dark bars on the tail. The spots and bars gradually disappear as the animal matures. The annual molt occurs in spring and is so gradual as to be hardly noticeable.

Males are larger than females. Measurements range from 150 to 275 cm (59–108 in); tail, 66 to 82 cm (26–32 in); and hind foot, 24 to 29 cm (9.5–11.4 in). Weights vary from 36 to 103 kg

EASTERN MOUNTAIN LION

(80–227 lb) (Hall and Kelson 1959). A mountain lion that was killed in New Hampshire in 1853 was 254 cm (100 in) long and weighed 90 kg (198 lb); and a Vermont male taken in 1875 was 221 cm (87 in) long and weighed 50 kg (110 lb) (Silver 1957).

**Distribution.** The original distribution of the mountain lion included both North and South America from southern Quebec to Vancouver Island on the Pacific coast and south to Patagonia in South America. The species once had the widest distribution of any American mammal. It was equally at home in an extraordinary variety of conditions. It occurred from tidelands to deep coniferous forests, and from areas of extreme cold at elevations of 11,000 feet in California to the hot, arid desert area of Yuma, Arizona and in most humid areas in the tropical forests, swamps, and canebrakes of Central and South America. Since 1900 the species has been extirpated east of the Mississippi River except for a declining population in Florida and New Brunswick. The mountain lion still roams west of the 100th meridian in wilderness areas of most of the Rocky Mountain states.

The eastern mountain lion, or catamount, or panther, as it was usually called by the early New England settlers, was not especially common, but evidently it was troublesome to the settlers, for bounties were offered and many were killed. Allen (1942) reported that as early as 1694 Connecticut paid a bounty of twenty

shillings for each lion killed and paid four to five shillings as late as 1769, and that Massachusetts offered a bounty of forty shillings in 1742 and in 1753 increased the reward to four pounds.

In Maine, the mountain lion was rare throughout the state and probably never more than a straggler (Norton 1930). A mountain lion was killed in Sebago about 1845 (Cram 1901), one was killed in 1891 near Andover (Goodwin 1936), and another was trapped on the Maine side of Little Saint John Lake on the Maine-Quebec border in Somerset County in January 1938 (Wright 1961).

In New Hampshire, Dearborn (1927) reported that a mountain lion was killed on 1 November 1853 in the town of Lee. Stone and Cram (1902) stated that the species disappeared in northeastern New Hampshire about 1852. Jackson (1922) reported that a mountain lion was killed in the White Mountains about 1885 and that a pair of "cougars" existed in the state in the early 1920s along the east side of the Androscoggin River in the town of Cambridge as far as the southern shores of Lake Umbagog. Siegler (1971) reported that since about 1961 the New Hampshire Fish and Game Department has received numerous reports of sightings of mountain lions throughout the state, and some seemed so authentic that the state legislature passed a bill in 1967 stating: "No person shall, at any time, shoot, hunt, take, or have in his possession, any mountain lion or any part of the carcass thereof, taken in this state. However, this section shall not apply to a person acting in protection of his person or property."

In Vermont, the mountain lion was often hunted and killed before the Revolutionary War (Allen 1942). Thompson (1853) commented on the animal's occurrence in Vermont: "The mountain lions were formerly much more common in Vermont than at the present day, and have at times done much injury by destroying sheep and young cattle." The last recorded lion was killed in Barnard, Windsor County, on Thanksgiving Day in 1881 (Titcomb 1901). Osgood (1938) considered this cat to be extinct in Vermont by the late 1930s. The author wrote: "Since 1934 there have been repeated reports of panthers seen and heard in various parts of the state but to date none has been captured to secure the reward of $1000 offered by a local paper for a Vermont panther, dead or alive."

This species was known to range the greater part of Massachusetts during early colonization. On 20 April 1741, the *Boston Gazette* reported a strange creature that was called a "Cattamount" and was exhibited at the Greay Hound Tavern in Roxbury. It was "caught in the woods about 80 miles to the westward" and "It has a tail like a Lyon, its leggs are like a Bears, its Claws like an Eagle, its eyes like a Tyger, its countenance is a mixture of every Thing that is Fierce and Savage, he is exceedingly ravenous and devours all sorts of Creatures that he can come near; its Agility is surprising, it will Leap 30 Foot at one jump not withstanding it is but Three months old. Whoever inclines to see this Creature may come to the Place aforesaid, paying a Shilling each, shall be welcome for their Money." Emmons (1840) noted that the mountain lion was then extirpated in Massachusetts, and Crane (1931) cited an occurrence of the animal as late as 18 January 1926 in the commonwealth.

In Connecticut, Linsley (1842) wrote: "I saw a fine specimen, said to have been killed in the northern part of the State, exhibited in Mix's Museum some years since." Goodwin (1935) reported that the cat was not uncommon in suitable places in the northern part of Connecticut, especially in the mountainous country of Litchfield County.

There are almost no records for Rhode Island. Mearns (1900) listed the mountain lion as one of the wild mammals known to have inhabited the state during the historical period, and Allen (1942) mentioned that a mountain lion was killed in 1847 or 1848 in West Greenwich.

The eastern mountain lion may be considered extirpated in New England, since no specimens have been taken for more than seventy years. There are, however, occasional but persistent rumors and reports of the species' occurring in the wilder areas of New England. These stories often result from misidentification of animal tracks or sighting of catlike animals with long tails. The tracks of the lynx are much larger than those of the mountain lion. Large fishers seen fleetingly may be taken for mountain lions.

# Walrus

*Odobenus rosmarus* (Linnaeus)

[*Phoca*] *rosmarus* Linnaeus, 1758. *Systema naturae*, 10th ed., 1:38.
*Odobenus rosmarus*, Trouessart, 1897. *Catalogus mammalium tam viventium quam fossilium.* Berlin: R. Friedlander, Fasc. 1, pp. 1–218.
Type Locality: Arctic regions
Dental Formula:
  I 1/0, C 1/1, P 3/3, M 0/0 × 2 = 18
Names: Sea horse, sea cow, and Morse. The meaning of *Odobenus* = tooth-walker, and *rosmarus* = a Norwegian rossmaal or rossmaar, meaning "whale horse."

**Description.** The walrus is a massive, bulky pinniped with thick, tough, wrinkled, thinly haired skin, large tusks, prominent, stiff whiskers on the upper lip, and a short, thick neck. The moustache contains some 400 yellowish sensory whiskers, about 10 cm (4 in) long, which aid in locating and grasping food organisms. They become worn in old adults from the constant friction involved in feeding. The eyes are small, often bloodshot, and situated far back on the small, rounded head. There are no ear pinnae. The nostrils are crescent-shaped and set anterior to the short, flat, rounded, mobile muzzle. The upper lip is fleshy and protrudes over the mouth. The large front flippers are oarlike, and the hind flippers can be bent forward at the tarsus and turned under the body, helping the animal walk on land. Both front and hind flippers are thick and have five clawed digits. The tail is vestigial.

The walrus is distinguished from other pinnipeds by the pair of long tusks, present in both sexes but more slender in females. The tusks are greatly elongated canines that extend up to 63.5 cm (25 in) from the gum, and they may weigh about 5 kg (11 lb). They begin to protrude through the gums between 2 and 3 months after birth, and their growth parallels that of the body, at 1 inch per year in both sexes, although tusk growth in females is slightly slower than in males. Tusk growth in bulls decreases with advanced age, and in old bulls wear is greater than growth so that the tusks decrease in length. The tusks are used for defense, to obtain food, and to assist in climbing on ice floes or land (Loughrey 1959). Miller (1975) reported that walrus tusks are important in fighting. Walruses with large tusks tend to initiate agonistic interactions with smaller individuals having smaller tusks.

The skull is massive, without a postorbital process; the orbits are relatively small; the sagittal crest is absent; and the palate is concave below. The number and structure of teeth vary. Supernumerary canines are sometimes present. The rooted cheek teeth lack distinctive crowns and soon wear down to large, flat-surfaced pegs.

The sexes are colored alike, generally cinnamon to pinkish gray. Old walruses are much grayer and nearly naked, and the young are yellowish orange dorsally, becoming reddish orange ventrally and at the base of the limbs.

Males are nearly one-third larger than females. Measurements of adults range from 3 to 4 m (10–13 ft) long. Adult males weigh up to 1,500 kg (3,308 lb) and females 900 kg (1,985 lb) (Loughrey 1959).

**Distribution.** Walruses are circumpolar, generally found in shallow coastal waters, and rarely stray farther south than the ice packs. In recent historical times the walrus was abundant and widespread in the Atlantic Ocean. The animal occurred in large breeding herds as far south as Sable Island, off the coast of Nova Scotia, and at the Magdalen Islands in the Gulf of Saint Lawrence (Mansfield 1967). The distribution and populations of walruses have declined significantly as a result of overkilling and the warming climate since the Pleistocene (Manville and Favour 1960).

In New England the species occasionally strayed southward into the Gulf of Maine. One animal was taken at Plymouth Bay, Massachusetts, in December 1734. Fossil bones discovered include three records from Maine, in Gardiner, Portland, and Addison. Four instances of fossil walrus remains have been dredged or hooked up from the Gulf of Maine; one from Portsmouth, New Hampshire; and another from Gay Head, Martha's Vineyard, Massachusetts (Allen 1930). Walrus bones have also been recorded from Penobscot Bay, Maine (Palmer 1944), and from Bluehill Bay, Maine (Dow 1954). Fossil walrus bones, estimated to be upward of 15,000 years old, have been recorded as far south as Virginia (Manville and Wilson 1970).

WALRUS

WAPITI
OR AMERICAN ELK

# Wapiti or American Elk

*Cervus elaphus* Erxleben

*Cervus elaphus* Erxleben, 1777.
   *Systema regni animalis* . . . , p.
   305.
Type Locality: Eastern Canada, Prov-
   ince of Quebec
Dental Formula:
   I 0/3, C 1/1, P 3/3, M 3/3 × 2 = 34
Names: Elk, American red deer, and
   Canada stag. The Shawnee Indians
   called this mammal "wapiti," but
   the early settlers called it "elk"
   because it closely resembled the
   European red deer of that name.
   The animal is known best as elk in
   North America and as wapiti in
   Europe. In international scientific
   literature the animal is known as
   wapiti. The meaning of *Cervus* = a
   stag, and *elaphus* = a deer, relating
   to, or resembling the red deer.

**Description.** The wapiti is second only to the
moose in weight and is the handsomest of the
deer. The neck is maned, the ears are prominent,
and the tail is short. The muzzle is naked. The
anteorbital facial glands are prominent, but the
tarsal and interdigital glands are absent. The
metatarsal glands are situated below the back on
the outsides of the hind legs.

   Adult bulls have superb massive, widely
branching antlers. Each antler consists of a long,
round beam sweeping up and back from the
skull, usually bearing five to seven well-
developed, unpalmate tines. These include a
brow tine, a bez tine, and a trez tine, before the
longest is reached, beyond which is another
fork, its tines in the same anteroposterior plane.
The last posterior tine generally slants downward, and the brow tine extends out above

the muzzle. Spike bulls have smaller antlers
with fewer tines. Cows with antlers are uncommon. Bulls begin shedding their antlers in
March, though some shed them in Feburary.
Spike bulls normally carry their antlers through
April and even as late as June. In older bulls the
new antlers are usually fully grown and in the
velvet by late May or June. By late August or
early September most of the velvet is rubbed off.

   The skull is long and narrow with lacrimal
vacuity and fossa moderately large. The vomer
does not divide the posterior nares into two
chambers as in the white-tailed deer and woodland caribou. The base of the pedicel extends
conspicuously over the posterior part of the
orbital cavity. Small maxillary canines are present in both sexes.

   The sexes are essentially the same in color,

but the female is generally slightly darker than the male in winter. Color tone varies considerably with the individual, and there is slight seasonal color variation. In winter the male possesses a heavy fringe of long hairs on the back of the neck and throat, and the pelage consists of long, coarse guard hairs and woolly underfur. The general color of the body is tawny brown or brownish gray. The head is usually dark brown with a paler eye ring, and the chin is lighter brown with a black spot near the angle of the mouth. The ears have a black spot on the lower edge of the anterior surface. The neck, chest, and legs are much darker brown, almost blackish, and the underparts are darker than the back. The tail and rump patch are buffy or whitish.

The summer pelage, which is somewhat redder or more tawny, consists of short, stiff, sparse hairs that lie close to the skin. There seems to be little, if any, underfur. The tips of the blackish hoofs and dew claws are usually a lighter color. Calves are yellowish brown and spotted yellowish white until the first fall. Melanistics and albinos occasionally occur. There are two molts each year. Molting of the ragged winter coat takes place in spring, and the gradual transition from the summer to winter pelage begins in early August.

Bulls are larger than cows. Measurements range from 203 to 254 cm (80–100 in); tail, 10 to 13 cm (4–5 in); hind foot, 60 to 71 cm (24–28 in); and height at shoulder, 125 to 150 cm (49–59 in). There is a great variation in the weight of elk at the same age, at different seasons, and in different localities. The animals are heaviest during late summer. Seton (1929) reported that the average weight of an adult male is 317 kg (700 lb).

**Distribution.** In former days the wapiti was the most widely distributed of our hoofed game animals. It was found in most of temperate North America from parts of Canada south to northern Mexico, and from the Pacific coast to most of the Atlantic seaboard. The species has been extirpated throughout most of its original range since the settlement by Europeans. Today the species occurs in the wilder regions of western North America, chiefly in the Rocky Mountains. It has been reintroduced into some localities.

There are gaps in information on the distribution of the wapiti in New England during historical times. Hays (1871) reported that the species was found all along the coast from Canada to the Gulf of Mexico, and that as late as 1826 a few wapiti were killed in the region of Saranac, New York. Wyman (1868) reported that wapiti bones were found in shell heaps on Mount Desert Island, Maine. Carpenter and Siegler (1958) reported that a perfectly preserved naturally shed wapiti antler was found buried in a lime deposit in Lime Pond, Columbia Bridge, Coos County, New Hampshire. Allen (1942) reported that the wapiti must have occurred in western Vermont, since its antlers have been found in bogs there. Nelson (1916) remarked that this animal occurred in central Massachusetts, while Allen (1942) stated that it ranged through extreme western Massachusetts. Cahalane (1947) gave its original range as from the Berkshires in Massachusetts westward. In Connecticut, Goodwin (1935) supposed that the wapiti could have occurred in the state during the sixteenth and seventeenth centuries. Waters and Mack (1962a) examined wapiti bones excavated from an archaeological site in Scituate, Providence County, Rhode Island, and speculated that the bones could be as much as 6,000 or as little as 1,600 years old.

**Wapiti in New Hampshire.** A total of 60 wapiti were introduced into the Corbin Preserve Newport, Sullivan County, New Hampshire from northern Minnesota in 1891. The animals did well from the first and were reported to have increased to 90 after a year; at one time they were estimated at 1,000 animals (Manville 1964). Many wapiti perished during the severe winter of 1897, and in 1903 the estimated number was only 300 (Palmer 1910).

The following account of wapiti problems in New Hampshire is from Silver (1957). In 1903 the Corbin heirs gave 4 bulls and 8 cows to the State of New Hampshire, which were released at the base of Ragged Mountain, Merrimack County, by the Andover Fish and Game Club. By 1912 as many as 40 wapiti were seen in one herd and some damage to crops was reported. In time the wapiti were shot. In 1933 the Corbin officials gave the state a second gift of 2 bulls and 10 cows which were released on the Pillsbury Reservation, Sullivan County—then a state game refuge—in the towns of Washington and Goshen. By 1941 at least 60 wapiti ranged over several surrounding towns. To forestall extensive damage to crops, the legislature authorized the reduction of the herds by a maximum of 125 animals. During a two-day open season, 46 wapiti were taken by hunters. In the vicinity of Straw Hill, in Unity, Sullivan County, 11 wapiti caused some damage to crops in 1951. Two years later the herd had increased to 26 animals, costing the department $470 in damage payments. The legislature then passed a law authorizing the New Hampshire Fish and Game Department to capture and relocate, or destroy, wapiti causing damage to agriculture.

The department considered translocating the entire Unity herd to northern Coos County, but decided against it because of the extremely high cost and because the habitat in northern Coos had been considerably altered. Sites that were considered favorable were now marginal even for deer, and it takes approximately three times as much food to support an elk as a deer. Nothing was done, and by 1954 there were 32 elk in the Unity herd and the damage claim rose to $580. In 1955 the herd had been reduced by 18 animals by conservation officers and biologists. In that year it was estimated that between 20 and 30 elk remained at large in the state. There now appear to be no free-ranging wapiti in New England, except in the confines of the Blue Mountain Forest preserve, Sullivan County, New Hampshire. The wapiti can be considered a wild species of the past in New England.

# Eastern Woodland Caribou

*Rangifer tarandus caribou* (Gmelin)

[*Cervus tarandus*] *caribou* Gmelin, 1788. *Systema naturae*, 13th ed., 1:177.
*Rangifer tarandus caribou*, True, 1884. *Proc. U.S. Natl. Mus.* 7:592.
Type Locality: Eastern Canada, Province of Quebec
Dental Formula:
I 0/3, C 0/1 or 1/1, P 3/3, M 3/3 × 2 = 32 or 34
Names: American reindeer, caribou, pawer, and carribouck. The meaning of *Rangifer* = a reindeer or caribou, *tarandus* = the reindeer fabled to change color like the chameleon, and *caribou* comes from a Micmac word meaning "a shoveler," an allusion to the way the caribou paws away the snow in winter to get lichen.

**Description.** The stocky caribou is intermediate in size between the white-tailed deer and the wapiti. Caribou have a wide, heavy muzzle covered with thick, short hairs. The ears and tail are short. The neck is maned on the underside. Tarsal glands are present but metatarsal glands are absent.

Unlike other members of the deer family, most caribou females have antlers, though they are much smaller and more spikey than those of males. Caribou antlers are distinctly unlike those of other members of the deer family and vary markedly in shape and form. Antlers of adult bulls are composed of two heavy main beams, each reaching back from the forehead almost in a line with the rostrum, then sweeping upward with a more or less noticeable angle for nearly half its length and curving forward near the tip. There are a pair of brow tines, either one or both flattened into a broad vertical "shovel," often leading forward nearly to the nose; a pair of bez tines, longer than the brow tines and bearing fewer points near the tip; and usually a small point or back tine at the posterior angle of each main beam and near the tip of the main beam, often palmate, with a series of points projecting from the upper edge. The main beams are flattened rather than cylindrical. The brow

tines are not used as shovels to scrape away the snow in uncovering food as was once commonly believed.

The antlers of mature bulls are usually shed in November, and the older animals drop their antlers first. Young males may carry their antlers until late in April, and females carry theirs until mid-May or even June. Pregnant females often keep their antlers until about the time the calf is born.

In the skull, the posterior nares are divided by the high vomer; the lacrimal vacuity is large and the lacrimal fossae are small. The orbital cavities do not extend backward under the bases of the horn pedicels, and the width of the auditory bullae is about one-half the length of the bony tube leading to the meatus. The maxillary canines are often present in both sexes, and the small incisors are spatulate, decreasing in size from the inner ones outward. The lower molariform teeth are only about half the diameter of the upper ones.

The pelage is composed of long, coarse, hollow guard hairs and fine, woolly, oily underfur. The guard hairs are filled with air and help the caribou float. The underfur is excellent insulation in cold weather. The sexes are colored alike and show slight seasonal variation, but there are great individual color variations. In autumn the general coloration above is dark cinnamon brown, with grizzled white to yellowish on the neck and shoulders, often continuing as a lateral stripe along the sides of the body. The area below this stripe is darker than the back. The underparts are darkest on the chest, becoming grizzled posteriorly and shading into the white abdomen. The inner side of the thigh is a pale buff, and the rump and underside of the tail are white. The nose and the tip of the lower lip are white, with a slight greenish tinge possibly caused by staining from plants. There are a few white hairs about the eye, sometimes forming a definite eye ring, and there is a white rim above each hoof. In winter the coloration is somewhat paler or grayish brown. Immatures are colored much like the adults but have faint traces of white spotting. Color variants are rare.

Males are larger than females. Measurements range from 183 to 213 cm (72–84 in); tail, 10 to 15 cm (4–6 in); hind foot, 50 to 66 cm (20–26 in); and height at shoulder, 107 to 127 cm (42–50 in). Weights vary from 90.8 to 181.6 kg (200–400 lb).

**Distribution.** The woodland caribou is found in the northern forested borders from Alaska and Canada south to northern Maine, northern Minnesota, Idaho, and British Columbia. It probably has reached the Adirondack region of New York. It is nearly extinct over much of its range.

Caribou remains were excavated from a shell heap (kitchen midden) at Duxbury, Plymouth County, Massachusetts, in 1967. The Duxbury site may indicate a relatively recent southern range extension of the species along the northeastern seaboard. The estimated age of the bones is 2,000 years or less (Guilday 1968).

**Former Abundance.** In Maine, the reports of the Commissioners of Inland Fisheries and Game have yielded some interesting data on the former abundance of caribou. The first record of

EASTERN WOODLAND CARIBOU

the species was mentioned in the report of 1886 as follows: "Of caribou it is difficult to make any estimate of increase or decrease. The reports to us are of plenty and in all sections. We have heard of many being killed, but of all our game animals the caribou is the most capable of taking care of itself." The report for 1889–90 stated: "We think moose and caribou have made no increase. Caribou, being migratory in their habits, cannot be depended on, often being plenty one year and none the next." The report for 1895 stated: "We learn that the shipment of caribou for 1895 was double that of 1894. These shipments represent a larger per cent of the whole number of these animals taken than the shipment of deer, for but few of them were consumed by our people, or in the forest, and most of them passed through the express offices. The whole number of these animals shipped was one hundred and five for the current year. And it will also be remembered that in 1894 a person might lawfully take two caribou while in 1895 he could take but one."

Allen (1942) wrote: "In 1895 and again in 1896 the Bangor and Aroostook Railroad alone shipped out about 130 caribou each year that had been killed by visiting sportsmen, but after that the number in the State began to decline, owing in part probably to the eastward movement of the animals once more and in part to summer and winter killing in both Maine and the adjacent parts of Canada."

The commissioners' report of 1896 gave a bleak statement: "The caribou is fast disappearing, and will very soon be practically extinct, unless a closed time, for a series of years, is put on them, or more stringent laws enacted for their protection." The same report gave 239 caribou killed. The report for 1900 mentioned the closed season and the prohibition on killing caribou before 15 October 1905. The report for 1904 stated: "There is no indication that the caribou are returning or will ever return." The report for 1906 briefly stated: "There are no indications of any caribou in the state."

Although Allen (1942) reported that the last caribou in Maine was in the Mount Katahdin region in 1905, Lorenz (1917) stated that the last caribou sighting at Mount Katahdin was in 1908.

In New Hampshire the caribou was not a permanent resident in any area, though small bands of caribou have drifted in and out of the state. Caribou first appeared on the New Hampshire law books in 1878, when the hunting season for caribou was closed except in Coos County. Regulations were changed often over the next few years, and in 1891 the first bag limit was two caribou of either sex. Ten years later, hunting of caribou was prohibited. Near the year 1900 seventeen caribou were seen near the First Connecticut Lake, Coos County (Silver 1957). Spears (1893) reported that six woodland caribou and four barren ground caribou were introduced into the Corbin Preserve, Newport, Sullivan County in 1890, but all soon died for lack of suitable food.

**Reintroduction into Maine.** In December 1963, the Maine Department of Inland Fisheries and Game released 24 adult woodland caribou on the base and plateau of Mount Katahdin. The animals were imported from Newfoundland. Observations made during the following summer indicated that some caribou traveled 90 miles northeast from the release site (Dunn 1965a). Periodic aerial and ground searches conducted from 1 July 1964 to 30 June 1967 revealed that one caribou was seen at Big Pleasant Pond, Piscataquis County (Dunn 1965b); remains of one caribou were found in "T2–R12"; one doe was sighted in "T6–R8"; two animals with antlers were seen in Bernard Township; one doe was seen in Island Falls; another doe was seen in "T6–R7" (Dunn 1966); a doe was seen in Wellington; and a stag was seen in "T4–R9" (Dunn 1967). In all probability the species has been extirpated in New England.

Studies in Maine have shown that 85 percent of the caribou range is infected with the meningeal worm *Pneumostrongylus tenuis* of the white-tailed deer (Dunn, personal communication). Anderson and Strelive (1968) demonstrated that this parasite kills caribou by infecting the spinal cord and brain.

Although habitat destruction, especially that associated with burning, may be a major factor for the decline of woodland caribou over much of its range in eastern North America, it is possible that the meningeal worm of the white-tailed deer could be a factor in the decline of caribou herds in areas occupied by both caribou and white-tailed deer.

At the beginning of the Pleistocene epoch, almost two million years ago, increasing cold, damp weather, year after year, led to an enormous buildup of ice in the polar regions. During the exceptional rigor of this great ice age, many species of mammals died because they could not adapt. The Pleistocene fauna of North America was notable for the abundance of huge mammals which immigrated from Asia. Caribou, moose, musk oxen, mountain sheep, wolverine, bears, and other species flourished after they crossed by the Bering land bridge between Asia and North America. But the mightiest of all land mammals, the elephant-sized mastodon and woolly mammoth, were slated for extinction as the centuries slipped by. The mastodon, which preceded the woolly mammoth to America by millions of years, still throve there perhaps as late as 5,000 years ago, roaming the edge of the great ice sheets. These two magnificent lumbering giants, the majestic wapiti, the caribou, and the mighty bison all inhabited New England. Superficially the mastodons and mammoths seem alike, but they differ structurally, notably in the cheek teeth. The mastodons possessed low-crowned, bunodont cheek teeth, while the mammoths and present-day elephants have high-crowned, lophodont cheek teeth used for grinding.

The following records of traces of extinct and extirpated species from New England are from Hay (1923). (1) Mastodon: in Massachusetts—a molar tooth from Shrewsbury, Worcester County, in 1885; in Connecticut—a molar tooth from Cheshire, New Haven County, in 1828, bones from Farmington, Hartford County, in 1914, remains of a skeleton from Bristol, Hartford County, in 1885, and bones from Sharon, Litchfield County, in 1828. (2) Woolly mammoth: in Vermont—remains of bones from Mount Holly, Rutland County, in 1849. (3) Wapiti: in Vermont—an antler and part of an upper jaw from Woodbury, Washington County, in 1910; in Connecticut—a humerus and a tibia from the Quinnipiac Valley near New Haven, New Haven County, in 1875. (4) Bison: in Massachusetts—milk teeth of a calf from Orleans, on Cape Cod, in 1920.

## REFERENCES

Allen, Glover M. 1930. The walrus in New England. *J. Mamm.* 11(2):139–45.

——. 1942. Extinct and vanishing mammals of the western hemisphere with the marine species of all the oceans. *Amer. Comm. Intl. Wild Life Proct. Spec. Pub.* 11:1–629.

Allen, Joel A. 1876. Former range of New England mammals. *Amer. Nat.* 10(12):708–15.

Anderson, Roy C., and Strelive, Uta R. 1968. The experimental transmission of *Pneumostrongylus tenuis* to caribou (*Rangifer tarandus terraenovae*). *Canadian J. Zool.* 46(3):503–10.

Anderson, Rudolph Martin. 1947. Catalogue of Canadian Recent mammals. *Natl. Mus. Canada Bull.* 102:1–238.

Cahalane, Victor H. 1947. *Mammals of North America.* New York: Macmillan Co.

Carpenter, Ralph G., II, and Siegler, Hilbert R. 1958. A list of New Hampshire mammals and their distribution. *New Hampshire Fish and Game Dept. Bull.,* pp. 1–15 (mimeographed).

Commissioners of Inland Fisheries and Game (Maine). *Annual Repts.* 1886, 1889–90, 1895, 1896, 1904, 1906.

Cram, Gardner. 1901. Panthers in Maine. *Forest and Stream* 56(6):123.

Crane, Jocelyn. 1931. Mammals of Hampshire County, Massachusetts. *J. Mamm.* 12(3):267–73.

Cronan, John M., and Brooks, Albert. 1968. *The mammals of Rhode Island.* Wildlife Pamphlet no. 6. Providence: Rhode Island Department of Agriculture and Conservation, Division Fish and Game.

Dearborn, Ned. 1927. An old record of the mountain lion in New Hampshire. *J. Mamm.* 8(4):311–12.

Dow, Robert L. 1954. Walrus skull (Odobenus) from Bluehill Bay, Maine. *J. Mamm.* 35(3):444.

Dunn, Francis D. 1965a. Reintroduction of woodland caribou to Mt. Katahdin, Maine. *Trans. N.E. Fish and Wildl. Conf.*

——. 1965b. Caribou population studies. W-37-R-14, K-1. *P-R Job Completion Report, 1 July 1964 to 30 June 1965.*

——. 1966. Caribou population studies. W-37-R-15, K-1. *P-R Job Completion Report, 1 July 1965 to 30 June 1966.*

——. 1967. Caribou population studies. W-37-R-16, K-1. *P-R Job Completion Report, 1 July 1966 to 30 June 1967.*

Emmons, Ebenezer. 1840. *A report on the quadrupeds of Massachusetts*. Cambridge: Folsom, Wells, and Thurston.

Foote, Leonard E. 1946. A history of wild game in Vermont. *Vermont Fish and Game Serv. State Bull.* 11:1–55.

Goodwin, George Gilbert. 1935. The mammals of Connecticut. *Connecticut Geol. Nat. Hist. Surv. Bull.* 53:1–221.

———. 1936. Big game animals in the northeastern United States. *J. Mamm.* 17(1):48–50.

Green, Jacob. 1822. Curious instinct of the common hog (*Sus scrofa* Linn.). *Amer. Sci.* 4(2):309–10.

Guilday, John E. 1968. Archaeological evidence of caribou from New York and Massachusetts. *J. Mamm.* 49(2):344–45.

Hall, E. Raymond, and Kelson, Keith R. 1959. *The mammals of North America*. 2 vols. New York: Ronald Press.

Hay, Oliver Perry. 1923. The Pleistocene of North America and its vertebrated animals from the states east of the Mississippi River and from the Canadian provinces east of longitude 95°. *Carnegie Institute of Washington Publications* 322:1–499.

Hays, W. J. 1871. Notes on the range of some of the animals in America at the time of the arrival of the white men. *Amer. Nat.* 5:387–97.

Iljin, N. A. 1941. Wolf-dog genetics. *J. Genetics* 42(3):359–414.

Jackson, C. F. 1922. Notes on New Hampshire mammals. *J. Mamm.* 3(1):13–15.

Linsley, James H. 1842. A catalogue of the mammals of Connecticut, arranged according to their natural families. *Amer. J. Sci.* 43(7):345–54.

Lorenz, Annie. 1917. Notes on the Hepaticae of Mt. Katahdin. *Bryologist* 20:41–46.

Loughrey, Alan G. 1959. Preliminary investigation of the Atlantic walrus *Odobenus rosmarus rosmarus* (Linnaeus). *Canadian Wildl. Serv. Wildl. Manage. Bull.* 1(14):1–123.

Mansfield, A. W. 1967. Seals of arctic and eastern Canada. *Fish. Res. Bd. Canada, Bull.* 137:1–35.

Mansueti, Romeo. 1954. Mystery mink. *Nature Mag.* 47(4):185–86.

Manville, Richard H. 1964. History of the Corbin Preserve. *Smithsonian Inst. Pub.* 4581:427–46.

Manville, Richard H., and Favour, Paul G., Jr. 1960. Southern distribution of the Atlantic walrus. *J. Mamm.* 41(4):499–503.

Manville, Richard H., and Wilson, Jerald J. 1970. Fossil walrus from Virginia waters. *J. Mamm.* 51(4):810–11.

Mearns, Edgar A. 1900. The native mammals of Rhode Island. *Newport Nat. Hist. Soc. Circ.* 1:1–4.

Miller, Edward H. 1975. Walrus ethology. I. The social role of tusks and applications of multidimensional scaling. *Canadian J. Zool.* 53(5):590–613.

Morton, Thomas. 1637. *New English Canaan or New Canaan*. Amsterdam.

Nelson, Edward William. 1916. The larger North American mammals. *Natl. Geogr. Mag.* 30(5):385–472.

Osgood, Frederick L., Jr. 1938. The mammals of Vermont. *J. Mamm.* 19(4):435–41.

Palmer, Ralph S. 1944. Walrus remains from New England. *J. Mamm.* 25(2):193.

Palmer, T. S. 1910. Private game preserves and their future in the United States. *U.S. Dept. Agric. Bur. Biol. Surv. Circ.* 72:1–11.

Seton, Ernest Thompson. 1929. *Lives of game animals*. 4 vols. Garden City, N.Y.: Doubleday, Doran and Co.

Shufeldt, R. W. 1924. The skull of the wolverine. *J. Mamm.* 5(3):189–93.

Siegler, Hilbert R. 1971. The status of wildcats in New Hampshire. In *Proceedings of a Symposium on the Native Cats of North America*, in conjunction with the 36th North American Wildlife and Natural Resources Conference.

Silver, Helenette. 1957. *A history of New Hampshire game and furbearers*. Survey report no. 6. Concord: New Hampshire Fish and Game Department.

Spears, John R. 1893. The Corbin Game Park. *Ann. Rept. Smithsonian Inst. for 1891*, pp. 417–23.

Stanton, Don C. 1963. A history of the white-tailed deer in Maine. *Maine Dept. Inland Fish. and Game, Game Div. Bull.* 8:1–75.

Stone, Witmer, and Cram, William Everett. 1902. *American animals*. Garden City, N.Y.: Doubleday, Page and Co.

Thompson, Zadock. 1853. *Appendix to the history of Vermont, natural, civil and statistical*. Burlington, Vt.: Chas. E. Tuttle Co.

Titcomb, John W. 1901. Animal life in Vermont. *The Vermonter* 5:2.

Waters, Joseph H., and Mack, Charles W. 1962. Note on former range of (*Cervus canadensis*) in New England. *J. Mamm.* 43(2):266–67.

Wright, Bruce S. 1961. The latest specimen of the eastern puma. *J. Mamm.* 42(2):278–79.

Wyman, Jeffries. 1868. An account of some Kjoekkenmoeddings, or shell-heaps in Maine and Massachusetts. *Amer. Nat.* 1:561–84.

Young, Stanley P., and Goldman, Edward A. 1944. *The wolves of North America*. Washington, D.C.: American Wildlife Institute.

———. 1946. *The puma, mysterious American cat*. Washington, D.C.: American Wildlife Institute.

# Glossary

## SKULL TERMS

*Alisphenoid*   A winglike bone forming part of the lateral wall of the braincase and part of the posterior wall of the orbit. Frequently fused to the basisphenoid.

*Alveolus*   A pit or socket in which the tooth is set.

*Angle of dentary*   Posterior projection process arising below condyle of dentary.

*Anterior nares*   Anterior openings of nasal cavities.

*Anterior palatine foramina*   A pair of elongated openings in the palate just behind the incisors.

*Antorbital*   Anterior to the orbit.

*Antorbital fossa*   An extensive opening just in front of orbit, caused by vacuities between two or more facial bones.

*Auditory bulla*   An inflated bony capsule at the base of braincase enclosing the middle ear.

*Auditory meatus*   The external opening into the auditory bulla.

*Baleen*   Horny plates of epidermal origin in the upper jaw of whalebone whales.

*Basilar length*   The length of the skull from the anterior border of the foramen magnum to the posterior border of the alveolus of the first upper incisor.

*Basioccipital*   The bone forming the ventral margin of the foramen magnum; it fuses with the two exoccipitals and the supraoccipital to form the occipital bone.

*Basisphenoid*   A median ventral bone of the skull, lying anterior to the basioccipital and between the auditory bullae.

*Bicuspid*   Having or ending in two points, as bicuspid teeth.

*Brachydont*   Low-crowned; any tooth whose width exceeds the height of the crown above the alveolus.

*Braincase*   The part of the skull that houses the brain; also known as the cranium.

*Buccal*   Pertaining to the mouth.

*Bunodont*   Low-crowned, squarish, enamel-capped teeth with low, rounded cusps.

*Canine*   The long, stout, cone-shaped, pointed stabbing tooth next to the incisors; occasionally bladelike but sometimes small and similar to the teeth preceding it. Lower canine bites in front of upper.

*Carnassial*   Flesh-cutting; in the order Carnivora the term refers to the last pair of premolar teeth in the upper jaw and of the first pair of true molars in the lower jaw.

*Condyle*   A rounded and somewhat elongated process.

*Coronoid process*   Ascending projection from dentary for attachment of temporal muscles.

*Cusp*   A peak or rounded elevation on the crown of a tooth.

*Cuspidate*   With a cusp or cusps; ending in a point or points.

*Dentary*   One of a pair of dermal bones of the lower jaw.

*Dentin*   Ivorylike substance beneath the enamel, usually constituting the bulk of the tooth.

*Diastema*   A distinct space between two teeth.

*Diphyodont*   Having two successive sets of teeth, deciduous and permanent.

*Fenestrated*   Having perforations or openings, as in skulls of lagomorphs.

*Foramen* (pl., *Foramina*)   An aperture through a bone for passage of nerves or blood vessels.

*Foramen magnum*   Large opening in back of the skull through which the spinal cord passes.

*Fossa*   An irregular ditchlike depression, such as the nasal fossa in which one of the nostrils lies.

*Frontal*   One or two bones covering the anterior part of the braincase above the orbits.

*Glenoid fossa*   Surface on squamosal bone for articulation of the lower jaw. Also called mandibular fossa.

*Heterodont*   Having teeth differentiated for various functions.

*Homodont*   Having teeth essentially similar throughout a tooth row.

*Hyoid*   A bone situated between vertical parts of the lower jaw.

*Hypsodont*   Teeth with high crowns; usually rootless and ever-growing.

*Incisive foramen*   An opening, paired or single, in the anterior part of the palate behind the incisors.

*Incisor*   Nipping or chiseling tooth in front of the canines.

*Infraorbital foramen*   An opening through the zygomatic plate from the orbit to the side of the rostrum, for passage of nerves, blood vessels, and muscles.

*Interorbital constriction*   The narrowest part of the top of the skull, between the orbits.

*Interorbital width*   Smallest measurement at interorbital constriction.

*Interparietal*   Unpaired bone in the middle at the extreme back and top of the braincase.

*Jugal*   The connecting bone that forms the midsection of the zygomatic arch (absent in shrews).

*Keel*   A ridge; an elevated line.

*Lacrimal*   A small bone situated at the front of the orbit.

*Lambdoidal crest*   A transverse ridge across the posterior region of the skull near the posterior border between the supraoccipital and parietals.

*Lateral*   Pertaining to the side.

*Lingual*   Pertaining to the tongue.

*Loph*   A transverse ridge of enamel across a tooth.

*Lophodont*   Teeth whose crowns have a series of lophs.

*Malar*   Bone situated below the lacrimal, corresponding to the cheekbone of man.

*Mandible*   The lower jaw, composed of two dentary bones.

*Mandibular condyle*   The part at the rear of the mandible that articulates with the squamosal.

*Mandibular ramus*   One side of the lower jaw or mandible.

*Mastoid process*   Lateral and ventral projection at base of the braincase, back of the auditory bullae.

*Maxillae*   Principal bones of the upper jaw that carry the upper molar teeth.

*Meatus*   Opening from the auditory bulla.

*Molar*   One of the cheek teeth behind the incisors and canines: a molar tooth.

*Nasal*   One of a pair of bones medial and posterior to the external nares, forming the roof of the nasal cavity.

*Nasal septum*   A thin, median, vertical partition of bone that divides the nasal cavity into right and left halves.

*Occipital*   The compound bone surrounding the foramen magnum and bearing the occipital condyles.

*Occipital condyle*   Knoblike process of exoccipital on either side of the foramen magnum; articulates with the first vertebra.

*Occipital crest*   Ridge across the skull near the posterior border of the parietals.

*Occiput*   The back of the head.

*Occlusal surface*   The crown of a tooth; the grinding surface that faces against the tooth opposing it.

*Orbit*   The eye socket, composed of several bones.

*Palatal bridge*   The posterior border of the palate in many Microtinae, which seems to bridge over the two troughs or rows of foramina that pass forward from the bridge to the incisive foramina.

*Palate*   The roof of the mouth, composed of parts of the premaxillary, maxillary, and palatine bones.

*Palatine*   One of a pair of bones that form the posterior part of the palate and wall the anterior part of the interpterygoid fossa. Considered an extension of the palate.

*Palatine vacuity*   Irregular openings in or perforations of the palatal portion of the palatine bones.

*Paroccipital*   Ventral projection from exoccipital just posterior to the auditory bulla.

*Parietal*   One of a pair of bones roofing the posterior part of the braincase.

*Posterior nares*   Posterior openings of the nasal cavity.

*Postorbital bar*   A bar between the orbit and the temporal fossa, formed by the union of the two postorbital processes from the jugal and frontal.

*Postorbital process*   A projection from the frontal or jugal bone or both, partly separating the orbit from the temporal fossa.

*Premaxillae*   Paired bones situated in front of the maxillae.

*Premolar*   Tooth preceding or in front of the molar tooth; a premolar tooth.

*Process*   A bony projection or elevation; it may be articular or nonarticular.

*Pterygoid*   Thin, straplike bones attached to the palatine bones.

*Ramus*   The vertical posterior part of the dentary bone, composed of the coronoid process, condyle, and angle.

*Rostrum*   Portion of the skull anterior to the orbit.

*Sagittal crest*   A longitudinal median ridge dorsal to the braincase. Often formed by fusion of the temporal line.

*Selenodont*   Type of tooth with longitudinal crescentic ridges of enamel.

*Spine*   A sharp projection.

*Squamosal*   A fan-shaped bone on each side of the braincase above the auditory bulla.

*Supraorbital*   Referring to the top rim of the orbit.

*Supraorbital ridge*   A beadlike ridge bordering the orbit dorsally.

*Suture*   An immovable line of union between two bones.

*Temporal fossa*   The large space above the zygomatic arch, between the orbit and the lambdoidal crest.

*Tricuspid*   A tooth with three points.

*Tubercle*   A small and usually more or less pointed process.

*Tuberculate*   Having tubercles or small knoblike prominences.

*Tuberculosectorial*   Primitive tooth consisting of cusps arranged in asymmetrical triangles.

*Tuberosity*   A rough or obtuse process.

*Tympanic*   The bone that forms the auditory bulla.

*Unicuspid*   A tooth bearing only a single cusp. In shrews, applied to a row of three to five small teeth behind the enlarged pair of incisors.

*Vomer*   A paired or unpaired dermal bone of the anterior base of the braincase, just behind the premaxilla.

*Zygoma*   The zygomatic arch, the bony bar lateral to the orbit and temporal fossa, formed by the maxillary, jugal, and squamosal.

*Zygomatic arch*   The bony arch that forms the lateral border of the orbit and temporal fossa.

*Zygomatic plate*   The flattened, expanded part of the maxillae in front of the orbit from which the anterior part of the zygomatic arch arises.

## OTHER BIOLOGICAL TERMS

*Abdomen*   The belly.

*Adult*   A sexually mature animal capable of producing young.

*Altricial*   Pertaining to young born naked and helpless.

*Annulated*   Marked with rings of color.

*Aquatic*   Living in water.

*Arboreal*   Living in trees.

*Baculum*   The penis bone or os priapi of some mammals.

*Bez tine*   The first tine above the first or brow tine antler in artiodactyls.

*Browse*   Twigs or shoots, with or without attached leaves, of shrubs, trees, or woody vines; to eat these parts of plants.

*Browse line*   The line marking the height of trees to which cervids have eaten browse.

*Cover*   The area of ground covered by plants, living or dead, for refuge of an animal.

*Crepuscular*   Active at twilight.

*Cruising radius*   The distance between locations at which an animal occurs at various hours of the day, seasons, or years.

*Cycle*   Periodic fluctuation of density of a given species in a given area.

*Distal*   Situated away from the center or place of attachment.

*Distribution*   Geographic area occupied by an organism; synonym for range.

*Diurnal*   Active in daytime.

*Dorsal*   On or pertaining to the back.

*Emarginate*   Having the margin notched, as a leaf.

*Embryo*   Unborn young in early stages of development.

*Emigration*   Moving out of an area, usually without returning.

*Exotic*   Foreign, not native in an area; an organism introduced from a foreign area.

*Extinction*   Annihilation, complete destruction.

*Extirpation*   Eradication; complete destruction within a certain area.

*Falcate*   Curved like a sickle.

*Fauna*   The animal life of a region.

*Feral*   Having escaped from domestication or cultivation and living in the wild.

*Fetus*   Unborn young in later stages of development.

*Flora*   The plant life of a region.

*Forage*   Unharvested plants available for animal consumption.

*Fossorial*   Inhabiting burrows; adapted for digging or burrowing in ground.

*Furbearer*   A mammal sought for its fur.

*Game animal*   An animal sought as a trophy or for flesh, fur, or feathers.

*Habitat*   Natural environment in which an animal lives.

*Hibernate*   To pass the winter in a lethargic or torpid state.

*Immature*   A nonbreeding animal. Synonym of juvenile, young.

*Immigration*   Moving into an area which an animal has not previously occupied.

*Inguinal*   On or pertaining to the groin.

*Integument*   External coating or skin.

*Juvenile*   Synonym of immature, young.

*Longevity*   Maximum age attained under optimum environmental conditions.

*Mast*   Acorns, nuts, and certain pine cones used as food by animals.

*Median*   Being situated in the middle.

*Mesaxonic foot*   Type of foot in which the axis passes through the middle digit, which is larger than the others and symmetrical in itself.

*Mesic*   Characterized by moderate moisture conditions.

*Migration*   Moving at regular intervals.

*Monogamy*   One male mating with one female of the same species.

*Niche*   The habitat best supplying the needs of an organism.

*Nocturnal*   Active at night.

*Palmate*   Hand-shaped, resembling a hand with fingers spread.

*Paraxonic foot*   Type of foot in which the axis passes between the third and fourth digits, which are almost equally developed.

*Parturition*   Act of giving birth.

*Pelage*   The fur, hair, or woolly covering of a mammal.

*Pelagic*   Oceanic; living in the sea far from the coast.

*Pectoral*   On or pertaining to the chest.

*Pinna*   The fleshy external ear.

*Polygamy*   A male mating with several females or a female mating with several males of the same species.

*Precocial*   Pertaining to young with hair that are able to run about when newly born.

*Prehensile*   Adapted for grasping, as an opossum's tail.

*Proximal*   Situated toward the center or place of attachment.

*Range*   Geographic area occupied by an organism; used as a synonym for distribution.

*Species*   A population of organisms reproductively isolated; a distinct sort or kind of animal.

*Subadult*   An animal which has not bred but which externally resembles an adult.

*Subterranean*   Living below the surface of the ground.

*Terrestrial*   Living on the surface of the ground.

*Territory*   The defended part of the home range.

*Tragus*   The central lobe anterior to the pinna of the ear.

*Ventral*   On or pertaining to the underside.

*Yearling*   An animal over one but under two years old.

*Young*   A synonym for one or more immature animals.

# INDEX

## A

Aardwolves, 196
*Alces alces*, 9, 269
Ambergris, 168
American porcupines, 148
Amphibians, 1, 3, 220, 261
Amphipods, 253
Ants, 38, 93
ARCHAEOCETI, 161
ARTIODACTYLA, 5, 9, 259
*Arvicola*, 4

## B

BALAENIDAE, 190
*Balaenoptera*
  *acutorostrata*, 8, 185
  *borealis*, 8, 184
  *musculus*, 8, 187
  *physalus*, 8, 182
BALAENOPTERIDAE, 182
Bandicoots, 16
Bats, 1, 3, 44, 225, 228, 233
  big brown, 5, 48, 49, 52, 55–57
  hoary, 5, 54, 55, 60–61
  red, 5, 58–59
  silver-haired, 5, 52–53, 61
Bears, 1, 3, 207, 292; as predators, 122
  black, 8, 207, 208–11; as predators, 88, 106, 219, 260, 271
  polar, 207, 250, 256
Beaver, 6, 25, 83, 104–10, 150, 234
Bees, 35, 214
Beetles, 26, 30, 38, 46, 54, 59, 101, 129, 148
  bloodsucking, 106, 107
*Berardius*, 161
Birds: 93; as predators, 1, 3, 109, 112, 114, 119; as prey, 94, 98, 101, 103, 113, 115, 132, 139, 177, 200, 204, 207, 214, 218, 220, 223, 225, 228, 233, 236, 239, 241, 261. *See also* Hawks; Owls
  American coot, 228
  bank swallow, 86
  blue jay, 58
  common grackle, 45, 56
  crow, 52, 67, 102, 122
  great black-backed gull, 125
  gull, 122
  heron, 23, 122
  herring gull, 125
  jay, 122
  junco, 86
  mallard, 228
  red-winged blackbird, 45
  ruffed grouse, 218
  shrike, 23, 31, 122
  solitary vireo, 24
  sparrow, 86
  starling, 86
  waterfowl, 199, 239
  woodpecker, 52, 58, 60, 100, 102
  yellow-bellied sapsucker, 101
Bison, 259, 280, 292
*Blarina brevicauda*, 5, 30
Bobcats, 8; as predators, 17, 27, 31, 67, 74, 79, 84, 88, 92, 97, 100, 102, 106, 122, 127, 132, 150, 206, 213, 217, 219, 227, 232, 235, 237, 240–42, 260, 265
Bugs, 46, 54, 59
  June, 90
  sow, 32
Butterflies, 86, 148

## C

*Calanus*, 184
  *finmarchicus*, 185, 192
CAMELIDAE, 1
Camels, 1, 259
CANIDAE, 196
*Canis*
  *latrans*, 8, 196
  *latrans* var., 198
  *lupus lycaon*, 280
Capybara, 104
Caribou, 203, 265, 280, 292. *See also* Eastern woodland caribou
CARNIVORA, 4, 8, 196
*Castor canadensis*, 6
CASTORIDAE, 104
Catamount, 284
Caterpillars, 113, 115, 148
Cats, 2, 3; as predators, 16, 23, 31, 34, 36, 38, 45, 50, 56, 58, 67, 84, 92, 97, 100, 102, 112, 118, 119, 122, 126, 129, 130, 132, 138, 141, 144, 147, 216, 222, 226, 231, 237, 242
Cattle, 259
Centipedes, 24, 28, 32, 35, 37, 38, 113, 148
Cephalopods, 162, 172, 177, 179, 181, 190
*Ceratherium*, 259
CERVIDAE, 261
*Cervus*
  *dama*, 9, 274
  *elaphus*, 287
Cestodes, 17, 25, 31, 38, 40, 45, 58, 67, 74, 78, 84, 92, 97, 103, 112, 118, 119, 122, 127, 132, 135, 138, 144, 150, 199, 203, 206, 213, 220, 222, 224, 227, 232, 237, 240, 250, 252, 254, 256, 265, 272. *See also* Parasites
CETACEA, 1, 4, 6, 161
Chipmunks, 1, 83, 218, 223, 233, 239, 241
  eastern, 6, 84–87
CHIROPTERA, 4, 5, 44
*Chaetomys*, 148
*Choloepus*, 1

**The Johns Hopkins University Press**

This book was set in VIP Melior text and display type by Monotype Composition Company, from a design by Susan Bishop. It was printed on 80-lb. Paloma Matte paper and bound in Holliston Record Buckram cloth by The Maple Press Company.

**Library of Congress
Cataloging in Publication Data**

Godin, Alfred J.
    Wild Mammals of New England.

Includes bibliographies and index.
1. Mammals—New England. I. Title.
QL719.N35G63   599'.09'74   77-4785
ISBN 0-8018-1964-4